Working bibliography of the Golden Eagle and the genus AQUILA

by

Maurice N. LeFranc, Jr.

and

William S. Clark

with a Foreword by

Leslie H. Brown

Research Assistant

Dawn M. Logan

NATIONAL WILDLIFE FEDERATION
Scientific & Technical Series No. 7

Raptor Information Center
Institute for Wildlife Research
National Wildlife Federation
1412 16th St., N.W.
Washington, D.C. 20036

LeFranc, Maurice N., 1951-
Working bibliography of the golden eagle and the genus Aquila.

(National Wildlife Federation scientific and technical series ; no. 7)
Includes indexes.
1. Golden eagle--Bibliography. 2. Aquila--Bibliography. I. Clark, William S., 1937-
II. Title. III. Series.
Z5333.E2L43 1983 [QL696.F32] 016.598916 83-17369
ISBN 0-912186-49-6

Library of Congress Catalog Card Number 83-17369
Cataloging in Publication Data

TO
LESLIE H. BROWN

Dr. Leslie Brown has been referred to as "the world's best known expert on birds of prey." His tireless enthusiasm for natural history studies provided the impetus for completing over twenty-five books on a variety of topics including several works on British and African birds of prey. Dr. Brown is perhaps best remembered for "Eagles, Hawks and Falcons of the World" written in collaboration with Dr. Dean Amadon.

This book is dedicated to the memory of this accomplished researcher and writer who shared his knowledge of natural history through his mastery of the written word.

TABLE OF CONTENTS

NATIONAL WILDLIFE FEDERATION
1412 Sixteenth Street, N.W., Washington, D.C. 20036

PREFACE

The *Working Bibliography of the Golden Eagle and the Genus Aquila*, Number 7 in the National Wildlife Federation's Scientific and Technical Series, marks a third contribution in our bibliographic series, following the *Working Bibliography of Owls of the World* and the *Working Bibliography of the Bald Eagle*. This project was initiated as a result of the NWF's involvement and concern over the problems and threats facing the golden eagle in the western United States and other parts of the world from livestock conflicts, poisoning, and energy developments. The problems which affect the golden eagle also adversely affect other members of the genus *Aquila*. Accordingly the NWF has summarized the research and management information available on this genus.

An extensive review of published work is a fundamental requirement in research. In response to this need the Raptor Information Center has compiled this bibliography of the golden eagle and the other nine members of the genus *Aquila*, to provide researchers with access to more than 3,400 references.

With the recent inclusion of the Raptor Information Center (established in 1976 with the aid of a grant from Exxon Corporation) and the newly formed Feline Research Center into the Institute for Wildlife Research, the National Wildlife Federation will assume a more active role in the research and management of our wildlife resources. Included in these programs will be the continued dissemination of information to the managers of our natural resources.

JAY D. HAIR
Executive Vice President

August 1983

FOREWORD

This is the third in a series of bibliographies on birds of prey planned and published by the Raptor Information Center of the National Wildlife Federation, Washington, D.C., U.S.A. The first was concerned with all species of owls in the world; the second with the U.S. National Bird, the bald eagle. The present bibliography covers what is probably the best known genus of true or booted eagles -- i.e., those with tarsi feathered to the toes. It is not the largest genus numerically (the largest is the almost unknown tropical forest genus *Spizaetus*), but it is the second largest and least specialized genus of booted eagles.

The members of this genus occur almost worldwide except in South America. They inhabit all latitudes from the Arctic to the Equator, and all habitats from tundra and desert to tropical forest, though they are not typical of the last named habitat. They are most often birds of woodland, open grasslands and steppes, and mountainous country. The altitude record is for [the tawny eagle] *A. rapax nipalensis*, one specimen having been found dead on the South Col of Mt. Everest at 7,925 m; and there are golden eagles, *A. chrysaetos*, breeding on sea coast cliffs in the west of Scotland, feeding on puffins and manx shearwaters.

Aquila eagles have fascinated and inspired mankind from ancient times. Aristotle knew something about them, for he observed, in what was probably one of the earliest comments on territorial behavior in birds, that "One does not find two eagles on the same hill . . . ." Imperial Rome adopted the eagle as the standard of her legions, and one member of the genus, the imperial eagle, *A. heliaca*, was more recently the standard of the Austrian Empire. The eagle has long been known as the King of Birds and in most cases the bird everyone concerned was thinking about was a member of the genus *Aquila*, most probably the golden eagle. This is the most widespread large eagle in Europe, and often the only species occurring in the countries whence fables of eagles came. Eagles appear on many a gatepost and coat of arms; they are not always *Aquila* for some have bare tarsi. However, there can be little doubt that most people, thinking of an eagle, are thinking of a member of the genus *Aquila*.

Of the nine *Aquila* species, the golden eagle inhabits a colossal range. It is found from the arctic to the subtropics and perhaps the borders of the tropics in northern India and southern Sahara; from the Pacific coast of North America to Maine; and then again from the Western Isles of Scotland to Japan. It is without any question the most successful, adaptable, and widespread eagle of its size in the world. It is also possibly the best known of any, though aspects of its life have yet to be fully studied, notably, how often, how long, and exactly how it hunts. In some areas which it inhabits the golden eagle is threatened by human activities, notably game preserving in eastern Scotland, and increased access to mountains by large numbers of human beings. Not so very long ago it was threatened by wholesale and largely unjustified shooting by ranchers in the western United States. Very detailed food studies have shown that the eagle does very little harm to ranching interests but the ranchers are still hostile. In many other parts of its vast range it is relatively unmolested and may be preserved. Since there may be 200,000 or more in North America, it would be very surprising if its world population were much less than half a million, quite possibly more. However, although it is so well known in some parts of its range, it is too little known in others (for instance the interior of Siberia) for anyone to make much more than an informed guess.

Within the genus, the golden eagle forms a superspecies with three others: the African Verreaux's or black eagle, *A. verreauxi*, which is its closest relative; the Australian wedge-tailed eagle, *A. audax*; and more doubtfully the curious short-winged, long-tailed Gurney's eagle, *A. gurneyi* of New Guinea. Verreaux's eagle is not only the golden eagle's nearest relative but its closest rival with respect to available knowledge, for it has been very intensively studied within the last two decades by a group of amateurs in the Matopos Hills in Zimbabwe. It may well be the best known eagle in the world at the nest, one female having become so used to observers that she allowed Mrs. Valerie Gargett to take the temperature of her brood patch, at a range of three feet and in full view, with an ordinary clinical thermometer.

In the forties, Verreaux's eagle was the subject of what was then the finest study available of any of the world's eagles, by E. G. Rowe in Tanzania. This eagle can thus claim to have led the world in the study of eagles in many respects, but there is still much to learn about its habits away from the nest, and away from the particular localities where it has been studied.

The Australian wedge-tailed eagle is by comparison very little known, though extensive studies have been made which have not yet been published. It is the only eagle occurring in Australia other than one member of the genus *Haliaeetus* on the coast and one hawk-eagle of the genus *Hieraaetus*. Despite relentless and as usual quite undeserved persecution the wedge-tailed eagle has probably benefited from man's activities, in that he introduced the rabbit, *Oryctolagus cuniculus*, a new source of live prey, as well as domestic sheep. The latter, while providing a main cause for animosity, has also provided plentiful carrion on badly managed rangelands. There are no vultures in Australia, so this eagle partly fills the niche normally occupied by Old World vultures in Asia, Africa and southern Europe, and by the unrelated New World vultures in the Americas.

The fourth member of the superspecies, Gurney's eagle, is very little known. Apparently inhabiting tropical forests, it shares some of the characteristics (i.e. relatively short wings and a long tail) of the genera *Spizaetus* and *Stephanoaetus*, which inhabit forests elsewhere. Within this superspecies in the genus *Aquila*, then, there is a variation from species inhabiting the arctic to those inhabiting tropical forests, and from one of the most well known eagles in the world to one of the least known.

All four members of this superspecies group are big majestic birds, with a wingspan of 2 to 2.5 m, and weighing 3.8 to 5.7 kg. The eastern Asian race of the golden eagle, *A. c. daphanea*, known as the Berkut, is among the largest eagles in the world, and has been trained by expert falconers to catch and hold wolves or foxes until a horseman can overtake and dispatch the quarry. The other members of the genus are all considerably smaller; the smallest, Wahlberg's eagle, *A. wahlbergi*, weighing 800 to 1,200 g, about a quarter of a golden eagle's weight. It differs in several respects from all other members of the genus and there is some doubt whether it should be included in *Aquila* at all. However, it does not really "fit" any better elsewhere.

The imperial eagle is probably the rarest and most threatened of the genus. The western race, *A. h. adalberti*, now confined to Spain, is reduced to less than 100 pairs, and even the eastern nominate race is uncommon. This is probably because at one time this eagle bred in lowlands, now all taken up for intensive agriculture, so that the eagle's habitat preference directly competes with that of man. In such a contest the eagle is likely to lose, imperial standard or not. It is nearly as large and majestic as the golden eagle, and on the whole more handsome, but its habits are very different than those of the golden eagle, and the two are probably not very closely related. The rarity of the Spanish race, which some regard as a species, has led to intensive studies in recent years. Among these have been fascinating experiments in cross-fostering; inducing a barren female to accept a chick that would otherwise have little or no chance of survival in a brood of two, three, or occasionally even four.

The tawny eagle, which is usually amalgamated with the steppe eagle in a single species, *A. rapax*, with five subspecies, [Author's note: These are treated as two separate species in this bibliography. See Chapters 1 and 2.] inhabits low lying plains and steppes, even sub-deserts, but in drier country, habitat suited to stock raising or wildlife rather than to intensive agriculture. The largest race of the species is *A. r. nipalensis*, and is nearly as big as an imperial eagle. The races of tropical India and Africa are much smaller. They are among the only eagles that are at least partly commensal with man in parts of their range in North Africa and India. No doubt as a result, they are among the most successful of all eagles, and they eat anything from a termite to a dead elephant, and moreover pirate prey from other raptors smaller or even much larger than themselves. Their instinct to piracy seems compulsive, and since both races, *orientalis* and *nipalensis*, are piratical like the African *belisarius* and *rapax*, there is at present no good ground for separating these two species.

The other three species are all woodland or forest birds, and are relatively small. The greater and lesser spotted eagles, *A. clanga* and *A. pomarina*, breed in woodlands in eastern Europe and western Asia and migrate to the tropics in winter. The wintering ground of the lesser spotted eagle is in the *Brachystegia* woodlands of southern Africa, but no one knows how it reaches its destination as after it enters Africa via Suez it largely disappears. It is easily mistaken for and possibly overlooked among the much more numerous tawny and steppe eagles and some recent records may be of tawny rather than spotted eagles. The greater spotted eagle is believed to winter in the Sudan and Ethiopia, north of the Equator but there have been very few definite records in recent times. Both of these species have declined in their European range in the last few decades, but may be more numerous farther east.

Wahlberg's eagle, *A. wahlbergi*, the last and smallest of the genus, is confined to Africa and is an intra-tropical trans-equatorial migrant, breeding mainly in the southern hemisphere, and wintering on the scorching plains of the southern Sudan. There are a few possibly doubtful records of its breeding north of the Equator in west Africa, and it undoubtedly does breed in Ethiopia, in habitat otherwise atypical of its main haunts. It differs from all other members of the genus in laying only one egg. All others regularly lay two, and occasionally three or more. It has a whistling, high-pitched voice unlike that of any other *Aquila* species. For these reasons it is doubtfully placed herein, but it equally does not seem closely allied to typical members of the genus *Hieraaetus*, hawk-eagles, so it remains in *Aquila*, even if it does not quite fit.

Within this genus, then, there is a surprisingly large variety, from some of the world's largest eagles to one of the smallest, and from one of the most successful to one of the most threatened. The level of knowledge varies from very comprehensive in the golden eagle to almost nil in Gurney's eagle, and one knows too that there is a mass of available knowledge not yet published for some species. This variation is reflected in the number of references cited in this bibliography for the various *Aquila* species. The number of references cited is not an absolute measure of the state of knowledge, however, and if one were to make a direct comparison between the level of knowledge in, say, the golden and Wahlberg's eagles, there would be much less difference than first appears. The number of references cited reflects at least to some extent the number of people who have been sufficiently interested to write something about a species, and for how long. The golden eagle has been attracting interest since the beginning of the century and earlier, whereas Wahlberg's eagle is probably better known in South Africa than is the golden eagle in Italy.

The study of birds of prey in general, and of the genus *Aquila* among them, was given a great impetus by the fears for their survival resulting from the extensive usage of organochlorine pesticides in the period 1945-65. At an accelerating rate since 1960, new studies have been made, and a mass of new published literature has appeared on a wide variety of species, expecially in North America and Europe. The threat of extinction posed by the use of organochlorine pesticides, and the possible unknown effects upon the human race itself resulted in massive new efforts to learn more about birds of prey. The immediate threat of extinction may have receded (at least in developed countries, though there is reason to fear it still continues in underdeveloped countries), but the stimulus for research remains and work continues to pour in. Without the threat of extinction, there is no doubt that we would have known far less about birds of prey, including *Aquila* species, than we now do. At any rate, there is far more to read now than there was, even in 1970.

In these circumstances a good, clear, keyworded bibliography becomes of inestimable value. Any aspiring student has his work cut out to locate sources of information on almost any well-known species and anything that helps to cut out this hack-work (for that is what it is) can only speed the acquisition of new knowledge and improve the quality of the subsequent research. Writing as one who has spent months of his life struggling through the *Zoological Record*, and the table of contents of journals (time which could have been better spent), how grateful I would have been if such a bibliography as this had been available for me then. As Richard Clark and his co-authors said in the first of these comprehensive bibliographies, "Others have provided the water - we offer a cup." This and other such bibliographies should offer the student a "cup" to the wellspring of knowledge which will spare him the need to take a pick and shovel and dig the well himself.

Some bibliographies I have seen are laboriously compiled lists of references in alphabetical sequence by author. While that is all very well, and better than nothing, a list so prepared is often very difficult to find one's way about in, and therefore tends to be underused or even discarded. The same applies to those bibliographies which list references to a species by a series of numbers, 1-100 or more, without enabling the reader to distinguish between a mere casual mention and an important paper. Such bibliographies become extremely difficult to use in practice, and it is difficult to sort the wheat from the chaff. A good bibliography should tell the seeker after knowledge what species he is dealing with and what particular aspect of it, at that. It should also say where to find it, and who wrote it. The last detail, often used as the key in alphabetically listed bibliographies, is probably the least important, unless you happen to know something about the gentleman or lady concerned, and can assess whether the work is likely to be of good quality or absolute rubbish.

In this bibliography on the genus *Aquila*, one can find what one is seeking by looking under keywords which have been assigned to each reference by the authors. There is also an appendix which lists the references for each species in the genus. When there are literally thousands of entries on the golden eagle, that becomes very much more necessary than in the case of a comparatively little-studied species. At the very worst, it should now take only a matter of days to work through the bibliography on even a very well-known and more studied species or topic to find all you want. Ideally, the true seeker after knowledge should read everything that is written on his subject, for it often happens that a seemingly trivial note contains a nugget of information that is valuable when related to other contexts of which the writer himself was unaware. However, that may be too much to ask of most people, who will start by being discriminating and selecting what seems to concern them most. They may then find that they have to go back to the finer detail later, and will now find it relatively easy to do so.

The Raptor Information Center and the National Wildlife Federation are to be congratulated most heartily on producing these bibliographies which deserve to be, and without doubt will be, welcomed enthusiastically by students of raptor biology worldwide. Writing as one who has often had to dig the well himself, and perhaps never dug it deep enough to get clear water, I can only be grateful that the cup is now held out to me, so that I may drink with greater ease. I wish someone had done it before!

Leslie Brown
March 1980

CHAPTER 1.

INTRODUCTION

Scope. The intent of this bibliography is to identify and keyword all references pertinent to the genus *Aquila*. Some authors treat the genus as being comprised of only nine species, combining *Aquila nipalensis* and *Aquila rapax* into one species *A. rapax* with several subspecies (Stresemann and Amadon 1979). However, recent evidence (Amadon, pers. comm.) suggests that they are two separate species and therefore we have chosen to treat them as such in this bibliography. Listed below are the 10 species which are covered and their English names:

Scientific Name	English Name
Aquila pomarina	Lesser spotted eagle
A. clanga	Greater spotted eagle
A. rapax	Tawny eagle
A. nipalensis	Steppe eagle
A. heliaca	Imperial eagle
A. wahlbergi	Wahlberg's eagle
A. gurneyi	Gurney's eagle
A. chrysaetos	Golden eagle
A. audax	Wedge-tailed eagle
A. verreauxi	Verreaux's or Black eagle

We have attempted to review most of the North American literature, and the major foreign journals (in particular the ornithological literature) up to and including September 1982. Although more than 3,400 citations have been included in this bibliography, we recognize that some information may have been inadvertently omitted. For this reason we request that all users of this bibliography note any references which may have been missed, and send citations or copies of the articles to the Raptor Information Center. We also request any comments or suggestions which would improve the usefulness of this bibliography.

HOW THE BIBLIOGRAPHY WAS PRODUCED

Compilation of the Literature. This bibliography grew out of a synthesis of published reports and partial bibliographies that began with the identification of *Aquila* references from the bibliography by Olendorff and Olendorff (1968). References from Anderson (1979) and personal collections of citations from M. Collopy, M. Fuller, V. Gargett, G. Hickman, B.-U. Meyburg, M. Ridpath, P. Rose, and P. Steyn were also incorporated. V. Galushin provided citations from the Russian literature.

Literature searches were commissioned from the Biological Information Service, Los Angeles, California, and the Fish and Wildlife Reference Service, Denver, Colorado. The Snake River Birds of Prey Research Project provided us with golden eagle references from their files. R. Olendorff also provided us with citations on a variety of species collected since the publication of Olendorff and Olendorff (1968), and from his Raptor Management Information System (RMIS) (Olendorff 1982).

Requests were also sent to all state and federal wildlife agencies soliciting published or unpublished references on golden eagles. Reprint request cards were sent to individual authors to acquire previously identified articles. A general request for _Aquila_ articles or citations was also printed in several volumes of the Raptor Information Center's newsletter, _The Eyas_.

The majority of articles were identified and photocopied at the Smithsonian Institution's Museum of Natural History, Division of Birds Library. The Library of Congress, U.S. Department of the Interior Main Library, the American Museum of Natural History's ornithological library, and the library of the National Institute for Health were used for additional searches. In addition to the journal searches, the _Zoological Record_, and the _Biological and Agricultural Index_ were also examined.

Once the articles were collected they were keyworded using the dictionary of keywords compiled for the _Working Bibliography of the Bald Eagle_ (Lincer et al. 1979). These articles were cataloged and are kept on file at the Raptor Information Center, Vienna, Virginia.

Creation and Management of the Aquila Data Base. The large number of references identified in our initial search demonstrated the need for an efficient data management system. The National Wildlife Federation's Data Processing Department aided in the development of a computerized bibliographic data management system designed as an updatable information processing system that could be sorted or edited at any point in the bibliographic procedure.

The bibliographic data base program, written for the NWF's Omega 480 computer, was designed with a screen capture format programmed in CICS. A data form was developed for input of citations in the bibliographic capture program. The program allowed records to be sorted by author, journal, and/or keywords and provided editorial and search capabilitites.

A master computer tape was created and was used to print the bibliography. Informatics, Inc. (Rockville, MD) provided processing and production computer programs which were used to edit and format the data for the printing of Chapters 3 through 7. The formatted production tape was used to create the camera ready copy for use by Port City Press, Inc. (Washington, D.C.) who printed the book.

ORGANIZATION OF THE BIBLIOGRAPHY

The bibliography is organized into seven chapters and one appendix. Chapter 3, the Master List of Citations, is comprised of a listing of all citations compiled alphabetically by author, and chronologically within each author. An article without a specified author is listed as "Anonymous" unless it is a publication by a particular organization to which the work would most appropriately be credited (i.e. U.S. Department of Agriculture, U.S. Department of the Interior). Each citation listed in Chapter 3 has been assigned a unique "Master Number" by which it is referred to in Chapters 4, 5 and 6.

We followed the *CBE Style Manual* (Council of Biology Editors, Third Ed., 1972) to list the citations in the Master List in Chapter 3, and for creating abbreviations. Journal titles were abbreviated using the *Biosis List of Serials*.

Chapter 4, the Permuted List of Keywords, is comprised of the primary keywords assigned to each reference listed in Chapter 3. Secondary keywords, which are species and geographic indicators, and keyword descriptors are listed after the primary keywords. These keyword strings refer by Master Number to specific citations listed in Chapter 3. Each Master Number listed in Chapter 4 is followed by the associated primary and secondary keywords. Therefore, in reviewing this section of the bibliography for specific information, the reader does not have to rely on a single keyword, but has supplementary additional keywords which further describe the contents of an article.

For each citation (as listed in Chapter 3) pertinent primary and secondary keywords have been assigned. Each primary keyword assigned to an article is placed on the left in the keyword string so that it appears first in the sequence under its respective alphabetic position. For example, a hypothetical citation, Master Number 1483 would appear in Chapter 3 as:

1483 Mearns, A.F. 1980. Golden eagle nesting in Idaho. Auk 80(3):56-61.

The primary and secondary keywords hypothetically assigned to this article could be:

Primary: Electrocution, Artificial Nest, Mortality
Secondary: *A. chrysaetos*, Idaho

In Chapter 4 this citation would be listed three times, as follows:

Under the A's: 1483 Artificial Nest • Electrocution • Mortality • *A. chrysaetos* • Idaho

Under the E's: 1483 Electrocution • Artificial Nest • Mortality • *A. chrysaetos* • Idaho

Under the M's: 1483 Mortality • Artificial Nest • Electrocution • *A. chrysaetos* • Idaho

As shown in the example, the secondary keywords (*A. chrysaetos*, Idaho) never appear to the left in the sequence, although they are always included as they provide additional information.

Chapter 5 is an Index of Master Numbers from Chapter 3 categorized by species and Chapter 6 is an Index of Master Numbers categorized by geographic location discussed in the reference. Chapter 7 is a list of records of golden eagle occurrences in North America, sorted geographically. Citations in Chapter 7 are not included in Chapter 3. Appendix A is a dictionary of the primary keywords as referenced in Chapter 4.

ACKNOWLEDGEMENTS

As with any project of this scope primary thanks must go to those individuals too numerous to mention, who responded to our requests for information and who expressed an interest in this project. We would like to acknowledge the cooperation of the various state wildlife agencies, U.S. Fish and Wildlife Service, U.S. Forest Service, U.S. Bureau of Land Management National Park Service, Fish and Wildlife Reference Service, U.S. Department of the Interior Main Library, Library of Congress, and the Smithsonian Institution's Museum of Natural History. Specific thanks go to G. Watson who coordinated our activities through the Division of Birds, Smithsonian Institution.

We would like to thank the following individuals who provided us with their own collections of *Aquila* citations early in the bibliographic process; G. Allen, S.B. Anderson, M. Collopy, M. Fuller, V. Galushin, V. Gargett, G. Hickman, R. Knight, B.-U. Meyburg, M. Ridpath, P. Rose, P. Saurola, and P. Steyn. Additionally, we would like to thank M. Kochert and K. Steenhof of the Snake River Birds of Prey Research Project and M.A. Jenkins of the Denver Service Center for their assistance. Various university libraries cooperated by providing interlibrary loan services.

Staff of the NWF provided the necessary assistance to make this bibliography a success. We would like to thank the Executive Staff of NWF for allowing us the freedom to complete this endeavor. We would like to express our appreciation to D. Miller and P. Parenteau for their support and assistance throughout this project. Specifically, we would like to acknowledge the assistance of the people in the various NWF deparments, including: M.A. Smith and T.B. Isom, Art Department; A. Thomas and P. Von Herbulis and Data Entry staff for keying citations; J. Dietz, W. Foley, M. McCarty, C. Walls and the staff of the Data Processing Department for computer assistance; A. Bohannan and J. Shore of the NWF Library for assistance with interlibrary loans; S. Artis, J. Rodriquez, and J. Weiser of the Word Processing Department for manuscript preparation; and to the staff of the Raptor Information Center, past and present for their assistance; K. Cline, R. Hudson, E. Junkins, R. Kennedy, K. Kirkpatrick, B. Millsap, L. Muse, M. Pramstaller, and G. Smith.

Additional thanks go to T. Roman, R. Blasick, J. Oster and J. Cray of Informatics, Inc. and C. Weigand of Port City Press, Inc. for their assistance in the production of this book.

Special thanks are due to the following individuals who provided advice and guidance, and whose past and present endeavors have influenced the completion of this project; D. Amadon, R. Clark, J. Lincer, R. Olendorff and W. Spofford. The late Leon Kelso provided much help by identifying eastern European and Russian citations, and by offering his services as a translator.

LITERATURE CITED

Anderson, S.R.B. 1979. Bibliography on the golden eagle. Raptor Research Foundation. Report.

Council of Biology Editors. 1972. CBE Style Manual. Third ed. American Institute of Biological Sciences, Washington, D.C. 292 pp.

Lincer, J.L., W.S. Clark and M.N. LeFranc, Jr. 1979. Working Bibliography of the Bald Eagle. NWF Scientific/Technical Series No. 2. 244 pp.

LITERATURE CITED (Cont.)

Olendorff, R.R. 1982. Raptor Management Information System: Raptor Data Base. U.S.D.I. Bureau of Land Management. 19 July 1982. 282 pp.

Olendorff, R.R. and S.E. Olendorff. 1968. An Extensive Bibliography on Falconry, Eagles, Hawks, Falcons, and other Diurnal Birds of Prey. Published by Authors.

Stresemann, E. and D. Amadon. 1979. Order Falconiformes. Pages 271-425. *in*: E. Mayr and E.W. Cottrell, eds. Check-list of Birds of the World. Vol. I, Second ed. Museum of Comparative Zoology, Cambridge, Mass.

CHAPTER 2.

INTRODUCTION TO THE GENUS *AQUILA*

The genus *Aquila* contains 10 closely related species (nine according to some authors who lump *Aquila rapax* and *A. nipalensis* into the single species *A. rapax*). We have treated all 10 as species in this bibliography. Because all members of the genus have their tibiotarsi feathered to the toes, they often are referred to as "booted" eagles. All taxonomic groupings and range information is based on Stresemann and Amadon (1979) unless otherwise noted.

Aquila pomarina - Lesser Spotted Eagle

A. p. pomarina - Europe from eastern Germany, Poland and the Balkan countries; Turkey through the U.S.S.R. to the Moscow and Ukraine regions. It winters in the Middle East and northeastern Africa to Rhodesia and occasionally southwest Africa.

A. p. hastata - Bengal, India south to Tamil Nadu, and Bangladesh.

The lesser spotted eagle occurs in habitat of mixed open and wooded landscape, from swampy lowlands to completely dry mountain woodlands. Its primary prey are small mammals, but some amphibians are taken in times of rodent shortages (Meyburg 1973). The plumage of the adult is uniformly dark and the immature has spots on the upper wing coverts, although smaller and less extensive than in the greater spotted eagle. The two spotted eagles can be separated in flight by primaries and secondaries darker than underwing coverts on the lesser spotted eagle and the reverse for the greater spotted eagle.

The breeding biology of this species, including the "Cain/Abel" conflict between the two nestlings, has been well documented (Meyburg 1974). The migration routes of the lesser spotted eagle are now becoming somewhat clearer through field work being conducted in the Middle East. There are 632 references to *A. pomarina* in the bibliography.

Aquila clanga - Greater Spotted Eagle

The distribution of the greater spotted eagle extends from eastern Europe through central Asia. The winter range is south into northwest Africa, the Middle East, and northern India. Its preferred breeding habitat is wet woodlands near water.

The greater spotted eagle is another small species of *Aquila*. It differs from the larger members of the genus by its round, not oval nostrils, a trait shared with its close relative the lesser spotted eagle and with *A. wahlbergi*. These two species are very similar and are difficult to differentiate in the field. The adult plumage is dark black, the immature somewhat lighter. The immature has white spots on the upper wing coverts which account for its English name. The bibliography contains 428 references to the greater spotted eagle.

Aquila rapax - *Tawny Eagle*

A. r. vindhiana - India, Bangladesh, Burma and Pakistan.

A. r. belisarius - Morocco, Tunisia, Sudan, southern Arabia, northern Nigeria and northern Kenya.

A. r. rapax - Angola, southern Zaire, and Kenya south to the Cape Province.

The tawny eagle is one of the smaller of the *Aquila* eagles. It has been reported to be a scavenger and pirate, but also capable of killing large prey. Plumage is variable but generally is dark brown and unstreaked in adults. It is generally sedentary throughout its range, but some local movement occurs in relation to wet and dry seasons in Africa (Thiollay 1977). It is often found in close association with humans.

The discussion under *A. nipalensis* explains our consideration of these as two separate species. There are 285 references to the tawny eagle in this bibliography.

Aquila nipalensis - Steppe Eagle

A. n. orientalis - Steppes of Rumania and Russia, east to the steppes north of the Caspian and Aral Seas. Winters from the Near East, Iraq, and Arabia to eastern Africa to Rhodesia.

A. n. nipalensis - Steppes of central Asia, from the Aral Sea across Mongolia to northern China - winters to India, Burma, and southern China.

The steppe eagle is almost as large as the golden eagle but is not as robust and has smaller feet. The diet of the steppe eagle consists primarily of small mammals, however, termites are taken in great numbers on the wintering grounds in Africa.

As discussed earlier, some authorities consider the steppe eagle and the tawny eagle the same species. A thorough review of all the points involved in this argument would not be appropriate in this chapter. We have based our decision to separate the two on comments made by Brooke et al. (1972) and Amadon (pers. comm.). These are:

1. There are great differences in the juvenal and immature plumages of steppe and tawny eagles.

2. The steppe eagle does not attain full adult plumage until its seventh year, the tawny eagle does so in its fifth.

3. The steppe eagle is migratory, the tawny is sedentary.

4. The hallux is longer in the female steppe eagle than in the male, but not in the tawny eagle.

Jankowitz (1976) adds an additional anatomical difference in that the gape of the steppe eagle is proportionately longer than in the tawny eagle. The mouth extends past the eye on the steppe eagle but ends at the eye in the tawny eagle. Other authors who have worked with these species concur with species separation (Christensen et al. 1972). However, Brown and Amadon (1968) and Cramp (1980) chose to combine the two into the one species, *A. nipalensis*.

The migration of great numbers of the western race, *A. n. orientalis*, through the Middle East in the fall and spring constitutes one of the great ornithological wonders of the region. The eastern race, *A. n. nipalensis*, is not well known and differs from the western race by being larger, by the adults having darker plumage, and by having a preference for habitat in rocky hills in lieu of steppes (Dementiev et al. 1966). There are 310 references to this species in the bibliography.

Aquila heliaca - Imperial Eagle

A. h. adalberti - Spain, Portugal and the upper peninsula of Morocco.

A. h. heliaca - Southeastern Europe, Russia, Turkey, Cyprus, Iran, and Chinese Turkestan. It winters south to northeastern Africa, Iraq, India, southeastern China and northern Indo-China.

The imperial eagle is slightly smaller than the golden eagle. It is a resident of lowland plains and foothills and preys upon small mammals. Its plumage is similar to that of the golden eagle except that the nape of the adult is more extensively covered with very pale golden feathers. The back contains two white areas on the scapulars which are much more prominent in the Spanish race (*A. h. adalberti*). The adult tail is bicolored and the head and neck are longer than in the golden eagle. Immature breast plumage is a pale yellow and is streaked on the nominate race and unstreaked on the Spanish race.

The northern breeders and immatures of *A. h. heliaca* are partially migratory in the fall. The Spanish race formerly bred in northwest Africa but there are no breeding records for the past 20 years. There are 468 references to the imperial eagle in the bibliography.

Aquila wahlbergi - Wahlberg's Eagle

Wahlberg's eagle is an African species which is widespread south of the Sahara to northern Cape Province. It is the smallest *Aquila* eagle, with rounded nostrils, as in the spotted eagles. The Wahlberg's eagle lays only one egg, somewhat atypical of the *Aquila* eagles.

The taxonomic position of *A. wahlbergi* within the genus is questioned by some authors. Smeenk (1974) stated that it was not closely related to the lesser spotted eagle, but was more similar to the booted eagle, *Hieraaetus pennatus*, based on plumage and vocalizations. Jollie (1957) suggested that on the basis of size, the presence of a small crest absent on all other *Aquila* species, and dimorphic plumage, *A. wahlbergi* may be more closely related to *Spizaetus* or *Hieraaetos* than to *Aquila*. There are 167 references included in this bibliography.

Aquila gurneyi - Gurney's Eagle

This species is the least studied member of the genus. Its known range is restricted to the western parts of New Guinea and some of the eastern islands of Indonesia.

The Gurney's eagle is a forest species and like other forest raptors has relatively short, broad wings and a long tail. Some taxonomists include it in the genus *Spizaetus*, but Jollie (1957), using feather structure and molt as criteria, has placed it in *Aquila*.

It is considered one of the large robust *Aquila* eagles. Little is known of its feeding habits. Brown and Amadon (1968) stated that in their opinion it is not closely related to the wedge-tailed eagle, the member of the genus with the nearest range. There are 18 references included in this bibliography.

Aquila chrysaetos - Golden Eagle

A. c. homyeri - Spain, Sardinia, northeastern Africa, south to Mauritania, Egypt, Sinai, Turkey, Syria, Arabia Caucasus and Iran.

A. c. chrysaetos - British Isles, Lapland, northern Russia south to the Pyrenees, Alps, Balkan Peninsula, east to western Siberia and northern Kirghiz Steppes intergrading with *A. c. canadensis* in the Altai Mountains and central Siberia.

A. c. canadensis - Kamchatka, Altai; eastern Asia east of the range of *A. c. chrysaetos*, south to northern Mongolia and Manchuria; North America, in the west, south to Durango, Mexico and sparingly in the east.

A. c. daphanea - Transbaicalia, Hazara, Pakistan; mountains of Asia south of *A. c. chrysaetos* and *A. c. canadensis*, from Russian Turkestan east to northern China, south to Afghanistan, Himalayas and southwestern China.

A. c. japonica - Mountains of Korea and Japan, south in winter.

The golden eagle is the most widespread of the genus *Aquila*. The plumage of the immature golden eagle is darker than other members of *Aquila*. Almost all individuals have the characteristic "golden" nape from which the English name is derived.

This species has been more thoroughly studied than the other *Aquila* species and there are 2305 references to it in this bibliography. The major food source for the golden eagle throughout its range is lagomorphs. The occassional predation upon lambs and other small livestock by golden eagles has resulted in considerable controversy over their protection. The extent of this predation is not well documented and it is believed to be caused primarily by sub-adult and immature birds.

Experiments with golden eagles have shown that in the absence of wind and on flat terrain, an adult bird has difficulty lifting an object half its weight from the ground (see Chapter 4, keyword "Lift"). This suggests a maximum prey size of less than 2 kg. Of course these eagles will feed upon the carcasses of much larger animals including adult domestic sheep and mule deer (*Odocoileus hemionus*).

Aquila audax - Wedge-tailed Eagle

A. a. audax - Australia, southern New Guinea.

A. a. fleayi - Tasmania.

The wedge-tailed eagle is a fairly common resident in Australia and Tasmania. It has been recorded in New Guinea where it is likely a vagrant from Australia. This species was formerly classified in a monotypic genus *Uroaetus*, but was placed into *Aquila* because of similarities of feather structure and molt with other members of the genus (Amadon 1978).

Analysis of prey from the nests of the wedge-tailed eagle show that rabbits are its primary prey. Additional prey includes other small mammals, birds and occassional reptiles (Ridpath 1969). The wedge-tailed eagle is widespread in Australia and relatively common in uninhabited areas despite mass persecution by shooting and poisoning coupled with a now defunct bounty system. This species does not have full legal protection in all areas of Australia. There are 86 references to *A. audax* included in this bibliography.

Aquila verreauxi - Black Eagle

The black (or Verreaux's eagle as it is often called) is widespread in the mountainous areas of eastern Africa, ranging from South Africa northward to Sinai, Israel and the Arabian Peninsula. This species has been studied extensively and 172 references to it are included in this bibliography.

Its primary prey are two species of hyrax (*Procavia capensis* and *Heterohyrax brucei*) which inhabit the rock outcrops around its nest site (Gargett 1980). Other species of small mammals have also been recorded as prey items. There are few animal damage problems associated with this species.

An interesting feature of its breeding biology is the "Cain/Abel" struggle in which the first hatched chick inevitably kills its younger sibling (Gargett 1978). This behavior, in which two eggs are laid but only one nestling survives, is reported in other species in the genus, but only in this species and the lesser spotted eagle is the outcome usually inevitable. In other species, for example the golden eagle, the mortality of a sibling may result from starvation.

The plumage of the black eagle is markedly different from the other *Aquila* eagles. Adults are jet black with large white patches on the back and the wings. The juvenal plumage is much lighter with a streaked breast and a chestnut mantle. Like all of the large, robust *Aquila* eagles, the black eagle is a graceful and skillful flyer.

LITERATURE CITED

Amadon, D. 1978. Remarks on the taxonomy of some Australasian raptors. Emu 78(3):115-118.

Brooke, R.K., J.H. Grobler, M.P. Stuart Irwin and P. Steyn. 1972. A study of the migratory eagles *Aquila nipalensis* and *A. pomarina* (Aves: Accipitridae) in southern Africa, with comparative notes on other large raptors. Occas. Pap. Natl. Mus. of Rhodesia 1972 B5:61-112.

Brown, L.H. and D. Amadon. 1968. Eagles, Hawks and Falcons of the World. McGraw-Hill Book Co., N.Y. 945 pp.

Christensen, S., B.P. Nielsen, R.F. Porter and I. Willis. 1972. Flight identification of European raptors. Part 3. *Aquila* and white-tailed eagles. Brit. Birds 65(2): 52-78.

Cramp, S. (ed.). 1980. Handbook of the Birds of Europe, the Middle East and North Africa. Vol. II. Oxford Univ. Press. 695 pp.

Dementiev, G.P., N.A. Gladkov, E.P. Spangenberg and A.M. Sudilovskaya. 1966. Birds of the Soviet Union. Vol. I. Smithsonian Inst. and Natl. Sci. Found., Israel; Program for Scientific Translation. 704 pp.

Gargett, V. 1978. Sibling aggression in the black eagle in the Matopos, Rhodesia. Ostrich 49(2):57-63.

__________. 1980. The black eagles of the Matopos. Zimbabwe Sci. News 14(7):155-159.

Jankowitz, M. 1976. Tawny or steppe eagle? Bokmakierie 28(3):64-65.

Jollie, M. 1957. Comments on the bird genus _Aquila_ and its occurrence in New Guinea and Australia. _Nova Guinea_ (New Ser.) 8(2):179-181.

Meyburg, B.-U. 1973. Studies of less familiar birds: 172. Lesser spotted eagle. Brit. Birds 66:439-447.

_____________. 1974. Sibling aggression and mortality among nestling eagles. Ibis 116:224-228.

Ridpath, M.G. 1969. The wedge-tailed eagle. Aust. Nat. Hist. 16(7):209-212.

Smeenk, C. 1974. Comparative-ecological studies of some east African birds of prey. Ardea 62(1/2):1-97.

Stresemann, E. and D. Amadon. 1979. Order Falconiformes. Pages 271-425. _in_: E. Mayr and G.W. Cottrell, eds. Check-list of Birds of the World. Vol. I. Second Ed. Museum of Comparative Zoology, Cambridge, Mass.

Thiollay, J.-M. 1977. L'importance des hivernants palearctiques dans le peuplement de rapaces d'Afrique tropicale. Nos Oiseaux 34(2):59-64.

Chapter 3

Master List of Citations

Master No.	Citation
1	**Abbott, C.G.** *1924.* Period of incubation of the golden eagle. Condor 26(5):194.
2	**Abdulali, H.** *1968.* A catalogue of the birds in the collection of the Bombay Natural History Society- 3. Falconiformes. J. Bombay Nat. Hist. Soc. 65(3):696-723.
3	**Acar, B., M. Beaman and R.F. Porter (For D.H.K. Dernegi).** *1977.* Status and migration of birds of prey in Turkey. Pages 182-187. *in:* R.D. Chancellor, Ed. Proc. of the World Conf. on Birds of Prey. 1-3 October, 1975. Vienna. I.C.B.P.
4	**Adam, R.M.** *1873.* Notes on the birds of the Sambhur Lake and its vicinity. Stray Feathers 1(5):361-404.
5	**Adams, A.L.** *1864.* Notes and observations on the birds of Egypt and Nubia. Ibis 6(21):1-36.
6	**Adams, E.** *1899.* A day with the raptores. Condor 1(2):21-22.
7	**Adkisson, C.S., R.N. Conner, I.D. Prather and J.W. Via.** *1978.* Recent sightings of eagles in southwestern Virginia. Raven 49(2):32-33.
8	**Agafonov, A.V., D.S. Rezinko, A.A. Rozhkov and N.M. Semenov.** *1957.* K ekologii stepnogo orla. (Ecology of *Aquila nipalensis* .) Bull. MOIP 62(2):174-181. (In Russian, English Summ.)
9	**Agardi, E.** *1925.* Nidologische daten aus der Mecsekgegend. Aquila 32-33:283-284.
10	**Agardi, E.** *1951.* *Aquila pomarina* Brehm - lesser spotted eagle - in Mecsek Mountains. Aquila 55-58:281.
11	**Agardi, E.** *1964-1965.* Observations in Transnubia. Aquila 71-72:241-243.
12	**Aharoni, I.** *1931.* Brutbiologisches aus der Syrischen Wuste und dem Libanon. Beitr. Fortpfl.-Biology Vogel 7(6):222-226.
13	**Aharoni, J.** *1932.* Bemerkungen und erganzungen zu R. Meinertzhagens werk "Nicoll's Birds of Egypt." J. Ornithol. 80(3):416-424.
14	**Aiken, C.E.H.** *1928.* Notes on the golden eagle *(Aquila chrysaetos)* in Colorado. Auk 45(3):373-374.
15	**Aitken, A.D., Jr.** *1935.* A golden eagle's nest. Bird-Lore 37(2):105-108.
16	**Albarda, H.** *1897.* Uber *Pomarinus* als Artname. J. Ornithol. 45:170-171.
17	**Albrektsson, T., R. Knape and M. Sandstrom.** *1969.* Ornitologisk inventering i Padjelanta Nationalpark 1966, 1967 och 1968. (The bird fauna of the Padjelanta National Park, Swedish Lapland.) Fauna Och Flora 64(1):20-30.
18	**Alcorn, G.D.** *1971.* Key to nest and eggs of the birds of the state of Washington. Univ. of Puget Sound, Occas. Pap. No. 42. 37 pp.
19	**Alcorn, G.D.** *1971.* Check list - birds of the state of Washington. Univ. of Puget Sound, Occas. Pap. No. 41. p. 155-199.
20	**Alexander, F.M.** *1930.* Notes on the birds of south central Kansas. Wilson Bull. 42(4):241-244.
21	**Alexander, H.G.** *1952.* Identifying birds of prey in the field. Bull. Brit. Ornithol. Club 72(6):55-61.
22	**Alford, J.R., III and E.G. Bolen.** *1972.* A note on golden eagle talon wounds. Wilson Bull. 84(4):487-489.
23	**Ali, S.** *1945.* The Birds of Kutch. Oxford Univ. Press, London. 175 pp.
24	**Ali, S.** *1949.* Indian Hill Birds. London. 240 pp.
25	**Ali, S.** *1961.* Book of Indian birds. 6th ed. Bombay. 209 pp.
26	**Ali, S. and S.D. Ripley.** *1978.* Handbook of the birds of India and Pakistan. Vol. 1. Oxford Univ. Press, Bombay and New York.
27	**Alippi, N.** *1919.* Gli uccelli di comparasa accidentale in Italia e il loro valore per lo studio delle migrazioni. Riv. Ital. Ornitol. 5:31-65.
28	**Allain, M.** *1946.* Capture d'un aigle fauve *Aquila c. chrysaetos* (L.) dans le Finistere. Oiseau Rev. Fr. Ornithol. 16:173.
29	**Allain, M.** *1947.* Seconde capture d'un aigle fauve *Aquila c. chrysaetos* (L.) dans le Finistere. Oiseau Rev. Fr. Ornithol. 17(1):128.
30	**Allan, P.F. and P.R. Sime.** *1943.* A hawk census on Texas panhandle highways. Wilson Bull. 55(1):29-39.
31	**Alldredge, A.W. and P.W. Stahlecker.** *1974.* Impact of project Rio Blanco on cliff-nesting hawks and eagles. J. Colo.-Wyoming Acad. Sci. 7(5):58-59.
32	**Allen, B.A.** *1979.* Determination of status and management of the golden eagle. New York Dept. Env. Conserv., Proj. No. E-1-3., Study No. VIII-11. 4 pp.
33	**Allen, C.A.** *1882.* Golden eagle's nest. Ornithol. and Ool. 7(23):181.
34	**Allen, G.T., R.L. Knight and M.V. Stalmaster.** *1980.* An annotated bibliography of bald and golden eagles in Washington. Washington Dept. of Game. Report. 66 pp.
35	**Allen, J.A.** *1869.* Notes on some of the rarer birds of Massachusetts. Am. Nat. 3(10):505-519.
36	**Allen, J.A.** *1886.* A revised list of the birds of Massachusetts. Bull. Am. Mus. Nat. Hist. 7(15):221-271.
37	**Allen, J.A.** *1905.* Report on the birds collected in northeastern Siberia by the Jesup North Pacific Expedition, with field notes by the collectors. Bull. Am. Mus. Nat. Hist. 21(13):216-257.
38	**Allen, J.A.** *1907.* The types of the North American genera of birds. Bull. Am. Mus. Nat. Hist. 23(16):279-384.
39	**Allen, J.A.** *1909.* The habitat groups of North American birds in the American Museum of Natural History. Auk 26(2):165-174
40	**Allen, J.A. and W. Brewster.** *1883.* List of birds observed in the vicinity of Colorado Springs, Colorado, during March, April, and May, 1882. Bull. Nuttall Ornithol. Club 8(4):189-198.

Master No.	Citation
41	**Allen, J.C.** *1939.* Ecology and management of Nelson's bighorn on the Nevada Mountain ranges. Trans. of the 4th North American Wildlife Conf., Wildlife Manage. Inst. p. 253-256.
42	**Allen, P.M.** *1978.* Steppe eagle in Nairobi National Park. E. African Nat. Hist. Soc. Bull. (Nov./Dec.):136-137. (Title Only)
43	**Alleon, A.** *1886.* Memoire sur les oiseaux dans la Dobrodja et la Bulgarie. Ornis 2:397-428.
44	**Allert, O.B.** *1939.* Notes on certain raptores in Allamakee, Clayton and Dubuque Counties, Iowa. Iowa Bird Life 9(3):34-36.
45	**Alleva, E. and E. Visalberghi.** *1977.* Uccelli rapaci di Palestina. Riv. Ital. Ornitol. 47(4):257-272. (English Summ.)
46	**Allouse, B.E.** *1960.* Birds of Iraq. Ar-Rabitts, Baghdad. 276 pp.
47	**Alsop, F.J., III.** *1979.* Population status and management considerations for Tennessee's threatened and endangered bird species. East Tennessee State Univ., Rep. to Tennessee Wildl. Res. Agency. 200 pp.
48	**Amadon, D.** *1978.* Remarks on the taxonomy of some Australasian raptors. Emu 78(3):115-118.
49	**Amadon, D. (Chairman).** *1975.* Sticky problems of hawk identification: a panel discussion. Pages 118-136. *in:* M. Harwood, Ed. Proc. of the North American Hawk Migration Conf., Syracuse, N.Y. 1974. Hawk Migration Assoc. of N. America.
50	**Ambruster, J.H.** *1920.* An eagle's cunning. Nature Study Rev. 16(9):383-384.
51	**Andersen, G.** *1967.* Sommeriagttagelse af kongeorn *(Aquila chrysaetos)* . Dansk Ornithol. 61(3/4):183-188.
52	**Andersen, T,. J.B. Brunn and B.P. Nielsen.** *1973.* Rapport fra sjaeldenhedsudvalget for 1972. (Report of the rarities committee for 1972.) Dansk Ornithol. 67(3/4):139-143.
53	**Andersen, W.C. and G.R. Craig.** *1976.* Raptor population and productivity studies in southeastern Colorado. Colorado Div. of Wildlife. Proj. No. W-124-R-4, Wk. Pl. 1, Job 3. pp. 209-211.
54	**Anderson, A.** *1871.* Notes on the raptorial birds of India. Proc. Zool. Soc. London 1871:675-690.
55	**Anderson, A.** *1872.* Additional notes on the birds of northwestern India. Proc. Zool. Soc. London 1872:619-623.
56	**Anderson, A.** *1872.* Notes on the raptorial birds of India. Proc. Zool. Soc. London 1872:68-81.
57	**Anderson, A.** *1872.* Voracity of the imperial eagle. Zoologist (2nd Ser.) 7:3331-3332.
58	**Anderson, A.** *1875.* On the nidification of certain Indian birds. Part IV. Ibis (3rd Ser.) 5(18):199-204.
59	**Anderson, A.** *1876.* Corrections of and additions to "Raptorial birds of north-western India." Proc. Zool. Soc. London 1876:310-317.
60	**Anderson, A.E.** *1956.* Unusual behavior of a golden eagle in southeastern New Mexico. Auk 73(2):287-288.
61	**Anderson, D.W. and J.J. Hickey.** *1972.* Eggshell changes in certain North American birds. Pages 514-540. *in:* Proc. 15th Int. Ornithol Conf.
62	**Anderson, J.** *1971.* Eagle habitat requirements and forest management. Pages 56-59. *in:* Proc. of the Fish and Wildlife Habitat Training Conf., Nov. 10 - Dec. 2. Eugene, OR. U.S.D.A. Forest Service, Region 6.
63	**Anderson, N.A.** *1940.* Mountain goat study, progress report. Washington Dept. of Game, Biology Bull. No. 2.
64	**Anderson, R.J. and A.M. Bruce.** *1980.* A comparison of selected bald and golden eagle nests in western Washington. Pages 117-120. *in:* R.L. Knight, G.T. Allen, M.V. Stalmaster and C.W. Servheen, Eds. Proc. of the Washington Bald Eagle Symposium. 14-15 June. Seattle, Wa. Wash. Dept. of Game.
65	**Anderson, R.M.** *1907.* The birds of Iowa. Proc. of the Davenport Acad. Sci. 11:125-417.
66	**Anderson, W.W.** *1975.* Pole changes keep eagles flying. Trans. and Distrib. (Nov.):28-31.
67	**Andersson, C.J.** *1872.* Notes on the birds of Damava Land. John Van Voorst, London. 394 pp.
68	**Anfrie, E.** *1913.* A propos des aigles criards observes en France. Oiseau Rev. Fr. Ornithol. 3(55):169-172.
69	**Anfrie, E.** *1917.* Les anomalies et les variations individuelles parmi les oiseaux faisant partie de notre collection. Oiseau Rev. Fr. Ornithol. 5(100/101):113-116.
70	**Angele, T.** *1910. Aquila clanga* Pall. in Oberosterreich erlegt. Ornithol. Jahrbuch 21(1/2):60.
71	**Angus, W.C.** *1866.* Osprey and golden eagle in Aberdeenshire. Zoologist 24:497.
72	**Anonymous.** *1847.* Occurrence of the golden eagle near Hungerford. Zoologist 5:1695.
73	**Anonymous.** *1851.* Eagles. Bentley's Misc. 30:115-128.
74	**Anonymous.** *1873.* Deutsche ornithologische gesellschaft in Berlin. J. Ornithol. 21(4):455-460.
75	**Anonymous.** *1873.* Battle in the air (between "large brown eagles'). Forest and Stream 1(Sept.):71. (Title Only)
76	**Anonymous.** *1874.* Singular capture of an eagle. Zoologist (2nd Ser.) 9:3953.
77	**Anonymous.** *1883.* Golden eagle. Ornithol. and Ool. 8(3):24.
78	**Anonymous.** *1883.* The age of an eagle. Zoologist (3rd Ser.) 7:422-423.
79	**Anonymous.** *1886.* IX. Jahresbericht (1884) des ausschusses fur beobachtungsstationen der vogel deutschlands. J. Ornithol. 34(2):129-387.
80	**Anonymous.** *1887.* X. Jahresbericht (1885) des ausschusses fur beobachtungsstationen der vogel deutschlands. J. Ornithol. 35(4):337-615.
81	**Anonymous.** *1888.* Bericht uber die September-Sitzung. J. Ornithol. 36(1):99-115.
82	**Anonymous.** *1889.* A terrier carried off by an eagle. Ornithol. and Ool. 14(5):78.
83	**Anonymous.** *1895.* Escape of a caged eagle. Zoologist (3rd Ser.) 19:380.
84	**Anonymous.** *1895.* Escape of a caged eagle. Zoologist (3rd Ser.) 19:434.
85	**Anonymous.** *1895.* Escape of a caged eagle. Zoologist (3rd Ser.) 19:455.
86	**Anonymous.** *1895.* The golden eagle. Knowledge (Nov. 1):268-270.
87	**Anonymous.** *1896.* Golden eagle. Ornithol. and Ool. 8(3):24.
88	**Anonymous.** *1900.* Untitled. Ibis 6(23):530-531.
89	**Anonymous.** *1904.* The golden eagle. Birds and Nature 15(1):10-13.
90	**Anonymous.** *1905.* Ueber einen kampf zwischen steinadler und gemse. Ornithol. Beob. 4(1):15-16.
91	**Anonymous.** *1911.* Kills eagle attacking girls. Oologist 28(3):52.

Master List of Citations (Cont.)

Master No.	Citation
92	**Anonymous.** *1912.* Aigle criard *(Aquila maculata,* Gm.). Gerfaut 2(4):64.
93	**Anonymous.** *1912.* Aigle royal. Ornithol. Beob. 9(12):235.
94	**Anonymous.** *1912.* Le vol de l'aigle. Gerfaut 2(1):12-13.
95	**Anonymous.** *1915.* The B.O.U. list of British birds. Brit. Birds 8(12):278-286.
96	**Anonymous.** *1915.* Additions and corrections to the hand-list of British birds. Brit. Birds 9(1):2-10.
97	**Anonymous.** *1917.* Weight of an eagle's brain. Wilson Bull. 29(3):164-165.
98	**Anonymous.** *1922.* Vom steinadler. Ornithol. Beob. 19(8):127.
99	**Anonymous.** *1922.* Einen steinadler erlegt im Engadin. Ornithol. Beob. 20(2):28.
100	**Anonymous.** *1922.* Adler am Alpstein. Ornithol. Beob. 19(8):126.
101	**Anonymous.** *1924.* Adlerhorste im Glarnerland 1923. Ornithol. Beob. 21(6):90.
102	**Anonymous.** *1926.* A closed season on golden eagle eggs. Condor 28(4):184.
103	**Anonymous.** *1927.* Filming of the golden eagle. Spectator 138(April 9):639. (Title Only)
104	**Anonymous.** *1927.* Collision with small plane. Condor 29:172.
105	**Anonymous.** *1927.* Fifth report of the Committee on the nomenclature and records of occurrences of rare birds in the British Islands. Ibis 3(2):310-313.
106	**Anonymous.** *1929.* A check-list of the birds of the Trans-Nzoia District of Kenya Colony. Part I. Bateleur 1(4):115-135.
107	**Anonymous.** *1930.* Secrets of eagles' flight revealed by camera. Pop. Mech. 54(2):190.
108	**Anonymous.** *1930.* Scientific observation and collecting. Bateleur 2(1):20-23.
109	**Anonymous.** *1932.* Wrong eagle on Washington coin. Lit. Digest 114(13):26.
110	**Anonymous.** *1932.* Pleistocene eagles. Science 76(1977):10.
111	**Anonymous.** *1932.* Golden eagle at home. Country Life Jan. 2:5-6. (Title Only)
112	**Anonymous.** *1934. Aquila pomarina,* Brehm, lesser spotted eagle. Ostrich 5(2):85.
113	**Anonymous.** *1935.* In Sudwest-Afrika uberwinterende schreiadler *(Aquila pomarina).* Ornithol. Monatsber. 43(1):26.
114	**Anonymous.** *1935.* Eagle nonsense again. Bird-Lore 37(6):456.
115	**Anonymous.** *1935.* Eagle nonsense again. Bird-Lore 37(6):456.
116	**Anonymous.** *1936.* Cattura di aquile in Sicilia. Riv. Ital. Ornitol. (2nd Ser.) 6(4):291-292.
117	**Anonymous.** *1936.* Fake eagle stories. Nature Mag. 27(2):106.
118	**Anonymous.** *1936.* Bundesgesetz uber jagd und vogelschutz. Ornithol. Beob. 33(12):195-206.
119	**Anonymous.** *1939.* The golden eagle. Nature Mag. 32(6):333.
120	**Anonymous.** *1939.* Eagles of other lands. Nature Mag. 32(6):334-336.
121	**Anonymous.** *1939.* Panhandle hunters mount fast autos and go eagle gunning. Nature Mag. 32(6):340.
122	**Anonymous.** *1942.* Legal protection for hawks and owls. Illinois Audubon Bull. 41:14.
123	**Anonymous.** *1942.* Eagle flight. Animal King. 45(3):67-70.
124	**Anonymous.** *1943.* On carrion habits of our eagles, July 18th, 1942: Hluhluwe Game Reserve. Ostrich 14(3):180.
125	**Anonymous.** *1945.* Rings um den steinadler. Ornithol. Beob. 42(4/5):70-72.
126	**Anonymous.** *1949.* Cranes and eagles. Blue Jay 7(4):16.
127	**Anonymous.** *1951.* Display of common buzzard. Brit. Birds 44(12):411-412.
128	**Anonymous.** *1955.* Recoveries. Ring 1(5):86.
129	**Anonymous.** *1956.* Golden eagle. Raven 27(1/2):16-17.
130	**Anonymous.** *1956.* Golden eagle killed on Shenandoah Mountain. Va. Wildl. 17(3):26.
131	**Anonymous.** *1959.* Banding in the Alps. Nat. Hist. 68(4):183-185.
132	**Anonymous.** *1962.* Avifauna van Nederland. Ardea 50(1/2):1-103.
133	**Anonymous.** *1963.* Materialy po awifauny Polski. (Contributions to the avifauna of Poland). Acta Ornithol. 7(9):253-304. (English Summ.)
134	**Anonymous.** *1963.* Wahlberg's eagle a migrant. Bokmakierie 15(2):17.
135	**Anonymous.** *1964.* Golden eagle survey. Brit. Birds 57(7):308.
136	**Anonymous.** *1965.* The case for the golden eagle. Natl. Parks Mag. 39(213):10-11.
137	**Anonymous.** *1965.* The case of the substitute golden eagle. Scott. Birds 3(6):282.
138	**Anonymous.** *1967.* What happens to the birds? Birds Illus. 13(2):62.
139	**Anonymous.** *1967.* Recoveries. Ring 5(50/51):18-20.
140	**Anonymous.** *1967.* Liste des especes d'oiseaux observes en Belgique et Leurs formes geographiques. Gerfaut 57(3/4):365-465.
141	**Anonymous.** *1968.* Hawks and eagles need more protection. S.C. Wildl. 15(2):15.
142	**Anonymous.** *1968.* Migrant eagle at sea. Bokmakierie 20(4):100.
143	**Anonymous.** *1968.* Close-up on the birds of prey. Animal King. 71(6):13-15.
144	**Anonymous.** *1969.* Wedge-tailed eagle. Birds of the World, I.P.C. Magazines Ltd., London.
145	**Anonymous.** *1969.* Zur freilandbestimmung Europaischer adlerarten. Ornithol. Mitt. 21(10):208-212.
146	**Anonymous.** *1970.* Golden eagles nest in man-made tree. Reclamation Era (August):5.
147	**Anonymous.** *1970.* Longevity of birds of prey and owls in captivity. Int. Zoo Yearb. 10:36-37.
148	**Anonymous.** *1970.* Secretary Hickel halts issuance of 'blanket" permits to kill golden eagles. Raptor Res. News 4(2):36-37.
149	**Anonymous.** *1971.* A slaughtering of eagles. Audubon (Sept.):72-73.
150	**Anonymous.** *1971.* Recovery round-up. Aust. Bird Band. 9(3):65-68.
151	**Anonymous.** *1971.* The loss of eagles in Wyoming: How one Federal agency replied to the protests. Audubon Bull. 160(Dec.):13-14.
152	**Anonymous.** *1972.* Illegale einfuhr von geraubten falken und steinadlern. Ornithol. Mitt. 24(24):178.
153	**Anonymous.** *1972.* Wrist slap given for eagle killing. Cons. News 37(24):2-3.
154	**Anonymous.** *1972.* Story about 509-50217. N.J. Outdoors 22(11):8-9.

Master No.	Citation
155	**Anonymous.** *1972.* Recovery round-up. Aust. Bird Band. 10(2):40-42.
156	**Anonymous.** *1973.* Wedge-tailed eagle survey. R.A.O.U. Newsl. 14:2.
157	**Anonymous.** *1973.* Birds of the Opalton area. Emu 1:21-22.
158	**Anonymous.** *1973.* Untitled. Mass. Wildlife 24(2):1.
159	**Anonymous.** *1973.* Massive "accidental" eagle kill laid to trappers. Cons. News 38(13):2-4.
160	**Anonymous.** *1974.* A checklist of birds of the Tennessee Valley. T.V.A. Div. of Forestry, Fisheries and Wildl. Development. 5 pp.
161	**Anonymous.** *1974.* Eagle killers. Defenders 49(3):250.
162	**Anonymous.** *1974.* Recovery round-up. Aust. Bird Band. 12(3):65-67.
163	**Anonymous.** *1974.* Recovery round-up. Corella 12(4):82-84.
164	**Anonymous.** *1974.* Protection of birds of prey and owls. I.C.B.P. Ann. Report 1974:11-12.
165	**Anonymous.** *1974.* Panhandle hunters mount fast autos and go eagle gunning. Nature Mag. 32(6):340.
166	**Anonymous.** *1974.* A rehabilitation program for raptorial birds in Minnesota. Defenders 49(1):28-30.
167	**Anonymous.** *1976.* Goldens eagles killed by poisoned baits. Nature Canada (1976). p. 22.
168	**Anonymous.** *1976.* 1976 Annual Report. Oregon Fish and Wildlife, Wildlife Div. 166 pp.
169	**Anonymous.** *1976.* Powerline electrocutions - hazards made safer. Cons. News 41(21):8-10.
170	**Anonymous.** *1977.* No sympathy for the eagle slayer. Cons. News 42(7):10-11.
171	**Anonymous.** *1977.* Some recent records of Palearctic birds in Kenya and Tanzania. Scopus 1(2):39-43.
172	**Anonymous.** *1977.* 1976 raptor mortality. Utah Div. of Wildl. Resources, Wildl. Report (March 14). p. 2.
173	**Anonymous.** *1977.* Eagles in Arkansas. Arkansas Audubon Soc. Quart. Newsl. 22(1):1
174	**Anonymous.** *1977.* Eagles, goshawks, fledge successfully thanks to loggers. Wash. Wildl. 29(3):19.
175	**Anonymous.** *1977.* Recovery round-up. Corella 1(4):84-87.
176	**Anonymous.** *1977.* With curved beak and claw. Audubon 79(4):13-49.
177	**Anonymous.** *1978.* Krasnaya Kniga Kazakhskoi SSR. (Red Data Book of Kazakhstan SSR.) Kainar, Alma-Ata. 206 pp.
178	**Anonymous.** *1978.* Krasnaya Kniga SSSR. (Red Data Book of the USSR.) Lesnaya Promyshlennost, Moscow. 460 pp. (In Russian)
179	**Anonymous.** *1978.* Krasnaya Kniga Moldavskoi SSR. (Red Data Book of Moldavian SSR.) Kartya Moldovenyagke, Kishinev. 120 pp.
180	**Anonymous.** *1978.* Colour-tagged Wahlberg's eagles. Scopus 2(1):28.
181	**Anonymous.** *1979.* Color-marked golden and bald eagles. N. Am. Bird Bander 4(3):115.
182	**Anonymous.** *1979.* Recovery round-up. Corella 3(1):13-14.
183	**Anonymous.** *1980.* A field guide for hawks seen in the northeast. New England Hawk Watch Study. Manuscript.
184	**Anonymous.** *1981.* Golden eagles. Royal Soc. for the Protect. of Birds. Report. 4 pp.
185	**Anonymous.** *1981.* Pictures from the life of golden eagles at Nahal Gilo. Torgos 1(2):44.
186	**Anonymous.** *1981.* The eagle who saved jews in a soviet prison from hunger. Torgos 1(2):58.
187	**Ansell, A.R. and W.E. Smith.** *1980.* Raptor protection activities of the Idaho Power Company. Pages 56-70. *in:* R.P. Howard and J.F. Gore, Eds. Proc. of A Workshop on Raptors and Energy Developments. 25-26 January, Boise, Id. Bonneville Power Admin., U.S. Fish and Wildlife Service, Idaho Power Comp., and Idaho Chapter of the Wildl. Soc.
188	**Antal, L., J. Fernbach, J. Mikuska, I. Pelle and L. Szlivka.** *1969.* Register of birds of the autonomous province of Vojvodina. Larus 23:69-118. (English Trans.)
189	**Anthony, A.J.** *1979.* Patagial tagging of Wahlberg's eagles. Safring 8(1):28-30.
190	**Araujo, J.** *1973.* Algunas presas interesantes de Strigiformes y Accipitriformes. Ardeola 19(1):24-25.
191	**Araujo, J.** *1973.* Unos datos sobre nidos de *Aquila chrysaetos* en Caceres. Ardeola 19(2):455.
192	**Araujo, J.** *1974.* Falconiformes del Guadarrama Suroccidental. Ardeola 19(2):257-278.
193	**Archer, G. and E.M. Goodman.** *1937.* The Birds of British Somaliland and the Gulf of Aden. (Vol. 1). Gurney and Jacobson, London. 285 pp.
194	**Argyle, F. and B. Gal.** *1978.* Birds in Eilat, spring 1978. Society for the Protection of Nature, Eilat, Israel. 17 pp.
195	**Arizona Game and Fish Dept.** *1976.* Non-game investigations. Wildlife surveys and investigations. Ariz. Game and Fish Dept., Spec. Perf. Report Proj. No. W-53-R-27. 71 pp.
196	**Arizona Game and Fish Dept.** *1976.* Non-game investigations. Wildlife surveys and investigations. Ariz. Game and Fish Dept., Spec. Perf. Report Proj. No. W-53-R-26. 52 pp.
197	**Arnault, C.** *1930.* Un nid d'aigles fauves. Oiseau Rev. Fr. Ornithol. 11(4):216-219.
198	**Arnell, W.B.** *1971.* Prey utilization by nesting golden eagles *(Aquila chrysaetos)* in central Utah. M.S. Thesis, Brigham Young Univ., Provo, UT. 53 pp.
199	**Arnold, J.R.** *1937.* Birds of the Coalinga area, Fresno County, California. Condor 39(1):31-35.
200	**Arnold, L.W.** *1948.* "Eagle attacks boy." The true story behind the distorted newspaper scoop. Audubon 50(4):256-257.
201	**Arnold, L.W.** *1954.* Golden eagle and its economic status. U.S.D.I. Fish and Wildlife Service. Circ. No. 27.
202	**Arredondo, O.** *1976.* The great predatory birds of the Pleistocene of Cuba. Smithsonian Contrib. to Paleobiology 27:169-187.
203	**Ash, J.S.** *1981.* Stork and raptor migration in south Nyana, Kenya. Scopus 5(2):51-52.
204	**Ashmole, N.P., R.G.B. Brown and R.P. Campbell.** *1957.* La migration a Gavarnie en automne 1955. Alauda 25(2):94-115. (English Summ.)
205	**Astley, H.D.** *1916.* The greater spotted eagle in Herefordshire. Brit. Birds 9(8):217.
206	**Atkinson, L.** *1947.* Nest and young of black eagle. Ostrich 18(2):196.
207	**Atkinson, W.L.** *1899.* Capture of a rabbit by a golden eagle. Condor 1(3):50-51.
208	**Attiwill, A.R.** *1972.* Birds breeding in Naracoorte District, 1941-1971. S. Aust. Ornithol. 26(3):59-64.

Master No. **Citation**

209 **Attwood, R.** *1979.* An unusual concentration of predators at Tea Tree Gully. S. Aust. Ornithol. 28(3):82.

210 **Audubon, J.J.** *1875.* The golden eagle. Oologist 1(10):68-69.

211 **Auger, H., R. Carrick, S. Gordon, W.E. Higham, E.J. Hosking and C.E. Palmar.** *1951.* Some recent photographic studies of the golden eagle. Brit. Birds 44(12):404.

212 **Aumann, G.** *1958.* Schelladler *(Aquila clanga* Pall.) bei Coburg. Ornithol. Mitt. 10(6):114.

213 **Aumann, T.** *1982.* The legal status of raptors in Australasia. Aust. Raptor Assoc. News 3(1):3-9.

214 **Ausobsky, A., Jr. and K. Mazzucco Jr.** *1964.* Die brutvogel des landes Salzburg und ihre vertikal-verbreitung. Egretta 7(1/2):1-49.

215 **Austin, O.L.** *1932.* Birds of Newfoundland, Labrador. Cambridge, Mass. 233 pp.

216 **Austin, O.L., Jr.** *1948.* The Birds of Korea. Bull. Mus. Comp. Zool. 101(1):3-301.

217 **Austin, O.L., Jr.** *1961.* Birds of the world. Golden Press, N.Y. 316 pp.

218 **Austin, O.L., Jr. and N. Kuroda.** *1953.* The birds of Japan, their status and distribution. Bull. Mus. Comp. Zool. 109(4):279-637.

219 **Austruy, J.-C. and J.-M. Cugnasse.** *1981.* L'aigle royal *Aquila chrysaetos chrysaetos,* dans le Massif central. (The golden eagle in the central Massif.) Nos Oiseaux 36(4):133-142.

220 **Averin, Y.V., I.M. Ganya and G.A. Uspenski.** *1971.* Ptitsi Moldavia, Tom 2. (Birds of Moldavia, Vol. 2.) Shtinitsa, Kishinev. p. 1-236.

221 **Aymar, G.C.** *1935.* Bird Flight. Dodd, Mead.

222 **Ayres, T.** *1878.* Additional notes on the ornithology of Transvaal. Ibis 2(7):281-301.

223 **Ayres, T.** *1879.* Additional notes on the ornithology of Transvaal. Ibis 3(1):285-300.

224 **Ayres, T.** *1884.* Additional notes on the ornithology of Transvaal. Ibis 2(7):217-233.

225 **Ayres, T.** *1886.* Additional notes on the ornithology of Transvaal. Ibis 4(15):282-298.

226 **Baboss, J.** *1955.* Ornithological observations in eastern Hungary. Aquila 59-62:465.

227 **Back, R.** *1952.* Anmarkningsvarda fageliakttagelser fran sodra Osterbotten. (Some remarkable bird-notes from southern East Bothnia.) Ornis Fenn. 29(2):67-69. (English Summ.)

228 **Backus, R.F.** *1918.* Catching an eagle by hand. Oologist 35(5):77-78.

229 **Badman, F.J.** *1979.* Birds of the southern and western Lake Eyre drainage. S. Aust. Ornithol. 28(2):29-54.

230 **Baehrmann, U.** *1974.* (Comparing osteometrical investigations of body skeletal elements and extremities of some diurnal birds of prey from Accipitridae, Pandionidae and Falconidae families.) Zool. Abhand. 33(3):33-62. (In German, English Summ.)

231 **Baer, K.P., C.Q. Heumier, R.G. Kinghorn, D.H. Luchtel and L.C. Goldman.** *1963.* Report of investigations of the eagle problem in southwest Texas and New Mexico. U.S.D.I. Bureau of Sport Fisheries and Wildlife, Unpublished Rep. 22 pp. (Title Only)

232 **Bagge, P. and J. Tiussa.** *1957.* Kiljukotka, *Aquila clanga* Pall., Parikkalan Siikalahdella. (Observation of the spotted eagle at Parikkala.) Ornis Fenn. 34(4):136.

233 **Baglien, J.W.** *1975.* Biology and habitat requirements of the nesting golden eagle in southwestern Montana. M.S. Thesis. Montana State Univ., Bozeman, MT. 53 pp.

234 **Baha, S.** *1981.* Observations on raptor migration in Suez. Torgos 1(2):21.

235 **Bailey, A.M. and R.J. Niedrach.** *1939.* Filming the golden eagle. Am. Forests 45(9):446-449,476-477.

236 **Bailey, B.H.** *1918.* The raptorial birds of Iowa. Iowa Geological Survey, Bull. No. 6. 238 pp.

237 **Bailey, E.P.** *1975.* Discovery of a golden eagle nest on the Alaska Peninsula. Condor 77(2):207-208.

238 **Bailey, F.M** *1905.* Breeding notes from New Mecixo. Condor 7(2):39-40.

239 **Bailey, H.B.** *1912.* Notes on birds breeding in the mountains of Virginia. Auk 29(1):79-80.

240 **Baillie, J.L.** *1955.* The golden eagle nesting in the Gaspe Peninsula, Quebec. Can. Field-Nat. 69(1):13-14.

241 **Baily, W.S.** *1930.* Eggs of *Aquila rapax* and *Aquila heliaca.* Bull. Brit. Ool. Assoc. 2(24):143.

242 **Baker, E.C.S.** *1928.* The fauna of British India including Ceylon and Burma. Birds. Vol. 5. Taylor and Francis, London. 470 pp.

243 **Balasch, J., S. Musquera, L. Palacios, M. Jimenez and J. Palomeque.** *1976.* Comparative hematology of some Falconiforms. Condor 78(2):258-259.

244 **Baldridge, F.A.** *1977.* Raptor nesting survey of southern San Diego County, Spring 1977; with an analysis of impacts of power lines. U.S.D.I. Bureau of Land Management. Report. 41 pp.

245 **Baldwin, M.** *1975.* Birds of Inverell District, NSW. Emu 75(3):113-120.

246 **Baldwin, S.P., S.C. Kendeigh and R.W. Franks.** *1932.* The protection of hawks and owls in Ohio. Ohio J. Sci. 32(5):403-424.

247 **Ball, V.** *1874.* On the avifauna of the Chutia (Chota) Nagpur Division, S.W. frontier of Bengal. Stray Feathers 2(405):355-440.

248 **Ballance, D.K. and S.L.B. Lee.** *1961.* Notes on autumn migration at the Bosphorous and in the Aegean. Ibis 103(2):195-204.

249 **Balsac, H. de.** *1929.* Voyages d'etudes ornithologiques dans la region Nord du Sahara. Oiseau Rev. Fr. Ornithol. 8(184/185):433-434.

250 **Balsac, H.H. de** *1952.* Rythme sexuel et fecondite chez les oiseaux du nord-ouest de l'Afrique. Alauda 20(4):213-242.

251 **Balsac, H.H. de.** *1926.* Contribution a l'ornithologie de l'Algerie Septentrionale, exploration ornithologique de l'Ouarensis. Oiseau Rev. Fr. Ornithol. 10(211):446-474.

252 **Balthasar, V.** *1949-1950.* Ptactvo gottwaldovskeho (zlinskeho) regionu. Sylvia 11-12(1):1-35. (In Czech., French Summ.)

253 **Baltosser, W.H.** *1975.* A survey for peregrine falcons and other birds in the Gila Wilderness, New Mexico, in summer 1975. N. Mexico Dept. of Game and Fish, Unpublished Report. 2 pp.

254 **Bamberg, O.** *1903.* Uber die erlegung eines lammergeiers und das besteigen eines steinadlerhorstes. Z. Ool. 13(6):81-88.

255 **Bamberg, O.** *1905.* Nachrichten aus Turkestan. Z. Ool. 15(9):134-139.

Master No.	Citation
256	**Bangs, O. and A. Loveridge.** *1933.* Reports on the scientific results of an expedition to the southwestern highlands of Tanganyika Territory. III. Birds. Bull. Mus. Comp. Zool. 75(3):143-221.
257	**Bangs, O. and J.L. Peters.** *1928.* Birds collected by Dr. Joseph F. Rock in western Kansu and eastern Tibel. Bull. Mus. Comp. Zool. 68(7):313-381.
258	**Bank, P.** *1915.* Ein schreiadler *(Aquila naevia)* in der provinz Hannover verschossen. Vogelwelt 35(5):236-237.
259	**Bannerman, D. and J. Bannerman.** *1953.* A second journey to the Moroccan Sahara (in 1952) and over the Great Atlas. Ibis 95(1):128-139.
260	**Bannerman, D.A.** *1910.* On a collection of birds made in northern Somaliland by Mr. G.U. Bury. Ibis (9th Ser.) 4(14):291-327.
261	**Bannerman, D.A.** *1930.* The Birds of Tropical West Africa. Crown Agents for the Colonies, London. 376 pp.
262	**Bannerman, D.A. and W.M. Bannerman.** *1958.* Birds of Cyprus. Oliver and Boyd, Ltd., Edinburgh. 384 pp.
263	**Bannerman, D.A. and W.M. Bannerman.** *1971.* Handbook of the Birds of Cyprus and Migrants of the Middle East. Oliver and Boyd, Ltd., Edinburgh. 237 pp.
264	**Bannerman, D.A. and J.A. Vella-Gaffiero.** *1976.* Birds of the Maltese Archipelago. Museums Dept., Malta. 550 pp.
265	**Banzhaf, W.** *1927.* Ornithologische beobachtungen von der kurischen nehrung aus dem jahre 1926. J. Ornithol. 75(3):443-452.
266	**Banzhaf, W.** *1939.* Zu: schelladler *(Aquila clanga* Pall.) auf der Monne bei Stettin. Ornithol. Monatsber. 47(1):17-18.
267	**Barber, D.** *1976.* The continuing slaughter of birds of prey in Britain. Brit. Birds 69(5):186-187.
268	**Barbour, R.W., C.T. Peterson, D. Rust, H.E. Shadowen and A.L. Whitt, Jr.** *1973.* Kentucky Birds, A Finding Guide. Univ. Press of Kentucky, Lexington. 306 pp.
269	**Barker, E.S.** *1947.* Golden eagle kills deer. New Mexico 25(12):29. (Title Only)
270	**Barlow, C.** *1897.* The golden eagle. *Aquila chrysaetos.* Avifauna 1(3):34-36.
271	**Barlow, C.** *1898.* After the golden eagle. Osprey 2:82-84.
272	**Barlow, J.W.** *1845.* Occurrence of the golden eagle in Cheshire. Zoologist 3:1022.
273	**Barnes, H.E.** *1893.* On the birds of Aden. Ibis (6th Ser.) 5(17):57-84.
274	**Barnes, R.M.** *1910.* Very unusual. Oologist 27(5):57.
275	**Barnes, R.M.** *1911.* Wanted - eggs. Oologist 28(8):140.
276	**Barnes, R.M.** *1916.* The eagle's flight. Oologist 33(1):1.
277	**Barnes, R.M.** *1918.* 349 1/2. Oologist 35(12):154-156.
278	**Barnes, R.M.** *1923.* Ingersoll's eagle eggs. Oologist 40(10):157,160.
279	**Barnes, R.M.** *1936.* An outrage. Oologist 53(6)78.
280	**Barnes, R.M.** *1940.* Scandalous! Oologist 57(9):107.
281	**Barnes, R.M.** *1940.* Another eagle killer. Oologist 57:132.
282	**Barnhill, M.V., III.** *1972.* A tentative DOS Delaware list. Delmarva Ornithol. 7(2):16-19.
283	**Barrington, R.M.** *1874.* Golden eagle at Powerscourt. Zoologist (2nd Ser.) 9:3952.
284	**Barrington, R.M.** *1915.* The last(?) Irish golden eagle. Irish Nat. 24(3):63.
285	**Barruel, P.** *1953.* Jusqu'ou habitent les oiseaux de nos pays? Nos Oiseaux 22(2):43-46.
286	**Bartelt, P.E.** *1976.* A raptor survey and trapping techniques in Clay County, South Dakota. S. Dakota Bird Notes 28(2):24-30,36. (Title Only)
287	**Barth, E.K.** *1958.* Rovfuglene i naturens husholdning. Sterna 3(1):1-12.
288	**Barth, E.K.** *1974.* Rodrev drept av kongeorn. (Red fox killed by a golden eagle.) Fauna 27(1):53-54. (English Summ.)
289	**Bassouls, G. and R. de Naurois.** *1959.* Peuplement et nidification des grands rapaces dans les Pyrenees Orientales. Osieau Rev. Fr. Ornithol. 29(3):218-220.
290	**Bassouls, M. and G. Petit.** *1959.* L'aigle royal gravement menace dans les Pyrenees-Orientales. Terre Et Vie (Suppl.):54.
292	**Bastyai, L.** *1952-1955.* Adatok nappali regadozomadaraink vedleserol. Aquila 59-62:161-165. (English and Russian Summ.)
291	**Bastyai, L.** *1957.* Data about birds of prey. Aquila 63-64:332.
293	**Bates, G.G.** *1976.* Breeding of sub-adult golden eagle. Bird Study 23(4):284.
294	**Batey, I.** *1907.* Wedge-tailed eagles and lambs. Emu 7(1):43-45.
295	**Bau, A.** *1904.* Mitteilungen. Meinungsaustausch. Z. Ool. 13(10):156-157.
296	**Bauer, E.A.** *1966.* Treasure bird. Field and Stream 70(12):61,118-121.
297	**Bauer, E.A.** *1976.* In the land of the raptors. Natl. Wildl. 14(3):12-17.
298	**Bauer, K.** *1952.* Ornithologische beobachtungen in den Leitha-Auen bei Zurndorf (Burgenland). J. Ornithol. 93(2):112-113.
299	**Bauer, K.** *1954.* Adler am Neusiedlersee. Ornithol. Mitt. 6(4):69-72.
300	**Bauer, K.** *1977.* Present status of birds of prey in Austria. Pages 83-85. *in:* R.D. Chancellor, Ed. Proc. of the World Conf. on Birds of Prey. 1-3 October, 1975. Vienna. I.C.B.P.
301	**Baum, J.** *1937.* Ornithologika pozorovani v Tatrach. (Ornithological notes from the Tatra Mountains.) Sylvia 2(2):29-33. (English Summ.)
302	**Baumgart, W.** *1971.* Wurgfalkenpaar zieht jungen steinadler auf. (Saker pair raises young golden eagle.) Falke 18(12):414-417.
303	**Baumgart, W.** *1971.* Beitrag zur kenntnis der greifvogel Bulgariens. Beitr. Vogelkd. 17(1):33-70.
304	**Baumgart, W.** *1979.* Zur signalfunktion von gefiedermerkmalen bei greifvogeln. Beitr. Vogelkd. 25(3/4):209-246.
305	**Baumgart, W.** *1980.* Steht der schreiadler unter zeitdruck? Falke 27:6-17.
306	**Baumgart, W. and D. Zenker.** *1969.* Steinadler und wurgfalke bei der beizjagd und haltung im Freiflug. Falke 16(12):416-425.
307	**Baxendale, F.R.S.** *1915.* Notes on the ornithology of Cyprus. Ibis (10th Ser.) 3(2):217-227.
308	**Baxter, C.J.** *1980.* Birds of Belair Recreation Park. S. Aust. Ornithol. 28(4):90-98.
309	**Beal, W.J.** *1867.* Novel way of shooting eagles. Am. Nat. 1(7):439.
310	**Beale, D.M. and A.D. Smith.** *1973.* Mortality of pronghorn antelope fawns in western Utah. J. Wildl. Manage. 37(3):343-352.
311	**Beaman, M. and C. Galea.** *1974.* The visible migration of raptors over the Maltese Islands. Ibis 116(4):419.

Master No. Citation

312 **Beaman, M. (ed.)** *1978.* Bird report No. 4. Ornithol. Soc. Turkey p. 1-126.

313 **Bean, J.R. and R.H. Hudson.** *1976.* Acute oral toxicity and tissue residues of thallium sulfate in golden eagles, *Aquila chrysaetos.* Bull. Env. Contam. Toxicol. 15(1):118-121.

314 **Beaton, R.J.** *1951.* Hawk migration at South Mountain. Atlantic Nat. 6(4):166-168.

315 **Beck, H.H.** *1924.* The Pennsylvania German names of birds. Auk 41(2):288-295.

316 **Beck, R.H.** *1901.* A season with the golden eagles of Santa Clara Co., Cal. Condor 3(3):59-64.

317 **Becsy, L.** *1971-1972.* Angaben zur ernahrung des kaiseradlers. Aquila 78-79:236. (German Trans.)

318 **Becsy, L. and A. Keve.** *1977.* The protection and status of birds of prey in Hungary. Pages 125-129. *in:* R.D. Chancellor, Ed. Proc. of the World Conf. on Birds of Prey. 1-3 October, 1975. Vienna. I.C.B.P.

319 **Bedichek, R.** *1948.* Golden eagle, airplane hater. Sci. Digest 23(3):56-60.

320 **Bednall, D.K.** *1959.* A spotted eagle in Nairobi? J. E. African Nat. Hist. Soc. 23(3):144.

321 **Bee, R.G. and J. Hutchings.** *1942.* Breeding records of Utah birds. Great Basin Nat. (3/4):61-85.

322 **Beebe, F.L.** *1975.* Field studies of the Falconiformes of British Columbia (vultures, hawks, falcons, eagles). British Columbia Prov. Mus. Nat. Hist., Occas. Pap. No. 17. 163 pp.

323 **Beebe, F.L.** *1976.* Hawks, Falcons and Falconry. Hancock House Publ., Seattle. 320 pp.

324 **Beecham, J.J. and M.N. Kochert.** *1975.* Breeding biology of the golden eagle in southwestern Idaho. Wilson Bull. 87(4):506-513.

325 **Beecham, J.J., Jr.** *1970.* Nesting ecology of the golden eagle in southwestern Idaho. M.S. Thesis. Univ. of Idaho, Moscow, ID. 48 pp.

326 **Beesley, J.S.S.** *1973.* The breeding seasons of birds in the Arusha National Park, Tanzania. Bull. Brit. Ornithol. Club 93(1):10-20.

327 **Behle, W.H.** *1958.* The birds of the Raft River Mountains, northwestern Utah. Univ. of Utah Biology Ser. 11(6):1-40.

328 **Behrends, O.** *1948.* Rovfuglenes ruter paa foraarstraekket over Als. (The routes of birds of prey on their spring migration over the island of Als.) Dansk Ornithol. 42(1):11-24. (English Summ.)

329 **Beidleman, R.G.** *1949.* Guide to the birds of prey of Colorado. Univ. of Colorado Museum, Leaflet No. 6. 24 pp.

330 **Belcher, M.** *1961.* Birds of Regina. Saskatchewan Nat. Hist. Soc., Spec. Publ. No. 3. 76 pp.

331 **Beldi, M.** *1952-1955.* Madarelofordulasi adatok erdelybol. Aquila 59-62:229-231. (English Summ.)

332 **Bell, A.** . Common Australian birds. New York. 229 pp.

333 **Bell, R.K.** *1979.* A golden eagle stays three months in southwestern, Pa. Redstart 46(4):148-150.

334 **Belyakov, V.V.** *1964.* O chiselennosti i raspredelenii dnevnykh khistchnykh ptits na territorii Kalininoskoi oblasti. (Number and distribution of diurnal birds of prey on a territory of Kalinin dist.) Uchenye Zaoiski Kalinin. Gos. Pedagog. Inst. 31:20-40.

335 **Bendire, C.** *1877.* Notes on some of the birds found in southeastern Oregon, particularly in the vicinity of Camp Harney, from November, 1874, to January, 1877. Proc. Boston Soc. Nat. Hist. 19:109-149.

336 **Benington, A.** *1954.* Golden eagle breeding in Co. Antrim. Brit. Birds 47(8):272.

337 **Bennett, E.V.** *1977.* 1976 Nebraska nesting survey. Nebr. Bird Rev. 45(1):3-5.

338 **Bennett, E.V.** *1978.* The first eleven years of breeding bird surveys in Nebraska. Nebr. Bird Rev. 46(3):38-62.

339 **Bennett, E.V.** *1978.* 1977 Nebraska nesting survey. Nebr. Bird Rev. 46(1):13-16.

340 **Bennett, E.V.** *1980.* 1978 Nebraska nesting survey. Nebr. Bird Rev. 48(2):39-43.

341 **Benson, C.W.** *1940.* Further notes on Nyasaland birds (with particular reference to those of the Northern Province). Ibis 4(2):257-298.

342 **Benson, C.W.** *1942.* Additional notes on Nyasaland birds. Ibis 6(2):197-224.

343 **Benson, C.W.** *1944.* Notes from Nyasaland. Ibis 86(Oct.):445-480.

344 **Benson, C.W.** *1945.* Notes on the birds of southern Abyssinia. Ibis 87(July):366-400.

345 **Benson, C.W. and F.M. Benson.** *1947.* Some breeding records from southern Nyasaland. Ool. Rec. 21(4):1-9.

346 **Benson, C.W. and R.S. Pitman.** *1966.* Further breeding records from Zambia (formerly Northern Rhodesia). Bull. Brit. Ornithol. Club 86(2):21-33.

347 **Benson, C.W., R.K. Brooke, R.J. Dowsett and M.P.S. Irwin.** *1970.* Notes on the birds of Zambia. Arnoldia 4(40):1-59.

348 **Benson, C.W., R.K. Brooke and C.J. Vernon.** *1964.* Bird breeding data for the Rhodesias and Nyasaland. Occas. Pap. Natl. Mus. S. Rhodesia 27(4):30-105.

349 **Benson, P.C.** *1980.* A study of large raptor electrocution and powerpole utilization in six western states. Pages 36-40. *in:* R.P. Howard and J.F. Gore, Eds. Proc. of A Workshop on Raptors and Energy Developments. 25-26 January, Boise, Id. Bonneville Power Admin., U.S. Fish and Wildlife Service, Idaho Power Comp., and Idaho Chapter of the Wildl. Soc.

350 **Benson, S.V., W.M. Irving, C. McDowell, C. Higginbottom and P.B. Lind.** *1961.* Birds seen in Yugoslavia. Larus 14:159-161. (English Trans.)

351 **Bent, A.C.** *1937.* Life Histories of North American Birds of Prey. Order Falconiformes (Part 1). Smithsonian Inst. U.S. Natl. Mus. Bull. 167. 409 pp.

352 **Benton, A.H.** *1966.* Wires, poles and birds. Pages 390-395. *in:* A. Stefferud (ed.) Birds in Our Lives. U.S. Bureau of Sport Fisheries and Wildlife, Wash., D.C.

353 **Bentz, P.G.** *1976.* Imperial eagle *Aquila heliaca* in Skane, 1975. Anser 15(3):211-216. (Title Only)

354 **Bereszynski, A.** *1970.* Contributions to the avifauna of Poland. VI. Acta Ornithol. 12(2):25-349 (In Polish, English Summ.)

355 **Beretzk, P.** *1955.* Ujabb adatok a szegedifeherto madarvilagahoz, 1949-1953. Aquila 59-62:217-227. (English and Russian Summ.)

356 **Berger, A.** *1913.* Schreiadler auf der Feldgrillenjagd. Aquila 20:527-528.

357 **Bergman, G.** *1961.* The food of birds of prey and owls in Fenno-Scandia. Brit. Birds 54:307-320.

Master No.	Citation
358	**Bergman, G.** *1977.* Birds of prey: the situtation in Finland. Pages 96-103. *in:* R.D. Chancellor, Ed. Proc. of the World Conf. on Birds of Prey. 1-3 October, 1975. Vienna. I.C.B.P.
359	**Bernath, E.-L.** *1936.* Notes sur l'avifaune des iles Baleares et Pityuses. Oiseau Rev. Fr. Ornithol. (New Ser.) 6(1):117-131.
360	**Berndt, R.** *1939.* Steinadler, *Aquila chrysaetos* (L.) bei Steckby in Anhalt. Ornithol. Monastber. 47(1):15
361	**Berndt, R. and R. Reinecke.** *1961.* Steinadler, *Aquila chrysaetos* L., im sudostlichen Niedersachsen. (Golden eagle in southern Niedersachsen.) Ornithol. Mitt. 13(5):91-93.
362	**Berney, F.L.** *1903.* The wedge-tailed eagle in North Queensland. Emu 3(2):123-124.
363	**Berney, F.L.** *1906.* Eagles and rats. Emu 6(1):13.
364	**Bernhardt, A.** *1965.* Golden eagle - *Aquila chrysaetos* (L.). Acta Ornithol. 9:128. (English Trans.)
365	**Bernhoft-Osa, A.** *1973.* Hvor tungt bytte klarer rovfuglene a fly avsted med. (Upon a clarification of the difficulties of raptors to cross flyways). Sterna 12(2):73-83. (German Summ.)
366	**Bernis, F.P.** *1973.* Algunos datos de alimentacion y depredacion de Falconiformes y Estrigiformes Ibericas. Ardeola 19(2):225-248.
367	**Bernis, F.P.** *1974.* Mas sobre fenologia de reproduccion y status de *Aquila chrysaetos* en Iberia. Ardeola 19(2):447-454.
368	**Berns, V.D.** *1979.* Golden eagle nest on Kodiak Island, Alaska. Condor 81(2):218.
369	**Berthet, G.** *1936.* Notes sur quelques oiseux du Chablais. Alauda 8(2):228-257.
370	**Berthet, G.** *1936.* Une aigle fauve *Aquila chrysaetos* dans les Monts du Lyonais. Alauda 8(3/4):495.
371	**Beruldsen, G** *1956.* Hawks of Yorke Peninsula. S. Aust. Ornithol. 22(2):23-25.
372	**Besson, J.** *1964.* L'aigle royal *(Aquila chrysaetos)* dans le Alpes du sud. Alauda 32(1):45-50.
373	**Besson, J.** *1967.* Notes sur la reproduction de quatre couples d'aigles royaux dans les Alpes Maritimes en 1966. Alauda 35(1):49-61.
374	**Besson, J.** *1969.* Un aigle royal *(Aquila chrysaetos)* de la variante "Barthelemyi" dans les Basses-Alpes. Alauda 37(3):258-260.
375	**Besson, J.** *1972.* Chasses statiques de longue duree chez l'aigle royal et le Circaete Jean-Le-Blanc. Alauda 40(1):100-101.
376	**Bethlenfalvy, E.** *1952.* Hrdlickazahradna, orol sklny a bocian biely pod vysokymi tatrami. Sylvia 14:135-136.
377	**Bethlenfalvy, E.** *1955.* Golden eagle feeding on ants. Aquila 59-62:451.
378	**Beyer, G.E., A. Allison and H.H. Kopman.** *1908.* List of the birds of Louisiana. Part V. Auk 25(4):439-448.
379	**Bianchi, E., L. Martire and A. Bianchi.** *1969.* Gli uccelli della provincia di Varese (Lombardia). Riv. Ital. Ornitol. 39(4):384-401.
380	**Biddulph, J.** *1881.* The birds of Gilgit. Stray Feathers 9(5/6):301-366.
381	**Biddulph, J.** *1881.* On the birds of Gilgit. Ibis (4th Ser.) 5(17):35-102.
382	**Bietsch, B.** *1886.* Vungefunder todter goldadler. Vogelwelt 11(4):103.
383	**Biggs, H.C.** *1978.* South West African contribution to raptor measurement and plumage study. Safring 7(2):2-8.
384	**Biggs, R. and H. Biggs.** *1978.* Comparative portraits of steppe and tawny eagles. Bokmakierie 30(3):77.
385	**Bigland, C.H.** *1957.* Isolation of *Erysioelothrix rhusipathiac* from a golden eagle. Can. J. Comp. Med. and Vet. Sci. 21(8):290-291. (Title Only)
386	**Bijleveld, M.** *1974.* Birds of Prey in Europe. MacMillan Press Ltd., London. 263 pp.
387	**Bild, U.** *1973.* Am horst des kaiseradlers. Falke 20(12):432.
388	**Bille, R.-P.** *1980.* Remarquable comportement du grand corbeau. Nos Oiseaux 35(5):227-231.
389	**Binford, L.C.** *1979.* Fall migration of diurnal raptors at Pt. Diablo, California. West. Birds 10(1):1-16.
390	**Birchetti, P.** *1977.* Rapporto tra nidificazione e massima altimetria relativo ad alcune specie nelle Alpi centrali (Lombardia). Riv. Ital. Ornitol. (2nd Ser.) 47(2):114-118.
391	**Bischoff, A.I., J.A. Azevedo and L.A. Woods.** *1971.* Pesticides investigation. Current fish and wildlife problems. Calif. Dept. of Fish and Game. Proj. No. FW-1-R-8, Job I-9. 7 pp.
392	**Bisgard, M.** *1934.* Kongeorne, *Aquila chrysaetos,* i Krengerupskovene, Fyn. Dansk Ornithol. 28(3/4):118-142.
393	**Bishop, S.C.** *1928.* The golden eagle *(Aquila chrysaetos)* in Columbia Co., New York. Auk 45:212.
394	**Black, C.T.** *1937.* Additional Illinois golden eagle records. Auk 54(3):385-388.
395	**Blackford, J.L.** *1956.* Duel in the sky. Nature Mag. 49(7):345-347.
396	**Blackwelder, E.** *1919.* Notes on the summer birds of the upper Yukon region, Alaska. Auk 36(1):57-64.
397	**Blagosklonov, K.N.** *1960.* (Birds of the Kandalaksha Reserve and vicinities of Belomorskaya Biological Station of Moskow University.) Trudy Kandalakshsk. Gos. Zopovednika. 2:5-104. (In Russian)
398	**Blair, H.M.S.** *1936.* On the birds of east Finmark. Part II. Ibis 6(3):429-459.
399	**Blair, H.M.S.** *1965.* On the oology of the Spanish imperial eagle *(Aquila heliaca adalberti)* . Ool. Rec. 39(1):19-20.
400	**Blair, H.M.S.** *1968.* Some old Spanish records. Ool. Rec. 42(4):64-66.
401	**Blake, E.A.** *1962.* The golden eagle. Birds Illus. 7(9):258-261.
402	**Blake, E.R.** *1953.* Birds of Mexico. Chicago. 674 pp.
403	**Blakiston, T. and H. Pryer.** *1878.* A catalogue of the birds of Japan. Ibis (4th Ser.) 2(7):209-250.
404	**Blanchan, N.** *1932.* Birds that hunt and are hunted. Doubleday & Mc Clure Co., N.Y. 359 pp.
405	**Blancou, L.** *1939.* Contribution a l'etude des oiseaux de l'Oubangui-Chari Occidental. Oiseau Rev. Fr. Ornithol. 9(1):58-88.
406	**Blandamer, J.S. and P.J.K. Burton.** *1979.* Anatomical specimens of birds in the collections of the British Museum (Natural History). Bull. Brit. Mus. Nat. Hist. Zool. 34(4):125-180.
407	**Blanford, W.T.** *1894.* On the scientific names of the imperial and spotted eagles and on the generic names of Bonelli's eagle and the black eagle. Ibis 6(22):283-289.
408	**Blasius, R.** *1862.* Die adler. J. Ornithol. 11:76-106.
409	**Blasius, R.** *1862.* Beobachtungen uber die brut- und zugverhaltnisse der vogel bei Braunschweig. J. Ornithol. 11:37-75.
410	**Blasius, R.** *1896.* Ornis von Malta und Gozo, und den umliegenden Inseln. Ornis 8:139-211.
411	**Blezard, E. and R. Laidler.** *1964.* Buzzard as golden eagle food. Scott. Nat. 71:97-98. (Title Only)

Master No. **Citation**

412 **Blignaut, J.** *1958.* Some observations on a pair of black eagles in the Jansenville District. Bokmakierie 10(1):12.

413 **Blondel, J.** *1962.* Observation d'un aigle imperial ? dans les Monts des Ksours. Alauda 30(1):69.

414 **Blondel, J.** *1967.* Reflexions sur les rapports entre predateurs et proies chez les rapaces. I. Les effets de la predation sur les populations de proies. Terre Et Vie 1:5-62.

415 **Blyth, E.** *1863.* Catalogue of the birds of India, with remarks on their geographical distribution. Ibis 5(17):1-31.

416 **Boag, D.A.** *1977.* Summer food habits of golden eagles in southwestern Alberta. Can. Field-Nat. 91(3):296-298.

417 **Boardman, G.A.** *1881.* The golden eagle in New Brunswick. Bull. Nuttall Ornithol. Club 6(1):58-60.

418 **Bocage, B., du.** *1876.* Die im Museum zu Lissabon befindlichen vogel der Westafrikanischen besitzungen Portugals. J. Ornithol. 24(3):285-317.

419 **Bock, P.** *1857. Aquila clanga* in Preussen beobachtet. J. Ornithol. 5(4):296.

420 **Boehm, E.F.** *1452.* Some ecological factors affecting bird life in the region of the north-west bend of the Murray, S.A. Emu 52(4):297-300.

421 **Boehm, E.F.** *1955.* Critical aspects of bird conservation of South Australia. S. Aust. Ornithol. 21(8):98-100.

422 **Boeker, E.L.** *1970.* Use of aircraft to determine golden eagle, *Aquila chrysaetos,* nesting activity. Southwest. Nat. 15(1):131-143.

423 **Boeker, E.L.** *1974.* Status of golden eagle surveys in the western states. Wildl. Soc. Bull. 2(2):46-49.

424 **Boeker, E.L.** *1976.* Golden eagle population study. U.S.D.I. Fish and Wildlife Service, Denver Wildlife Res. Center. Rep. 6 pp.

425 **Boeker, E.L. and E.G. Bolen.** *1972.* Winter golden eagle populations in the Southwest. J. Wildl. Manage. 36(2):477-484.

426 **Boeker, E.L. and P.R. Nickerson.** *1975.* Raptor electrocutions. Wildl. Soc. Bull. 3(2):79-81.

427 **Boeker, E.L. and T.D. Ray.** *1971.* Golden eagle population studies in the Southwest. Condor 73(4):463-467.

428 **Boeker, E.L., M.A. Jenkins and L.D. Crowley.** *1978.* Golden eagle population studies in the western states. U.S.D.I. Fish and Wildlife Service, Denver Wildl. Res. Center, Ann. Prog. Rep. - 1978. p. 38-42.

429 **Boetticher, H., von.** *1927.* Kurze uebersicht uber die raubvogel und eulen Bulgariens. Anz. Ornithol. Ges. Bayern 17(4):535-549.

430 **Boetticher, H., von.** *1919.* Ornithologische beobachtungen in der Muss-Alla-Grappe (Rila-Gebirge) 1916-1919. J. Ornithol. 67(3):233-257.

431 **Bojnotti, S.** *1959.* Catture rare o interessanti de uccelli ricevuti per la preparazione dall'Aprile 1958 all'Aprile 1959. Riv. Ital. Ornitol. 30(2):138-139.

432 **Bokai, B.** *1955.* Ornithological data from the Pilis Mountains, north of Budapest. Aquila 59-62:459-460.

433 **Bolander, L.P. and C.A. Bryant.** *1929.* Flying defense of a golden eagle. Condor 31(6):253.

434 **Bolen, E.G.** *1975.* Eagles and sheep: a viewpoint. J. Range Manage. 28(1):11-17.

435 **Bolivar, J.C.** *1970.* Premieres donnees sur l'ecologie hivernale des vertebres de la cordilliere Cantibrique. (Preliminary paper on the winter ecology of vertebrates in the Cantabrian Mountains.) Alauda 38(2):126-149. (English Summ.)

436 **Bolt, B.F.** *1912.* Capture of a golden eagle at Kansas City, Mo. Auk 29(1):102.

437 **Bonaparte, C.** *1824.* Observations on the nomenclature of Wilson's Ornithology. J. Acad. Nat. Sci. Philadelphia 3(2):340-371.

438 **Bonnin, J.M. and R.C. Angove.** *1980.* Wedge Island. S. Aust. Ornithol. 28(4):104-106.

439 **Booth, E.S.** *1952.* Ecological distribution of the birds of the Blue Mountains region of southeastern Washington and northwestern Oregon. Walla Walla College, Publ. No. 7. p. 65-107.

440 **Booth, E.S.** *1971.* Birds of the west. Fourth Ed., Outdoor Pictures, Anacortes, Wa.

441 **Borg, K.** *1962.* Predation on roe deer in Sweden. J. Wildl. Manage. 26(2):133-136.

442 **Borrer, C.D., F.C.R. Jourdain and E.C. Stuart Baker. (Eds.).** *1926.* Untitled. Bull. Brit. Ool. Assoc. 1(9):89-95.

443 **Boshoff, A.F.** *1979.* Variation in the Afrikaans vernacular for some eagles in the Cape Province. Bokmakierie 31(3):22-23.

444 **Boshoff, A.F.** *1980.* Some socio-economic aspects of a bird of prey questionnaire survey. S. Afr. J. Wildl. Res. 10:71-81.

445 **Boshoff, A.F. and C.J. Vernon.** *1980.* The distribution and status of some eagles in the Cape Province. Ann. Cape Prov. Mus. Nat. Hist. 13:107-132.

446 **Bosmans, R. and F. Moreaux.** *1977.* Birds of the Ethiopian Begemdir Province in August and September 1973-1974. Gerfaut 67(4):395-412.

447 **Boubier, M.** *1935.* Rapport sur les travaux de la Commission pour l'Unification des Noms Francais des Oiseaux. Oiseau Rev. Fr. Ornithol. 5(2):189-213.

448 **Bouchner, M.** *1951.* Orel volavy *(Aquila clanga)* na Rickansu. (The spotted eagle *(Aquila clanga)* in the vicinity of Prague.) Sylvia 13(4):136-137. (In Czech., English Summ.)

449 **Bouillot, M.** *1970.* Notes sur l'avifaune specifique de certain biotopes de Basse-Ardeche. Alauda 38(3):174-185.

450 **Boutinot, S.** *1955.* Faune ornithologique du Vermandois (Region de Saint-Quentin). Oiseau Rev. Fr. Ornithol. 25(4):256-294.

451 **Bouvier, G.** *1973.* La frequence des Mallophages chez les oiseaux de Suisse. Nos Oiseaux 32:9-15.

452 **Bowles, A.H.** *1877.* Golden eagle near Killarney. Zoologist (3rd Ser.) 1:25.

453 **Bowles, J.H.** *1921.* Breeding dates for Washington birds. Murrelet 2(2):8-12.

454 **Bowles, J.H.** *1925.* A study of bird stomachs and their contents. Murrelet 6:27-31.

455 **Bowles, R.V.** *1967.* Lesser spotted eagle on migration. J. E. African Nat. Hist. Soc. 26(2):87.

456 **Boxberger, L., von.** *1934.* Beitrage zur brutvogelfauna der Provinz Malaga. J. Ornithol. 82(2):185-209.

457 **Boxberger,. L. von.** *1938.* Zunahme des steinadlers in Oesterreich. Beitr. Fortpfl.-Biology Vogel 14(3):109.

458 **Boyd, J.M.** *1955.* Golden eagle at St. Kilda. Brit. Birds 48(10):454.

459 **Boyer, R.H.** *1948.* Fawns escape eagle's attack. Yosemite Nat. Notes 27(11):129-130.

460 **Boyles, A.K.** *1901.* Eagle vs. hounds. Am. Ornithol. 1(3):46.

461 **Bradfield, R.D.** *1932.* Unusual experiences with birds. Ostrich 3(1):13-16.

462 **Brakefield, T.** *1980.* Eagle in the sky. Sports Afield (Feb.):55-57,120,122.

Master No. **Citation**

463 **Brandolini, A.** *1935.* Anno 1934. Osservioni interessanti l'avifauna ravenate. Riv. Ital. Ornitol. 5(4):292.

464 **Brandon, T.** *1938.* Nesting of the wedge-tailed eagle *(Uroaetus audax)* in 1937, etc. S. Aust. Ornithol. 14(5):117-121.

465 **Brandon, T.** *1938.* Bird notes from Wilmington District. S. Aust. Ornithol. 14(8):218-221.

466 **Brandon, T.** *1948.* Bird notes from Wilmington. S. Aust. Ornithol. 19(1):3-5.

467 **Brandt, H., O. Lassoe, P. Thomsen and S.K. Visherijule.** *1970.* Iagttagelse af skrigeorn *(Aquila pomarina/clanga),* sandsynligvis lille, ved skagen i pasken 1968. Dansk Ornithol. 64(3):193-199.

468 **Brauner, A.** *1907.* Uber *Aquila orientalis* Cab. im taurischen goovernement (Sudrussland). Ornithol. Jahrbuch 18(1/2):24-27.

469 **Bravery, J.A.** *1970.* The birds of Atherton Shire, Queensland. Emu 70(2):49-63.

470 **Brawley, K.C.** *1977.* Domestic sheep mortality during and after tests of several predator control methods. M.S. Thesis. Univ. of Montana, Missoula, MT. 69 pp.

471 **Breckenridge, W.J.** *1946.* Bird hawks, eagles, and vultures. Cons. Vol. 9(50):16-22.

472 **Bredahl, W.** *1924.* Ein steinadler in Pommern erlegt. Ornithol. Monatsber. 32(1):19-20.

473 **Breninger, G.F** *1890.* The golden eagle. *Aquila chrysaetos.* Ornithol. and Ool. 2(1):17-21.

474 **Brett, J.J. and A.C. Nagy.** *1973.* Feathers in the Wind; the Mountain and the Migration. Hawk Mountain Sanctuary Assoc., Kempton, PA. 70 pp.

475 **Brewster, W.** *1924.* The Birds of Lake Umbagog Region of Maine. Harvard College, Cambridge, Mass. 620 pp.

476 **Brewster, W.** *1925.* The birds of the Lake Umbagog region of Maine. Bull. Mus. of Comp. Zool. 66(Part 2):213-402.

477 **Brichetti, P.** *1973.* Gli uccelli del bresciano. Riv. Ital. Ornitol. 43(4):519-649.

478 **Briggs, J.J.** *1843.* Notes on the capture and appearance of some of our rarer British birds, in the County of Derby. Zoologist 1:178-180.

479 **Briggs, S.A.** *1957.* Safer passage for hawks in Pennsylvania. Atlantic Nat. 12(6):293-296.

480 **Britton, H.A.** *1978.* E.A.N.H.S. nest record scheme: July 1976-December 1977. Scopus 1(5):132-141.

481 **Britton, H.A.** *1979.* E.A.N.H.S. nest record scheme: 1978. Scopus 2(5):126-132.

482 **Britton, P.L.** *1970.* Birds of the Balovale District of Zambia. Ostrich 41(3):145-190.

483 **Britton, P.L.** *1970.* Some non-passerine bird weights from East Africa. Bull. Brit. Ornithol. Club 90(5):142-144.

484 **Britton, P.L.** *1978.* The Anderson Collection from Tanzania. Scopus 2(4):77-85.

485 **Brodezky, E.** *1972.* I found a nest of a golden eagle. Saleit 1:271-272. (In Hebrew)

486 **Brodtkorb, R.** *1964.* Flying Free. Rand Mac Nally and Comp., Chicago. 141 pp.

487 **Broekhuysen, M.H., G.J. Broekhuysen and J. Martin.** *1974.* Six black eagles and a missed chance. Bokmakierie 26(4):82-83.

488 **Brokehuysen, G.J. and R. Liversidge.** *1950.* Members of the Cape Bird Club ring the first black eagle *Pteroaetus verreauxi.* Ostrich 21(1):35-36.

489 **Brooke, A.B.** *1873.* Notes on the ornithology of Sardinia. Ibis 3(10):143-155.

490 **Brooke, R.** *1973.* Steppe eagle in S.A. and its identification. Bokmakierie 25(4):86-88.

491 **Brooke, R.K.** *1964.* Avian observations on a journey across central Africa and additional information on some of the species seen. Ostrich 35(4):277-292.

492 **Brooke, R.K.** *1965.* Notes, chiefly distributional, on some birds in the Kafue National Park. Puku, Occas. Pap. Dept. Game and Fisheries, Zambia. No. 3. pp. 59-65.

493 **Brooke, R.K.** *1968.* Notes on some birds of Mozambique. Ostrich 39(3):180-184.

494 **Brooke, R.K.** *1972.* An out of season breeding record for *Aquila wahlbergi* from Zambia. Bull. Brit. Ornithol. Club 92(3/4):97.

495 **Brooke, R.K.** *1974.* Steppe eagle in Africa. Brit. Birds 67(9):393.

496 **Brooke, R.K.** *1974.* The African southern limits of the steppe eagle in winter. Bull. Brit. Ornithol. Club 94(2):62.

497 **Brooke, R.K., I.C. Cannell and R.D. Jeffery.** *1971.* New distributional records of raptors in western Angola. Bull. Brit. Ornithol. Club 91(6):165-167.

498 **Brooke, R.K., J.H. Grobler, M.P. Stuart Irwin and P. Steyn.** *1972.* A study of the migratory eagles *Aquila nipalensis* and *A. pomarina* (Aves:Accipitridae) in southern Africa, with comparative notes on other large raptors. Occas. Pap. Natl. Museum of Rhodesia, 1972 B5:61-114.

499 **Brooker, M.** *1973.* Wedge-tailed eagle survey. R.A.O.U. Newsl. 15:4.

500 **Brooker, M.G.** *1974.* Field observations of the behavior of the wedge-tailed eagle. Emu 74:39-42.

501 **Brooker, M.G. and M.G. Ridpath.** *1980.* The diet of the wedge-tailed eagle, *Aquila audax,* in western Australia. Aust. Wildl. Res. 7:433-452.

502 **Brooker, M.G., M.G. Ridpath, A.J. Estbergs, J. Bywater, D.S. Hart and M.S. Jones.** *1978.* Bird observations on the north-western Nullarbor Plain and neighbouring regions, 1967-1978. Emu 79(4):176-187.

503 **Brooks, A.** *1933.* Eagles, hawks and vultures. Natl. Geog. 64(1):64-95.

504 **Brooks, M.** *1934.* The golden eagle in West Virginia. Auk 51(3):370.

505 **Brooks, W.E.** *1868.* Remarks on various stages of plumage of *Aquila imperialis, A. fluvescens* and *A. noevia.* Ibis 4(15):349-353.

506 **Brooks, W.E.** *1870.* Comments on *Aquila naevia.* Ibis 6(22):288-291.

507 **Brooks, W.E.** *1872.* On the imperial eagles of India. Proc. Zool. Soc. London 1872:502-504.

508 **Brooks, W.E.** *1872.* Imperial eagles of India. Proc. Asiatic Soc. Bengal pp. 64-65. (Title Only)

509 **Brooks, W.E.** *1873.* Notes upon some of the Indian and European eagles. Stray Feathers 1(5):325-331.

510 **Brooks, W.E.** *1873.* Notes upon some of the Indian and European eagles. Stray Feathers 1(4):290-294.

511 **Brooks, W.E.** *1873.* Notes upon some of the Indian and European eagles. Stray Feathers 1(6):463-464.

512 **Brooks, W.E.** *1874.* Notes on some European and Asiatic eagles. Ibis 4(13):84-87.

513 **Brooks, W.E.** *1875.* Notes on the "spotted eagle', *Aquila navia.* Stray Feathers 3(4):304-313.

514 **Brooks, W.E.** *1876.* *Aguila maculata,* Melin *(Aguila clanga,* Pallas), the spotted eagle. Stray Feathers 4(4-6):268-278.

515 **Brooks, W.E.** *1876.* A few ornithological notes and corrections. Ibis 6(24):499-504.

516 **Brooks, W.E.** *1879.* Ornithological observations in Sikhim, the Punjab, and Sind. Stray Feathers 8(6):464-489.

Master No.	Citation
517	**Brooks, W.E.** *1884.* A few ornithological notes and corrections. Ibis 2(7):234-240.
518	**Brooks, W.E.** *1885.* Stray ornithological notes. Ibis 3(2):380-389.
519	**Brosset, A.** *1956.* Les oiseaux du Maroc Orientale la Mediterranee a Berguent. Alauda 24(3):161-205.
520	**Brosset, A.** *1957.* Les oiseaux de la steppe de Berguent. Remarques particulieres sur leurs migrations. Alauda 25(3):196-208.
521	**Brosset, A.** *1967.* Fecondite potentialle et fecondite reele chez les rapaces des regions Pre-Sahariennes. Terre Et Vie 1967 (1):63-75.
522	**Broun, M.** *1935.* The hawk migration during the fall of 1934, along the Kittatinny Ridge in Pennsylvania. Auk 52(3):233-248.
523	**Broun, M.** *1939.* Fall migrations of hawks at Hawk Mountain, Pennsylvania, 1934-1938. Auk 56(4):429-441.
524	**Broun, M.** *1947.* Golden eagle captures red-shouldered hawk. Auk 64(2):317-318.
525	**Broun, M.** *1948.* Golden eagles at Hawk Mountain. Nature Mag. 41:490-491,500.
526	**Broun, M.** *1960.* Hawks Aloft: The Story of Hawk Mountain. Kutztown Publ. Co., Kutztown, Pa. 222 pp.
527	**Broun, M.** *1966.* What future for birds of prey? Audubon 68(5):331-334,341-342.
528	**Broun, M. and B.V. Goodwin.** *1943.* Flight-speeds of hawks and crows. Auk 60(4):487-492.
529	**Brouwer, G.A. and G.C.A. Junge.** *1945.* Waarnemingen van broedvogels en trekvogels in 1943. Ardea 33(3/4):136-222.
530	**Brouwer, G.A. and G.C.A. Junge.** *1946.* Waarnemingen van broedvogels en trekvogels in 1944 en 1945. Ardea 34(4):315-393.
555	**Brown, P.** *1978.* The birds of Tasmania. Avic. Mag. 84(1):32-50.
531	**Brown, C.E.** *1928.* Longevity of birds in captivity. Auk 45(3):345-348.
532	**Brown, L. and D. Amadon.** *1968.* The hunting hawk. Audubon 70(5):38-53.
533	**Brown, L.H.** *1946.* A golden eagle census? Country Life 100(2597):766-767.
534	**Brown, L.H.** *1952.* On the biology of the large birds of prey of the Embu District, Kenya Colony. Ibis 94(4):577-620.
535	**Brown, L.H.** *1955.* Supplementary notes on the biology of the larger birds of prey of Embu District, Kenya Colony. Ibis 97(1):38-64.
536	**Brown, L.H.** *1958.* What controls the number of golden eagles? Country Life 124(3215):434,437.
537	**Brown, L.H.** *1966.* Observations on some Kenya eagles. Ibis 108:531-572.
538	**Brown, L.H.** *1969.* Status and breeding success of golden eagles in northwest Sutherland in 1967. Brit. Birds 62(9):345-363.
539	**Brown, L.H.** *1969.* Verreaux's eagle. Birds of the World, I.P.C. Magazines, Ltd. London.
540	**Brown, L.H.** *1970.* Some factors affecting breeding in eagles. Ostrich (Suppl. 8):157-167.
541	**Brown, L.H.** *1971.* African Birds of Prey. Houghton Mifflin, Boston. 320 pp.
542	**Brown, L.H.** *1974.* A record of two young reared by Verreaux's eagle. Ostrich 45(2):146-147.
543	**Brown, L.H.** *1975.* The status of East African and Ethiopian raptors. Pages 177-184. *in:* J.R. Murphy, C.M. White and B.E. Harrell, Eds. Population Status of Raptors. Proc. of the Conf. on Raptor Conserv. Techniques. 22-24 March, 1973. Fort Collins, Co. Raptor Res. Rep. No. 3.
544	**Brown, L.H.** *1976.* British Birds of Prey. William Collins Sons and Co., Ltd., London. 400 pp.
545	**Brown, L.H.** *1976.* Eagles. Arco Publishing Co., New York. 96 pp.
546	**Brown, L.H.** *1976.* Birds of Prey: Their Biology and Ecology. A. + W. Publishers, Inc., New York. 256 pp.
547	**Brown, L.H.** *1977.* Eagles of the World. Universe Books, New York. 244 pp.
548	**Brown, L.H.** *1977.* The status of, and threats to, diurnal and nocturnal birds of prey in East Africa and Ethiopia. Pages 15-28. *in:* R.D. Chancellor, Ed. Proc. of the World Conf. on Birds of Prey. 1-3 October, 1975. Vienna. I.C.B.P.
549	**Brown, L.H.** *1978.* African birds of prey: an assessment of our current knowledge. Pages 2-12. *in:* Proc. of the Symposium on African Predatory Birds. 29 August- 1 September, 1977. Pretoria. Northern Transvaal Ornithol. Soc.
550	**Brown, L.H.** *1978.* Wahlberg's eagle in northeastern Africa and the first record for Somalia. Scopus 2(1):19-20.
551	**Brown, L.H. and D. Amadon.** *1968.* Eagles, Hawks, and Falcons of the World. McGraw-Hill Book Co., N.Y. 945 pp.
552	**Brown, L.H. and A. Watson.** *1964.* The golden eagle in relation to its food supply. Ibis 106(1):78-100.
553	**Brown, L.H., V. Gargett and P. Steyn.** *1977.* Breeding success in some African eagles related to theories about sibling aggression and its effects. Ostrich 48(3/4):65-71.
554	**Brown, O.** *1938.* Golden eagle controls ground squirrels in Yosemite Valley. Yosemite Nat. Notes 17(12):159.
556	**Brown, V., H.G. Weston, Jr. and J. Buzzell.** *1973.* Handbook of California Birds. Naturegraph Publ., Healdsburg. 224 pp.
557	**Browning, B.** *1962.* Hell on wings. Guns and Hunt. 5(1):16-19,78-79.
558	**Brozek, C.** *1976.* "African" birds in Poland. Bokmakierie 28(4):87-90.
559	**Bruce, D.** *1971.* Bird of Jove. Ballantine Books, New York. 283 pp.
560	**Bruderer, B.** *1976.* Unter welchen Umstanden greifen steinadler *Aquila chrysaetos* flugzeugean? Ornithol. Beob. 73(1):29-30.
561	**Bruderer, B. and W. Thonen.** *1977.* Liste rouge des especes d'oiseaux menacees et rares en Suisse. Nos Oiseaux 34(Suppl.):1-36.
562	**Brunel, J. and J.-M. Thiollay.** *1969.* Liste preliminaire des oiseaux de Cote-d'Ivoire. Alauda 37(3):230-254.
563	**Bruner, L.** *1901.* Birds that nest in Nebraska. Ann. Report of Nebraska State Board of Agric. for 1900. Nebr. Ornithol. Union. p. 240-253.
564	**Brunn, J.B., J.R. Jacobsen, E. Kramshoj and B.P. Nielsen.** *1972.* Rapport fra sjaeldenhedsudvalget for 1971. (Report for the Rarity Committee for 1971.) Dansk Ornithol. 66(3/4):123-131.
565	**Bruns, E.H.** *1970.* Winter predation of golden eagles and coyotes on pronghorn antelopes. Can. Field-Nat. 84(3):301-304.
566	**Bruns, H.** *1975.* Zur bestimmung der Europaischen greifvogel im Fluge. Ornithol. Mitt. 27(12):227-231.
567	**Bruun, B.** *1971.* Birds of Europe. New York. 320 pp.
568	**Bry, E.** *1963.* Narrative of the eagle. N.D. Outdoors 26(2):16-19.
569	**Bryant, C.A.** *1930.* Golden eagles. Gull 12(12):1.
570	**Bryant, H.C.** *1929.* Sparrow hawk pesters golden eagle. Yosemite Nat. Notes 8(7):76.

Master No.	Citation
571	**Bryant, W.E.** *1889.* A catalogue of the birds of lower California, Mexico. Proc. of Calif. Acad. Sci. 2:237-320.
572	**Buchheister, C.W.** *1961.* A bill to save both eagles. Audubon 63(5):258-259.
573	**Buchheister, C.W.** *1962.* Eagle slaughter is shameful thing. Audubon 64(3):148.
574	**Bucknill, J.A.** *1910.* On the ornithology of Cyprus. Part II. Ibis (9th Ser.) 4(13):1-47.
575	**Bucknill, J.A.** *1911.* A further contribution to the ornithology of Cyprus. Ibis 5(20):632-656.
576	**Bucknill, J.A.** *1913.* A third contribution to the ornithology of Cyprus. Ibis (10th Ser.) 1(1):2-14.
577	**Bue, I.G. and B.A. Nelson.** *1949.* Pheasants and winter cover. S. Dakota Dept. Game, Fish and Parks. Proj. 19-D-2. 17-R-3. 43 pp.
578	**Buechner, H.K.** *1950.* Life history, ecology, and range use of the pronghorn antelope in Trans-Pecos, Texas. Am. Midland Nat. 43(2):257-354.
579	**Buell, A.W.** *1954.* Eagle through the windshield. Nat. Hist. 63(8):381.
580	**Buettikofer, J.** *1948.* The golden eagle - we must not let it die out. Pro Natura Bale 1:65-76. (Title Only)
581	**Buhler, A.** *1964.* Steinadler baut einem baumhorst im Suldtal BE. Ornithol. Beob. 61(1):46-47.
582	**Bundy, G.** *1976.* The birds of Libya. British Ornithol. Union Checklist No. 1. 102 pp.
583	**Bundy, G. and J.H. Morgan.** *1969.* Notes on Tripolitanian birds. Bull. Brit. Ornithol. Club 89(5):139-144.
584	**Burckhardt, D.** *1953.* Ruckmeldung eines beringten steinadlers aus dem Ausland. Ornithol. Beob. 50(2):48-49.
585	**Bureau, L.** *1911.* Reponses a la question relative a l'age augquel un oiseau male est capable de feconder. Oiseau Rev. Fr. Ornithol. 2:479.
586	**Burgh, G.W.H., vos.** *1913.* The hawks of southern Wisconsin and northern Illinois. Oologist 30(2):30-32.
587	**Burnier, J. and R. Hainard.** *1954.* En presence de l'aigle royal. Nos Oiseaux 22(8/9):178-186.
588	**Burns, F.L.** *1921.* Comparative periods of nestling life of some North American Nidicolae. Wilson Bull. 33(1/2):4-15,90-99.
589	**Burton, M.** *1952.* Golden eagles and other giants. Illus. London News 220(5890):4249
590	**Busch, D.E.** *1979.* Successful golden eagle nest in southwest Nebraska. Nebr. Bird Rev. 47(3):57.
591	**Busse, H.** *1968.* Unfer vogel atlas. 35. Der schelladler. Falke 15(11):394-395.
592	**Busse, H.** *1973.* Unser vogel atlas. 89. Der Wahlbergadler. Falke 20(5):178-179.
593	**Butler, A.L.** *1905.* A contribution to the ornithology of Egyptian Soudan. Ibis 5(19):301-401.
594	**Butler, A.L.** *1908.* A second contribution to the ornithology of the Egyptian Soudan. Ibis 2(6):205-263.
595	**Butler, E.A.** *1875.* Notes on the avifauna of Mount Aboo and Northern Guzerat. Stray Feathers 3(6):437-500.
596	**Butler, E.A.** *1880.* A tentative catalogue of the birds of the Deccan and South Mahratta Country. Stray Feathers 9(5/6):367-442.
597	**Buttikoler, J.** *1945.* Der kanton schwyz schutztden steinadler. Ornithol. Beob. 42(2):31.
598	**Buturlin, S.A.** *1932.* Name of the Asiatic golden eagle. Bull. Brit. Ornithol. Club 52(361):149-150.
599	**Buxton, E.C.** *1861.* Golden eagle nesting in Scotland. Zoologist 19:7463.
600	**C., J.R.** *1886.* Nesting of the golden eagle. Ornithol. and Ool. 11(6):85.
601	**Cabler, C.** *1964.* Golden eagle in eastern Arkansas. Migrant 35(1):21.
602	**Cadbury, C.J.** *1980.* Silent death. The destruction of birds and mammals through the deliberate misuse of poisons in Britain. Royal Soc. for the Protection of Birds. Report. 24 pp.
603	**Cade, T.J.** *1953.* A synoptic bibliography of Alaskan ornithology. Univ. of Calif., Publ. Zool. (Title Only)
604	**Cade, T.J. and L. Greenwald.** *1966.* Nasal salt secretion in Falconiform birds. Condor 68(4):338-350.
605	**Cain, C.** *1936.* Notes on the wedge-tailed eagle, *Uroaetus audax*, in the Farina District, S.A., during 1934-1935. S. Aust. Ornithol. 13(6):169-170.
606	**Cain, S.A., J.A. Kadlec, D.L. Allen, R.A. Cooley, M.G. Hornocker, A.S. Leopold and F.H. Wagner.** *1972.* Predator control-1971. Report to the Council on Environmental Quality and the Department of the Interior. Advisory Comm. on Predator Control, Inst. for Environmental Quality, Univ. of Michigan, Ann Arbor, Mich. 207 pp.
607	**Cairns, J.S.** *1889.* The summer birds of Buncombe County, North Carolina. Ornithol. and Ool. 14(2):17-23.
608	**Calaby, J.H.** *1951.* Notes on the little eagle; with particular reference to rabbit predation. Emu 51(1):33-56.
609	**Caldwell, J.** *1936.* Recent eagle occurences. Migrant 7(2):46.
610	**Call, M.** *1979.* Habitat management guide for birds of prey. U.S.D.I. Bureau of Land Management, Tech. Note TN-338. 70 pp.
611	**Call, M.W.** *1978.* Nesting habitats and surveying techniques for common western raptors. U.S.D.I. Bureau of Land Management, Tech. Note TN-316. 115 pp.
612	**Callahan, P.S.** *1974.* The Manificent Birds of Prey. Holiday House, N.Y. 190 pp.
613	**Callison, C.H.** *1971.* A massacre of eagles Audubon 73(4):94-95.
614	**Camenzind, F.J.** *1968.* Nesting ecology and behavior of the golden eagle in west central Utah. M.S. Thesis. Brigham Young Univ., Provo, UT. 42 pp.
615	**Cameron, A.C.** *1952.* A what height do birds fly? Emu 52(2):86-88.
616	**Cameron, E.S.** *1905.* Nesting of the golden eagle in Montana. Auk 22:158-167.
617	**Cameron, E.S.** *1907.* The birds of Custer and Dawson Counties, Montana. Auk 24(3):241-270.
618	**Cameron, E.S.** *1908.* Observations on the golden eagle in Montana. Auk 25(3):251-268.
619	**Cameron, E.S.** *1908.* Nesting of the golden eagle in Montana. Country Life 21(623):834-836.
620	**Cameron, R.A.D., L. Cornwallis, M.J.L. Percival and A.R.E. Sinclair.** *1967.* The migration of raptors and storks through the Near East in autumn. Ibis 109(4):489-501.
621	**Campbell, J.M.** *1956.* A note on winter foraging of the golden eagle in Yakima County. Murrelet 37(1):11-12.
622	**Campbell, J.M.** *1960.* Nesting of the golden eagle in the central Brooks Range of Arctic Alaska. Condor 62(4):298.
623	**Campbell, J.S.** *1934.* The golden eagle in northern Louisiana. Wilson Bull. 46(2):116.

Master No.	Citation
624	**Campbell, J.S.** *1934.* Golden eagle in north Louisiana. Auk 51(3):370-371.
625	**Campbell, R.W., M.G. Shepard and R.H. Drent.** *1972.* Status of birds in the Vancouver area in 1970. Syesis 5:137-167.
626	**Cantoni, J.** *1960.* Note sur la capture d' *Aquila pomarina* Brehm en Tunisie. Alauda 28(3):230-232.
627	**Capiluppi, M.** *1971.* Notizie ornitologiche dalla provincia di Reggio Emilia (1967-1970). Riv. Ital. Ornitol. (2nd Ser.) 41(2):122-126.
628	**Carnie, S.K.** *1954.* Food habits of nesting golden eagles in the coast ranges of California. Condor 56(1):3-12.
629	**Carpelan, J.** *1927.* Zur brutbiologie des steinadlers und fischadlers. Vogelwelt 3(2):47-48. (Title Only)
630	**Carpenter, N.K.** *1919.* Evidence that many birds remain mated for a number of years. Condor 21(1):28-30.
631	**Carr, W.H.** *1944.* Eagles wild and tame. Nat. Hist. 53(10):442-447.
632	**Carroll, C.J.** *1916.* Extermination of the golden eagle in Ireland. Brit. Birds 9(9):251-252.
633	**Carter, T.** *1888.* Notes from western Australia. Zoologist (3rd Ser.) 12:28-30.
634	**Casey, R.G.** *1956.* Eagle hunt. Flying 59(6):28,83-84.
635	**Cassidy, J.** *1938.* Australian bird-beauties. Avic. Mag. 3(5):130-136.
636	**Castan, R.** *1963.* Notes de Tunisie (Region de Gabes). Alauda 31(4):294-303.
637	**Castellani, C., G. Siracusano, R. Parri and G. Rolandi.** *1973.* Osservazioni ornitologiche sul Massiccio del Pollino. Riv. Ital. Ornitol. 43(1):135-146.
638	**Caterini, F.** *1941.* Gli uccelli del Pisano. Riv. Ital. Ornitol. (2nd Ser.) 11(4):137-149.
639	**Caterini, F.** *1953.* Aquila reale *(Aquila chrysaetus).* Riv. Ital. Ornitol. 23(4):153-155.
640	**Cawkell, E.M.** *1965.* Notes on Gambian birds. Ibis 107(4):535-540.
641	**Cawkell, E.M. And. R.E. Moreau.** *1963.* Notes on birds in the Gambia. Ibis 105(2):156-177.
642	**Cayley, N.** *1973.* What Bird Is That? A Guide to the Birds of Australia. Sydney, Australia. 359 pp.
643	**Cederholm, C.-G. and M. Sylven.** *1972.* Aigle criard. Alauda 40(4):398.
644	**Cejchan, A.** *1948.* O vyskytu dravcu Accipitres v okoli Adrspachu. (The occurrence of Accipitres in Adrspach.) Sylvia 9-10(3):70-71. (In Czech., English Summ.)
645	**Cerny, W.** *1951.* Orel volavy *(Aquila clanga)* na Ostravsku. (The spotted eagle *(Aquila clanga)* in the vicinity of Prague.) Sylvia 13(4):137-138. (In Czech., English Summ.)
646	**Chace, L.W.** *1936.* Golden bald eagle. Oologist 53(10):134-135.
647	**Chalker, J.R.** *1886.* Nesting of the golden eagle. Ornithol. and Ool. 11(6):85.
648	**Chalker, J.R.** *1887.* A trip after golden eagles eggs. Ornithol. and Ool. 12(6):86-88.
649	**Chalker, J.R.** *1888.* A trip to San Benito County after golden eagle's eggs. Ornithol. and Ool. 13(8):119-121.
650	**Chamberlain, E.B.** *1957.* Golden eagles killed in the Carolinas. Chat 21(2):44.
651	**Chambers, F.** *1903.* The Tennessee eagles. Independent 55(2829):495-496.
652	**Chance, E.** *1927.* Eggs of the golden eagle. Bull. Brit. Ool. Assoc. 1(9):93-94.
653	**Chancellor, R.D. (ed.)** *1977.* Proceedings of the World Conference on Birds of Prey. 1-3 October, 1975. Vienna, Austria. International Council for Bird Preservation. 434 pp.
654	**Chapman, A.** *1888.* Winter notes in Spain. Ibis 6(24):444-461.
655	**Chapman, F.M.** *1912.* Color Key to North American Birds. D. Appleton and Comp., NY. 336 pp.
656	**Chapman, F.M.** *1932.* Handbook of Birds of Eastern North America. D. Appleton and Co., N.Y. and London. 581 pp.
657	**Charriere, A.-L.** *1913.* Observations ornithologiques dans l'Arrondissement de Setif (Algerie) (1895-1900). Oiseau Rev. Fr. Ornithol. 3(52/53):135-139.
658	**Chase, R.W.** *1916.* Untitled. Bull. Brit. Ornithol. Club 37:55.
659	**Chase, R.W.** *1917.* Abnormal golden eagle's eggs. Brit. Birds 11(5):119.
660	**Chasen, F.N.** *1921.* Field notes on the birds of Macedonia. With special reference to the Struma Plain. Ibis 3(2):185-227.
661	**Chavigny, J., de. and N. Mayaud.** *1932.* Sur l'avifaune des Acores. Alauda 4(4):416-441.
662	**Cheesman, R.E. and W.L. Sclater.** *1935.* On a collection of birds from north-western Abyssinia. Ibis 5(1):151-191.
663	**Cheltsov-Bebutov, A.M.** *1970.* Pernatye khistchniki i okhotnichie khozyaistvo. (Raptors and game management.) Trans. 9th Int. Cong. Game Biologist, Moscow. p. 449-457. (In Russian, English Summ.)
664	**Cheng, T.-H.** *1955.* A distribution list of Chinese birds. Volume I. Non-passeriformes. Transl. of Monogr., Chung-Kuo Niao-Lei Fen-Pu Mo-Lu, Peiping. Joint Publ. Res. Service ,NTIS, Va. 259 pp.
665	**Chernel, S.** *1909.* Beitrage zur nahrungsfrage unserer carnivoren vogelwelt. Aquila 16:145-155.
666	**Chernel, S.** *1918.* Nomenclator avium regni Hungarie. Aquila 25(Suppl.):1-76.
667	**Chessex, C.** *1948.* Journees de Fionnay 1948. Nos Oiseaux 19(11):287-288.
668	**Chessex, C.** *1949.* Lacher d'un aigle royal *Aquila chrysaetos.* Nos Oiseaux 20(6):135-136.
669	**Chessex, C.** *1963.* "Nos Oiseaux" fete son premier demi-siecle d'activite. Nos Oiseaux 27(2):34-45.
670	**Cheylan, G.** *1973.* Les deplacements de la niverolle *Montifringilla nivalis* et son hivernage en France meridionale. Alauda 41(3):213-226. (English Summ.)
671	**Cheylan, G.** *1973.* Notes sur la competition entre l'aigle royal *Aquila chrysaetos* et l'aigle Bonnelli *Hieraetus fasciatus.* (Notes on competition between golden and Bonnelli's eagles.) Alauda 41(3):203-212.
672	**Cheylan, G.** *1974.* Biogeographie d'un Mediterraneene: la Sainte-Victoire (Bouches-du-Rhone). Alauda 42(1):57-78. (English Summ.)
673	**Cheylan, G.** *1977.* La place trophique de l'aigle de Bonnelli *Hieraetus fasciatus* dans les biocenoses Mediterraneenes. (The trophic place of Bonnelli's eagle in Mediterranean biocenoses.) Alauda 45(1):1-15. (English Summ.)

Master No. **Citation**

674 **Chiavetta, M.** *1977.* Diurnal birds of prey ringed in Europe and in North Africa, shot in Italy from 1969 to 1973. Considerations on effects of shooting and status of the Italian population. Pages 272-274. *in:* R.D. Chancellor, Ed. Proc. of the World Conf. on Birds of Prey. 1-3 October, 1975. Vienna. I.C.B.P.

675 **Childs, J.L.** *1905.* Taking a set of golden eagle *(Aquila chrysaetos).* Warbler (2nd Ser.) 1(2):43.

676 **Chisholm, E.C.** *1936.* Birds of the Pilliga scrub. Emu 36(1):32-38.

677 **Choate, E.A.** *1973.* The Dictionary of American Bird Names. Gambit, Boston. 261 pp.

678 **Chogakkal, N.** *1974.* Check-list of Japanese birds. Fifth and Rev. Ed. Gakken Co., Ltd. Tokyo. 364 pp.

679 **Christensen, N.H.** *1960.* Ornithologiske iagttagelser ved Atbara, Nord-Sudan.(Ornithological observations at Atbara, North-Sudan.) Dansk Ornithol. 54(3):144-161. (English Summ.)

680 **Christensen, N.H.** *1962.* Observationer af fugle og noter om fugletraek i Iraq, 1958. Dansk Ornithol. 56(1):56-81.

681 **Christensen, N.H. and L.H. Sorensen.** *1961.* Efterastraek af rovfugle i Danmark. (The autumn migration of raptors in Denmark.) Dansk Ornithol. 55(3):113-136. (English Summ.)

682 **Christensen, S.** *1967.* Observationer af fugle og noter om fugletraek pa Cypern 1964-1965. Dansk Ornithol. 61(1):40-55.

683 **Christensen, S.** *1968.* Kejserorn *(Aquila heiliaca)* iagttaget pa Amager 1965. (Imperial eagle *(Aquila heliaca)* observed on the island Amager in 1965.) Dansk Ornithol. 62(1):95-101.

684 **Christensen, S. and E. Kramshoj.** *1965.* Iagttagelser af mindre almindelige fugle pa sydvestamager. Dansk Ornithol. 59(2):103-106. (English Summ.)

685 **Christensen, S., B.P. Nielson, N.H. Christensen and L.H. Sorensen.** *1968.* Feltbestemmelse of orne af slaegten *Aquila.* (Field identification of *Aquila* eagles. Dansk Ornithol. 62(1):68-94.

686 **Christensen, S., B.P. Nielsen, R.F. Porter, and I. Willis.** *1972.* Flight identification of European raptors. Part 3. *Aquila* and white-tailed eagles. Brit. Birds 65(2):52-78.

687 **Christison, P.** *1949.* Golden eagle versus wildcat. Scott. Nat. 61(2):121-122.

688 **Christoleit, T.** *1904.* Ueber ein vorkommendes schelladlers *(A. maculata).* Vogelwelt 29(7):309-311.

689 **Christoleit, W.** *1903.* Der schreiadler *(Aquila naevia)* . Deutsche Jager-Zeitung 41:144. (Title Only)

690 **Christopher, E.** *1972.* Massacre at Jackson's Canyon; slaughtering of eagles and other predators. Out. Life 149:39-41. (Title Only)

691 **Christy, M.** *1892.* Spotted eagle in Essex. Zoologist (3rd Ser.) 16:76.

692 **Chubb, E.C.** *1909.* On the birds of Bulawayo, Southern Rhodesia. Ibis 3(9):140-172.

693 **Cieslak, M.** *1977.* Prawdopodobne gniazdowanie orlika grubudziobego *(Aquila clanga)* w rezerwacie jata w nadlesnictwie Lukow. (Probable nesting of spotted eagle in Lukow forest range.) Notaki Ornitol. 18(1/2):54.

694 **Clancy, P.A.** *1956.* First report of the S.A.O.S. list Committee. Ostrich 27(4):173-183.

695 **Clark, W.S.** *1974.* Cape May Point raptor banding station - 1973 results. E.B.B.A. News 37(1):51-64.

696 **Clarke, G.** *1967.* Birds of Para Wirra National Park. S. Aust. Ornithol. 24(7):119-134.

697 **Clarke, W.E.** *1884.* Field-notes from Slavonia and Hungary, with an annotated list of the birds observed in Slovonia. Ibis (5th Ser.) 2(6):125-148.

698 **Clawson, M.M.** *1948.* Golden eagle attacks young deer. Yosemite Nat. Notes 27(9):107-109.

699 **Cleland, J.B.** *1946.* Wedge-tailed eagle *(Uroaetus audax)* attacked by five kestrels. S. Aust. Ornithol. 18(1):1.

700 **Cleland, J.B.** *1956.* Birds as they affect the health and well-being of man in Australia. Emu 56(3):183-193.

701 **Coates, T.** *1901.* Notes from Point Cloates, N.W. Australia. Zoologist (4th Ser.) 5:255-262.

702 **Cofer, H.H.** *1939.* A summary of information concerning the hawks, eagles and vultures of Missouri, including a bulletin for public use. M.A. Thesis, Univ. of Missouri, Columbia, Mo. 55 pp.

703 **Coffey, B.B.** *1937.* Golden eagle in Mississippi. Migrant 8(2):35.

704 **Coffey, J.W.** *1965.* Golden eagle captured in Mitchell County, North Carolina. Chat 29(2):52-53.

705 **Coheleach, G.** *1972.* The way to extinction - Spanish imperial eagle. Audubon 74(1):32-33.

706 **Coheleach, G.** *1973.* Under the spell of the golden eagle. Natl. Wildl. 12(1):12-16.

707 **Cohen, D.A.** *1893.* The golden eagle in captivity. Nidiologist 1(2):28.

708 **Colby, W.E.** *1921.* Golden eagles in the Sierra. Sierra Club Bull. 11(2):206-207.

709 **Cole, W.** *1891.* Spotted eagle shot at Leigh. Essex Nat. 5:218-219.

710 **Colebrook-Robjent, J.F.R. and T.O. Osborne.** *1973.* Nesting of Verreaux's eagle *Aquila verreauxi* in Zambia. Bull. Brit. Ornithol. Club 93(4):166.

711 **Collier, C.** *1904.* The birds of the island of Raasay. Ibis 4(16):490-512.

712 **Collman, J.R. and J.P. Croxall.** *1967.* Spring migration at the Bosphorous. Ibis 109(3):359-372.

713 **Collopy, M.W.** *1978.* The bioenergetic requirements for growth and maintenance of raptors common to the Snake River Birds of Prey Natural Area. Contract Final Rep., U.S.D.I. Bureau of Land Management, Idaho State Office.

714 **Collopy, M.W.** *1980.* Food consumption and growth energetics of nestling golden eagles. Ph.D. Dissertation. Univ. of Michigan, Ann Arbor, MI. 202 pp.

715 **Comble, J., de la.** *1956.* Capture d'un aigle criard *Aquila clanga* Pallas en Saone-et-Loire. Alauda 24(1):70.

716 **Compton, H.O.** *1958.* The effects of predation on pronghorn antelope in south central Oregon. M.S Thesis. Oregon State Coll., Corvallis, OR.

717 **Conder, P.** *1977.* Legal status of birds of prey and owls in Europe. Pages 189-193. *in:* R.D. Chancellor, Ed. Proc. of the World Conf. on Birds of Prey. 1-3 October, 1975. Vienna. I.C.B.P.

718 **Condon, H.T.** *1955.* Aboriginal bird names - South Australia. S. Aust. Ornithol. 21(6/7):74-88.

719 **Condon, H.T.** *1959.* Bird protection in South Australia. S. Aust. Ornithol. 23(1/2):3-6.

Master No.	Citation
720	**Condon, H.T.** *1962.* A handlist of the birds of South Australia. S. Aust. Ornithol. 23(6/8):85-150.
721	**Condon, H.T. and D. Amadon.** *1954.* Taxonomic notes on Australian hawks. Records of the S. Aust. Mus. 112(2):189-249. (Title Only)
722	**Congreve, W.M.** *1929.* Some notes from south-western Transylvania and Banat of New Rumania. Ibis 5(3):443-491.
723	**Congreve, W.M.** *1933.* Destruction of birds of prey in the British Islands. Ool. Rec. 13(2):46.
724	**Congreve, W.M.** *1935.* Further notes from south-western Transylvania. Ibis (13th Ser.) 5(3):564-577.
725	**Connolly, G.E.** . Prey remains at a golden eagle, *Aquila chrysaetos,* nest near Hopland, California. Calif. Fish and Game 62(1):85-86.
726	**Conway, W.G.** *1962.* Weights of 11 living eagles and vultures at the New York Zoological Park. Auk 79:274-275.
727	**Cook, A.J.** *1893.* Birds of Michigan. Michigan Agric. Expt. Stat. Bull. 94. 148 pp.
728	**Cooke, M.T.** *1941.* Returns from banded birds: recoveries of some banded birds of prey. Bird-Band. 12(4):150-160.
729	**Cooke, W.W.** *1884.* Bird nomenclature of the Chippewa indians. Auk 1(3):242-250.
730	**Cookman, A.** *1909.* Notes on the birds of California. Oologist 26(12):221-222.
731	**Coomber, R.** *1977.* Golden eagle repeatedly catching sticks in flight. Brit. Birds 70(9):391-392.
732	**Cooper, A.B.** *1969.* Golden eagle kills red deer calf. J. Zool. 158(2):215-216.
733	**Cooper, J.E.** *1972.* Some haematological data for birds of prey. Raptor Res. 6(4):133-136.
734	**Cooper, J.E.** *1973.* Use of the steroid anaesthetic CT 1341 in birds. Vet. Record 92(18):474-479.
735	**Cooper, J.E.** *1974.* Metomidate anesthesia of some birds of prey for laparotomy and sexing. Vet. Record 94(19):427-440.
736	**Cooper, J.E.** *1975.* The role of vaccination in the maintenance of captive birds of prey. Raptor Res. 9(1/2):21-26.
737	**Cooper, J.E.** *1980.* An eagle at the College. Annals of the Royal College of Surgeons of England. 62:240-241.
738	**Cooper, R.P.** *1972.* Birds of the Mopoke Plains. Aust. Bird Watcher 4(6):171-181.
739	**Copley, H.** *1944.* Occasional notes (tawny eagle). J. E. African Nat. Hist. Soc. 17(5/6):395.
740	**Cordeaux, W.H.** *1851.* Occurrence of the golden eagle *(Aquila chrysaetos)* in Herefordshire. Zoologist 9:3027.
741	**Corkhill, P.** *1980.* Golden eagles on Rhum. Scott. Birds 11(2):33-43.
742	**Cornish, D.M. and R. Ellis.** *1974.* Birds of the Mulligan River and adjacent Simpson Desert Area. S. Aust. Ornithol. 26(7):164-165.
743	**Corti, V.A.** *1956.* Die vogel des Kantons Wallis. Ornithol. Beob. 53(5/6):121-171.
744	**Cory, C.B.** *1909.* The birds of Illinois and Wisconsin. Field Mus. Nat. Hist. - Zoology 9:1-764.
745	**Cottam, C. and C.S. Williams.** *1939.* Food and habits of some birds nesting on islands in Great Salt Lake. Wilson Bull. 51(3):150-155.
746	**Coues, E.** *1884.* Ornithophilologicali. Auk 1(1/2):49-58,140-144.
747	**Coues, E.** *1892.* Nesting of the golden eagle in Arizona. Auk 9(2):201.
748	**Coues, E.** *1927.* Key to North American birds. Page Comp., Boston. 1152 pp.
749	**Couey, F.M.** *1944.* Golden eagle vs. big horn sheep. Mont. Wildl. Bull. 1(1):2-3. (Title Only)
750	**Courtney, C.H., D.J. Forrester and F.H. White.** *1977.* Epornitic of pox in a wild bird population. J. Am. Vet. Med. Assoc. 171(9):993-994.
751	**Coverley, H.W.** *1933.* Nesting notes from Portugal. Ibis (13th Ser.) 3(4):782-785.
752	**Coward, T.A.** *1969.* Birds of the British Isles and Their Eggs. Frederick Warne + Co., Ltd., London. 359 pp.
753	**Cowden, J.** *1976.* For the Love of an Eagle. David McKay Co., Inc., N.Y.
754	**Cox, J.B.** *1973.* Birds of the Mannum Area. S. Aust. Ornithol. 26(5):103-114.
755	**Cox, J.B. and L.P. Pedler.** *1977.* Birds recorded during three visits to the far north-east of South Australia. S. Aust. Ornithol. 27(7):231-250.
756	**Craig, G.** *1976.* Raptor population and productivity studies in northeastern Colorado. Colorado Div. of Wildlife, Proj. No. W-124-R-48, Wk. Pl. 1, Job No. 2. p. 199-206.
757	**Craig, G.R.** *1976.* Statewide raptor populations and characteristics studies. Colorado Div. of Wildl., Proj. No. W-124-R-4, Wk. Pl. 1, Job No. 1. p. 177-197.
758	**Craig, G.R.** *1977.* Statewide raptor populations and characteristics studies. Colo. Div. of Wildl. Game Res. Rep., April 21. 9 pp.
759	**Craig, G.R.** *1979.* Bald and golden eagle nesting studies. Colo. Dept. of Fish and Game, Job Prog. Rep. W-124-R, Job 1. p. 81-85.
760	**Craig, G.R.** *1979.* Bald and golden eagle winter population surveys. Colo. Dept. of Fish and Game, Job Prog. Rep. W-124-R, Job No. 2. p. 87-93.
761	**Craig, J.** *1973.* The eagles, hawks and falcons of Colorado. Colo. Outdoors 22(2):24-33.
762	**Craig, T.H.** *1978.* A car survey of raptors in southeastern Idaho 1974-1976. Raptor Res. 12(1/2):40-45.
763	**Craig, T.H.** *1979.* The raptors of the Idaho National Engineering Laboratory site. U.S. Dept. of Energy, Radiol. and Environ. Sci. Lab. ID 0-12089.
764	**Craighead, F.C., Jr.** *1976.* Progress toward tracking migrating raptors by satellite. Raptor Res. 10(4):112-120.
765	**Craighead, J.** *1979.* The eagles speak. Frontier (July-August):42-44.
766	**Craighead, J. and F. Craighead.** *1940.* In quest of the golden eagle. Natl. Geog. 77(5):692-710.
767	**Craighead, J., C. Craighead and D. Craighead.** *1967.* Sharing the lives of wild golden eagles. Natl. Geogr. 132(3):420-439.
768	**Cramp, S.** *1964.* Predators and toxic chemicals. Pages 53-58. *in:* Proc. of the Working Conf. on Birds of Prey and Owls. 10-12 April, 1964. Caen.
769	**Cramp, S. and I.J. Ferguson-Lees.** *1963.* The birds of the Danube delta and their conservation. Brit. Birds 56(9):323-339.
770	**Cramp, S. (ed.)** *1980.* Handbook of the birds of Europe, the Middle East and North Africa. Volume II. Oxford Univ. Press. 695 pp.

Master No. **Citation**

771 **Cramp, S., P.J. Conder and J.S. Ash.** *1967.* The risk to bird life from chlorinated hydrocarbon pesticides. Angew. Ornithol. 2(1):15-22.

772 **Crawford, D.N.** *1972.* Birds of Darwin area, with some records from other parts of Northern Territory. Emu 72(4):131-148.

773 **Cripps, J.R.** *1878.* First list of the birds of Furreedpore, eastern Bengal. Stray Feathers 7(3/4):238-315.

774 **Cruon, R. and J. Vielliard.** *1975.* Notes d'ornithologie Francaise. Alauda 43(1):1-21.

775 **Csato, J., v.** *1881.* Beobachtungen uber den schreiadler. *Aquila naevia,* Br. Mitt. Ornithol. Vereines Wein 5(7):52.

776 **Csato, J., v.** *1891. Aquila clanga* Pall. und *A. pennata* Gm. in der Umgebung von Nagy-Enyed brutend. Ornithol. Jahrbuch 2(2):49-52.

777 **Csato, J., von** *1881.* Beobachtungen uber den schreiadler, *Aquila naevia,* Br. Mitt. Ornithol. Vereines Wein 5(6):43-44.

778 **Csato, J., von.** *1894. Aquila orientalis* Cab. in Siebenburgen. Aquila 1:168.

779 **Csornai, R.** *1943.* Serbische und kroatische namen der vogel der Bacska. Aquila 50:394-402.

780 **Csornai, R., L. Szlivka and L. Antal.** *1958.* Adatok a Bacska es Banat Madarvilagahoz. (Data on the ornis of Batchka and Banat.) Aquila 65:234-239. (English Summ.)

781 **Cugnasse, J.M.** *1977.* Mise en reserve de nourriture chez les rapaces. (Hoarding of food by raptors.) Alauda 45(2/3):241-242.

782 **Cunningham, E.B.** *1970.* A golden eagle harrassing bighorn sheep. Can. Field-Nat. 84(2):183.

783 **Curry-Lindahl, K.** *1961.* Conservation and predation problems of birds of prey in Sweden. Brit. Birds 54(8):297-306.

784 **Curry-Lindahl, K.** *1972.* Let them live, a world wide survey of animals threatened with extinction. Wm. Morrow, NY. 394 pp.

785 **Curtis, E.L. and R.C. Miller.** *1938.* The sclerotic ring in North American birds. Auk 55(2):225-243.

786 **Cvitanic, A.** *1961.* Ornithological notes from Middle Dalmatia. Larus 15:125-150. (English Trans.)

787 **Cvitanic, A.** *1968.* The relationships between intestine and body length and the nutrition of some bird species. Larus 21-22:181-190. (English Trans.)

788 **Czarnecki, Z.** *1962.* Ptaki jeziora Goplo. (The birds of Goplo Lake.) Acta Ornithol. 6(11):181-194. (English Summ.)

789 **D' Ostilio, D.O.** *1954.* Nesting status and food of the golden eagle in northern Colorado. M.S. Thesis. Univ. of Colorado, Boulder, CO. 48 pp.

790 **D'Andria, A.** *1967.* Courtship of wedge-tailed eagles. Emu 67(1/2):55-56.

791 **Daguin, F.** *1913.* La verite sur les oiseaux de proie. Oiseau Rev. Fr. Ornithol. 3(51):114-117.

792 **Dahlbeck, N.** *1942.* Havsorn *(Haliaetus albicilla* L.) och kungsorn *(Aquila chrysaetos* L.) som hackfaglar i Sverige 1940-41. Var Fagelvarld 1(1):1-6. (German Summ.)

793 **Dal Nero, V.** *1923.* Elenco degli uccelli piu notevoli catturati negli anni 1922-23 nel Veronese e preparati per decorazione. Riv. Ital. Ornitol. 6(2):43-44.

794 **Dale, F.H.** *1936.* Eagle "control" in northern California. Condor 38(5):208-210.

795 **Dandl, J.** *1955.* Imperial eagle at Dunabogdany. Aquila 59-62:437.

796 **Danecki, J.** *1965.* Golden eagle - *Aquila chrysaetos* (L.). Acta Ornithol. 9:128. (English Trans.)

797 **Danford, C.G.** *1878.* A contribution to the ornithology of Asia Minor. Ibis 2(5):1-35.

798 **Danford, C.G.** *1880.* A further contribution to the ornithology of Asia Minor. Ibis 4(13):81-99.

799 **Danford, C.G. and J.A. Harvie Brown.** *1875.* The birds of Transylvania. Ibis (3rd Ser.) 5(19):291-312.

800 **Danielson, K. and E.L. Larson.** *1982.* Avledningsbeteendehos kungsorn *Aquila chrysaetos.* (Distraction display by a golden eagle.) Var Fagelvarld 41(2):108.

801 **Danilov, O.V.** *1976.* Khistchnye ptitsy i sovy Baraby i severnoi Kulundy.) (Birds of prey and owls of Barara and North Kulunda.) Nauka, Novosibirsk. p. 1-158.

802 **Dankin, N.S.** *1975.* (Biocenotic and economic significance of raptors in the plains of Zeya and Bureya Rivers and measures for their protection and attraction.) Zhivotnyi Mir Dalnego Vostoka, Blagoveshchensk. p. 63-65. (In Russian)

803 **Danko, S.** *1971.* Ornithologische beobachtungen in Mazedonien, mit besonderer Berucksichtigung der greifvogel. Ornithol. Mitt. 23(1):9-18.

804 **Dantoft, O., P.L. Hansen and B. Zobbe.** *1971.* Systematisk liste over forarets fugle. Dansk Ornithol. 65(4):148-163.

805 **Darling, F.F.** *1934.* Speed of a golden eagle's flight. Nature 134(3383):325-326.

806 **Daut, C.** *1902.* Raum fur alle hat die erde. Ornithol. Beob. 1(3):19-21.

807 **Daut, C.** *1906.* Der steinadler in der Schweiz. Ornithol. Beob. 5(6):87-90.

808 **Daut, C.** *1906.* Zum schutze des steinadlers. Ornithol. Beob. 5(10):155-159.

809 **Daut, C.** *1908.* Weiteres uber den steinadler in der Schweiz. Ornithol. Beob. 6(12):177-183.

810 **Daut, K.** *1910.* Neueres vom steinadler in der Schweiz. Ornithol. Beob. 7(4):59-60.

811 **Daut, K.** *1914.* Der steinadler in der Schweiz. Ornithol. Beob. 11(5):97-107.

812 **Daut, K.** *1923.* Der steinadler, *Aquila chr. chrysaetos.* Ornithol. Beob. 21(3):33-36.

813 **Daut, K.** *1928.* Vom steinadler. Ornithol. Beob. 26(17):4-7.

814 **David-Beaulieu, A.** *1949.* Les oiseaux de la Province de Savannakhet. Oiseau Rev. Fr. Ornithol. 19(2):153-194.

815 **Davidson, C.S. and C.E. Wenden.** *1878.* A contribution to the avifauna of the Deccan. Stray Feathers 7(1/2):68-95.

816 **Davidson, I.** *1978.* Flight identification of southern African raptors. Bokmakierie 30(2):43-48.

817 **Davidson, J.** *1882.* Rough list of the birds of western Khandesh. Stray Feathers 10(4):279-327.

818 **Davidson, R.C.** *1955.* Peregrine "stooping" at golden eagle. Irish Nat. J. 11(11):309.

819 **Davie, O.** *1882.* Eagles in Ohio. Ornithol. and Ool. 7(18):142.

820 **Davie, O.** *1882.* Capture of the golden eagle *(Aquila chrysaetos chrysaetos)* near Columbus, O. Bull. Nuttall Ornithol. Club 7(2):123.

821 **Davies, S.A.** *1905.* On the birds of the upper Muonio River. Ibis 5(17):67-85.

822 **Davies, S.J.J.F.** *1974.* The interpretation of the function of avian display. Emu 74(1):1-4.

Master No. **Citation**

823 **Davis, C.A., P.E. Sawyer, J.P. Griffing and B.D. Borden.** *1974.* Bird populations in a shrub-grassland area, southeastern New Mexico. New Mexico Agric. Expt. Station, Bull. 619. 29 pp.

824 **Davis, L.I.** *1972.* A Field Guide to the Birds of Mexico and Central America. Austin, Texas. 300 pp.

825 **Davis, R., Jr.** *1845.* Occurrence of the smaller eagle *(Aquila naevia)* in Ireland. Zoologist 3:1190.

826 **Davis, W.B. and B.P. Glass.** *1951.* Notes on eastern Chinese birds. Auk 68(1):86-91.

827 **Dawson, B.** *1974.* Letting them go. Santa Barbara Museum of Nat. Hist., Museum Talk 48(2):45-48.

828 **Dawson, W.L.** *1897.* A preliminary list of the birds of Okanogan County, Washington. Auk 14(2):168-182.

829 **Day, J.W.** *1961.* Strange tales of the golden eagle. Country Life 129(3333):108-109.

830 **De S. Disney, H.J.** *1979.* Royal Australian Ornithologists Union Pilot Atlas Scheme. Corella 2(Suppl.):97-163.

837 **Dean, W.R.J.** *1975.* Martial eagles nesting on high tension pylons. Ostrich 46(1):116-117.

838 **Deane, C.D.** *1962.* Irish golden eagles and a link with Scotland. Brit. Birds 55(7):272-274.

839 **Deane, R.** *1880.* Additional cases of albinism and melanism in North American birds. Bull. Nuttall Ornithol. Club 5(1):25-30.

840 **Deane, R.** *1905.* A hitherto unpublished letter of John James Audubon. Auk 22:170-175.

841 **Dearborn, N.** *1909.* Catalogue of a collection of birds from British East Africa. Field Mus. Nat. Hist. Ornithol. Series 1(4):141-190.

842 **Debus, S.J.** *1978.* Notes on wedge-tailed eagle behaviour. Corella 2(3):54-55.

843 **Deditius, C.** *1885.* Die vogel des gouvernements St. Petersburg von Eug. Buchner. J. Ornithol. 33(2):196-210.

844 **Deditius, K.** *1897.* Die vogel der westchinesischen Provinz Gan-su. J. Ornithol. 45(1):57-75.

845 **Deignan, H.G.** *1945.* The birds of northern Thailand. Smithsonian Inst., U.S. Natl. Mus. Bull. 186. 616 pp.

846 **Dejonghe, J.-F.** *1980.* Analyse de la migration prenuptiale des rapaces et des cigognes au Cap Bon (Tunisie). Oiseau Rev. Fr. Ornithol. 50(2):125-147.

847 **Dekker, D.** *1970.* Migration of diurnal birds of prey in the Rocky Mountain foothills west of Cochrane, Alberta. Blue Jay 28(1):20-24.

848 **Delacour, J.** *1929.* On the birds collected during the fourth expedition to French-Indo China. Ibis (12th Ser.) 5(2):193-220.

849 **Delacour, J. and P. Jabouille.** *1940.* Liste des oiseaux de l' Indochine Francaise. Oiseau Rev. Fr. Ornithol. 10(1/2):89-220.

850 **Delareulle, R.R.** *1945.* Eagle antics in Contra Costa County. Gull 30(5):17.

851 **Delibes, M. and J. Calderon.** *1977.* Golden eagle fledging three young. Brit. Birds 70(3):118-119.

852 **Delibes, M., F. Amores and J. Calderon.** *1975.* Tamano de la puesta y mortalidad entre los pollos del aguila real iberica *(Aquila chrysaetos homeyeri).* Donana Acta Vertebrata 2(2):179-191.

853 **Delibes, M., J. Calderon and F. Hiraldo.** *1975.* Seleccion de presa y alimentacion en Espana del aguila real *(Aquila chrysaetos)* . Ardeola 21(Special):285-303.

854 **Delmas, A.** *1911.* Catalogue des oiseaux observes dans l'Aveyron. Oiseau Rev. Fr. Ornithol. 2:151-156.

855 **Delme-Radcliffe, H.** *1922.* Weight carrying power of the golden eagle. Brit. Birds 15(9):217.

856 **Demandt, C.** *1933.* Uber das vorkommen des steinadlers als brutvogel in den Ostalpen. Beitr. Fortpfl.-Biology Vogel 9(3):99.

857 **Demandt, C.** *1938.* Vom steinadler in Voralberg. Beitr. Fortpfl.-Biology Vogel 14(2):68.

858 **Demandt, C.** *1938.* Zunahme des steinadlers in Oesterreich? Beitr. Fortpfl.-Biology Vogel 14(4):148-149.

859 **Demandt, C.** *1939.* Der steinadler als brutvogel in Vorarlberg. (The golden eagle breeding in Vorarlberg.) Beitr. Fortpfl.-Biology Vogel 15(1):14-16.

860 **Demandt, C.** *1944.* Vom steinadler in Vorarlberg. Beitr. Fortpfl.-Biology Vogel 20(2/3):64-65.

861 **Demandt, C.** *1954.* Schrei- und schelladler als irrgaste im rheinisch-westfalischen Grenzgebiet. Vogelwelt 75(6):233-236.

862 **Demartis, A.M.** *1974.* Oiseaux observes l'ete en Corse. Oiseau Rev. Fr. Ornithol. 44(1):28-39. (English Summ.)

863 **Demenchuk, G.** *1976.* Zapiski sovremennogo sokolnika. (Notes of a modern falconer.) Okhotn. Prostory. 33:74-82.

864 **Dementiev, G.P.** *1933.* Systema avium rossicarum. I. Ordo Accipitres (Oiseaux de proie Diurnes). Oiseau Rev. Fr. Ornithol. 3(3):457-518.

865 **Dementiev, G.P.** *1935.* Contribution a l'avifaune de la Siberie N.E. Alauda 7(2):153-169.

866 **Dementiev, G.P.** *1935.* Changements de noms neccessaires. Alauda 7(2):253-254.

867 **Dementiev, G.P.** *1936.* Le vol a l'aigle au Turkestan. (The flight of the eagle in Turkestan.) Oiseau Rev. Fr. Ornithol. 6(3):361-365.

868 **Dementiev, G.P.** *1945.* La fauconnerie en Russie, esquisse historique. Oiseau Rev. Fr. Ornithol. 15:10-39.

869 **Dementiev, G.P.** *1961.* Okhota s lovchimi ptitsami. (Falconry.) Okhota I Okhotn. Khozyaistvo. 1:59.

870 **Dementiev, G.P.** *1962.* Okhota s lovchimi ptitsami, ikh soderzhanie i dressirovka. (Hunting with raptors. their care and training.) Okhota I Okhotn. Khozyaistvo. 1:28-30.

871 **Dementiev, G.P., N.A. Gladkov, E.P. Spangenberg and A.M. Sudilovskaya.** *1966.* Birds of the Soviet Union. Vol. 1. Smithsonian Inst. and Natl. Science Foundation, Israel Program for Scientific Translation. 704 pp.

872 **Dennis, R.H.** *1971.* Scottish bird report 1970. Scott. Birds 6(7):347-402.

873 **Dennis, R.H., R.H. Hogg, A. Brown, R.W. Forrester and D.J. Bates.** *1978.* Scottish bird reports 1976-1977. Scott. Birds 10(3/4):73-153.

874 **Deppe, H.-J.** *1974.* Discoveries of the golden eagle *(Aquila chrysaetos)* in southern parts of the North Sea and Baltic Sea coast regions. Abh. Verhand. Natur. Hamburg 18/19:43-68. (Title Only)

875 **Deppe, H.-J.** *1974.* Zum vorkommen des steinadlers *(Aquila chrysaetos)* in Mitteleuropaischen tiefebene zwischen ems un memel. (Occurrence of golden eagle in northern plains of middle Europe.) Vogelwelt 95(6):201-227.

876 **Deppe, H.-J.** *1977.* Zum zug des steinadlers *(Aquila chrysaetos)* im nordlichen Mitteleuropa. Ornithol. Mitt. 29(8):167-169.

877 **Desfayes, M.** *1965.* Gains recents de l'avifaune valaisanne. Nos Oiseaux 28(2/3):50-53.

878 **Desfayes, M.** *1966.* En Valais: les changements en une decennie. Nos Oiseaux 28(7):169-172.

Master No. **Citation**

879 **Despott, G.** *1917.* Notes on the ornithology of Malta. Ibis 5(3):281-349,466-526.

880 **Despott, G.** *1933.* Ornitologia delle Isole Maltesi. Riv. Ital Ornitol. (2nd Ser.) 3(1):1-15.

881 **Detmers, E.** *1912.* Studien zur avifauna der Emslande. J. Ornithol. 60(1):1-68.

882 **Devillers, P.** *1976.* Projet de nomenclature Francaise des oiseaux du monde. 2. Anhimides aux Otidides. (Project of French names of birds of the world. 2. Anhimidae to Otididae.) Gerfaut 66(3/4):391-421.

831 **DeBoer, L.E.M.** *1976.* The somatic chromosome complements of 16 species of Falconiformes (Aves) and the karyological relationships of the orders. Genetica 46(1):77-113.

832 **DeGroot, D.S.** *1928.* Record sets of eggs of California raptores. Condor 30(6):360-361.

833 **DeLima, J. and L. Wagner.** *1959.* Notes on nesting of *Aquila chrysaetos* in the Serra do Marao. Publ. Inst. Zool. " Dr. Augusto Nobre" Faculidad Ciencia Porto. 66:7-19. (Title Only)

834 **DeMay, I.S.** *1942.* An avifauna from indian kitchen middens at Buena Vista Lake, California. Condor 44(5):228-230.

835 **DeVilliers, J.S.** *1958.* A report on the bird life of the Kalahari Gemsbok National Park. Koedoe 1:143-161.

836 **DeWitt, J.B.** *1963.* Studies on pesticide-eagle relationships. Audubon 65(1):30-31.

883 **Dharmakumarsinhji, R.S.** *1954.* Birds of Saurashtra, India. Times of India Press, Bombay. 561 pp.

887 **Dick, J.A. and M.B. Fenton.** *1979.* Tool using by a black eagle? Bokmakierie 31(1):17.

888 **Dickinson, L.E.** *1957.* Utilities and birds. Audubon 59(2):54-55,86-87.

889 **Diesselhorst, G.** *1971.* Liste der eigenen greifvogelbeobachtungen im September 1970. Orntihol. Mitt. 23(4):72-73. (English Summ.)

890 **Diez, M.G.** *1960. Aquila heliaca.* Ardeola 6(2):380.

891 **Dini, G.** *1934.* Cattura di un' aquila anatraia maggiore *(Aquila maculata).* Riv. Ital. Ornitol. (2nd Ser.) 4(1):47.

892 **Dixon, C.** *1882.* Notes on the birds of the Province of Constatine, Algeria. Ibis 6(24):550-579.

893 **Dixon, C.H.** *1913.* The nesting history of a pair of golden eagles. Oologist 30(3):34-47,49.

894 **Dixon, J.B.** *1911.* The golden eagle. Oologist 28(8):126-129.

895 **Dixon, J.B.** *1937.* The golden eagle in San Diego County, California. Condor 39(2):49-56.

896 **Dixon, J.E.W** *1968.* Prey of large raptors. Ostrich 39(3):203-204.

897 **Dixon, J.S.** *1933.* Three magpies rob a golden eagle. Condor 35(4):161.

898 **Dixon, J.S.** *1933.* Red fox attacked by a golden eagle. J. Mammal. 14(3):257.

899 **Dixon, J.S.** *1934.* A study of the life history and food habits of mule deer in California. Calif. Fish & Game 20(3):181-282.

900 **Dixon, J.S.** *1938.* Birds and mammals of the Mt. Mc Kinley National Park. U.S. National Park Service. Fauna Series No. 3.

884 **DiCarlo, E.A.** *1956.* Notizie ornitologiche. Aquila capit muscam? Riv. Ital. Ornitol. (2nd Ser.) 26(4):198-200.

885 **DiCarlo, E.A.** *1958.* Risultati di ricerche ornitologiche sulle montagne d'Abruzzo. Riv. Ital. Ornitol. (2nd Ser.) 28(3/4):145-217.

886 **DiCarlo, E.A.** *1972.* Gli uccelli del Parco Nazionale d' Abruzzo. Riv. Ital. Ornitol. (2nd Ser.) 42(1/2):1-160.

901 **Dobay, L., von.** *1934.* Beitrage zur biologie der schrei- und schelladler. Kocsag 7:31-39. (Title Only)

902 **Doerr, P.D. and J.H. Enderson.** *1965.* An index of abundance of the goshawk in Colorado in winter. Auk 82(2):284-285.

903 **Dolbik, M.S. and A.M. Dorofeev.** *1978.* Redkie i ischezayuschie ptitsi Belorussi. (Rare and endangered birds in Belorussia.) Uradzhai, Minsk. p. 1-200.

904 **Dolgushin, I.A. (ed.)** *1962.* Ptitsi Kszakhstana. Tom 2. (Birds of Kazakhstan, Vol. 2.) Alam-Ata. 425 pp.

905 **Domaniewski, J.** *1934.* Sprawozdanie z dzialalnosci Stacji Badania wedrowek ptakow za rok 1933. Acta Ornithol. 1(11):321-364. (French Summ.)

906 **Domaniewski, J. and B. Kreczmer.** *1936.* Sprawozdanie z dzialalnosci Stacji badania wedrowek ptakow za rok 1934. Acta Ornithol. 1(16):467-510.

907 **Donald, C.H.** *1919.* The birds of prey of the Punjab. Part II. J. Bombay Nat. Hist. Soc. 26(2):629-655.

908 **Donald, C.H.** *1952.* The flight of eagles. J. Bombay Nat. Hist. Soc. 50(4):839-844.

909 **Donegal.** *1861.* A domesticated golden eagle. Zoologist 19:7381.

910 **Dovrat, E.** *1979.* Raptor migration 17 km east of Tel-Aviv, autumn 1979. Cnaf-Renanim 12:15-22.

911 **Dovrat, E.** *1982.* Summary of 5 years of fall migration at Kfar Kassem. Torgos 2(1):63-116. (English Summ.)

912 **Dowsett, R.J.** *1979.* Recent additions to the Zambian list. Bull. Brit. Ornithol. Club 99(3):94-98.

913 **Dowsett, R.J. and F. Dowsett-Lemaire.** *1980.* The systematic status of some Zambian birds. Gerfaut 70(2):151-199.

914 **Drake, C.F.T.** *1867.* Notes on the birds of Tangier and eastern Morocco. Ibis 3(12):421-430.

915 **Dresser, H.E.** *1872.* Remarks on exhibition of skins of various eagles *(Aquila).* Proc. Zool. Soc. London p. 863-865.

916 **Dresser, H.E.** *1873.* On certain species of *Aquila* . Proc. Zool. Soc. London 1873:514-517.

917 **Dresser, H.E.** *1875.* Notes on Severtzoff's "Fauna of Turkestan" (Turkestanskie Jevotnie). Ibis 5(17):96-112.

918 **Dresser, H.E.** *1885.* Species of British-killed spotted eagles determined. Zoologist (3rd Ser.) 9:230-231.

919 **Dresser, H.E.** *1905.* An oological journey to Russia. Ibis 5(18):149-158.

920 **Drew, F.M.** *1885.* On the vertical range of birds in Colorado. Auk 2(1):11-18.

921 **Dubke, K.H.** *1964.* Golden eagle on Roan Mountain. Migrant 35(3):58-59.

922 **Dubois, A.** *1898.* Nouvelles observations sur la faune ornithologique de Belgique. Ornis 9:127-131.

923 **Dunajewski, A.** *1938.* Badania nad ptakami Wolynia. Acta Ornithol. 2(17):335-411.

924 **Dunlop, E.B.** *1912.* On incubation. Brit. Birds 5(12):322-327.

925 **Dunstan, T.C.** *1977.* Types and uses of radio packages for North American Falconiform and Strigiform birds. Pages 30-39. *in:* F.M. Long, Ed. First Int. Conf. on Wildlife Biotelemetry. 27-29 July, 1977. Univ. of Wyoming, Laramie.

926 **Dupond, C.** *1932.* Capture d'un aigle dore *Aquila chr. chrysaetus* (L.) en Belgique. Gerfaut 22(1):20-22.

927 **Dupond, C.** *1944. Aquila clanga* Pallas. Aigle criard. Gerfaut 34(3/4):77-78.

928 **Dupond, C.** *1946. Aquila chr. chrysaetus* Linne. Aigle royal. Gerfaut 36(4):240.

929 **Dupond, M.C.** *1932. Aquila chr. chrysaetus* (L.) - aigle dore. Gerfaut 22(3):138.

Master No.	Citation
930	**Dupuy, A.** *1969.* Catalogue ornithologique du Sahara Algerien. Oiseau Rev. Fr. Ornithol. 39(2):140-160.
931	**Dupuy, A. and J. Verschuren.** *1978.* Note sur les oiseaux, principalement aquatiques, de la region du Parc National du Delta du Saloum (Senegal). Gerfaut 68(3):321-345.
932	**Dupuy, A.-R.** *1976.* Donnees nouvelles concernant la reproduction de quelques especes aviennes au Senegal. Oiseau Rev. Fr. Ornithol. 46(1):47-62.
933	**Duse, A.** *1932.* Dall' osservatorio ornitologico cel Garda. Riv. Ital. Ornitol. (2nd Ser.) 2(2):78-82.
934	**Dutcher, W.** *1893.* Notes on some rare birds in the collection of the Long Island Historical Society. Auk 10(3):267-277.
935	**Dwight, J., Jr.** *1893.* Summer birds of Prince Edward Island. Auk 10(1):1-15.
936	**Dyche, L.L.** *1905.* The golden eagle *(Aquila chrysaetos).* Trans. Kansas Acad. of Science 19:179-181.
937	**Dyck, J., J.R. Jacobsen, E. Kramshoj and J. Rabol.** *1970.* Rapport fra sjaeldenhedsudvalget med oversigt over godkendte forekomster 1965-1969 (Report of the Rarity Committee for 1965-69.) Dansk Ornithol. 64(3):126-151.
938	**Dyrcz, A.** *1964.* Ptaki strachocina pod Wroclawiem - porownanie stanu dzisiejszego ze stanem z okolo 1890 roku. Acta Ornithol. 8(7):303-311.
939	**Dyrez, A., J. Okulewicz, L. Tomialojc and J. Witkowski.** *1974.* Breeding avifauna of the Biebrza Marshes and adjacent territories. Acta Ornithol. 13(10):185-252. (English Trans.)
940	**Eates, K.R.** *1937.* The distribution and nidification of the greater spotted eagle *(Aquila clanga* Pallas) in Sind. J. Bombay Nat. Hist. Soc. 39(2):403-405.
941	**Eaton, R.L.** *1976.* Golden eagle *(Aquila chrysaetos).* Pages 82-118. *in:* Marine Shoreline Fauna of Washington. Vol. II. Wash. Dept. of Game, Coastal Zone Environ. Studies. Rep. No. 3.
942	**Eaton, W.F.** *1935.* Eagles. Bird-Lore 37(2):136.
943	**Eckert, J.** *1974.* Birds of Streaky Bay. S. Aust. Ornithol. 26(6):138-141.
944	**Eddy, R.J.** *1959.* The wedge-tailed eagle in Bendigo, Mandurang and Big Hill Ranges. Aust. Bird Watcher 1(1):19-22.
945	**Edge, C.N.** *1939.* Hawk Mountain Sanctuary Association first annual report. Hawk Mt. Sanctuary Assoc., Publ. No. 1. 4 pp.
946	**Edge, R.** *1945.* Eagles in wonderland including a brief description of the McGuffey eagle. Hawk Mt. Sanctuary Assoc., Publ. No. 10. 7 pp.
948	**Edwards, P.** *1975.* First Pleistocene record of the golden eagle from the central Great Plains. Nebr. Bird Rev. 43(4):84.
947	**Edwards, C.C.** *1969.* Winter behavior and population dynamics of American eagles in western Utah. Ph.D. Dissertation. Brigham Young Univ., Provo, UT. 142 pp.
949	**Eggeling, W.J.** *1964.* The Nature Conservancy and its work: with mainly ornithological examples. Scott. Birds 3(2):59-72.
950	**Eichler, W.** *1958.* Bird lice: a biological test for ornithologists. Emu 58(1):1-4.
951	**Eifrig, C.W.G.** *1905.* The golden eagle *(Aquila chrysaetos)* near Ottawa. Auk 22:310.
952	**Elgood, J.H., C.H. Fry and R.J. Dowsett.** *1973.* African migrants in Nigeria. Ibis 115(1):1-45.
953	**Elliott, C.C.H. and M.J.F. Jarvis.** *1970.* Fourteenth ringing report. Ostrich 41(1/2):1-117.
954	**Elliott, R.R.** *1948.* Antelope surveys and investigations. Colo. Dept. of Game and Fish Quart. Prog. Rep., Proj. No. W-40-R-29. p. 63-69.
955	**Ellis, D.H.** *1973.* Behavior of the golden eagle: an ontogenic study. Ph.D. Diss. Univ. of Montana, Missoula, Montana.
956	**Ellis, D.H.** *1975.* First experiments with capturing golden eagles by helicopter. Bird-Band. 46(3):217-219.
957	**Ellis, D.H.** *1979.* Development of behavior in the golden eagle. Wildlife Monographs No. 70. 94 pp.
958	**Ellis, D.H. and C.H. Ellis.** *1975.* Color marking golden eagles with human hair dyes. J. Wildl. Manage. 39(2):445-447.
959	**Ellis, D.H. and J.R. Varney.** *1973.* A fully automated egg for telemetering adult attentiveness and incubation temperatures. Raptor Res. 7(3/4):73-77.
960	**Ellis, D.H., J.G. Goodwin, Jr. and J.R. Hunt.** *1978.* Wildlife and electric power transmission. Pages 81-104. *in:* Effects of Noise on Wildlife. Proc. of the 9th Int. Conf. on Acoustics. 1977. Madrid. Academic Press.
961	**Ellis, D.H., D.G. Smith and J.R. Murphy.** *1969.* Studies on raptor mortality in western Utah. Great Basin Nat. 29(3):165-167.
962	**Ellis, R.** *1958.* Strange behaviour of two wedge-tailed eagles. S. Aust. Ornithol. 22(5/6):69.
963	**Ellis, R.** *1960.* Notes on the breeding of the wedge-tailed eagle. S. Aust. Ornithol. 23(3):39-41.
964	**Elosegui, J.** *1974.* Informe preliminar sobre alimentacion de aves rapaces en Navarra Provicias Limitrofes. Ardeola 19(2):249-256.
965	**Elwes, H.J.** *1870.* A list of the birds of Turkey. Ibis 6(21):59-77.
966	**Embury, L.** *1940.* The odyssey of Uncle Sam's eagle. Nat. Hist. 45(5):276-279.
967	**Emde, E.** *1968.* Schelladler *(Aquila clanga)* in Hessen. Ornithol. Mitt. 20(11):239.
968	**Enderson, J.H.** *1965.* Roadside raptor count in Colorado. Wilson Bull. 77(1):82-83.
969	**Enderson, J.H., F.A. Colley and J. Stevenson.** *1970.* Aerial eagle count in Colorado. Condor 72(1):112.
970	**Engel, A.** *1915.* Etude d'une collection d'oiseaux rapportes de l'Inde. Oiseau Rev. Fr. Ornithol. 4(72):57-61.
971	**Engel, R.L. and T.A. Vaughan.** *1966.* A coyote-golden eagle association. J. Mammal. 47(1):143.
972	**Engels, W.L.** *1941.* Wing skeleton and flight of hawks. Auk 58(1):61-69.
973	**England, M.D.** *1974.* A further review of the problem of "escapes." Brit. Birds 67(5):177-197.
974	**English, W. and C.W. Servheen.** *1977.* A rehabilitation program for bald and golden eagles. J. N.A.F.A. 16:25-27.
975	**Ennion, H.E.** *1962.* Notes on birds seen in Aden and the western Aden Protectorate. Ibis 104(4):560-562.
976	**Enomoto, Y.** *1922.* (Method of the flight of *Aquila chrysaetos* .) Tori 3(12/13).
977	**Eprecht, W.** *1945.* Ornithologische notizen aus den Gebirgen des Gotthardgebietes. Ornithol. Beob. 42(9/10):121-136.
978	**Erard, C.** *1961.* L'aigle royal dans la Marne. Alauda 29(3):230-231.
979	**Erard, C.** *1961.* Quelques especes interessantes de la foret de Darbey (Vosges). Alauda 29(1):56-58.

Master No. **Citation**

980 **Erard, C. and F. Larigauderie.** *1972.* Observations sur la migration prenuptiale dans l'ouest de la Libye (Tripolitaine et plus particulierement Fezzan.) Oiseau Rev. Fr. Ornithol. 42(2/3):81-169.

981 **Erard, C., G. Jarry and R. Millo.** *1980.* Aigle pomarin dans les Bouches-du-Rhone. Oiseau Rev. Fr. Ornithol. 50(2):164-165.

982 **Erhard, H.** *1922.* Das wiederauftreten des steinadlers in den Berchtesgadner Alden. Ornithol. Monatsber. 30(1):12-13.

983 **Erhardt, T.** *1972.* Attempted breeding of the golden eagle *(Aquila chrysaetos canadensis)* in captivity. Raptor Res. News 5(5/6):147-149.

984 **Eriksson, K. and R. Saarenoksa.** *1955.* Kiljukotka, *Aquila clanga* Pall., nahty Espoossa. Ornis Fenn. 32(2):65.

985 **Erola, R., E. Hietanen and R. Niemi.** *1970.* Havaintoja paivapetolintujen muutosta ja esiintymisesta Ita-Hameessa. (Observations on the migration of birds of prey in eastern Hame.) Ornis Fenn. 47(1):36-39. (English Summ.)

986 **Ertl, G.** *1901.* Nidologia et oologia. I. Comitat Lipto. Aquila 8:165-172.

987 **Etchecopar, R.D. and F. Hue.** *1967.* The birds of North Africa from the Canary Islands to the Red Sea. Oliver + Boyd, Edinburgh and London. 612 pp.

988 **Etchecopar, R.D. and F. Hue.** *1978.* Les oiseaux de Chine. Editions Du Pacifique, Tahiti. 585 pp.

989 **Etoc, G.** *1913.* Les aigles criards. Oiseau Rev. Fr. Ornithol. 3(50):88-90.

990 **Evans, H.A.** *1963.* Falconry for You. (Third Ed.) John Gifford, Ltd., London. 206 pp.

991 **Evans, P.R. and W.V. Flower.** *1967.* The birds of the Small Isles. Scott. Birds 4(6):404-445.

992 **Evans, S.C.** *1888.* Nest of the golden eagle. *(Aquila chrysaetos canadensis).* Ornithol. and Ool. 13(6):90-91.

993 **Everett, M.** *1977.* The golden eagle. William Blackwood and Sons, Ltd., Edinburgh. 60 pp.

994 **Everett, M.J.** *1969.* Golden eagle calling. Scott. Birds 5(5):297.

995 **Everett, M.J.** *1971.* The golden eagle survey in Scotland in 1964-1968. Brit. Birds 64(2):49-56.

996 **Everett, M.J.** *1971.* Eagle survey. Birds 3:193.

997 **Everett, M.J.** *1971.* Golden eagle numbers in Scotland. Brit. Birds 64:511.

998 **Everett, M.J.** *1976.* Birds of Prey. Orbis Publ. Co., London. 128 pp.

999 **Everett, M.J.** *1978.* Birds of prey on Royal Society for the Protection of Birds Reserves. Pages 16-22. *in:* T.A. Geer, Ed. Bird of Prey Management Techniques. British Falconers' Club.

1000 **Ey, A.E.** *1944.* Birds breeding in the Millicent District. S. Aust. Ornithol. 17(4):32-37.

1001 **Eylmann, E.** *1911.* Die vogelwelt der kolonie Sudaustralien. J. Ornithol. 59(1):93-148.

1002 **Eyre, L. and D. Paul.** *1973.* Raptors of Utah. Utah Division of Wildlife Resources. 76 pp.

1003 **F., J.** *1965.* Observations on the breeding sites of the golden eagle in the Scottish highlands. Ool. Rec. 39(3):7-10.

1004 **Fables, D., Jr.** *1955.* Annotated List of New Jersey Birds. Urner Ornithological Club. 95 pp.

1005 **Farkas, T.** *1951.* Bird-faunistical and coenological researches at the Lake of Solymar. Aquila 55-58:148-158. (English Trans.)

1006 **Farman, C.** *1869.* On some of the birds of prey of central Bulgaria. Ibis 5(1):199-204.

1007 **Fasce, P.** *1979.* Censimento e dati sulla riproduzione dell' Aquila reale *(Aquila chrysaetos)* nelle Alpi occidentali Italiane e nell'appennino settentrionale. Riv. Ital. Ornitol. 49(1):34-39. (English Summ.)

1008 **Fatio, V. and T. Studer.** *1889.* Catalogue des oiseaux de la Suisse. Dept. Fed. De L'Industrie Et De L'Agriculture. 816 pp.

1009 **Fatton, M.-F.** *1963.* L'aigle et le petit chamois. Nos Oiseaux 27(3):109.

1010 **Favero, L.** *1933.* Catture de aquile nel Veneto. Riv. Ital. Ornitol. 3(3):186-187.

1011 **Favero, L.** *1935.* Aquila clanga Pallas. Riv. Ital. Ornitol. 5(2):110-111.

1012 **Favero, L.** *1935.* *Aquila clanga* Pallas. Riv. Ital. Ornitol. 8(2):110-111.

1013 **Favero, L.** *1942.* Comparsa dell'aquila imperiale in Italia. Riv. Ital. Ornitol. (2nd Ser.) 12(1):62-63.

1014 **Favero, L.** *1952.* Notizie ornitologiche, *Aquila chrysaetos chrysaetos* (L.). Riv. Ital. Ornitol. 22(1):35-39.

1015 **Favero, L.** *1954.* Aquila anatraia maggiore. Riv. Ital. Ornitol. 24(2):139.

1016 **Fawks, E.** *1972.* One day eagle count February 19 or 20, 1972. Iowa Bird Life 42(2):48-49.

1017 **Fawks, E.** *1976.* One day bald eagle count - February 14, 1976. Ill. Audubon Bull. No. 177. 2 pp.

1018 **Feeny, P.O., R.W. Arnold and R.S. Bailey.** *1968.* Autumn migration in the South Caspian region. Ibis 110(1):35-86.

1019 **Fentzloff, C.** *1978.* The breeding and release of the white-tailed sea eagle *(Haliaeetus albicilla)* as performed by the Deutsche Greifenwarte burg Guttenberg. Pages 97-103. *in:* T.A. Geer, Ed. Bird of Prey Management Techniques. British Falconers' Club.

1020 **Ferguson-Lees, I.J.** *1963.* Changes in the status of birds of prey in Europe. Brit. Birds 56(4):140-148.

1021 **Fernandez-Cruz, M.** *1968.* Activadades del centro de migracion de la Sociedad Espanola de Ornitologia. Ardeola 14:5-27.

1022 **Fernandez-Cruz, M.** *1968.* Capturas de aves anilladas en Espana: informe n. 11/12 (Anos 1967-1968.) Ardeola 14:29-87.

1023 **Fernandez-Cruz, M.** *1974.* Activades del centro de migracion de la sociedad Espanola de Ornitologia. Ardeola 20:5-35.

1024 **Fernandez, J.A.** *1976.* Donana: fortress for wildlife. Int. Wildl. 6(2):48-55.

1025 **Fernandez, J.A.** *1978.* Donana, Spain's wildlife wilderness. Taplinger Publ. Comp., New York.

1026 **Festetics, A.** *1957.* Ornithological observation in 1955 in Csorvas. Aquila 63-64:362.

1027 **Fevold, H.R. and J.J. Craighead.** *1958.* Food requirements of the golden eagle. Auk 75(3):312-317.

1028 **Field, M. and W. Rayner.** *1972.* Hawk Cliff raptor banding station: second annual report, 1972. Ont. Bird Band. 8(3):44-60.

1029 **Field, M. and W. Rayner** *1974.* Hawk Cliff raptor banding station: third annual report, 1973. Ont. Bird Band. 9(2):26-58.

1030 **Field, M. and W. Rayner.** *1978.* Hawk Cliff raptor banding station: sixth annual report, 1976. Ont. Bird Band. 12(1):1-27.

1031 **Fimreite, N., R.W. Fyfe and J.A. Keith.** *1970.* Mercury contamination of Canadian prairie seed eaters and their avian predators. Can. Field-Nat. 84(3):269-276.

1032 **Finch-Davies, C.G.** *1919.* South African hawks. Ibis 1(3):559-561.

1033 **Finch-Davies, C.G.** *1920.* Notes on South African Accipitres. Ibis 2(3):608-629.

1034 **Finch-Davies, C.G.** *1920.* Notes on South African Accipitres. Ibis 2(2):409-438.

1035 **Finlay, E. (Ed).** *1968.* Hawks and eagles need more protection. S. Carolina Wildl. 15(2):15.

Master No. Citation

1036 **Finley, W.L.** *1905.* The golden eagle of Mission Ridge. Country Cal. 1(1):41-46.

1037 **Finley, W.L.** *1906.* The golden eagle. Condor 8(1):5-11.

1038 **Firmin, A.** *1945.* Verreaux's eagle breeding in Kenya. J. E. African Nat. Hist. Soc. 18(1/2):95.

1039 **Fischer, W.** *1973.* Die greifvogel dererde. Falke 20(3):78-87.

1040 **Fischer, W.** *1976.* Steinadler, kaffern- und keilschwanzadler. Die Neue Brehm-Bucherei, A. Ziemsen Verlag, Wittenberg Lutherstadt. 215 pp.

1041 **Fischer, W., D. Zenker and W. Baumgart.** *1975.* Ein beitrag zum bestand und zur ernahrung des steinadlers *(Aquila chrysaetos)* auf der Balkanhalbinsel. (Occurrence and feeding of the golden eagle on the Balkan Peninsula.) Beitr. Vogelkd. 21(3/4):275-287.

1042 **Fisher, A.K.** *1893.* The hawks and owls of the United States in their relation to agriculture. U.S.D.A., Div. of Ornithol. and Mammal., Bull. No. 3. 210 pp.

1043 **Fisher, C.D., E. Lindgren and W.R. Dawson.** *1972.* Drinking patterns and behavior of Australian desert birds in relation to their ecology and abundance. Condor 74(2):111-136.

1044 **Fisher, G.C.** *1908.* A golden eagle taken in Alabama. Wilson Bull. 20(1):55.

1045 **Fisher, H.I.** *1946.* Adaptations and comparative anatomy of the locomotor apparatus of New World vultures. Am. Midland Nat. 35(3):545-727.

1046 **Fisher, J. and R.T. Peterson.** . The World of Birds. Doubleday, New York. p. 25.

1047 **Fitch, E.A.** *1890.* Golden eagle laying at Woodham Walter. Essex Nat. 4:124.

1048 **Fitch, E.A.** *1912.* Golden eagle *(Aquila chrysaetos).* Zoologist (4th Ser.) 16:34-35.

1049 **Fitch, H.S., H.A. Stephens and R.O. Bare.** *1973.* Road counts of hawks in Kansas. Kansas Ornithol. Soc. Bull. 24(4):33-35.

1050 **Fitch, H.S., F. Swenson and D.F. Tillotson.** *1946.* Behavior and food habits of the red-tailed hawk. Condor 48(5):205-237.

1051 **Fitter, R. (ed.)** *1973.* Book of British Birds. Drive Publ. Ltd., London. 471 pp.

1052 **Fitzner, R.E.** *1978.* Behavioral ecology of the Swainson's hawk *(Buteo swainsoni)* in southeastern Washington. Ph.D. Dissertation. Washington St. Univ., Pullman, WA.

1053 **Flasar, J.** *1951.* Orel kralovsky *(Aquila heliaca* Sav.) v Ceskoslovensku. Sylvia 13(3):89-93. (French Summ.)

1054 **Fleay, D.** *1951.* A wedge-tailed eagle for Christmas. Animal King. 54(6):162-169,188-189.

1055 **Fleay, D.** *1952.* Further adventures of Horatia. Animal King. 55(1):22-23.

1056 **Fleay, D.** *1952.* With a wedge-tailed eagle at the nest. Emu 52(1):1-17.

1057 **Fleming, J.H.** *1907.* Birds of Toronto, Canada. Brit. Birds 24(1):71-89.

1058 **Fleming, R.L. and M.A. Traylor.** *1961.* Notes on Nepal Birds. Fieldiana: Zoology 35(8):443-487.

1059 **Fleming, R.L., R.L. Fleming, Jr. and L.S. Bangdel.** *1976.* Birds of Nepal. Avalok Publ., Kathmandu. Nepal.

1060 **Flinn, W.L.** *1957.* A guide to the eagles, hawks and falcons of Colorado. Colo. Outdoors 6(2):14-18.

1061 **Floericke, C.** *1891.* Zur ornis der Bartschniederung. J. Ornithol. 39(3):275-292.

1062 **Flook, D.R. and D.C. Thomas.** *1962.* An observation of a golden eagle dominating coyotes. Can. Field-Nat. 76(2):123.

1063 **Flower, S.S.** *1938.* Further notes on the duration of life in animals. IV. Birds. Proc. Zool. Soc. London 108(12):195-235.

1064 **Fluckiger, A.** *1971.* Uhu und Steinadler. Jager in Wald und Fels. Schweizer Verlagshaus. Zurich. 287 pp.

1065 **Ford, H. and D. Paton.** *1976.* Birds of Para Wirra Recreation Park: changes in status over 10 years. S. Aust. Ornithol. 27(3):88-95.

1066 **Ford, H.S. and J.S. Alcorn.** *1964.* Observations of golden eagle attacks on coyotes. Condor 66(1):76-77.

1067 **Ford, J. and E.H. Sedgwick.** *1967.* Bird distribution in the Nullabor Plain and Great Victoria Desert region, western Australia. Emu 67(1/2):99-124.

1068 **Ford, J.R. and P.S. Stone.** *1957.* Birds of the Kellerberrin/Kwolyn District, western Australia. Emu 57(1):9-21.

1069 **Formosof, A.N.** *1934.* Birds of prey and rodents. Zool. Zhur. 13(4):664-700.

1070 **Fortenberry, D.K.** *1974.* Resources inventory Yukon Planning Region. Migratory birds, raptors, and endangered species. Joint Federal-State Land Use Planning Comm., Resource Planning Team. 40 pp.

1071 **Fortenberry, D.K.** *1974.* Resources inventory South Central Planning Region. Migratory birds, raptors, and endangered species. Joint Federal-State Land Use Planning Comm., Resource Planning Team. 15 pp.

1072 **Fortenberry, D.K.** *1974.* Resources inventory Southwest Planning Region. Migratory birds, raptors, and endangered species. Joint Federal-State Land Use Planning Comm., Resource Planning Team. 35 pp.

1073 **Fortenberry, D.K.** *1974.* Resources inventory Arctic Planning Region. Migratory birds, raptors and endangered species. Joint Federal-State Land Use Planning Comm., Resource Planning Team. 20 pp.

1074 **Foschi, T.** *1959.* E ritornata l'aquila reale in Romagna. Riv. Ital. Ornitol. 29(2):73-74.

1075 **Foster, J.W.** *1976.* Project Babe. NABB 1(4):179.

1076 **Francois, J.** *1975.* Contribution a la connaissance de l'avifaune d'Afrique du Nord. Alauda 43(3):279-292.

1077 **Frank, L.G. and J.E. Cooper.** *1974.* Further notes on the use of CT 1341 in birds of prey. Raptor Res. 8(1/2):29-32.

1078 **Frank, L.G., R.M. Jackson, J.E. Cooper and M.C. French.** *1977.* A survey of chlorinated hydrocarbon residues in Kenyan birds of prey. E. African Wildl. J. 15(4):295-304.

1079 **Frauca, H.** *1967.* Kings of the Australian skies. Birds Illus. 13(1):2-5.

1080 **Freiherr, C.** *1898.* Beitrage zur avifauna Tunesiens. J. Ornithol. 46(3):377-497.

1081 **Freiherr, C.** *1904.* Beitrage zur vogelfauna Nordostafrikas. J. Ornithol. 52(2):137-244.

1082 **Fresenius, R.** *1886.* Ornithologische miscellen. Monats. Deutschen Vereins Zum Schutze Vogelwelt 11(4):88-93.

1083 **Fretzdorff, A.-M., E. Schuz and G. Weitzel.** *1966.* Uber den Eisengehalt von vogel-Burzeldrusen. J. Ornithol. 107(2):225-228.

1084 **Frey, E.S.** *1940.* Hawk notes from Sterret's Gap. Auk 57(2):247-250.

1085 **Frey, H. and W. Walter.** *1972.* Steinadler im Waldviertel. Egretta 15(2):64-66.

1086 **Friedman, H.** *1937.* Bird bones from archeological sites in Alaska. J. Wash. Acad. Sci. 27(10):431-438.

Master No.	Citation
1087	**Friedmann, H.** *1930.* Birds collected by the Childs Frick Expedition to Ethiopia and Kenya Colony. Part 1. Non-passeres. Smithsonian Inst., U.S. Natl. Mus. Bull. 153. 561 pp.
1088	**Friedmann, H. and A. Loveridge.** *1937.* Notes on the ornithology of tropical east Africa. Bull. Mus. Comp. Zool. 81(1):1-413.
1089	**Froggatt, W.W.** *1920.* The wedge-tailed eagle. Aust. Zool. 1(8):252-254.
1090	**Frome, M.** *1971.* The eagle killings and their end result. Field and Stream 76(8):44,79-81.
1091	**Frossard, J.** *1945.* Les aigles du Val de Bagnes II. Nos Oiseaux 18:21-34.
1092	**Frossard, J.** *1945.* Les aigles du Val de Bagnes I. Nos Oiseaux 18:1-7.
1093	**Fry, C.H.** *1964.* The ecological distribution of birds in northern Guinea Savanna, Nigeria. Ostrich (Suppl. 6):335-356.
1094	**Fuggles-Couchman, N.R. and H.F.I. Elliott.** *1946.* Some records and field-notes from north-eastern Tanganyika Territory. Ibis 88:327-347.
1095	**Fuhrer, L., von.** *1900.* Der prachtadler *(Aquila fulvescens* Gray) in Nord-Albanien erlegt. Ornithol. Jahrbuch 11(1):25-28.
1096	**Fuller, M.R., P.T. Redig and G.E. Duke.** *1973.* Raptor rehabilitation and conservation in Minnesota. Loon 45(2):50-55.
1097	**Fuller, M.R., P.T. Redig and G.E. Duke.** *1974.* Raptor rehabilitation and conservation in Minnesota. Raptor Res. 8(1/2):11-19.
1098	**Fyfe, R. and J.W. Grier.** *1972.* Canadian Wildlife Service involvement with birds of prey. Trans. 36th Federal-Provincial Wildlife Conf. p. 69-78.
1099	**Fyfe, R.W.** *1959.* Golden eagles nesting in Saskatchewan. Blue Jay 17(3):110-111.
1100	**Fyfe, R.W.** *1976.* Status of Canadian raptor populations. Can. Field-Nat. 90(3):370-375.
1101	**Fyfe, R.W.** *1977.* Status of Canadian raptor populations. Pages 34-39. *in:* R.D. Chancellor, Ed. Proc. of the World Conf. on Birds of Prey. 1-3 October, 1975. Vienna. I.C.B.P.
1102	**Fyfe, R.W. and R.R. Olendorff.** *1976.* Minimizing the dangers of nesting studies to raptors and other sensitive species. Canadian Wildl. Service, Occas. Pap. No. 23. 17 pp.
1103	**Gabrielson, I.N.** *1922.* Some hawks of Harney Valley, Oregon. Condor 24(1):33-34.
1104	**Gabrielson, I.N. and S.G. Jewett.** *1924.* Notes on the birds of western North Dakota. Auk 41(2):296-303.
1105	**Gabrielson, I.N. and F.C. Lincoln.** *1959.* The Birds of Alaska. Stackpole, Harrisburg, PA. 922 pp.
1106	**Gabrielson, I.N., H.A. Hochbaum, R.A. McCabe, D.A. Munro, R. Pough and C. Cottam.** *1962.* Report to the American Ornithologists' Union by the Committee on Bird Protection, 1961. Auk 79(3):463-478.
1107	**Gabrielson, I.N., H.A. Hochbaum, D.A. Munro, R.A. McCabe, R. Pough and C. Cottam.** *1963.* Report of the Committee on Bird Protection, 1962. Auk 80(3):352-364.
1108	**Gaeta, L.** *1971.* Reports on a golden eagle breeding project. Raptor Res. News 5:19-21.
1109	**Gal, J.K.** *1958.* Contributions to the knowledge of the ornis of the Bihar-Mountains. Aquila 65:217-223. (English Summ.)
1110	**Gal, J.K.** *1959.* Data on the bird-life of the Kisbanyahavas (Baisoara) and Belavara (Scarisoara - Belioara). Aquila 66:225-232.
1111	**Galbreath, D.S. and R. Moreland.** *1953.* The chukar partridge in Washington. Washington Dept. of Game, Biology Bull. No. 1.
1112	**Gall, B.C. and N.W. Longmore.** *1978.* Avifauna of the Thredbo Valley, Kosciusko National Park. Emu 78(4):189-196.
1113	**Gallagher, M.D. and M.R. Brown.** *1982.* Golden eagle *Aquila chrysaetos* breeding in Oman. Bull. Brit. Ornithol. Club 102(1)41-42.
1114	**Galushin. V.M.** *1970.* Khistchnye ptitsi. (Birds of prey.) Lesnaya Promyshlennost, Moscow. p. 1-136.
1115	**Galushin. V.M.** *1977.* Recent changes in the actual and legislative status of birds of prey in the USSR. Pages 152-159. *in:* R.D. Chancellor, Ed. Proc. of the World Conf. on Birds of Prey. 1-3 October, 1975. Vienna. I.C.B.P.
1116	**Galushin, V.M.** *1961.* Absolyutnyi uchet chislennosti dnevnykh khistchnyka ptits v raione Oskogo zapovednika. (Survey of diurnal birds of prey in a region of Oka State Reserve.) Sovestchanie Po Voprosam Organizatsii I Metodam Ucheta Resurov Fauny Nazemnykh Pozvonochnykh, Tezisy Dokl., Moscow. p. 146-147.
1117	**Galushin, V.M.** *1962.* Bolshoi podorlik doliny reki Oki i ego vozdeistvie na chislennost nekotorykh ptits. (*Aquila clanga* in Oka Valley and its influence upon populations of some birds.) Uchenye Zapiski Mosk. Lenin Pepogog. Inst. N. 186. p. 115-151.
1118	**Galushin, V.M.** *1963.* (Character of variability and factors determining sizes of hunting areas of birds of prey.) 5 Pribalt Ornithol. Konf., Tezisy Doklady, Tartu. p. 36-40. (In Russian)
1119	**Galushin, V.M.** *1967.* Popytka orientirovochnogo rascheta obstchei chislennosti khistchnykh ptits Moskovxkol oblasti.) (Attempt of preliminary estimation of total number of birds of prey in Moscow.) Zhivotnoe Naselenie Moskvy I Podmoskovia, Moscow. p. 30-32. (In Russian)
1120	**Galushin, V.M.** *1967.* (Number of birds of prey in the European part of the USSR.) Materialy 3 Zool. Konf. Pedagog. Inst., Volgograd. p. 406-408. (In Russian)
1121	**Galushin, V.M.** *1970.* Ecological and economic effects of birds of prey in the central region of the European part of Russia with special reference to the "index of predatory pressure." Pages 166-174. *in:* Proc. 11th Tech. Meeting. Int. Union Conserv. of Nature. Vol. 1.
1122	**Galushin, V.M.** *1970.* A quantitative estimation of predatory bird's pressure upon game birds' populations in the central region of the European part of the USSR. Trans. 9th Int. Cong. Game Biologists, Moscow. p. 553-562.
1123	**Galushin, V.M.** *1971.* The legal position of birds of prey in the U.S.S.R. Bull. of the Int. Council for Bird Pres., London. p. 107-111.
1124	**Galushin, V.M.** *1971.* (Number and territorial distribution of birds of prey in the European center of the U.S.S.R.) Trudy Oskogo Gos. Zapovednika 8:5-132. (In Russian)
1125	**Galushin, V.M.** *1976.* Orel-berkut. (*Aquila chrysaetos.*) Ochota I Ochoton: Khozyaistvo. 4:13.

Master No. Citation

1126 **Galushin, V.M.** *1977.* Pernatye khistchniki. (Raptors.) Broshyura, Minist. Selsk. Khozyaistva SSSR, Moscow. p. 1-6. (In Russian)

1127 **Galushin, V.M.** *1977.* Opyt obzora problemy: khistchnye ptitsy i sovremennaya sreda. (Review of a problem: birds of prey and modern environment.) Adaptivnye Osobennosti Evolyutsia Ptits, Moscow. p. 78-88.

1128 **Galushin, V.M.** *1978.* Recent outlines of birds of prey ranges in the European part of the USSR. Page 53. *in:* Proc. of the 17th Int. Ornithol. Cong., 4-11 June. Berlin. Symposia and Spec. Interest Groups. Abstract.

1129 **Galushin, V.M.** *1981.* Changes in population status and nest range distribution of Falconiforms in the USSR since 1950. Raptor Res. 15(1):4-11.

1130 **Galushin, V.M. and A.A. Inozemtsev.** *1970.* Opyt otsenki khistchnichestva nekotorykh ptits iz otryadov Falconiformes i Passeriformes. (Estimation of predatory pressure of some birds of Falconiformes and Passeriformes.) Uchenye Zapiski Mosk. Lenin Gos. Pedagog. Inst. 272:82-90. (In Russian)

1131 **Ganier, A.F.** *1940.* Tall tales about eagles. Migrant 11(4):99-102.

1132 **Ganier, A.F.** *1940.* Notes on Tennessee birds of prey. Migrant 11(1):1-4.

1133 **Ganier, A.F.** *1953.* Eagles at the Duck River Refuge. Migrant 24(4):83-84.

1134 **Ganier, A.F.** *1973.* The wild life met by Tennessee's first settlers. Migrant 44(1):58-74.

1135 **Ganya, I.M.** *1964.* Ekologo-faunisticheskaya kharakteristika dnevnykh khistchnykh ptits Moldavii. (Ecological and faunistic characteristics of diurnal birds of prey in Moldavia.) Voprosy Ekologii I Praktich. Znachenia Ptits I Mlekopitayustchikh Moldavii. 2:34-54.

1136 **Ganya, I.M.** *1975.* Vliyanie antropogennykh faktorov na ornithofauno Moldavii. (Influence of anthropogenic factors on avifauna of Moldavia.) Ecologia Ptits I Mlekopitayustchikh Moldavii, Kishinev. p. 39-63.

1137 **Gardonyi, G.G.** *1958.* Ornithological observations from the Borzsony Mountains (40 miles N. of Budapest). Aquila 65:361-362. (English Trans.)

1138 **Gargett, V.** *1965.* Death of a black eagle in the Matopos. Honeyguide 50:9-10. (Title Only)

1139 **Gargett, V.** *1967.* Black eagle experiment. Bokmakierie 19(4):88-90.

1140 **Gargett, V.** *1969.* A black eagle's survival trek. Bokmakierie 21(4):78.

1141 **Gargett, V.** *1970.* Black eagle survey, Rhodes Matopos National Park: a population study, 1964-1968. Ostrich (Suppl. 8):397-414.

1142 **Gargett, V.** *1970.* Black eagle experiment no. 2. Bokmakierie 22(2):32-35.

1143 **Gargett, V.** *1971.* Nature's way. Bokmakierie 23(1):16-17.

1144 **Gargett, V.** *1971.* Some observations on black eagles in the Matapos, Rhodesia. Ostrich (Suppl. 9):91-124.

1145 **Gargett, V.** *1972.* Observations at a black eagle nest in the Matopos, Rhodesia. Ostrich 43:77-108.

1146 **Gargett, V.** *1972.* Black eagle *Aquila verreauxi* population dynamics. Ostrich 43:177-178.

1147 **Gargett, V.** *1972.* Black eagles of the Matopos. Afr. Wild Life 26(1):12-17.

1148 **Gargett, V.** *1973.* Marking black eagles in the Matopos. Honeyguide 76:26-31.

1149 **Gargett, V.** *1974.* "Turns the eggs?" Bokmakierie 26(4):90-91.

1150 **Gargett, V.** *1975.* The spacing of black eagles in the Matopos, Rhodesia. Ostrich 46(1):1-44.

1151 **Gargett, V.** *1977.* A 13-year population study of the black eagles in the Matopos, Rhodesia, 1964-1976. Ostrich 48(1/2):17-27.

1152 **Gargett, V.** *1977.* A black eagle lays her second egg. Bokmakierie 28(3):83-84.

1153 **Gargett, V.** *1978.* Sibling aggression in the black eagle in the Matopos, Rhodesia. Ostrich 49(2):57-63.

1154 **Gargett, V.** *1978.* Black eagles in protected and unprotected habitats. Pages 96-102. *in:* Proc. of the Symposium on African Predatory Birds. Pretoria. Northern Transvaal Ornithol. Soc.

1155 **Gargett, V.** *1978.* Eagles are individuals. Afr. Wildl. 33(2):6-7.

1156 **Gargett, V.** *1980.* The black eagles of the Matopos. Zimbabwe Sci. News 14(7):155-159.

1157 **Gargett, V.** *1980.* Black eagle survey. Rhodes Matapos National Park. 17th Annual Report-1980. Ornithol. Soc. of Zimbabwe - Matabeleland Branch. Report. 2 pp.

1158 **Gargett, V.** *1981.* Black eagle survey. 18th annual report. Ornithol. Assoc. of Zimbabwe-Matabeleland Branch. 3 pp.

1159 **Garzon, J.** *1968.* Las rapaces y otras aves de la Sierra de Gata. Ardeola 14:97-130.

1160 **Garzon, J.** *1972.* Especies en peligro: el aquila imperial. Adena 4:8-11. (Title Only)

1161 **Garzon, J.** *1973.* Contribucion al conocimiento de la cronolgia de la reproducion de *Aquila chrysaetos* en Espana. Ardeola 19(2):457.

1162 **Garzon, J.** *1974.* Contribucion al estudio del status, alimentacion y proteccion del las Falconiformes en Espana central. Ardeola 19(2):279-330.

1163 **Garzon, J.** *1974.* Datos sobre un nido de *Aquila heliaca adalberti* con 4 huevos. Ardeola 20:347-348.

1164 **Garzon, J.** *1977.* Birds of prey in Spain, the present situation. Pages 159-170. *in:* R.D. Chancellor, Ed. Proc. of the World Conf. on Birds of Prey. 1-3 October, 1975. Vienna. I.C.B.P.

1165 **Garzon, J. and J. Araujo.** *1972.* El clima su posible influencia sobre las aves de presa (Falconiformes y Strigiformes) en Espana Central (Primavera 1971). Ardeola 16:193-213.

1166 **Gauci, C. and J. Sutlana.** *1971.* Systematic list for 1967-1970. Il-Merrill 5:1-24.

1167 **Gaukrodger, D.W.** *1924.* The way of an eagle. Emu 24(1):3-12.

1168 **Geary, N.** *1932.* Notes on the wedge-tailed eagle. Emu 31(4):288-289.

1169 **Gebruder.** *1877.* Zur ornis der dobrudscha. J. Ornithol. 25(1):59-69.

1170 **Geilikman, B.O.** *1975.* (Influence of feeding characteristics of Accipitridae on some characteristics of their mode of life and extremities.) Zool. Sbornik. Inst., Armenian SSR. 16:65-77. (In Russian)

Master List of Citations (Cont.)

Master No. **Citation**

1171 **Geilikman, B.O. and A.K. Unanyan.** *1974.* O gnezdovanii malogo podorlika *(Aquila pomarina* Brehm) v Armenii. (Nesting of *Aquila pomarina* in Armenia.) Biology Zhur. Armenii, Armian SSR. 27(5):100-102.

1172 **Genelly, R.E.** *1978.* Observations of the Australian kestrel on northern tablelands of New South Wales, 1975. Emu 78(3):137-144.

1173 **Gengler, J. and E. Kawelin.** . Die vogel von Koselsk and Umgebung. Ornithol. Jahrbuch 20(5/6):165-191.

1174 **Gentz, K.** *1965.* Am horst des schreiadlers. Falke 12(12):412-420.

1175 **Gentz, K.** *1966.* Wie alt werden vogel? Falke 13(3):190-195.

1176 **Gentz, K.** *1967.* Uber die brutbiologie des schreiadlers. (On the breeding biology of the lesser spotted eagle.) Ornitologia 8:294-298.

1177 **Gentz, K.** *1975.* Die schreiadlervorkommen in den brandenburgischen Bezirken 1952 bis 1972. Falke 22(2):52-57.

1178 **George, V.** *1969.* Zur ernahrung des steinadlers, *Aquila chrysaetos,* in der Sahara. J. Ornithol. 110(4):502.

1179 **Gepp, B.C. and A.J. Fife.** *1975.* Birds seen in forest reserves in South Australia. S. Aust. Ornithol. 27(1):12-17.

1180 **Gerell, R.** *1965.* Imperial eagle *(Aquila heliaca)* observed in Scania in the autumn 1964. Var Fagelvarld 24(4):403. (Title Only)

1181 **Gerick, K.** *1929.* Ein schreiadler-horst mit 2 jungen. Ornithol. Monatsber. 37(1):18-19.

1182 **Germain, M., J. Dragesco, F. Roux, and H. Garcin.** *1973.* Contribution a l'ornithologie du Sud-Cameroun. I. Non-Passeriformes. Oiseau Rev. Fr. Ornithol. 43(2):119-182.

1183 **Geroudet, P.** *1945.* L'aigle royal dans la region de Geneve. Nos Oiseaux 18:45-47.

1184 **Geroudet, P.** *1949.* L'aigle royal *(Aquila chrysaetos)* dans les cantons du Valais, de Vaud et de Fribourg en 1948. (The golden eagle in the districts of Valais, Vaud and Fribourg in 1948.) Nos Oiseaux 20(1):1-6.

1185 **Geroudet, P.** *1958.* Apercus ornithologiques sur la Yougoslavie. II. Les marais de l'Obedska Bara et de Kupinovo. Nos Oiseaux 24(8/9):213-220.

1186 **Geroudet, P.** *1958.* Apercus ornithologiques sur la Yougoslavie. V. Les bois et marais pres de Sid. Nos Oiseaux 24(11):269-272.

1187 **Geroudet, P.** *1958.* Apercus ornithologiques sur la Yougoslavie. I. Fruska gora. Nos Oiseaux 24(7):184-193.

1188 **Geroudet, P.** *1962.* Letter concerning aircraft-eagle collision. Nos Oiseaux 26(12):320.

1189 **Geroudet, P.** *1968.* Contribution a l'ornithologie du Val Ferret, Valais. Nos Oiseaux 29(7):177-198.

1190 **Geroudet, P.** *1971.* Declin alarmant des rapaces diurnes en Allemagne federale. Nos Oiseaux 31(1):29.

1191 **Geroudet, P.** *1972.* Contribution a l'ornithologie du Grande Paradis. Nos Oiseaux 31(12):269-296.

1192 **Geroudet, P.** *1976.* Une aire d'aigle royal desertee par la faute d'un inconscient. Nos Oiseaux 33(8):356.

1193 **Geroudet, P. and Y. Reverdin.** *1971.* L'aigle royal attaque par une pie. Nos Oiseaux 31(4):92.

1194 **Geyr, H.F.** *1915.* Ornithologische beobachtungen im Komitat Syrmien. J. Ornithol. 63(1):70-111.

1195 **Ghidini, A.** *1906.* Steinadler bei Genf. Ornithol. Beob. 5(4):62.

1196 **Ghidini, A.** *1913.* "*Vultur monachus* ', "*Gyps fulvus* " ed "*Aquila fulva* " nelle Alpi nel 1912. Riv. Ital. Ornitol. 2(3):198-199.

1197 **Ghidini, A.** *1913.* Le aquile nel Ticino. Riv. Ital. Ornitol. 2(2):117.

1198 **Ghidini, G.** *1914.* Aquile ed avoltoi nelle Alpi. Riv. Ital. Ornitol. 3(1/2):82-83.

1199 **Gibet, L.A.** *1960.* (Number of birds of prey and their relation to rodents in the Steppe zone of north Kazakhstan.) Ornitologia 3:278-291. (In Russian)

1200 **Gibson, D.D. and G.V. Byrd.** *1973.* Alaska region. Am. Birds 27(5):905-907.

1201 **Giglioli, H.H.** *1881.* Notes on the avifauna of Italy. Ibis (4th Ser.) 5(18):181-222.

1202 **Gilbert, H.A.** *1925.* Fledging period of golden eagle. Brit. Birds 19(5):133.

1203 **Gilbert, H.A.** *1926.* Lifting power of the golden eagle. Brit. Birds 19(10):259-260.

1204 **Gilbert, H.A.** *1926.* Lifting power of the golden eagle. Brit. Birds 19(11):292.

1205 **Gilbert, H.A.** *1934.* Der steinadler in Schottland. (The golden eagle in Scotland.) J. Ornithol. 82(4):561-567.

1206 **Gilbertson, M. and L. Reynolds.** *1974.* DDE and PCB in Canadian birds, 1969 to 1972. Canadian Wildl. Service, Occas. Pap. No. 19. 18 pp.

1207 **Gill, H.B.** *1970.* Birds of Innisfail and Hinterland. Emu 70(3):105-116.

1208 **Gillman, A.R.** *1921.* Weight-carrying power of a golden eagle. Brit. Birds 15(1):24.

1209 **Gilpin, J.B.** *1872-1873.* On the eagles of Nova Scotia. Proc. and Trans. Nova Scotia Inst. Nat. Sci. 3(3):202-208.

1210 **Giol, A.** *1959.* Cattura di aquila anatraia maggiore. Riv. Ital. Ornitol. 29(1):50-52.

1211 **Giol, A.** *1961.* Cattura di un'aquila anatraia minore *Aquila p. pomarina,* Brehm in provincia di Treviso. Riv. Ital. Ornitol. 31(1):1-6.

1212 **Girtanner, U.** *1899.* Flanderei uber den steinadler *(Aquila fulva* (L.), *A. chrysaetus* (L.)). Vogelwelt 24(4):101-111.

1213 **Glegg, W.E.** *1914.* A list of the birds of Macedonia. Ibis (11th Ser.) 6(1):46-86.

1214 **Glegg, W.E.** *1941.* The birds of "L'Ile de la Camarque et la Petite Camarque." Ibis 5(4):556-610.

1215 **Glegg, W.E.** *1943.* The birds of L'Ile de la Camarque et la Petite Camarque. Early records. Ibis 85:300-307.

1216 **Glegg, W.E.** *1948.* An early oologist: John Drew Salmon, 1802-1859. Bull. Brit. Ornithol. Club 68(2):56-58.

1217 **Glenny, F.H.** *1941.* Presence of the ligamentum botalli in the golden eagle *Aquila chrysaetos* L., the red-tailed hawk, *Buteo borealis borealis* (Gmelin) and common pigeon *Columba livia.* Ohio J. Sci. 41(1):46-49.

1218 **Glenny, R.** *1950.* The golden eagle in Co. Louth. Irish Nat. J. 10(1):15-16.

1219 **Gloger.** *1860.* Die flugelkraft des steinadlers, *(Aquila fulva).* J. Ornithol. 8(6):468-469.

1220 **Glotov, I.N.** *1959.* Materialy po biologii bolshogo podorlika. *(Aquila clanga*). (Data on the biology of *Aquila clanga* .) Trudy Biology Inst., Sibirsk. Otdel, SSR. 5:167-170.

1221 **Glover, F.A. and L.G. Heugly.** *1970.* Golden eagle ecology in west Texas. Report to National Audubon Soc., Colorado Coop. Wildl. Res. Unit., Fort Collins, Co. 84 pp. (Title Only)

1222 **Gloyd, H.K.** *1925.* Field studies of the diurnal raptores of eastern and central Kansas. Wilson Bull. 37(3):133-149.

1223 **Glutz Von Blotzheim, U.N.** *1960.* Berich uber die brutperiode 1959. Ornithol. Beob. 57(2):90-107.

Master No.	Citation
1224	**Glutz Von Blotzheim, U.N.** *1964.* Hochstalter schweizerischer ringvogel. Ornithol. Beob. 61(4):106-127.
1225	**Glutz Von Blotzheim, U.N.** *1964.* Die Brutvogel der Schweiz. Verlag Aargauer Tagblatt AG, Aarau. 648 pp.
1226	**Glutz Von Blotzheim, U.N.** *1965.* Der steinadler-bestand der schweizer alpen. Ornithol. Beob. 62(2):60-61.
1227	**Glutz Von Blotzheim, U.N.** *1966.* Ergebnis der steinadler-Bestandesaufnahurel 1965. Ornithol. Beob. 63(2):51-52.
1228	**Glutz Von Blotzheim, U.N., K.M. Bauer and E. Bezzel.** *1971.* Handbuch der vogel Mitteleuropas. Part 4. Falconiformes. Akademische Verlagsgesellschaft, Frankfurt. 943 pp.
1229	**Goddard, T.R.** *1926.* Spotted eagle. Trans. Nat. Hist Soc. Northumberland, Durham and Newcastle (New Ser.) 6(2):241.
1230	**Godfrey, W.E.** *1956.* Some Canadian Birds. National Museum of Canada, Ottawa. 44 pp.
1231	**Goebbels, F., H. Kirchner and F. Moebert.** . Ornitholgische hilfstabellen. Vogelwelt 61(3/4):38-52.
1232	**Goebel, H.** *1879.* Ueber vogel des uman'schen kreises. J. Ornithol. 27(3):266-275.
1233	**Golddushko, B.Z.** *1961.* (Number of amphibia and reptilia and their role in the food of *Buteo buteo* and *Aquila pomarina* .) Fauna I Ecologia Nazemnykh Pozvonochnykh Belorussi, Minsk. p. 143-149. (In Russian)
1234	**Golodushko, B.Z.** *1958.* (Data on the food of *Buteo buteo* and *Aquila pomarina* in Belowiezhe.) Trudy Zapovedno-Okhotn. Khoz. Belowezhskaya Pustcha. 1:100-109. (In Russian)
1235	**Golodushko, B.Z.** *1958.* Materialy po ekologii malogo podorlika Belovezhskoi pustchi. (Data on ecology of *Aquila pomarina* in Belowezhie.) 1st Zool. Konf. Beloruss. SSR, Tezisy Doklady., Minsk. p. 34.
1236	**Golodushko, B.Z.** *1959.* (Data on the ecology of *Aquila pomarina* in the Belowiezhie Forest.) First Zool. Conf., Beloruss. SSR, Tezisy Doklad., Minsk. p. 34-35. (In Russian)
1237	**Golodushko, B.Z.** *1961.* O roli khistchnykh ptits v biotsenoze Belowezhskoi pustchi. (Role of birds of prey in a biocenosis of Belowiezhe.) Fauna I Ekologia Nazemnykh Pozvonochnnykh Belorussii, Minsk. p. 98-111.
1238	**Golodushko, B.Z.** *1961.* O pistchevykh otnosheniyakh ptits Belowezhskoi pustchi. (Trophic relations between birds of prey in Beloweziezhe.) Fauna I Ekologia Nazemuykh Pozvonochnykh Belorussii, Minsk. p. 112-132.
1239	**Golovanova, E.N. and Y.B. Pukinski.** *1969.* K voprosu o pistchevoi plastichnosti ptits. (Problem of feeding variability of birds.) 5th Vsesoyuzn. Ornithol. Konf. Askhabad. p. 171-174.
1240	**Goodacre, M.J. and D. Lack.** *1959.* Early breeding in 1957. Brit. Birds 52(3):74-83.
1241	**Goodrich, A.L.** *1946.* Birds in Kansas. Kansas State Board of Agric., Topeka. 340 pp.
1242	**Goodwin, D.** *1949.* Notes on the migration of birds of prey over Suez. Ibis 91(1):59-63.
1243	**Goodwin, G.A.** *1977.* Golden eagle predation on pronghorn antelope. Auk 94(4):789-790.
1244	**Gordon, S.** *1924.* The family life of a golden eagle. Country Life 56(1441):244-246.
1245	**Gordon, S.** *1927.* Days with the golden eagle. Williams and Norgate, London.
1246	**Gordon, S.** *1937.* The king of birds. Nat. Hist. 39(5):353-356.
1247	**Gordon, S.** *1939.* Letter. Nature 144(3657):942.
1248	**Gordon, S.** *1939.* Comments on nesting in the golden eagle. Nature 144:942.
1249	**Gordon, S.** *1944.* Early hatching of golden eagle. Brit. Birds 38(4):76-77.
1250	**Gordon, S.** *1958.* Golden eagle, fulmar and fox. Country Life 124(3213):330.
1251	**Gordon, S.** *1962.* When golden eagles bask in the sun. Country Life 131(3389):342.
1252	**Gordon, S.** *1964.* Golden eagles of the Hebrides. Country Life 136(3526):855-856.
1253	**Gordon, S.** *1966.* Transplanting a pair of eagles. Field (May 5):827. (Title Only)
1254	**Gordon, S.** *1968.* Courtship of the golden eagle. Birds 2(6):144-145.
1255	**Gordon, S. and A. Gordon.** *1925.* Some notes taken at the eyrie of a golden eagle. Brit. Birds 18(9):237-240.
1256	**Gordon, S.P.** *1904.* The golden eagle. Country Life 40(388):879-880.
1257	**Gordon, S.P.** *1904.* A young golden eagle. Country Life 40(390):959.
1258	**Gordon, S.P.** *1924.* Family life of a golden eagle. Country Life 56(Aug. 16):244-246. (Title Only)
1259	**Gordon, S.P.** *1924.* The eagle in a cage. Nineteenth Century 95(567):730-734.
1260	**Gordon, S.P.** *1926.* Letter (Lifting power of the golden eagle). Brit. Birds 19(11):291-292.
1261	**Gordon, S.P.** *1926.* The golden eagle. Nature 118(2967):377-379.
1262	**Gordon, S.P.** *1928.* Camera shots at a golden eagle. Field (Aug. 30):338. (Title Only)
1263	**Gordon, S.P.** *1929.* The golden eagle. Country Life 66(1697):114-118.
1264	**Gordon, S.P.** *1932.* Forty hours with the golden eagle. Field (July 16):90-91. (Title Only)
1265	**Gordon, S.P.** *1932.* The golden eagle at home. Country Life 71(1824):5-6.
1266	**Gordon, S.P.** *1935.* Golden eagle mating during incubation. Brit. Birds 29(2):59-60.
1267	**Gordon, S.P.** *1939.* Fledging period of the golden eagle. Brit. Birds 33(4):111-112.
1268	**Gordon, S.P.** *1939.* Golden eagles mating. Brit. Birds 32(12):398.
1269	**Gordon, S.P.** *1945.* A golden eagle's roosting time. Country Life 97(2507):209.
1270	**Gordon, S.P.** *1946.* Golden eagle running. Brit. Birds 39(11):344.
1271	**Gordon, S.P.** *1947.* Male golden eagle incubating. Brit. Birds 40(6):181-182.
1272	**Gordon, S.P.** *1949.* Golden eagle running. Brit. Birds 42(8):247-248.
1273	**Gordon, S.P.** *1955.* The golden eagle, king of birds. Citadel Press, New York, NY. 246 pp.
1274	**Gordon, S.P.** *1957.* Hebridean golden eagles. Country Life 121(3152):1191.
1275	**Gordon, S.P.** *1957.* Coition of golden eagles in Skye. Scott. Nat. 69(3):183-184.
1276	**Gore, M.E.J. and W. Pyong-Oh.** *1971.* The Birds of Korea. Royal Asiatic Soc., Korea Branch, Seoul. 450 pp.
1277	**Gorham, F.W. and A.C. Ivy.** *1938.* General function of the gall bladder from the evolutionary standpoint. Field Mus. of Nat. Hist., Zool. 22(3):159-213.

Master No.	Citation
1278	**Gorman, J.** *1981.* A new home for eagles. Discovery 2(9):72-73.
1279	**Gorodon, A.G. and M.J.P. Gregory** *1973.* Golden eagles rear three young. Scott. Birds 7(8):408,412-413.
1280	**Gouttenoire, G.** *1955.* Inventaire des oiseaux de Tunisie. Alauda 23(1):1-64.
1281	**Graham, R.** *1922.* "Eagle tricks." Oologist 39(3):43.
1282	**Graham, R.** *1922.* Eagle and fox terrier dog. Oologist 39(5):80.
1283	**Graham, R.** *1924.* Texas eagles. Oologist 41(7):87.
1284	**Grandjean, P.** *1971.* Kongeorn. (Golden eagle.) Dansk Ornithol. 71(3):147.
1285	**Grant, G.S.** *1970.* Golden eagle specimen from Buncombe County, N.C. Chat 34(2):48.
1286	**Grant, H.B.** *1915.* On a collection of birds from British East Africa and Uganda, presented to the British Museum by Capt. G.P. Cosens. Part II. Ibis 3(2):235-316.
1287	**Grayson, D.K.** *1976.* A note on the prehistoric avifauna of the lower Klamath Basin. Auk 93(4):830-833.
1288	**Green, W.R.** *1927.* The golden eagle nesting in Woods County, Oklahoma. Wilson Bull. 39(3):172.
1289	**Greenway, J.C., Jr.** *1973.* Type specimens of birds. Bull. Am. Mus. Nat. Hist. 150(3):207-346.
1290	**Greenwood, A.** *1977.* The role of disease in the ecology of British raptors. Bird Study 24(4):259-265.
1291	**Gregory, M.J.P. and C. Gregory.** *1969.* Apparent distraction display by golden eagle. Scott. Birds 5(5):282.
1292	**Grier, J.W.** *1972.* Congratulations to an absentee father. Natl. Wildl. 10(6):44-45.
1293	**Grier, J.W.** *1973.* Techniques and results of artificial insemination with golden eagles. Raptor Res. 7(1):1-12.
1294	**Grier, J.W., H.R. Postovit and J.D. Crawford.** *1978.* Little Missouri Breaks raptor studies. Dunn County, North Dakota. 1978 (and 1977) final report. North Dakota State Univ., U.S.D.I. Bureau of Land Management, Contract. 67 pp.
1295	**Griffee, W.E.** *1941.* Large sets of golden eagle eggs. Murrelet 22(2):40-41.
1296	**Griffin, W.W.** *1941.* Four unpublished golden eagle records from Georgia. Oriole 6(1):12.
1297	**Grimm, P. and B. Nammert.** *1978.* Der schreiadler als beuteschmarotzer. Falke 25(4):140.
1298	**Grinnell, G.B.** *1929.* Eagles' prey. J. Mammal. 10(1):83.
1299	**Grinnell, J.** *1909.* A collection of birds from forty-mile, Yukon Territory, Canada. Condor 11(6):202-207.
1300	**Grinnell, J., J. Dixon and J.M. Linsdale.** *1930.* Vertebrate natural history of a section of northern California through the Lassen Peak region. Univ. Calif. Press, Berkeley. 594 pp.
1301	**Grinnell, J. and A.H. Miller.** *1944.* The distribution of the birds of California. Pac. Coast Avifauna 27:1-608.
1302	**Griscom, L.** *1921.* Some notes on the winter avifauna of the Camargue. Ibis (11th Ser.) 3(4):595-608.
1303	**Gronvall, J.S.** *1948.* Storre skrikornens, *Aquila clanga* (Pall.), forekomst i Ostra Nyland. Ornis Fenn. 25(1):19. (English Summ.)
1304	**Gross, A.O.** *1968.* Albinistic eggs (white eggs) of some North American birds. Bird-Band. 39(1):1-6.
1305	**Grossman, M.L. and J. Hamlet.** *1964.* Birds of Prey of the World. Bonanza Books, New York. 496 pp.
1306	**Grote, H.** *1932.* Zwei junge im adlerhorst. Ornithol. Monatsber. 40(3):85-86.
1307	**Grote, H.** *1933.* Steinadlerhorst in Fischreiherkolonie. Beitr. Fortpfl.-Biology Vogel 9(3):101.
1308	**Grote, H.** *1933.* Zwei flugge junge bei *Aquila clanga* Pall. Beitr. Fortpfl.-Biology Vogel 9(5):188.
1309	**Grote, H.** *1936.* Beitrage zur biologie sudostrussischer Steppenvogel. Beitr. Fortpfl.-Biology Vogel 12(4):133-139.
1310	**Grote, H.** *1936.* Die zahl der jungenim stein- und kaiser adlerhorst Beitr. Fortpfl.-Biology Vogel 12(2):74.
1311	**Grupp, E.** *1977.* I helped the eagles. Ranger Rick 11(6):20-25.
1312	**Gruson, E.S.** *1972.* Words for Birds. Quadrangle Books, New York. 305 pp.
1313	**Gubler, W.** *1972.* Merkwurdiges flugspiel eines steinadlers. Ornithol. Beob. 69(1):46-47.
1314	**Gude, H.** *1924.* Uber *Aquila fluvescens* Gray. Falco 20(1):9-13.
1315	**Gullion, G.W.** *1957.* Two records of unprovoked attack by golden eagles. Condor 59(3):210-211.
1316	**Gunston, D.** *1971.* The golden eagle in England. Birds and Country 24(99):7-8.
1317	**Gurney, J.H.** *[?].* An eagle forty years old. Trans. Norwich Nat. Soc. 9:458.
1318	**Gurney, J.H.** *1859.* List of birds of prey received from Beyrout. Ibis 1(4):389-391.
1319	**Gurney, J.H.** *1859.* List of a collection of birds from the colony of Natal in south-eastern Africa. Ibis 1(3):234-251.
1320	**Gurney, J.H.** *1860.* Note on the eggs of the eared vulture and the wedge-tailed eagle. Ibis 2(6):171.
1321	**Gurney, J.H.** *1862.* Further note on the eggs of *Vultur auricularis* and *Aquila audax.* Ibis 4(15):207-208.
1322	**Gurney, J.H.** *1864.* Note on *Aquila barthelemyi,* Jaubert. Ibis 6(23/24):339-340.
1323	**Gurney, J.H.** *1871.* Remarks on *Aquila naevioides.* Ibis 1(2):247-248.
1324	**Gurney, J.H.** *1872.* Remarks on *Aquila clanga.* Ibis 2(7):327-330.
1325	**Gurney, J.H.** *1872.* Untitled. Ibis 2(8):472-473.
1326	**Gurney, J.H.** *1873.* Two specimens of *Aquila naevia.* Ibis (3rd Ser.) 3(12):422-423.
1327	**Gurney, J.H.** *1873. Aquila imperialis* and allies. Ibis 3(9):98-100.
1328	**Gurney, J.H.** *1877.* Notes on a "Catalogue of Accipitres in the British Museum, " by R. Bowdler Sharpe (1874). Ibis (4th Ser.) 1(2):209-236,325-333.
1329	**Gurney, J.H.** *1886.* Notes on the two spotted eagles inhabiting the European continent. Naturalist 127:45-47.
1330	**Gurney, J.H.** *1894.* Catalogue of the birds of prey (Accipitres and Striges), with the number of specimens in Norwich Museum. R.H. Porter, London. 56 pp.
1331	**Gurney, J.H.** *1895.* Alleged former nesting of golden eagle and goshawk in England. Zoologist (3rd Ser.) 19:450-451.
1332	**Gurney, J.H.** *1899.* On the comparative ages to which birds live. Ibis 5(17):19-42.
1333	**Gurney, J.H.** *1917.* A long-lived eagle. Ibis 5(2):273.
1334	**Gurney, J.H.** *1922.* On the sense of smell possessed by birds. Ibis 4(2):225-253.
1335	**Gurney, S.** *1860.* Child attacked by an eagle. Zoologist 1860:7142-7143.
1336	**Gyldenstolpe, N.** *1920.* A nominal list of the birds at present known to inhabit Siam. Ibis 2(4):735-780.

Master No.	Citation
1337	**Gyldenstolpe, N.** *1934.* The travels and collections of Johan August Wahlberg, 1810-1856: a pioneer naturalist in South Africa. Ibis (13th Ser.) 4(2):264-292.
1338	**Gyorffy, S.** *1947.* The golden eagle in the Carpates and in the Mountain Malra. Aquila 51-54:188.
1339	**Haagner, A.** *1908.* The South African birds of prey: their economic relations to man. S. African Ornithol. Union Bull. No. 1. p. 1-23.
1340	**Haagner, A.K.** *1945.* A list of the birds observed in Beira and neighbourhood - with some notes on habits, etc. Ostrich 16(1):32-43.
1341	**Haar, H.** *1971.* Schreiadler *(Aquila pomarina)* bei Ilz. Egretta 14(1):25-26.
1342	**Haartman, L., von.** *1974.* Finnish nest records. Ornis Fenn. 51(1):48-58.
1343	**Haas, G.** *1956.* Vorkommen und verhalten des schelladlers in Wurttemberg. Vogelwelt 77:22-24. (Title Only)
1344	**Habfurther, A.** *1955.* Beobachtungen am steinadler *(Aquila chrysaetos)* und kolkraben *(Corvus corax)* im Allgau. Ornithol. Mitt. 7(9):168-169.
1345	**Hadfield, G.** *1892.* Golden eagle in the Isle of Wight. Zoologist (2nd Ser.) 7:3309-3310.
1346	**Hadley, A.H.** *1932.* State laws relating to birds of prey. Hawk and Owl Society Bull. No. 1. p. 33-34.
1347	**Hafner, J.C. and M.S. Hafner** *1977.* The cryptic occipital spot in the Accipitridae (Falconiformes). Auk 94(2):293-303.
1348	**Haftorn, S.** *1958.* Populasjonsendringer, spesielt geografiske forskyvninger, i den norske avifauna de siste 100 ar. Sterna 3(3):105-137.
1349	**Hagen, W.** *1929.* Aus der Nordmark. Ornithol. Monatsber. 37(2):45-46.
1350	**Hailey, T.L.** *1979.* A handbook for pronghorn antelope management in Texas. Texas Parks and Wildl. Dept., FA Rep. Ser. No. 20. p. 59.
1351	**Hainard, R.** *1954.* Coup d'oeil sur la faune estivale de Loeche-les-Bains. Nos Oiseaux 22(11):245-251.
1352	**Hajek, V.** *1977.* Orel volavy *(Aquila clanga* Pall.) u Breclavi. (The spotted eagle in Breclav.) Zpravy Mos. 19(35):99-101. (In Czechoslovakian, German Summ.) (Title Only)
1353	**Hakansson, G. and S. Hogstrom** *1965.* Golden eagles *(Aquila chrysaetos)* breeding in Gotland in 1963. Var Fagelvarld 24:56-59. (In Swedish, English Summ.) (Title Only)
1354	**Hale, P.E.** *1939.* Rough notes on the birds of Hartley District, January 1936 to June 1938. Ostrich 10(1):28-35.
1355	**Haley, D.** *1979.* In Idaho: a refuge for the raptors. Pac. Search 13(3):3-9,41-44.
1356	**Hall, G.A.** *1971.* The list of West Virginia birds. Redstart 38(1):2-18.
1357	**Hall, L.S.** *1977.* Feral rabbits on San Juan Island, Washington. Northwest Sci. 51(4):293-296.
1358	**Hamerstom, F.** *1971.* Cloacal sexing of raptors. Auk 88(1):173-174.
1359	**Hamerstrom, F.** *1942.* "Control" of the golden eagle in Texas. Wilson Bull. 54(3):218.
1360	**Hamerstrom, F.** *1969.* Golden eagle and the rearing of redtails. Nat. Hist. 78(5):62-69.
1361	**Hamerstrom, F.** *1970.* An Eagle to the Sky. Iowa State Univ. Press, Ames. 142 pp.
1362	**Hamerstrom, F.** *1971.* Reports on a golden eagle breeding project. Raptor Res. News 5:23-24.
1363	**Hamerstrom, F.** *1972.* Birds of Prey of Wisconsin. Wisconsin Dept. of Natural Resources. 64 pp.
1364	**Hamerstrom, F.** *1977.* Introducing captive-reared raptors into the wild. Pages 348-353. *in:* R.D. Chancellor, Ed. Proc. of the World Conf. on Birds of Prey. 1-3 October, 1975. Vienna. I.C.B.P.
1365	**Hamerstrom, F.N.** *1943.* Golden eagles and bighorns. Wilson Bull. 55(1):64-65.
1366	**Hanerstrom, F., T. Ray, C.M. White and C.E. Braun.** *1975.* Conservation committee report on status of eagles. Wilson Bull. 87(1):140-143.
1367	**Hanko, M.** *1955.* Second occurrence of the tawny eagle in Hungary. Aquila 59-62:432.
1368	**Hanna, W.** *1926.* Destruction of golden eagle in California. Calif. Fish & Game 12(3):153.
1369	**Hanna, W.C.** *1924.* Weights of about three thousand eggs. Condor 26(4):146-153.
1370	**Hanna, W.C.** *1930.* Notes on the golden eagle in southern California. Condor 32(2):121-123.
1371	**Hanni, E.** *1930.* Die steinadler im Kanton Bern. Ornithol. Beob. 27(4):64.
1372	**Hanni, E.** *1933.* Gefrevelte steinadler. Ornithol. Beob. 30(8):142.
1373	**Hanny, M.** *1922.* Ausplunderung eines steinadlerhorstes. Ornithol. Beob. 20(2):28.
1374	**Hanselmann, H.** *1923.* Beringte steinadler. Ornithol. Beob. 20(10/11):167.
1375	**Hansen, E. and B.P. Nielsen.** *1975.* Rapport fra sjaeldenhedsudvalget for 1974. (Report of the Rarities Committee for 1974.) Dansk Ornithol. 69:135-139.
1376	**Hansen, E., S. Hansen and B.P. Nielsen.** *1974.* Rapport fra Sjaeldenhedsudvalget for 1973. (Report of the Rarities Committee for 1973.) Dansk Ornithol. 68:138-144.
1377	**Hansen, L.** *1962.* Fugle pa Lolland-Falster. Dansk Ornithol. 56(1):1-32.
1378	**Hansen, M.B.** *1968.* Efterarstraek pa Langeland 1962, 1963, og 1964. (Autumn migration on Langeland, 1962, 1963, and 1964.) Dansk Ornithol. 62(1):1-16. (English Summ.)
1379	**Hardey, J.** *1977.* Display of golden eagle carrying stick. Scott. Birds 9(6):300.
1380	**Hardy, R.** *1945.* Breeding birds of the pigmy conifers in the Book Cliff region of eastern Utah. Auk 62(4):523-542.
1381	**Hare, H.L.** *1915.* The birds of Philipstown, Cape Province, with notes on their habits. J. S. African Ornithol. Union 11(1):1-18.
1382	**Hare, H.L.** *1932.* Records of unusual habits of birds. Ostrich 3(3):82-90.
1383	**Hargrave, L.L.** *1939.* Bird bones from abandoned indian dwellings in Arizona and Utah. Condor 41(5):26-210.
1384	**Harkness, R. and C. Murdoch.** *1971.* Birds of Prey in the Field. H.F.G. Witherby, Ltd., London. 208 pp.
1385	**Harman, I.** *1963.* Eagle lore. Birds Illus. 9(6):169-170.
1386	**Harrington, M.** *1928.* He trained a wild eagle to eat out of his hand. Am. Mag. 106(3):62-63.
1387	**Harris, J.** *1975.* Investigation probes eagles' impact on stock and game production. Chihuahuan Desert Discovery 1:6-7.
1388	**Harris, M.P. and R. Price.** *1962.* Golden eagle attacking reindeer. Brit. Birds 55(1):42-43.

Master No. **Citation**

1389 **Harrison, J.C.** *1931.* Taking of a grouse by an eagle: episode of the nesting season in the Highlands. Field (Dec. 5):857. (Title Only)

1390 **Hartert, E.** *1892.* On the birds of East Prussia. Ibis 4(15):353-372.

1391 **Hartert, E.** *1896.* Uber *Aquila pomarina* Br. Z. Ool. 6:2-39 (Title Only)

1392 **Hartert, E.** *1903.* The birds of Batjan. Novitates Zool. 10:43-64.

1393 **Hartert, E.** *1915.* List of a small collection of birds from Hausaland, northern Nigeria. Novitates Zool. 22:244-266.

1394 **Hartert, E.** *1918.* Types of birds in the Tring Museum. Novitates Zool. 25:4-63.

1395 **Hartert, E.** *1923.* The hitherto known birds of Marocco. Novitates Zool. 30:91-146.

1396 **Hartert, E.** *1925.* Remarks on *Aquila rapax* and *A. nipalensis* and on the Dobrogea eagle. Bull. Brit. Ornithol. Club 46:30-36.

1397 **Hartert, E.** *1931.* Types of birds in the Tring Museum. Novitates Zool. 37:35-52.

1398 **Harting, J.E.** *1915.* Spotted eagle in Herefordshire. Field (Dec. 4):935-936. (Title Only)

1399 **Hartman, F.A.** *1946.* Adrenal and thyroid weights in birds. Auk 63(1):42-64.

1400 **Hartzler, J.E.** *1974.* Predation and the daily timing of sage grouse leks. Auk 91(3):532-536.

1401 **Harwin, R. and J. Harwin.** *1972.* Raptor territories at Cathedral Peak, Natal. Ostrich 43(2):73-76.

1402 **Harwin, R.M.** *1955.* Field identification of raptores - eagles. Bokmakierie 7(2):33-38.

1403 **Harwood, M.** *1973.* The View from Hawk Mountain. Charles Scribner's Sons, New York. 191 pp.

1404 **Hatch, D.R.M.** *1968.* Golden eagle hunting tactics. Blue Jay 26(2):78-80.

1405 **Hatch, J.H.** *1977.* The birds of Comet Bore (Ninety Mile Plain). S. Aust. Ornithol. 27(5):163-171.

1406 **Hatler, D.F.** *1974.* Foods at a golden eagle nest in central Alaska. Condor 76(3):356-357.

1407 **Hatsofe, O.** *1981.* Raptor ringing recoveries in Israel. Torgos 1(2):5-13.

1408 **Haugh, J.R.** *1966.* Some observations on the hawk migration at Derby Hill. Kingbird 16(1):15-16.

1409 **Haugh, J.R.** *1971.* Ecology and present status of cliff-nesting raptors and other birds in arctic Alaska. Alaska Dept. of Fish and Game, Report. 33 pp.

1410 **Haugh, J.R.** *1972.* Ecology and present status of birds and mammals in interior and arctic Alaska. Alaska Dept. of Fish and Game, Report. 24 pp.

1411 **Haugh, J.R. and T.J. Cade.** *1966.* The spring hawk migration around the southeastern shore of Lake Ontario. Wilson Bull. 78(1):88-110.

1412 **Hauri, R** *1962.* Wie verhalten sich wasservogel gegenuber einem steinadler? Ornithol. Beob. 59(3):86-87.

1413 **Hauri, R.** *1953.* Ueber den beutetransport beim steinadler. Ornithol. Beob. 50(3):93.

1414 **Hauri, R.** *1959.* Fuchs, gemse, steinbock und ihr verhalten zu einem steinadler. Ornithol. Beob. 56(1):28-29.

1415 **Hauri, R.** *1970.* Eine brut des steinadlers im Napfgebiet. Ornithol. Beob. 67(5/6):297.

1416 **Hauri, R.** *1973.* Schicksal von zweiten jungvoegeln bei steinadlerbruten. Ornithol. Beob. 70(5/6):278-279.

1417 **Hauri, R.** *1973.* Neuerliches bruten des steinadlers im Napfgebiet. Ornithol. Beob. 70(5/6):277-278.

1418 **Havre, G. van.** *1921.* A propos des aigles criards. *Aquila clanga* Pallas et *Aquila pomarina pomarina* Brehm. Gerfaut 11(3/4):73-80.

1419 **Hawes, C.A.** *1878.* A nesting of the golden eagle *(Aquila chrysaetos).* Oologist 3(11):86.

1420 **Hawker, D.M.** *1975.* Golden eagle eating fulmars. Brit. Birds 68(7):293.

1421 **Haycock, P.** *1966.* The wedge-tailed eagle. Birds Illus. 11(10):293.

1422 **Haydock, E.L.** *1949.* Some notes on the family Falconidae. Ostrich 20(1):28-30.

1423 **Hayman, P.** *1969.* Tawny eagles. Birds of the World, I.P.C. Magazines, Ltd., London.

1424 **Hayward, C.L.** *1967.* Birds of the upper Colorado River Basin. Brigham Young Univ. Sci. Bull. Biology Series 9(2):1-64.

1425 **Head, F.** *1892.* A tame golden eagle. Zoologist (3rd Ser.) 16:335.

1426 **Headstrom, R.** *1961.* Birds' nests: A Field Guide (Eastern). New York. 128 pp.

1427 **Hebard, F.V.** *1938.* Golden eagle in Georgia. Oriole 3(2):18.

1428 **Hedgren, S., G. Rudebeck, G. Stromberg, L. Svensson and U. Unger.** *1976.* Fornyad granskning av nagra aldre fynd av sallsynta faglar. (A revision of some older Swedish records of rare birds.) Var Fagelvarld 35(2):137-144. (English Summ.)

1429 **Heerden, J. van.** *1977.* Quelea predators. Bokmakierie 29(2):45.

1430 **Hefs, A.** *1919.* Schrei- und schelladler in der Schweiz. Ornithol. Monatsber. 27(1/2):9-10.

1431 **Hegdal, P.L., T.A. Gatz, K.A. Fagerstone, J.F. Glahn and G.H. Matschke.** *1979.* Hazards to wildlife associated with 1080 baiting for California ground squirrels. U.S.D.I. Fish and Wildlife Service, Denver Wildlife Research Center. Report. 59 pp.

1432 **Heinrici, P.** *1951.* Erganzende bemerkungen zu der arbeit: zwanzigjahrige beobachtungen uber den schreiadler *Aquila pomarina* von Victor Wendland. Vogelwelt 72:190-192. (Title Only)

1433 **Heinroth, O.** *1922.* Die beziehungen zwischen vogelewicht, eigewicht, gelegegewicht und brutdauer. (Relation between bird weight, egg weight, clutch weight and incubation.) J. Ornithol. 70(2/3):172-285.

1434 **Heintzelman, D.S.** *1967.* Bake Oven Knob autumn hawk migration observations, 1966 and 1967. Cassinia 50:26-27.

1435 **Heintzelman, D.S.** *1970.* The Hawks of New Jersey. New Jersey State Museum Bull. 13.

1436 **Heintzelman, D.S.** *1970.* Bake Oven Knob autumn hawk migration observations (1957 and 1969). Cassinia 52:37.

1437 **Heintzelman, D.S.** *1975.* Autumn-hawk Flights: The Migrations in Eastern North America. Rutgers Univ. Press, New Brunswick, N.J. 398 pp.

1438 **Heintzelman, D.S.** *1976.* A Guide to Eastern Hawk Watching. Keystone Books, Penn. State Univ. Press, University Park, PA. 99 pp.

1439 **Heintzelman, D.S.** *1978.* The 1976, 1977, and 1978 autumn hawk counts at Bake Oven Knob, Pennsylvania. Cassinia 57:19-20.

1440 **Heintzelman, D.S.** *1982.* Variations in utilization of the Kittatinny Ridge in eastern Pennsylvania in autumn by migrating golden eagles and bald eagles (1968-1981). Am. Hawkwatcher 1(3):1-4.

Master No. **Citation**

1441 **Heintzelman, D.S. and R. MacClay.** *1972.* The 1970 and 1971 autumn hawk counts at Bake Oven Knob, Pennsylvania. Cassinia 53:3-23.

1442 **Heintzelman, D.S. and R. MacClay.** *1973.* The 1972 autumn hawk count at Bake Oven Knob, Pennsylvania. Cassinia 54:3-9.

1443 **Heintzelman, D.S. and R. MacClay.** *1975.* The 1973 and 1974 autumn hawk counts at Bake Oven Knob, Pennsylvania. Cassinia 55:17-28.

1444 **Heintzelman, D.S. and R. MacClay.** *1976.* The 1975 autumn hawk count at Bake Oven Knob, Pennsylvania. Cassinia 56:15-21.

1445 **Heinzel, H., R. Fitter and J. Parslow.** *1977.* The Birds of Britain and Europe with North Africa and the Middle East. Wm. Collins Sons + Co. Ltd., London. 114 pp.

1446 **Heirman, G.** *1921.* Aquila clanga Pall. - Grand aigle criard. Gerfaut 11(1):49.

1447 **Helander, B.** *1977.* The white-tailed sea eagle in Sweden. Pages 319-329. *in:* R.D. Chancellor, Ed. Proc. of the World Conf. on Birds of Prey. 1-3 October, 1975. Vienna. I.C.B.P.

1448 **Helander, B.** *1978.* Feeding white-tailed sea eagles in Sweden. Pages 47-56. *in:* T.A. Geer, Ed. Bird of Prey Management Techniques. British Falconers' Club.

1449 **Hell, P.** *1958.* A few ornithological notes about raptorial-birds from western Slovakia. Aquila 65:342.

1450 **Hellmayr, C.E.** *1902.* Ubersicht der von Dr. A. Penther in Sudafrika gesammelten vogel. J. Ornithol. 50(2):210-237.

1451 **Hellmayr, C.E. and B. Conover.** *1949.* Catalogue of birds of the Americas. Part 1, Number 4. Field Mus. Nat. Hist. Zool. Ser. 13(1/4):1-358.

1452 **Helmstaedt, K.W.** *1961.* Faunistiche bemerkungen uber vogel der Slowakischen Hohen Tatra. J. Ornithol. 102(3):308-316.

1453 **Henderson, A.D** *1922.* An eagle on the woodpile. Oologist 39(4):61.

1454 **Henderson, A.D.** *1920.* Golden eagle and horned owl. Oologist 37(10):122.

1455 **Henderson, T.G.** *1889.* Strange capture of a golden eagle. Zoologist (3rd Ser.) 13:232-233.

1456 **Henm, T.** *1889.* Schreiadler. Vogelwelt 14(17):498.

1457 **Henninger, W.F.** *1903.* Golden eagle in middle southern Ohio. Auk 20:67.

1458 **Henninger, W.F. and L. Jones.** *1909.* The falcones of North America. Wilson Bull. 21(4):205-218.

1459 **Henry, C., P. Henry, J. Hesse and B. Lunais.** *1971.* Contribution a la connaissance de l'avifaune du Loir-et-Cher (regions de Blois et de la Sologne). Oiseau Rev. Fr. Ornithol. 41(2/3):94-116.

1460 **Henry, C.J.** *1939.* Golden eagle takes wounded horned owl. Auk 56(1):75.

1461 **Henty, C.F.** *1977.* Thermal soaring of raptors. Brit. Birds 70(11):471-475.

1462 **Henze, O.** *1949.* Ein deutsches steinadler Gelege. Vogelwelt 70(4):118.

1463 **Hepworth, W.G.** *1968.* Causes of antelope fawn mortality. Wyoming Dept. of Fish and Game, Proj. No. FW-3-R-14, Wk. Pl. 3, Job No. 3w. Final Rep. p. 59-66.

1464 **Heredia Armada, R.** *1973.* Nota sobre alimentacion y reproduccion de *Aquila chrysaetos* en Sobre y Toledo. Ardeola 19(2):459-460.

1465 **Heron, S.J.** *1970.* The birds of Middle Swan, western Australia. Emu 70(4):155-158.

1466 **Heron, S.J.** *1973.* Birds of the Orange District, NSW. Emu 73(1):1-8.

1467 **Herren, H.** *1962.* Steinadler und feldhase. Ornithol. Beob. 59(2):56-57.

1468 **Herren, H.** *1977.* The situation regarding birds of prey in Switzerland in 1975. Pages 178-182. *in:* R.D. Chancellor, Ed. Proc. of the World Conf. on Birds of Prey. 1-3 October, 1975. Vienna. I.C.B.P.

1469 **Herron, G.B. and P.A. Lucas.** *1978.* Population surveys, species distribution, and key habitats of selected nongame species. Nevada Dept. of Fish and Game, Proj. W-53-R, Study 1, Job 1&2. 14 pp.

1470 **Herron, G.B. and P.A. Lucas.** *1979.* Population surveys, species distribution, and key habitats of selected non-game species. Nevada Dept. of Wildlife, Proj. W-53-R, Study 1, Job 1&2. 18 pp.

1471 **Herron, G.B., S.J. Stiver and R. Turner.** *1980.* Population surveys, species distribution, and key habitats of selected nongame species. Nevada Dept. of Wildlife, Proj. W-53-R, Study 1, Job 1&2. 21 pp.

1472 **Hertig, W.** *1962.* Heringsmowe als adlerweute. Ornithol. Beob. 59(6):200.

1473 **Hess, A.** *1912.* Steinadler in den Schwyzeralpen. Ornithol. Beob. 9(5):82.

1474 **Hess, A.** *1912.* Der steinadler in der Schweiz. Zool. Beob. 11:1-9.

1475 **Hess, A.** *1918.* Le grand aigle criard *(Aquila clanga,* Pallas ou *Aquila maculata,* Gm.). Ornithol. Beob. 15(8/9):138.

1476 **Hess, A.** *1918.* Schreiadler, *Aquila naevia* Brisson. Ornithol. Beob. 15(6):90-91.

1477 **Hess, A.** *1919.* Plunderung eines adlerhorstes im Berner Oberland. Ornithol. Beob. 16(11):181-182.

1478 **Hess, A.** *1921.* Vogelschutzfragen. I. Der steinadler. Ornithol. Beob. 18(11):170-173.

1479 **Hess, A.** *1921.* Verbotener steinadlerfang. Ornithol. Beob. 18(9):140.

1480 **Hess, A.** *1922.* Zahlung der steinadlerhorst im Berner Oberland. Ornithol. Beob. 19(11):175.

1481 **Hess, A.** *1922.* Adlerabschuss. Ornithol. Beob. 20(3):41.

1482 **Hess, A.** *1922.* Steinadler und Schafzucht? Ornithol. Beob. 20(1):13.

1483 **Hess, A.** *1923.* Adler in Graubunden. Ornithol. Beob. 20(6):91-92.

1484 **Hess, A.** *1923.* Ein erlegter jungadler. Ornithol. Beob. 20(12):183-184.

1485 **Hess, A.** *1924.* Jagd duf adler. Ornithol. Beob. 21(7):107.

1486 **Hess, A.** *1924.* Erlegung eines steinadlers. Ornithol. Beob. 22(3):47.

1487 **Hess, A.** *1924.* Von einem steinadlerabschuss. Ornithol. Beob. 21(5):73-74.

1488 **Hess, A.** *1924.* Das flugbid des steinadlers, *Aquila chr. chrysaetos* (L.). Ornithol. Beob. 21(8):113.

1489 **Hess, A.** *1926.* Abschuss-statistik fur den bartgeier und steinadler in Graubunden. Ornithol. Beob. 23(12):221.

1490 **Hesse, E.** *1914.* Die vogel der Havellandischen Luchgebiete. J. Ornithol. 62(3):334-386.

1491 **Hesse, E.** *1915.* Einige bemerkenswerte belegstucke der deutschen ornis im Konigl. Zoolog. Museum Berlin. J. Ornithol. 63(4):569-603.

Master No.	Citation
1492	**Heuglin, T., von.** *1859.* List of birds observed and collected during a voyage in the Red Sea. Ibis 1(4):337-352.
1493	**Heyder, R.** *1914. Aquila clanga* Pall. in Sachsen erlegt. Ornithol. Monatsber. 22(1):1-11.
1494	**Heyder, R.** *1922.* Nachtrage zur ornis Saxonica. J. Ornithol. 70(2/3):137-172.
1495	**Hiatt, R.W.** *1944.* A raptor census in Montana. Am. Midland Nat. 31(3):684-688.
1496	**Hickey, J.J. and D.W. Anderson.** *1968.* Chlorinated hydrocarbons and eggshell changes in raptorial and fish eating birds. Science 162(3850):271-273.
1497	**Hickman, G.L.** *1971.* Escape responses and swimming abilities of nestling golden eagles. Auk 88(2):427.
1498	**Hickman, G.L.** *1971.* Prairie falcons and red-tailed hawks rearing young in inactive golden eagle nests. Condor 73(4):490.
1499	**Hickman, G.L.** *1971.* Duck band found in golden eagle casting. Bird-Band. 42(1):50.
1500	**Hickman, G.L.** *1972.* Aerial determination of golden eagle nesting status. J. Wildl. Manage. 36(4):1289-1292.
1501	**Higby, L.N.** *1975.* The eagle survey in Wyoming. Pages 97-102. *in:* J.R. Murphy, C.M. White and B.E.Harrell, Eds. Population Status of Raptors. Proc. of the Conf. on Raptor Conserv. Techniques. 22-24 March, 1973. Fort Collins, Co. Raptor Res. Rep. No. 3.
1502	**Higham, W.E.** *1951.* The most magnificent of birds, which was at one time threatened with extinction: the domestic life of the golden eagle revealed in a brilliant series of photographs. Illus. London News 219(5858):138-142.
1503	**Hilgert, C.** *1928. Aquila rapax belisarius* (Lev.). (Nordafrikanischer raubadler.) Beitr. Fortpfl.-Biology Vogel 4(1):24-25.
1504	**Hill, H.M. and T.H. Work.** *1947.* Protocalliphora larvae infesting nestling birds of prey. Condor 49(2):74-75.
1505	**Hill, N.P.** *1944.* Sexual dimorphism in the Falconiformes. Auk 61(2):228-234.
1506	**Hill, R.** *1968.* Australian Birds. New York. 281 pp.
1507	**Hilts, W.** *1971.* Where are eagles. Fur-Fish-Game (Dec.):4-5,23-25.
1508	**Hinman, R.A.** *1960.* Antelope populations in southwestern Utah, with special reference to golden eagle predation. M.S. Thesis. Utah State Univ., Logan, UT. 66 pp.
1509	**Hintovt, F.** *1940.* Notes on the ecology of steppe eagles. Rev. Microbiol. Parasitol. 19:322-331. (Title Only)
1510	**Hirsch, L.P. and L. Yager.** *1978.* Status and management of endangered avian species. New York Dept. of Env. Conserv., Proj. No. E-1-2. Study No. VIII-6. 5 pp.
1511	**Hirtz, M.** *1914.* Beitrage zur kenntnis der ornithofauna Croatica. Ornithol. Jahrbuch 25(1/2):1-16.
1512	**Hissa, R.** *1964.* Censuses of the nesting birds of the island Lapporen in Bjorko, W. Finland. Ornis Fenn. 41(3/4):69-81.
1513	**Hobbie, J.E. and T.J. Cade.** *1962.* Observations on the breeding of golden eagles at Lake Peters in northern Alaska. Condor 64(3):235-237.
1514	**Hobbs, J.N.** *1958.* Some notes on grebes. Emu 58(2):129-132.
1515	**Hobbs, J.N.** *1961.* The birds of south-west New South Wales. Emu 61(1):21-55.
1516	**Hobbs, J.N.** *1962.* Food of the wedge-tailed eagle. Emu 62(3):180.
1517	**Hock, R.J.** *1952.* Golden eagle versus red fox: predation or play? Condor 54(5):318-319.
1518	**Hocke, H.** *1891.* Von schreiadler. Z. Ool. 1:5-6,11. (Title Only)
1519	**Hodek, E.** *1881.* Steinadler - goldadler. Mitt. Ornithol. Vereines Wein 5(3):19-20.
1520	**Hodek, E.** *1881.* Steinadler-goldadler. Mitt. Ornithol. Vereines Wein 5(4):30-32.
1521	**Hoechlin, D.R.** *1974.* Behavioral ecology of nesting golden eagles *(Aquila chrysaetos)* in San Diego County, California. M.S. Thesis, San Diego State Univ.
1522	**Hoechlin, D.R.** *1976.* Development of golden eaglets in southern California. West. Birds 7(4):137-152.
1523	**Hoefs, M.** *1973.* Birds of the Kluane Game Sanctuary, Yukon Territory and adjacent areas. Can. Field-Nat. 87(4):345-355.
1524	**Hoesch, W.** *1936.* Beobachtungen am horst von *Aquila verreauxi.* Ornithol. Monatsber. 44(6):165-169.
1525	**Hoesch, W.** *1938.* Ornithologische beobachtungen auf einer Fahrt ins Etoscha-Gebeit (S.W. Afrika). J. Ornithol. 86(2):222-233.
1526	**Hoesch, W. and G. Niethammer.** *1940.* Die vogelwelt Deutsch-Sudwestafrikas namentlich des Damara- und Namalandes. J. Ornithol. 88(Special):84-365.
1527	**Hoffman, R., R. Faust, U. Weinandt and G.Hoffmann-Frezer.** *1976.* Chromosomenuntersuchungen un funf spezies der ordnng Falconiformes. Zool. Garten 46(1/2):99-107.
1528	**Hoffmann, G.** *1931.* Brut eines schelladlers *(Aquila clanga)* in Ostpreuben. Ornithol. Monatsber. 39(6):161-163.
1529	**Hoffmann, G.** *1931.* Brutet der schelladler, *Aquila clanga* (Pallas), doch in Ostpreuben? Ornithol. Monatsber. 39(4):115-116.
1530	**Hoffmann, G.** *1931.* Der kleine schreiadler. Aus Der Heimat 44:101-106. (Title Only)
1531	**Hoffmann, G.** *1932.* Der schelladler. Aus Der Heimat 45:259-262. (Title Only)
1532	**Hoffmann, G.** *1935.* Vom schelladler *(Aquila clanga)* in Ostpreuben. Ornithol. Monatsber. 43(1):25-26.
1533	**Hoffmann, G.** *1938.* Zur brutbiologie des kleinen schreiadlers. Aus Der Heimat 51:131-134. (Title Only)
1534	**Hoffmann, W.J.** *1885.* Bird names of the Selish, Pah-Uta and Soshoni indians. Auk 2(1):7-10.
1535	**Hofslund, P.B.** *1954.* The hawk pass at Duluth, Minnesota. Wilson Bull. 66(3):224.
1536	**Hogg, P.** *1950.* Some breeding records from the Anglo-Egyptian Sudan. Ibis 92(4):574-578.
1537	**Hogg, P.** *1974.* Trans-Saharan migration through Sarir, 1969-70. Ibis 116(4):466-476.
1538	**Hohn, E.O.** *1959.* Eagles and thunderbirds. Can. Audubon 21(5):146-149.
1539	**Holgersen, H.** *1957.* Bird-banding in Norway 1956. (Report No. 7). Sterna 2(5):137-184.
1540	**Holgersen, H.** *1959.* Ringmerkingsoversikt 1958. Sterna 3(7):269-313.
1541	**Holgersen, H.** *1961.* Ringmerkingsoversikt 1959. Sterna 4(5):177-220.
1542	**Holgersen, H.** *1967.* Ringmerkingsoversikt 1967. (Summary: bird-banding in Norway 1967.) Sterna 8(6):285-290.
1543	**Holgersen, H.** *1967.* Stavanger museums gjenfunn 1965. Sterna 7(7):305-330.
1544	**Holgersen, H.** *1969.* Stavanger Museums gjenfunn 1967-1968. (Stavanger Museum's recoveries of ringed birds 1967 and 1968.) Sterna 8(8):390-424.
1545	**Holgersen, H.** *1971.* Stavanger museums Gjenfunn 1969-1970. Sterna 10(3):175-224.

Master No.	Citation
1546	**Holgersen, H.** *1977.* Ringmerkingsoversikt 1975. (Summary: bird-ringing in Norway 1975.) Sterna 16(2):81-86.
1547	**Holman, F.C.** *1947.* Birds of Gold Coast. Ibis 89(Oct.):623-650.
1548	**Holmes, R.T. and C.P. Black.** *1973.* Ecological distribution of birds in the Kolomak River-Asinuk Mt. Region, Yukon-Kuskokwim Delta, Alaska. Condor 75(2):150-163.
1549	**Holroyd, G.L.** *1968.* Recent golden eagle sightings in eastern Ontario. Ontario Field Biology 22:30-33.
1550	**Holt, E.G.** *1935.* Golden eagle in Louisiana: A correction and a reaffirmation of a positive record. Auk 52(2):183.
1551	**Holt, J.B., Jr. and R. Frock, Jr.** *1980.* Twenty years of raptor banding on the Kittatinny Ridge. Hawk Mt. News 54:8-32.
1552	**Homeyer, E., von.** *1868.* Zur kenntniss der vogel ostsibiriens und des Amurlandes. J. Ornithol. 16(3):197-206.
1553	**Homeyer, E.F., von.** *1875.* Ueber die gruppe der schreiadler. J. Ornithol. 23(2):153-166.
1554	**Homeyer, E.F., von.** *1877.* Bemerkungen zur ornis Bulgariens mit rucksicht auf den bericht der gebruder sintenis und die reiseergebnisse von Dr. Finsch in Journ. f. Orn. 1859 p. 378. J. Ornithol. 25(1):69-74.
1555	**Homeyer, E.F., von.** *1881.* Notiz uber den schreiadler. Mitt. Ornithol. Vereines Wein 5(7):51.
1556	**Hoogerwerf, A.** *1964.* On birds new for New Guinea or with a larger range than previously known. Bull. Brit. Ornithol. Club 84(5):94-96.
1557	**Hore-Lacy, I.** *1964.* Birds in the You Yangs, Victoria: habitat selection and status. Emu 64(1):29-32.
1558	**Horsbrugh, C.B.** *1918.* Further notes on birds observed at Alix, Buffalo Lake, and Red Deer in the Province of Alberta, Canada, in 1915 and 1916. Ibis (10th Ser.) 6(3):477-496.
1559	**Hortling, I.** *1944.* Finlands ornar. Var Fagelvarld 3(1):33-38.
1560	**Horvath, L.** *1976.* Unusual record of lesser spotted eagle *(Aquila pomarina).* Aquila 83:299.
1561	**Hosken, J.H. and N. Hosken.** *1955.* The eagle and the cobra. Bokmakierie 7(1):24.
1562	**Hosking, E.** *1964.* British bird photographers. I. Seton Gordon. Brit. Birds 57(1):24-25.
1563	**Hosta, A.B., A. de Juan Monzon, M.M. Brieca Bombon and A.V. Palet.** *1977.* Sobra fenologica, status y distribucion de accipitriformes (Aves) en la Provinicia Barcelona. Misc. Zoologica 4(1):265-298. (English Summ.)
1564	**Hough, J.** *1950.* Watching Verreaux's eagle. Bokmakierie 3(1):6.
1565	**Houston, C.S.** *1965.* Raptores banded in 1964. Blue Jay 23(1):10-12.
1566	**Houston, C.S.** *1970.* Hawk and owl banding in Saskatchewan, Canada. Ring 62:5-8.
1567	**Hovel, H.** *1963.* Verreaux's eagle in Israel. Bull. Brit. Ornithol. Club 83(1):1-2.
1568	**Hovel, H.** *1969-1970.* Ornithological news from Israel. Aquila 76-77:195-196.
1569	**Hovette, C.** *1972.* Nouvelles acquisitions avifaunistiques de la Camarque. Alauda 40(4):343-352.
1570	**Howard, H.** *1929.* The avifauna of Emeryville shellmound. Univ. of Calif. Publ. in Zool. 32(2):301-394.
1571	**Howard, H.** *1930.* A census of the Pleistocene birds of Rancho La Brea from the collections of the Los Angeles Museum. Condor 32(2):81-88.
1572	**Howard, H.** *1932.* Eagles and eagle-like vultures of the Pleistocene of Rancho La Brea. Carnegie Inst. Washington, Contrib. to Paleontology. 82 pp.
1573	**Howard, H.** *1935.* A new species of eagle from Quartenary cave deposits in eastern Nevada. Condor 37(4):206-209.
1574	**Howard, H.** *1936.* Further studies upon the birds of the Pleistocene of Rancho La Brea. Condor 38(1):32-36.
1575	**Howard, H.** *1947.* An ancestral golden eagle raises a question in taxonomy. Auk 64(2):287-291.
1576	**Howard, H.** *1947.* A preliminary survey of trends in avian evolution from Pleistocene to recent time. Condor 49(1):10-13.
1577	**Howard, H.** *1971.* Quartenary avian remains from Dark Canyon Cave, New Mexico. Condor 73(2):237-240.
1578	**Howard, H. and L.M. Dodson.** *1933.* Bird remains from an indian shellmound near Point Mugu, California. Condor 35(6):235.
1579	**Howard, H. and A.H. Miller.** *1933.* Bird remains from cave deposits in New Mexico. Condor 35(1):15-18.
1580	**Howard, R.P. and J.F. Gore. (Eds.).** *1980.* A workshop on raptors and energy developments. Bonneville Power Admin., U.S. Fish and Wildlife Service, Idaho Power Comp., and Idaho Chapter of the Wildlife Soc. 125 pp.
1581	**Howard, R.P., L.O. Wilson and F.B. Renn.** *1976.* Relative abundance of nesting raptors in southern Idaho. Raptor Res. 10(4):120-128.
1582	**Howard, R.P., L.O. Wilson and F.B. Renn.** *1977.* Relative abundance of nesting raptors in the southern Idaho deserts. U.S.D.I. Bureau of Land Management. Report. 11 pp.
1583	**Howard, V.W., Jr. and R.E. Shaw.** *1978.* Preliminary assessment of predator damage to the sheep industry in southeastern New Mexico. New Mexico St. Univ. Agric. Expt. Stat., Res. Rep. 356. 8 pp.
1584	**Howland, S.** *1882.* Golden eagle's eggs. Ornithol. and Ool. 7(17):131.
1585	**Howland, S.** *1885.* Peculiarly marked eggs of the golden eagle. Ornithol. and Ool. 10(6):94-95.
1586	**Hudec, K. and W. Cerny.** *1977.* Fauna CSSR. Nakladatelstvi Ceskoslovenske Akademie Ved, Praha. 893 pp.
1587	**Hudson, G.E.** *1948.* Studies on the muscles of the pelvic appendage in birds II: The heterogeneous order Falconiformes. Am. Midland Nat. 39(1):102-127.
1588	**Hudson, L.W.** *1920.* Golden eagle at Porterville, California. Condor 22(2):72.
1589	**Hudson, R. (ed.)** *1975.* Threatened birds of Europe. Mac Millan London Ltd. and the Council of Europe. 128 pp.
1590	**Hue, F. and R.D. Etchecopar.** *1966.* Notes ornithologiques du Moyen-Orient. Oiseau Rev. Fr. Ornithol. 36(2):95-109.
1591	**Huey, L.M.** *1931.* Three noteworthy bird records from Barrow, Alaska. Condor 33(1):36-37.
1592	**Huey, L.M.** *1962.* Comparison of the weight-lifting capacities of a house finch and a golden eagle. Auk 79(3):485.
1593	**Hughes, C.G.** *1973.* Help for the golden eagle. Wildl. in N. Carolina 37(9):10.
1594	**Hughes, M.R.** *1970.* Relative kidney size in non-passerine birds with functional salt glands. Condor 72(2):164-168.
1595	**Hugues, A.** *1937.* Contribution a l'etude des oiseaux du Gard, de la Camarque et de la Lozere. Alauda 9(2):151-209.
1596	**Hulten, M.** *1960.* Observations ornithologiques faites au Grand-Duche de Luxembourg en 1958. Gerfaut 50(1):67-72.

Master No. **Citation**

1597 **Hume, A.** *1888.* Detailed list of species observed in Manipur, together with notices of all other species observed in Assam, Sylhet and Cachar. Stray Feathers 11(1/4):1-353.

1598 **Hume, A.O.** *1873.* Contributions to the ornithology of India. Stray Feathers 1(3):91-289.

1599 **Hume, A.O.** *1875.* A first list of the birds of Upper Peru. Stray Feathers 3(1):1-94.

1600 **Hume, A.O.** *1878.* Birds occurring in India, not described in Jerdon or hitherto in "Stray Feathers." Stray Feathers 7(3/4):320-451.

1601 **Hummel, G.** *1904.* Steinadler abseits vom Gebirge. Ornithol. Beob. 3(5):76.

1602 **Hunsicker, G.R.** *1972.* Nest behavior of the golden eagle, *Aquila chrysaetos,* in San Diego County, California. M.S. Thesis, Univ. of California, Riverside.

1603 **Hurley, G.** *1970.* Fall hawk migration along Peters Mountain in Monroe County, West Virginia. Redstart 37(3):82-86.

1604 **Hurley, G.** *1975.* Fall hawk migration along Peters Mountain, Part 2. Redstart 42(4):114-117.

1605 **Hurt, A.B.** *1961.* Golden eagle captured in Ashe County, N.C. Chat 25(1):15.

1606 **Huth, H.** *1968.* Zusammenstoss zwischen segelflugzeug und steinadler. Ornithol. Beob. 65(3):131.

1607 **Hyndman, C.C. and A.S. Hyndman.** *1972.* The shell pigment of golden eagle eggs. Condor 74(2):200-201.

1608 **Hynt, J.** *1979.* Poisons in the countryside. Scott. Birds 10(5):157-158.

1609 **Ijams, H.P.** *1931.* Two interesting notes from east Tennessee. Auk 48(2):267-268.

1610 **Ijzendoorn, E.J., van.** *1979.* Steenarend *Aquila chrysaetos* overwinterend in de Wieringermeer. (Golden eagle *Aquila chrysaetos* wintering in the Wieringermeer.) Dutch Bird. 1(5):15-16.

1611 **Ilani, G.** *1981.* Black eagle mystery. Israel Land and Nature 6(4):168.

1612 **Ilani, G.** *1981.* The golden eagles return to Jerusalem. Israel Land and Nature 6(4):168.

1613 **Ilyichev, V.D. and V.E. Fomin.** *1979.* Ornithofauna Bashkirii i eye izmenenia v dvadtsatom veke. (Avifauna of Bashkiria and its changes in the twentieth century.) Ornithologia 14:83-96.

1614 **Imhof, T.A.** *1962.* Alabama Birds. Univ. of Alabama Press, University, AL. 445 pp.

1615 **Imler, R.H.** *1936.* An annotated list of the birds of Rooks County, Kansas, and vicinity. Trans. Kansas Acad. Sci. 39:295-312.

1616 **Imler, R.H.** *1937.* Weights of some birds of prey of western Kansas. Bird-Band. 8(4):166-169.

1617 **Ingram, C.** *1926.* Notes on nestling birds. Brit. Birds 19(107):247-248.

1618 **Ingram, C.** *1959.* The importance of juvenile cannibalism in the breeding biology of certain birds of prey. Auk 76(2):218-226.

1619 **Insley, H. and D. Dugan.** *1973.* Buzzard and golden eagle feeding on other avian predators. Brit. Birds 66(7):310-311.

1620 **Irby, L.H.** *1861.* Notes on birds observed in Oudh and Kumaon. Ibis 3(1):217-251.

1621 **Irby, L.H.** *1879.* Notes on the birds of the Straits of Gibraltar. Ibis 3(11):342-346.

1622 **Irby, L.H.** *1883.* Notes on the birds of Santander, northern Spain. Ibis 1(2):173-190.

1623 **Iredale, T.** *1915.* The new B.O.U. List: more corrections. Ibis 3(2):388-390.

1624 **Iredale, T.** *1956.* Birds of New Guinea. Georgian House, Melbourne. 230 pp.

1625 **Iribarren, J.J.** *1977.* The present status of birds of prey in Navarra (Spain). Pages 381-382. *in:* R.D. Chancellor, Ed. Proc. of the World Conf. on Birds of Prey. 1-3 October, 1975. Vienna. I.C.B.P.

1626 **Irving, L.** *1960.* Birds of Anaktuvuk Pass, Kobuk, and Old Crow. Smithsonian Inst., U.S. Natl. Museum Bull. 217. 409 pp.

1627 **Irving, L. and S. Paneak.** *1954.* Biological reconnaissance along the Ahlasuruk River east of Howard Pass, Brooks Range, Alaska, with notes on the avifauna. J. Wash. Acad. Sci. 44(7):201-211.

1628 **Irwin, M.P.S. and C.W. Benson.** *1966.* Notes on the birds of Zambia: Part II. Arnoldia 2(37):1-23.

1629 **Iselin, H.** *1956.* Beobachtungen an einem steinadlerhorst. Ornithol. Beob. 53(2):25-28.

1630 **Iselin, H. and E. Hammerle.** *1960.* Beobachtungen am adlerhorst: der nahrungsbedarf zweier jungadler. Ornithol. Beob. 57(3):129-132.

1631 **Ismagilov, M.I.** *1962.* (Nesting period of *Aquila nipalensis* in central Kazakhstan.) 3rd Vsesoyuzn Ornithol Konf., Lvov. 2:3-4. (In Russian)

1632 **Ivanov, A.I. and B.K. Stegmann.** *1978.* (Concise key to U.S.S.R. birds.) Second Ed., Publisher Unknown. (In Russian)

1633 **Ivanov, V.G. and V.V. Dmitriev.** *1961.* Khistchnie ptitsi Kabardino-Balkarii. (Birds of prey of Kabardino-Balkaria.) Uchenye Zapiski Kabardino-Balkarsk. Gos. Univ. 10:161-173. (In Russian)

1634 **Ivanovic, B.** *1968.* Some ornithological observations on Skadar (Scutari) Lake. Larus 21-22:127-147. (English Trans.)

1635 **Ivanovski, V.V.** *1977.* K biologii berkuta *(Aquila chrysaetos* L.) v Belorusskom poozerie. (Biology of *Aquila chrysaetos* in northern Belorussia.) 7th Zsesoyuzn. Ornithol. Konf., Kiev. p. 215-216. (In Russian)

1636 **Ivanovsky-Ivanauskas, T.** *1927.* Zwei *Aquila chrysaetos* L. in Litauen erlegt. Ornithol. Monatsber. 35(2):54.

1637 **Jablonski, B** *1964.* Materialy do awifauny wschondniej czesci Niziny Maxowieckiej. Ptaki okolic Klembowa, pow. Wolomin. (Contributions to the knowledge of the Mazowsze Plain avifauna.) Acta Ornithol. 8(1):1-66. (English Summ.)

1638 **Jablonski, B.** *1976.* Estimation of bird abundance in large areas. Acta Ornithol. 16(2):23-76.

1639 **Jackman, S.M. and J.M. Scott.** *1975.* Literature Review of Twenty Three Selected Forest Birds of the Pacific Northwest. U.S.D.A. Forest Service, Portland, Or. 382 pp.

1640 **Jackson, F.J.** *1906.* On a collection of birds made by Mr. Geoffrey Archer during a journey to the Ruwenzori Range. Ibis (8th Ser.) 6(23):505-570.

1641 **Jackson, F.J.** *1938.* The Birds of Kenya and the Uganda Protectorate. Vol. 1. Publisher Unknown.

1642 **Jackson, H.D. and P. Steyn.** *1968.* Bird notes from the Nata sector of the Markarikari Pan, Botswana. Ostrich 39(1):3-8.

1643 **Jackson, S.W.** *1907.* Egg collecting and bird life of Australia. F.W. White, Printer, Sydney. 171 pp.

1644 **Jacobi, A.** *1915.* Weiteres vorkommen von *Aquila clanga* Pall. in Sachsen. Ornithol. Monatsber. 23(7/8):125.

1645 **Jacobsen, J.T., B.P. Nielsen and J. Rabol.** *1970.* Rapport fra sjaeldenhedsudvalget for 1970. (Report of the Rarity Committee for 1970.) Dansk Ornithol. 65(3):133-139.

1646 **Jacobsson, S.** *1955.* Kungsorn *(Aquila chrysaetos)* . Var Fagelvarld 14(3):187.

Master No. **Citation**

1647 **Jacquat, B.** *1976.* Betrachtliches alter eines freilebenden steinadlers *Aquila chrysaetos* . Ornithol. Beob. 73(1):28-29.

1648 **Jacquat, R.** *1977.* Age remarquable d'un aigle royal *Aquila chrysaetos* en liberte. Nos Oiseaux 34(1):31.

1649 **Jahn, H.** *1942.* Zur oekologie und biologie der vogel Japans. J. Ornithol. 90(1/2):1-302.

1650 **Jahn, L.R., R.C. Clement, W.W. Dykstra, R.A. McCabe, D.A. Poole and R.L. Rudd.** *1963.* Annual report of the Conservation Committee. Wilson Bull. 75(3):295-325.

1651 **James, R.D., P.L. McLaren and J.C. Barlow,** *1976.* Annotated checklist of the birds of Ontario. Royal Ontario Museum, Misc. Publ. 1. 75 pp.

1652 **Jankowitz, M.** *1976.* Tawny or steppe eagle? Bokmakierie 28(3):64-65.

1653 **Jankowitz, M.** *1978.* In search of the golden eagle. Bokmakierie 30(4):89-92.

1654 **Janossy, D.** *1960-1961.* Eine fossile vogelfauna aus den Mousterien-Schichten der Subalyuk-Hohle im Bukk-Gebirge (Nordostungarn). Aquila 67-68:175-188. (In German, Hungarian Summ.)

1655 **Janowski, K.** *1967.* On some birds observed in Niemodlin District, 1964-1965. Acta Ornithol. 10:187-198.

1656 **Jantzen, R.A.** *1961.* Bighorns and golden eagles. Pages 47-50. *in:* Proc. Fifth Ann. Meeting of the Desert Bighorn Council, 4-7 April 1961. Sonora, Mexico.

1657 **Jarman, H.** *1953.* The R.A.O.U. camp-out in central Australia, 1952. Emu 53(2):169-185.

1658 **Jeheber, P.** *1976.* Hawks over the Hudson Highlands. Kingbird 26(3):136-140.

1659 **Jenkins, D., A. Watson and G.R. Miller.** *1964.* Predation and red grouse populations. J. Appl. Ecol. 1(1):183-195.

1660 **Jenney, C.E.** *1901.* A golden eagle. Birds and Nature 9(4):152.

1661 **Jensen, J.K.** *1923.* Notes on the nesting birds of northern Santa Fe County, New Mexico. Auk 40(3):452-469.

1662 **Jensen, J.K.** *1928.* Nests and eggs of the golden eagle. Ool. Rec. 8(2):37-39.

1663 **Jensen, R.A.C.** *1972.* The steppe eagle *Aquila nipalensis* and other termite-eating raptors in South West Africa. Madoqua 1(5):73-76. (Title Only)

1664 **Jensen, R.A.C., M.K. Jensen, A.C. Pooley. T.P. Dutton, G.W. Schutte and H.E. Evans.** *1969.* Lesser spotted eagle in Zululand. Ostrich 40(1):24.

1665 **Jerdon, T.C.** *1871.* Supplementary notes to "The birds of India." Ibis 1(2):234-247.

1666 **Jesse, W.** *1897.* Birds' nesting in and around Lucknow. Additional notes taken in the season of 1896. Ibis (7th Ser.) 3(12):554-562.

1667 **Jesse, W.** *1901.* On a few additions to the birds of Lucknow. Ibis 1(4):597-607.

1668 **Jesse, W.** *1903.* A list of the birds of Lucknow. Ibis (8th Ser.) 3(9):49-81.

1669 **Jewett, S.G.** *1936.* Birds notes from Harney County, Oregon, during May 1934. Murrelet 17(2/3):41-47.

1670 **Jewett, S.G., W.P. Taylor, W.T. Shaw and J.W. Aldrich.** *1953.* Birds of Washington State. Univ. of Washington Press, Seattle. 767 pp.

1671 **Johansen, H.** *1957.* Die vogelfauna Westsibiriens. J. Ornithol. 98(3):262-278.

1672 **Johansen, H.** *1963.* Forelobig liste over fuglene pa Laeso. Dansk Ornithol. 57(3):143-154.

1673 **Johns, B.W.** *1977.* Golden eagle attempts to kill sandhill crane. Blue Jay 35(2):92-93.

1674 **Johnsgard, P.A.** *1979.* The breeding birds of Nebraska. Nebr. Bird Rev. 47(1):3-16.

1675 **Johnson, D. and J.H. Enderson.** *1972.* Roadside raptor census in Colorado - Winter 1971-72. Wilson Bull. 84(4):489-490.

1676 **Johnson, D.R. and W.E. Melquist.** *1973.* Unique, rare and endangered raptorial birds of northern Idaho; nesting success and management recommendations. Univ. of Idaho-U.S. Forest Service, Publ. No. R1-73-021.

1677 **Johnson, F.E.** *1879.* Capture of the golden eagle *(Aquila chrysaetos)* at Gravesend, L.I.. Bull. Nuttall Ornithol. Club 4(3):189.

1678 **Johnson, M.D.** *1959.* Bounty study. N. Dakota Game and Fish Dept., Proj. W-38-R-G, Job 11. 8 pp.

1679 **Johnson, N.K. and H.J. Peeters.** *1963.* The systematic position of certain hawks in the genus Buteo. Auk 80(4):417-446.

1680 **Johnson, W.B.** *1946.* Golden eagle's nest in eastern Washington. Murrelet 27(2):34.

1681 **Johnson, W.W.** *1949.* Texas eagle "chowser." Life (June 13):2,4,7,10,13.

1682 **Jollie, M.** *1957.* Comments on the bird genus *Aquila* and its occurrence in New Guinea and Australia. Nova Guinea (New Ser.) 8(2):179-181.

1683 **Jollie, M.** *1976.* A contribution to the morphology and phylogeny of the Falconiformes. (Part I.) Evol. Theory 1(9):285-298.

1684 **Jollie, M.** *1976.* A contribution to the morphology and phylogeny of the Falconiformes. (Part II.) Vol. Theory 2:115-208.

1685 **Jollie, M.** *1976.* A contribution to the morphology and phylogeny of the Falconiformes. (Part IV.) Evol. Theory 3:1-141.

1686 **Jollie, M.** *1976.* A contribution to the morphology and phylogeny of the Falconiformes. (Part III.) Evol. Theory 2:209-300.

1687 **Jollie, M.T.** *1943.* The golden eagle: its life history, behavior and ecology. M.S. Thesis. Univ. of Colorado, Boulder, CO. 227 pp.

1688 **Jollie, M.T.** *1947.* Plumage changes in the golden eagle. Auk 64(4):549-576.

1689 **Jollie, M.T.** *1957.* Avian anatomy and the anatomist. Condor 59(6):394-397.

1690 **Jones, H.** *1945.* Notes on some birds of the northern Rhodesia copperbelt. Ostrich 16(3):176-183.

1691 **Jones, P.V., Jr.** *1949.* Experimental management of antelope. Texas Game, Fish and Oyster Comm., F.A. Rep., Ser. No. 1-3. 31 pp.

1692 **Jonkel, G.M.** *1965.* South Dakota wintering eagle inventories. S.D. Bird Notes 17(3):61-62.

1693 **Jordania, R.G** *1961.* Ornitofauna Tbilisi i okolic. (Avifauna of Tbilisi and its vicinity). Acta Ornithol. 6(8):83-105. (In Russian, English Summ.)

1694 **Jordans, A., v.** *1928.* Steinadler in Westdeutschland. Ornithol. Monatsber. 36(3):84-85.

1695 **Jordans, A., von.** *1924.* Die ergebnisse meiner zweiten Reise nach Mallorca. III. Teil. J. Ornithol. 72(4):518-5369

1696 **Jordheim, S.** *1980.* Eagle/coyote incident. Blue Jay 38:47-48.

1697 **Jouard, H.** *1936.* Qu'en est-il des "aigles criards' de France? Alauda 8(2):198-219.

1698 **Jourdain, F.C.R.** *1910.* Aggressiveness of a buzzard in Westmoreland and a golden eagle in Invernesshire. Brit. Birds 4(5):152.

1699 **Jourdain, F.C.R.** *1911.* Extermination of the sea-eagle in Ireland. Brit. Birds 5(5):138-139.

Master No. **Citation**

1700 **Jourdain, F.C.R.** *1912.* Notes on the ornithology of Corsica. Ibis (9th Ser.) 6(21):63-82.

1701 **Jourdain, F.C.R.** *1913.* Corrections to the "Catalogue of collection of bird's eggs in the British Museum." Ibis 1(4):513-527.

1702 **Jourdain, F.C.R.** *1925.* Golden eagle in Berkshire. Brit. Birds 18(8):219-220.

1703 **Jourdain, F.C.R.** *1926.* Eggs of Palearctic eagles. Bull. Brit. Ool. Assoc. 1(9):89-95.

1704 **Jourdain, F.C.R.** *1926.* Status of the golden eagle in Ireland. Brit. Birds 19(8):211-212.

1705 **Jourdain, F.C.R.** *1928.* Brutdauer von *Aquila chrysaetos.* Ornithol. Monatsber. 36(6):175.

1706 **Jourdain, F.C.R.** *1930.* On the eggs of the golden eagle. Bull. Brit. Ool. Assoc. 2(24):142.

1707 **Jourdain, F.C.R.** *1934.* Moltoni on Italian ornithology. Ibis (13th Ser.) 4(2):414.

1708 **Jourdain, F.C.R., H.M. Wallis and F.R. Ratcliff.** *1915.* Notes on the bird-life of eastern Algeria. Ibis (10th Ser.) 3(1):133-169.

1709 **Jozefik, M.** *1969.* (Changes in avifauna of Bieszczady Mountain as the result of civilization factors influenced in the last quarter of the century.) Notatki Ornitol. 10(2/3):34-39. (In Russian)

1710 **Juillard, M., J.-C. Praz, A. Etournard and P. Beaud.** *1978.* Donnees sur la contamination des rapaces de Suisse romande et de leurs oeufs par les biocides organochlores, les PCB et les metaux lourds. Nos Oiseaux 34(5):189-206. (English Summ.)

1711 **Junge, G.C.A.** *1952.* Trekvogels 1948 t/m 1951. Ardea 40(1/2):83-93.

1712 **Junge, G.C.A.** *1954.* Waarnemingen van broedvogels en trekvogels in 1952 en 1953. Ardea 42(3/4):314-328.

1713 **Junge, G.C.A.** *1955.* Waarnemingen van broedvogels en trekvogels in 1954. Ardea 43(4):248-260.

1714 **Jussl, F. and T. Randla.** *1971.* Kaljukotkas ja merikotkas Eestis 1970. Eesti Loodus 14:222-223.

1715 **Kaiser, W. and H. Zimmerman.** *1969.* Ornithologische notizen aus der Lewitz (1968). Falke 16(5):166-173.

1716 **Kalitsch, L., von.** *1935.* Eine ornithologische studienreise nach Ostpreuben. Beitr. Fortpfl.-Biology Vogel 11(5):176-180.

1717 **Kalmbach, E.R., R.H. Imler and L.W. Arnold.** *1966.* The American eagles and their economic status, 1964. U.S.D.I. Fish and Wildlife Service. 35 pp.

1718 **Kappes, W., H. Riesch, J. Poltz and B. Diekhaus.** *1965.* Steinadler auf dem Durchzug in Hamburg. Ornithol. Mitt. 17(4):86-87.

1719 **Karaman, S.L.** *1948.* Ornithofauna of Vardar Park in Skopje. Larus 2:81-85. (English Trans.)

1720 **Karaman, S.L.** *1949.* Ornithofauna of the Shopska Kotlina Basin. Larus 3:155-220. (English Trans.)

1721 **Karczewski, Z.** *1965.* Golden eagle, *Aquila chrysaetos* (L.) in Masuria. Acta Ornithol. 9(1):155-156.

1722 **Karczewski, Z., J. Kopton and L. Tomialojc.** *1964.* Wycieczka ornitolgiczna na jezioro Druzno. (An ornithological trip to Lake Druzno (distr. Elblag.) Acta Ornithol. 8(8):317-323. (English Summ.)

1723 **Kasparson, G.R.** *1958.* (Food of some diurnal birds of prey in Latvia SSR.) Zool. Zhur. 37(9):1389-1396. (In Russian, English Summ.)

1724 **Kasparson, G.R.** *1970.* Sovremennoe sostoyanie fauny khistchnykh ptits Latvii. (Recent status of raptor fauna in Latvia.) 7th Pribalt. Ornithol. Konf., Materialy 2:99-103.

1725 **Kawata, K.** *1973.* Note on an American golden eagle *Aquila chrysaetos canadensis* hatched in captivity. Int. Zoo. Yearb. 13:114-115.

1726 **Kawata, K.** *1974.* Further notes on the breeding of the American golden eagle, *Aquila chrysaetos canadensis,* at Topeka Zoological Park. Zool. Garten 44(4):211-215.

1727 **Kayser, C.** *1915.* Ein unter eigentumlichen umstanden gefangener steinadler, *Aquila chrysaetos* (L.). Vogelwelt 35(5):224-225.

1728 **Kayser, C.** *1917.* Uber das fruhere horsten des steinadlers im Riesengebirge. Ornithol. Monatsber. 25(10):137-141.

1729 **Kazama, T.** *1973.* On the distribution and protection of *Aquila chrysaetos* in Niigata Prefecture. Yamashina Inst. for Ornithol. Misc. Reports 7(1):104-111

1730 **Keith, J.A. and I.M. Gruchy.** *1972.* Residue levels of chemical pollutants in North American birdlife. Pages 438-452. *in:* E.H. Voous, Ed. Proc. of the XVth Int. Ornithol. Cong., 30 August - 5 September, 1970. The Hague.

1731 **Kelemen, A., B.J. Kiss and I. Sterbetz.** *1976.* Observations on the autumn migration in Dobrudzha. Aquila 83:267-279.(In Hungarian, English Summ.)

1732 **Kelleher, J.V. and W.F. O'Malia.** *1971.* Golden eagle attacks a mallard. Auk 88(1):186.

1733 **Kemp, A. and M. Kemp.** *1974.* Observations on the buffalo weaver. Bokmakierie 26(3):55-58.

1734 **Kemp, A. and J. Mendelsohn.** *1975.* What colour is Wahlberg's eagle? Bokmakerie 27(4):72-74.

1735 **Kemp, A.C.** *1974.* The distribution and status of the birds of the Kruger National Park. Koedoe Suppl. Natl. Park Board of Trustees, Pretoria. 352 pp.

1736 **Kemp, A.C.** *1977.* Some marking methods used on a variety of southern African raptors. Safring 6(1):38-43.

1737 **Kemp, A.C. and M.I. Kemp.** *1977.* The status of raptorial birds in the Transvaal Province of South Africa. Pages 28-34. *in:* R.D. Chancellor, Ed. Proc. of the World Conf. on Birds of Prey. 1-3 October, 1975. Vienna. I.C.B.P.

1738 **Kennedy, A.W.M.C.** *1874.* Notes on the avifauna of the Desert of Sinai and of the Holy Land. Ibis 4(14):107-112.

1739 **Kennedy, C.A.** *1948.* Golden eagle kills bighorn lamb. J. Mammal. 29(1):68-69.

1740 **Kennedy, C.H.** *1917.* Some factors in the nesting habit of birds. Condor 19(3):87-93.

1741 **Kennedy, P.L.** *1980.* Raptor baseline studies in energy development. Wild. Soc. Bull. 8(2):129-135.

1742 **Kennell, J.** *1971.* Group acts to stop eagle kill. Pac. Search 5(6):6.

1743 **Kerlinger, P. and M. Bennett.** *1977.* Hawk migration at Oneonta. Kingbird 27(2):74-79.

1747 **Keve, A.** *1948-1951.* A magyar madartani intezet 1933-1950. Evi Madarjelolesei. Aquila 55-58:89-107.

1744 **Keve, A.** *1955.* Ornithological notes from the "postza' of Kunkapolnas on the Hortobagy. Aquila 59-62:461-464.

1745 **Keve, A.** *1955.* Uncommon visitors around the Lake Balaton in the years 1952-54. Aquila 59-62:439-440.

1746 **Keve, A.** *1959.* XXI. Gyuruzesi jelentes. (21st bird-banding report.) Aquila 66:201-210.

1748 **Keymer, I.F.** *1972.* Diseases of birds of prey. Vet. Record 90(21):579-594.

Master No. **Citation**

1749 **Khakhloff, V.A.** *1936.* Les oiseaux de la steppe de Kouznetzk et du Salair. Gerfaut 26(3):201-211.

1750 **Kharchenko, V.I. and V.A. Minoranski.** *1967.* (Recent distribution of *Aquila nipalensis* in the European part of USSR.) Zool. Zhur. 46(6):958-960. (In Russian)

1751 **Khusainov, A.** *1958.* O znachenii khistchnykh zverey i ptits v Alakulskom ondatrovom khozyaistve. (Significance of predatory mammals and birds in Alakul ondatra management area.) Trudy Inst. Zool. AN Kazakh. SSR. 10:206-240. (In Russian)

1752 **Kiff, L. and K. Axelson.** *1977.* A raptor survey of the Angeles National Forest. Audubon Imprint 2(4):1-7.

1753 **Kikkawa, J., I. Hore-Lacy and J. Le Gay Brereton.** *1965.* A preliminary report on the birds of the New England National Park. Emu 65(2):139-143.

1754 **Kingery, H.E. and W.D. Graul (Eds.).** *1978.* Colorado bird distribution latilong study. Colorado Field Ornithol. 58 pp.

1755 **Kiraly, I.** *1931.* A kis bekaszo *(Aquila pomarina* Brehm) feszkelese hazankban egykor es most. (On the past and present breeding of *Aquila pomarina* in Hungary.) Kocsag 3:89-95. (In Hungarian, German Summ.)

1756 **Kirk, D.** *1975.* California raptor survey. Calif. Dept. of Fish and Game, Nongame Wildlife Investigations. Proj. W-54-R-7, Job III- 1.0. 8 pp.

1757 **Kish, F.** *1970.* Egg laying and incubation by American golden eagles *Aquila chrysaetos canadenis* at Topeka Zoo. Int. Zoo. Yearb. 10:26-29.

1758 **Kish, F.** *1972.* First breeding by American golden eagles *Aquila chrysaetos canadensis* at Topeka Zoo. Int. Zoo Yearb. 12:136-138.

1759 **Kittenberger, K.** *1958.* My ornithological collecting expeditions in East-Africa I. Aquila 65:13-37.

1760 **Kiyosu, Y.** *1937.* (On the breeding habits of *Aquila chrysaetos japonica* in Hondo.) Tori 9(44):301-306. (In Japanese)

1761 **Kiyosu, Y.** *1952.* The Birds of Japan. Vol. II Cypseli to Limicolae. Published in Japan. 785 pp.

1762 **Kjaer, T.** *1975.* Mellem rovfugle og storke ved Bosporous. Dansk Fugle 27(3/4):92-99.

1763 **Klammet, G.** *1952.* Mein adlertagebuch. Columba 4(1/2):3-7.

1764 **Klebenow, D.A. and K. McAdoo.** *1976.* Predation on domestic sheep in northeastern Nevada. J. Range Manage. 29(2):96-100.

1765 **Kleinschmidt, O.** *1900. Aquila fluvescens* Gray? in Europa. Ornithol. Monatsber. 8(5):65-66.

1766 **Kleinschmidt, O.** *1924.* Bemerkungen uber schreiadler. Falco 20(1):13-14.

1767 **Kleinschmidt, O.** *1931.* Verwechselung von kaiser- und steppenadler. Falco 27(2):11.

1768 **Knauthe, R.** *1891.* Steinadler. Vogelwelt 16(6):183.

1769 **Knight, C.W.R.** *1927.* The golden eagle. Country Life 61(1576):516-518.

1770 **Knight, C.W.R.** *1943.* All British Eagle. Hodder & Stoughton, London.

1771 **Knight, R.L.** *1978.* Ecology of the Great Basin Canada goose on the Columbia River in north-central Washington. M.S. Thesis. Univ. of Washington, Seattle, WA.

1772 **Knight, R.L. and A.W. Erickson.** *1978.* Marmots as a food source of golden eagles along the Columbia River. Murrelet 59(1):28-30.

1773 **Knight, R.L., J.B. Athearn, J.J. Brueggman and A.W. Erickson.** *1979.* Observations on wintering bald and golden eagles on the Columbia River, Washington. Murrelet 60(3):99-105.

1774 **Knight, T.** *1977.* Watching raptors in winter. Wash. Wildl. 29(3):9-13.

1775 **Kochert, M.N.** *1972.* Population status and chemical contamination in golden eagles in southwestern Idaho. M.S. Thesis, Univ. of Idaho, Moscow, Idaho. 115 pp.

1776 **Kochert, M.N.** *1973.* Density, food habits, and reproductive performance of raptors in the Snake River Birds of Prey Natural Area. U.S.D.I. Bureau of Land Management, Snake River Birds of Prey Research Proj., Ann. Prog. Rep. 1973. 54 pp.

1777 **Kochert, M.N.** *1980.* Golden eagle reproduction and population changes in relation to jackrabbit cycles: implications to eagle electrocutions. Pages 71-86. *in:* R.P. Howard and J.F. Gore, Eds. Proc. of A Workshop on Raptors and Energy Developments. 25-26 January. Boise, Id. Bonneville Power Admin., U.S. Fish and Wildlife Service, Idaho Power Comp. and Idaho Chapter of the Wildl. Soc.

1778 **Kochert, M.N., A.R. Bammann, R.P. Howard, J.H. Doremus, M. De Late and D.D. Donohue.** *1975.* Reproductive performance, food habits and population dynamics of raptors in the Snake River Birds of Prey Natural Area. *in:* Snake River Birds of Prey Research Proj. Ann. Rep., U.S.D.I. Bureau of Land Management, Boise, ID. p. 1-50.

1779 **Kochert, M.N., A.R. Bammann and K. Steenhof.** *1979.* Reproductive performance, food habits and population dynamics of raptors in the Snake River Birds of Prey Natural Area. U.S.D.I. Bureau of Land Management, Snake River Birds of Prey Res. Proj., Annual Report. p. 1-31.

1780 **Kochert, M.N., A.R. Bammann, J.H. Doremus, M. De Late and J. Wyatt.** *1976.* Reproductive performance, food habits and population dynamics of raptors in the Snake River Birds of Prey Natural Area. *in:* Snake River Birds of Prey Research Proj. Ann. Rep., U.S.D.I. Bureau of Land Management, Boise, ID. p. 1-56.

1781 **Kochert, M.N., A.R. Bammann, K. Steenhof, J.H. Doremus, M. De Late, J. Oakley and T. Hamer.** *1977.* Reproductive performance, food habits and population dynamics of raptors in the Snake River Birds of Prey Natural Area. *in:* Snake River Birds of Prey Research Proj. Ann. Rep., U.S.D.I. Bureau of Land Management, Boise, ID. p. 1-39.

Master No.	Citation
1782	**Kochert, M.N., A.R. Bammann and K. Steenhof.** *1980.* Reproductive performance, food habits and population dynamics of raptors in the Snake River Birds of Prey Natural Area. (Study 1.) *in:* Snake River Birds of Prey Research Proj. Ann. Rep., U.S.D.I. Bureau of Land Management, Boise, ID. p. 1-21.
1783	**Koelz, W.** *1937.* Notes on the birds of Spiti, a Himalayan Province on the Punjab. Ibis 1(1):86-104.
1784	**Koenig, A.** *1892.* Zweiter beitrag zur avifauna von Tunis. J. Ornithol. 40(4):329-416.
1785	**Koenig, A.** *1907.* Die falconiden Aegyptens. J. Ornithol. 55(3):391-469.
1786	**Koepert, O.** *1896.* Die vogelwelt des Herzogtums Sachsen-Altenburg. J. Ornithol. 44(2):217-248.
1787	**Koersveld, E., van.** *1954.* Rarae aves. Ardea 42(3/4):281-313.
1788	**Koford, C.B.** *1958.* Prairie dogs, whitefaces and blue grama. Wildl. Monogr. No. 3. 78 pp.
1789	**Kohl, I.** *1959.* Bird victims of a hailstorm. Aquila 66:325.
1790	**Kolazy, J.** *1881.* Die vogelparsiten. Mitt. Ornithol. Vereines in Wein 5(1):89-90.
1791	**Kolesnikov, A.D.** *1960.* O khistchnykh ptitsakh lesov Onpropetrovstchiny district. (Birds of prey in Onepropetrovsk district.) Nauchn. Zapiski Onepropetrovsk Univ. 62:29-349 (In Russian)
1792	**Kolesnikov, A.D.** *1979.* Khistchnye ptitsi i sovy lesov yugo-vostoka Ukrainy. (Birds of prey and owls of forests in southeastern Ukraine.) Ornithologia 14:77-82.
1793	**Koller, A.** *1892.* Steinadler *(A. fulva)* in Oberosterreich geschossen. Ornithol. Jahrbuch 3(5):205.
1794	**Kollibay, P.** *1881. Vultur Cinereus* und *Aquila Fulva* inSchlesien. Ornithol. Centralblatt 6(5):34.
1795	**Kondla, N.G., H.W. Pinel, C.A. Wallis and C.R. Wershler.** *1973.* Avifauna of the Drumheller Area, Alberta. Can. Field-Nat. 87(4):377-393.
1796	**Koning, F.J.** *1976.* Notes on the birds of prey in the Indus Valley. J. Bombay Nat. Hist. Soc. 73(3):448-455.
1797	**Kostin, Y.V.** *1978.* Ptitsi i zveri Kryma. (Birds and mammals of Crimea.) Tavria, Simferopol. p. 1-110.
1798	**Kovshar, A.F.** *1966.* Ptitsi Talasskogo Alatau. (Birds of Talass Ala Tau.) Trudy Gos. Zapovednika Aksu-Dzhabagly. 3:1-435. (In Russian)
1799	**Kozlov, N.P.** *1959.* O poleznykh khistchikakh. (Useful raptors.) Priroda 7:50-52. (In Russian)
1800	**Kozlova, E.V.** *1932.* The birds of south-west Transbaikalia, northern Mongolia, and central Gobi. Part II. Ibis 2(3):405-438.
1801	**Kozlova, E.V.** *1975.* Ptitsi zonalnykh stepei i pustun tsentralnoi Azii. (Birds of zonal steppe and deserts in central Asia.) Nauka, Leningrad. p. 1-252.
1802	**Krampitz, H.E.** *1956.* Die brutvogel Siziliens. J. Ornithol. 97(3):310-334.
1803	**Krampitz, H.E.** *1958.* Wieteres uber die brutvogel Siziliens. J. Ornithol. 99(1):39-58.
1804	**Kratzig, H.** *1943.* Beitrage zur vogelkunde der Krim. J. Ornithol. 91(2/3):268-285.
1805	**Kraus, M. and P. Conradty.** *1965.* Zur kenntnis der vogelwelt Nordgriechenlands. Anz. Ornithol. Ges. Bayerns 7(4):475-485.
1806	**Krienke, W.** *1929.* Observations on some of the raptores breeding in the Beatrice District, S. Rhodesia. Ool. Rec. 9(2):31-35.
1807	**Krpan, M.** *1953.* Contribution to the study of the avifauna of the immediate environs of Split. Larus 6-7:113-115. (English Trans.)
1808	**Krpan, M.** *1954.* The beginnings of ornithology in Dalmatia. Larus 8:77-86. (English Trans.)
1811	**Krpan, M.** *1955-1956.* Birds of the Mosor Mountain and its environs. Larus 9-10:107-122.
1809	**Krpan, M.** *1959.* Contribution to the study of the birds of the environs of Split. Larus 12-13:61-83. (English Trans.)
1810	**Krpan, M.** *1960.* Notes on the behavior of birds. Larus 14:131-137. (English Trans.)
1812	**Krumm, K.** *1953.* Golden eagle decimates prairie dog colony. S. Dakota Bird Notes 5(3):48.
1813	**Kruper, T.** *1852.* Die adler pommerns. Naumannia 2(1):61-76.
1814	**Kruper, T.** *1853.* Die adler pommerns. Naumannia 3:39-46.
1815	**Krylov, D.G.** *1965.* (Character of distribution of birds of prey on Sarydzhazskie syrtys in central Tyan Shan.) Ornitologia 7:203-208. (In Russian)
1816	**Kubantsev, B.S.** *1959.* (Steppe raptors and the role of rodents in their feeding in the steppe of west Kazakhstan.) Uchenye Zapiski Stalingrad Pedagog. Inst. 9:179-180. (In Russian)
1817	**Kucera, L.** *1970.* Die vogel des mittleren Bohmerwaldes (Sumava). Ornithol. Mitt. 22(11):223-242.
1818	**Kuchin, A.P.** *1959.* K biologii bolshogo podorlika v Biiskoi lesostepi. (Biology of *Aquila clanga* in Byisk forest-steppe.) 2 Vsesoyuzn. Ornithol. Konf., Tezisy Dolk., Chast 3, Moscow. p. 75-80. (In Russian)
1819	**Kuchin, A.P.** *1961.* (Ecology of some raptors between the rivers Biya and Katun.) Zool. Zhur. 40(5):730-735. (In Russian, English Summ.)
1820	**Kuhk, R.** *1968.* (The status and protection of diurnal birds of prey in West Germany and future requirements.) Angew. Ornithol. 3(2):59-65. (In German)
1821	**Kulczycki, A.** *1964.* Wystepowanie nicktorych gatunkow ptakow w Bieszezadach w latach 1957-1963. Acta Ornithol. 8(7):324-325.
1822	**Kumari, E.** *1966.* The utilization and protection of the wild birds of Estonia. Ornithol. Researches in Estonia. p. 34-39.
1823	**Kumerloeve, H.** *1958.* Ein adler der *clanga/pomarina* - Gruppe bei Osnabruck. Ornithol. Mitt. 10(3):53.
1824	**Kumerloeve, H.** *1961.* Notes on a small bird-collection of A.E. Day in Beirut (Lebanon). Alauda 29(2):146.
1825	**Kumerloeve, H.** *1966.* Liste systematique revisee des especes d'oiseaux de Turquie. Alauda 34(3):165-186. (English Summ.)
1826	**Kumerloeve, H.** *1968.* Recherches sur l'avifaune de la Republique Arabe Syrienne essai d'un apercu. Alauda 36(1/2):1-26.
1827	**Kumerloeve, H.** *1972.* Liste comparee des oiseaux nicheurs de Turquie Meridional. Syrie et Liban. Alauda 40(4):353-366.
1828	**Kumerloeve, H. and G. Niethammer.** *1935.* Beitrage zur kenntnis der avifauna kleinasiens (Paphlagonien-Galatien). J. Ornithol. 83(1):25-75.
1829	**Kumlien, L. and N. Hollister.** *1903.* Birds of Wisconsin. Bull. Wisc. Nat. Hist. Soc. 3:1-143.
1830	**Kummerloeve, H. and G. Niethammer.** *1935.* Einige beobachtungen uber fruhjahrsdurchzug am Bosporus. Vogelzug 6(1):6-10.
1831	**Kuntzendorff, J.** *1932.* Importations et elevages d'oiseaux en Allemagne. Oiseau Rev. Fr. Ornithol. 2(1):176-189.

Master No.	Citation
1832	**Kunz, W.** *1958.* Ein schelladler *(Aquila clanga)* im Kreise Besenbruck. Ornithol. Mitt. 10(7):134.
1833	**Kurth, H. and D. Kurth.** *1968.* Ornithologische beobachtungen am Neusiedlersee. Ornithol. Mitt. 20(1):13-14.
1834	**Kurth, H. and D. Kurth.** *1967.* Rastender schelladler *(Aquila clanga)* bei Hamburg. Ornithol. Mitt. 19(8):173.
1835	**Kuschel, M.** *1895.* Abrifs einer beschreibung von vogeleiern der athiopischen Ornis. J. Ornithol. 43(1):80-98.
1836	**Kutschin, A.** *1959.* Zur biologie des schelladlers in der Bijsker Waldsteppe. Thesen 2 Ornithol. Allunions-Konf. 3. p. 75-76. (Title Only)
1837	**Kuyt, E.** *1965.* Golden eagle. Blue Jay 23(4):191.
1838	**Kuyt, E.** *1967.* Two banding returns for golden eagle and peregrine falcon. Blue Jay 25(2):70.
1839	**Kuyt, E.** *1974.* Golden-eagles attack white-fronted geese. Blue Jay 32(4):227-228.
1840	**L'Hermite, J.** *1936.* Complement a la "Contribution a l'etude ornithologique de la Provence." Alauda 8(3/4):316-325.
1841	**L'Hermite, J.L.** *1920.* L'aigle imperial a Marseille. Oiseau Rev. Fr. Ornithol. 6(140):153-154.
1842	**L'Hermitte, J.** *1915.* Contribution a l'etude ornithologique de la Provence. Oiseau Rev. Fr. Ornithol. 4(79):161-166.
1843	**La Touche, J.D.** *1892.* On birds collected or observed in the vicinity of Foochow and Swatow in south-eastern China. Ibis (6th Ser.) 4(16):477-503.
1844	**La Touche, J.D.** *1907.* Field-notes on the birds of Chinkiang, Lower Yangtse Basin. Part III. Ibis (9th Ser.) 1(1):1-30.
1845	**La Touche, J.D.** *1920.* Notes on the birds of north-east Chihli, in North China. Part II. Ibis (11th Ser.) 2(4):880-920.
1846	**La Touche, J.D.** *1922.* On the birds of the Province of Hupeh in Central China. Ibis (11th Ser.) 4(3):445-471.
1847	**La Touche, J.D.** *1934.* A handbook of the birds of eastern China. Vol. II. Taylor and Francis, London. 566 pp.
1849	**Labuschagne, R.J.** *1959.* Birds of the Kalahari Gemsbok National Park. Ostrich (Suppl.) 3:86-95.
1850	**Labutin, Y.V.** *1958.* Materialy po ornithofaune khistchnikov Yanskogo pluskogoria. (Data on raptor avifauna of Yana plateau.) Nauchn. Soobstchenia Yakutsk. Filiala Sibirsk. Otdel. An SSSR. 1:161-166. (In Russian)
1851	**Labutin, Y.V.** *1962.* Materialy po ekologii berkuta v mezhkhrebtovoi depressii basseina Yany. (Data on the ecology of *Aquila chrysaetos* in a region of Yana basin between ridges.) 3 Vsesoyuzn. Ornithol. Konf., Materialy, 2 Kniga, Lvov. p. 67-68. (In Russian)
1852	**Labutin, Y.V.** *1963.* Deyatelnost cheloveka i chislennost nekotorykh vidov khistchnykh ptits Verkhoyanie. (Man's activity and the number of some raptors in Verkhoyansk.) Problemy Okhrany Prirody Yakutii, Yakutsk. p. 141-143. (In Russian)
1853	**Labutin, Y.V.** *1974.* (Characteristics of territorial and trophic relations between populations of some birds of prey and owls in northeastern Yakutia.) Biology Problemy Severa, 6 Sympos., Tezisy Doklady, Yakutsk. 1:133-138. (In Russian)
1854	**Lacey, H.** *1912.* Additions to birds of Kerrville, Texas. Auk 29(2):254.
1855	**Lack, D.** *1940.* Courtship feeding in birds. Auk 57(2):169-178.
1856	**Lack, D.** *1946.* Competition for food by birds of prey. J. Anim. Ecol. 15(2):123-129.
1857	**Lack, P.C., W. Leuthold and C. Smeenk.** *1980.* Check-list of the birds of Tsavo East National Park, Kenya. J. E. African Nat. Hist Soc. 170:1-25.
1858	**Laferrere, M.** *1955.* Un apparition insolite: l'aigle royal en Dombes. Alauda 23(4):286-287.
1859	**Lakhanov, Z.** *1969.* Materialy po biologii berkuta- *Aquila chrysaetos fulva* L. - v yugo-zapadnykh Kyzylkumakh. (Data on the ecology of *Aquila chrysaetos* in southwestern Kyzyl-Kum.) Materialy 25 Nauchn Konf. Prof. - Prepod. Sostava Biology Facult. p. 12-14. (In Russian)
1860	**Lambert, A.** *1957.* A specific check list of the birds of Greece. Ibis 99(1):43-68.
1861	**Lambrecht, K.** *1912.* Die fossilen vogel Ungarns. Aquila 19:288-320.
1862	**Lambrecht, K.** *1917.* Die ausbildung und geschichte der Europaischen vogelwelt. Aquila 24:209-221.
1863	**Langhorne, E.J.B.** *1978.* The golden eagle *Aquila chrysaetos* . Hawk Trust Ann. Report 9:23-27.
1864	**Lankin, P.M.** *1976.* K raspredeleniyu stepnogo orla v nizoviyakh Syrdari. (Distribution of *Aquila nipalensis* in lower Syr Darya.) Ornitologia 12:235-236. (In Russian)
1865	**Lano, A.** *1897. Aquila chrysaetos* in central Minnesota. Auk 14:317.
1866	**Lano, A.** *1922.* Golden eagle *(Aquila chrysaetos)* and porcupine. Auk 39(2):258-259.
1867	**Lansell, G.L.** *1937.* A few more notes on wedge-tailed eagles and notes on ravens. Emu 36(3):246-247.
1868	**Lardy, M.E.** *1980.* Raptor inventory and ferruginous hawk breeding biology in southeastern Oregon. M.S. Thesis. Univ. of Idaho, Moscow, ID. 48 pp.
1869	**Larsen, F.** *1966.* Forarstreak af rovfugle ved koge sonakke med iagttagelse af stor skrigeorn *(Aquila clanga),* 1962. (The breeding birds on Nord Fuglay in northern Norway.) Dansk Ornithol. 59(3/4):261-267.
1870	**Lashmar, A.F.C.** *1946.* Birds noted on the eastern portion of Kangaroo Island. S. Aust. Ornithol. 18(3):28-30.
1871	**Laszlo, M.** *1951.* Occurrences of *Aquila ch. chrysaetos* L. - golden eagle. Aquila 55-58:281.
1872	**Laubmann, A.** *1950.* Der steinadler. Columba 2(2):41.
1873	**Laurent, G. and B. Mouillard.** *1939.* Les oiseaux de la region de Saint-Die. Alauda 11(2/3/4):104-174.
1874	**Lavauden, L. and H. de Balsac.** *1925.* Sur un aigle de la collection d'hamonville et sur la question des petits aigles. Oiseau. Rev. Fr. Ornithol. 9(192):81-88.
1875	**Laver, H.** *1891.* Spotted eagle *(Aquila naevia)* at Elmstead and Leigh: an addition to the Essex fauna. Essex Nat. 5:218-219.
1876	**Laver, H.** *1891.* Capture of a spotted eagle near Colchester. Zoologist 1891:470.
1877	**Lavery, H.J. and N. Hopkins.** *1963.* Birds of the Townsville District of Northern Queensland. Emu 63(3):242-252.
1878	**Layard, E.L.** *1869.* Further notes on South African ornithology. Ibis 5(17):68-79.
1879	**Laycock, G.** *1973.* Saving western eagles from traps and zaps. Audubon 75(5):133.
1848	**LaFontaine, A.R. and J.H. Fowler.** *1976.* Golden eagle preys on osprey. Auk 93(2):390.
1885	**Lea, D. and W.R.P. Bourne.** *1975.* The birds of Orkney. Brit. Birds 68(7):261-283.
1886	**Leakey, L.S.B.** *1944.* Eggs of *Aquila verreauxi.* J. E. African Nat. Hist. Soc. 18(1/2):95-96.

Master No.	Citation
1887	**Lebeurier, E. and J. Rapine.** *1934.* Ornithologie de la Basse-Bretagne. Oiseau Rev. Fr. Ornithol. 4(4):659-702.
1888	**Lechner, F.** *1967.* Steinadler, *Aquila chrysaetos,* kropft im Fluge. Anz. Ornithol. Ges. Bayern 8(1):68-69.
1889	**Lee, J.M., Jr.** *1980.* Raptors and the BPA Transmission system. Pages 41-55. *in:* R.P. Howard and J.F. Gore, Eds. Proc. of A Workshop on Raptors and Energy Developments. 25-26 January. Boise, Id. Bonneville Power Admin., U.S. Fish and Wildlife Service, Idaho Power Comp. and Idaho Chapter of the Wildl. Soc.
1890	**Lee, R.D.** *1954.* First report of the poultry bug from a golden eagle's nest, with new locality records. Econ. Entomol. 47(6):1144.
1891	**Lees, R.H.** *1935.* Observations on the breeding habits of some Accipitres in northern Rhodesia. Ostrich 6(2):89-94.
1892	**Lees, S.G.** *1968.* Notes on the nesting of Wahlberg's eagle *Aquila wahlbergi.* Ostrich 39(3):192-193.
1893	**Lees, S.G.** *1968.* The birds of Mutemwe. Ostrich 39(1):28-29.
1894	**Lefevre, R.H.** *1929.* Birds of China. Auk 46(4):494-501.
1895	**Legendre, M.** *1927.* L'ornithologie dans le departement de la Sarthe. Oiseau Rev. Fr. Ornithol. 11(215):98-115.
1896	**Lehti, R.W.** *1947.* The golden eagle attacking antelope. J. Wildl. Manage. 11(4):348-349.
1897	**Leinster, C.** *1971.* Sheepmen vs. eagles- slaughter in the sky. Life 71(8):36-37.
1898	**Lemoyne, A.** *1886.* Notes on some birds of the Great Smoky Mountains. Ornithol. and Ool. 11(10):147-148.
1899	**Leopold, A.** *1942.* "Control" of the golden eagle in Texas. Wilson Bull. 54:218.
1900	**Leopold, A.S. and T.O. Wolfe.** *1970.* Food habits of nesting wedge-tailed eagles, *Aquila audax,* in south-eastern Australia. CSIRO Wildl. Res. 15(1):1-17.
1901	**Leshem, Y.** *1979.* "Guarding forces" of golden eagles. Saleit 7(11):7-10. (In Hebrew)
1902	**Leshem, Y.** *1979.* How we protected a nest of the golden eagle. Teva Vaaretz 21:206-212. (In Hebrew)
1903	**Leshem, Y.** *1980.* Golden eagles in our backyard. Israel- Land and Nature 5(2):70-75.
1904	**Leshem, Y.** *1981.* Israel's raptors. The Negev and Judean Desert. Hawk Trust, Ann. Rep. 11:30-35.
1905	**Leuthold, W. and B. Leuthold.** *1972.* Blutschnabelweber *(Quelea quelea)* als beute von greif- und stelzvogln. Vogelwarte 26(4):352-354.
1906	**Leveque, R.** *1960.* Observaciones sobre aguila imperial en Donana, Abril-Mayo 1959. Ardeola 6(2):381.
1907	**Leveque, R. and R. Prodon.** *1976.* A propos de la collection d'oiseaux sauvy deposee av Laboratoire Arago. Alauda 44(4):431-439.
1908	**Leverkuhn, P.** *1907.* Kaiseradler und aasgeier am Horst. Pages 218-230. *in:* Proc. 4th Int. Ornithol. Cong., London. 1905.
1909	**Levin, A.S. and B.M. Gubin.** *1978.* Gnezdovye nakhodki bolshogo podorlika i zelenushki v Kazakhstane. (Nesting of *Aquila clanga* and *Chlorix chlorix* in Kazakhstan.) Materialy Konf. Molodych Uchenych Inst. Zool., Alma Ata. p. 35-37.
1910	**Lewis, A.D.** *1981.* Raptor and stork migration at Namanga, southern Kenya. Scopus 5(3):83-84.
1911	**Lewis, A.D. and D.J. Pearson.** *1981.* Stork and raptor migration in south Nyanza, Kenya. Scopus 5(2):51.
1880	**LeFur, R.** *1975.* Notes sur l'avifaune Algerienne. Alauda 43(3):317-319.
1881	**LePrince, P., G. Dandrifosse and E. Schoffeniels.** *1979.* The digestive enzymes and acidity of the pellets regurgitated by raptors. Biochem. Syst. Ecol. 7(3):223-227.
1882	**LeRoi, O.** *1923.* Die ornis der Sinai-Halbinsel. J. Ornithol. 71(1):28-95.
1883	**LeSouef, A.S.** *1925.* Size of the wedge-tailed eagle *(Uroaetus audax).* Emu 25(1):40.
1884	**LeSouef, D.** *1900.* Reports. Victorian Nat. 17(3):39.
1912	**Lheritier, J.N.** *1975.* Les rapaces diurnes du Parc National des Cevennes. Mem. Trav. E.P.H.E., Inst. Montpeller No. 2. 143 pp.
1913	**Lichtenstein, H.** *1853.* Eine seltene varietat des schreiadlers, *Falco naevius* Lin. J. Ornithol. 1:69-72.
1914	**Ligon, J.S.** *1961.* New Mexico birds, and Where to Find Them. Univ. of New Mexico Press. 360 pp.
1915	**Likhachev, G.N.** *1957.* Ocherk gnezdovania krupnykh dnevnykh khistchnykh ptits v shirokolistvennom lesu. (Description of nesting of large raptors in a deciduous forest.) Trudy 2 Pribalt. Ornithol. Konf. p. 308-336.
1916	**Lilford, L.** *1865.* Notes on the ornithology of Spain. Ibis 1(2):166-177.
1917	**Lilford, L.** *1866.* Notes on the ornithology of Spain. Ibis 2(6):173-187.
1918	**Lilford, L.** *1866.* Notes on the ornithology of Spain. Ibis (New Ser.) 2(8):377-392.
1919	**Lincoln, F.C.** *1925.* Some results of bird banding in Europe. Auk 42(3):358-388.
1920	**Lindeman, G.V.** *1971.* Ptitsi iskusstvennykh lesnykh nasazhdeni v glinistoi polupustyne severnogo Prikaspiya. (Birds of artificial forests in the semi-desert north of the Caspian Sea.) Zhivotnye Iskusstvennykh Lesnykh Nasazhdennii V Glinistoi Polupustyne, Moscow. p. 120-151.
1921	**Lindeman, G.V.** *1977.* Stepnol orel v Volzhsko-Uralskom mezhdurechie. (*Aquila nipalensis* between the Volga and Ural Rivers.) 7th Vsesoyuzn Ornithol. Konf., Materialy, Kiev. 2:226-228. (In Russian)
1922	**Linsdale, J.M.** *1930.* Problems of bird conservation in California. Condor 32(2):105-115.
1923	**Linsdale, J.M.** *1931.* Facts concerning the use of thallium in California to poison rodents - its destructiveness to game birds, song birds and other valuable wildlife. Condor 33(3):92-106.
1924	**Lintia, D.** *1916.* Materialien zur avifauna Serbiens. Aquila 23:74-162.
1925	**Lintia, D.** *1951.* *Aquila h. heliaca* Sav. - imperial eagle-breeding in Banat. Aquila 55-58:281.
1926	**Lintner, G.A.** *1883.* Capture of the golden eagle at Albany, N.Y. Bull. Nuttall Ornithol. Club 8(2):123-124.
1927	**Lish, J.W.** *1974.* Apparent predation by golden eagle at great blue heron colony. Bull. Oklahoma Ornithol. Soc. 7(1):4.
1928	**Lish, J.W.** *1975.* Status and ecology of bald eagles and nesting of golden eagles in Oklahoma. M.S. Thesis. Oklahoma State Univ., Stillwater, OK. 99 pp.

Master No.	Citation
1929	**Litvinov, N.I. and T.N. Gagina.** *1977.* Ptitsi ostrova Olkhon. (Birds of Olkhon Island.) Ecologia Ptits Vostochnoi Sibiri, Irkutsk. p. 176-188.
1930	**Llewellyn, L.C.** *1973.* The use of snares for bird trapping. Aust. Bird Band. 11(2):30-32.
1931	**Lloyd, G.** *1970.* Birds of Prey. Grosset and Dunlap Publ., New York. 159 pp.
1932	**Lloyd, J.H.** *1873.* On the birds of the Province of Kattiawar in western India. Ibis 3(12):397-421.
1933	**Lobachev, V.S.** *1959.* K biologii khistchnyh ptits Zaaraliya. (Biology of raptors eastward of the Aral Sea.) 2nd Vsesoyuzn Ornithol. Konf., Tezisy Dokl., Moscow. 3:31-32.
1934	**Lobachev, V.S.** *1960.* Materialy po biologii orlamogilnika v severo-vostochnom Priaralie. (Data on the biology of *Aquila heliaca* northeast of the Aral Sea.) Ornithologia 3:306-324.
1935	**Lobachev, V.S.** *1961.* O prostranstvennom razmestchenii vida vnutri areala i rasprostranenii orla-mogilnika. (Range and territorial distribution of *Aquila heliaca* .) 3rd Vsesoyuzn Nauchn. Konf. Molodykh Uchenykhbiologov, 1 Moscow Univ. p. 40-43.
1936	**Lobachev, V.S.** *1967.* Pitanie orla-mogilnica v severo-vostochnom Priaralie. (Food of *Aquila heliaca* northeast of the Aral Sea.) Ornithologia 8:366-369.
1937	**Lobkov, E.C.** *1978.* Kratkii obzor ptits tsentralnykh raionov vostochnoi Kamchatki. (Brief summary of birds in central areas of east Kamchatka.) Sbornik Trudov Zool. Museya, Mosk. Univ. 17:52-72.
1938	**Lock, R.A.** *1972.* Population trends of non-game species. Nebraska Game and Parks Comm., Proj. F-71, Job F4. 6 pp.
1939	**Lock, R.A.** *1974.* Population surveys of raptors. Nebraska Game and Parks Comm., Proj. W-15-R-30, Job No. 02. 25 pp.
1940	**Lock, R.A.** *1975.* Population surveys of raptors. Nebraska Game and Parks Comm., Proj. W-15-R-31, Job No. 02. 25 pp.
1941	**Lock, R.A.** *1978.* Raptor research. Nebraskaland 56(6):22-25.
1942	**Lockhart, J.M.** *1976.* The food habits, status and ecology of nesting golden eagles in Trans-Pecos region of Texas. M.S. Thesis. Sul Ross State Univ., Alpine, TX. 65 pp.
1943	**Lockhart, J.M.** *1976.* The food habits, status, and ecology of nesting golden eagles in the Trans-Pecos region of Texas. Chihuahuan Desert Res. Inst., Report. 38 pp.
1944	**Lockhart, J.M. and M.N. Kochert.** *1978.* Effects of visual markers and telemetry devices on the nesting success of golden eagles. Draft Manuscrpit. 14 pp.
1945	**Lockhart, J.M., T.P. McEneaney and R.L. Phillips.** *1978.* The effects of coal development on the ecology of birds of prey in southeastern Montana and northern Wyoming. Pages 9-20. *in:* U.S.D.I. Fish and Wildlife Service, Denver Service Center. Annual Prog. Rep. 1978.
1946	**Lockie, J.** *1973.* Wildlife and land use in highland Scotland. Oryx 12(1):53-63.
1947	**Lockie, J.D.** *1964.* The breeding density of the golden eagle and fox in relation to food supply in Western Ross, Scotland. Scott. Nat. 71:67-77. (Title Only)
1948	**Lockie, J.D.** *1967.* A field study of the effects of chlorinated hydrocarbons on golden eagles in west Scotland. Trans. of the 7th Int. Cong. on Game Biology, Beograd-Ljubljana, 1965. p. 275-278. (Title Only)
1949	**Lockie, J.D. and D.A. Ratcliffe.** *1964.* Insecticides and Scottish golden eagles. Brit. Birds 57(3):89-101.
1950	**Lockie, J.D. and D. Stephen.** *1959.* Eagles, lambs and land management on Lewis. J. Anim. Ecol. 28(1):43-50.
1951	**Lockie, J.D., D.A. Ratcliffe and R. Balharry.** *1969.* Breeding success and organo-chlorine residues in golden eagles in west Scotland. J. Appl. Ecol. 6(3):381-389.
1952	**Lodge, R.B.** *1908.* Experiences with eagles and vultures in the Carpathians. Aquila 15:268-273. (English Trans.)
1953	**Lodge, R.B.** *1909.* Eagles in the Carpathians. Country Life 25(627):41-44.
1954	**Lodge, R.B.** *1913.* Among the eagles. Wild Life 1(1):6-11.
1955	**Lofberg, L.M.** *1935.* Random notes on raptors at Florence Lake, California. Condor 37(3):171-173.
1956	**Lombard, A.L.** *1965.* Notes sur les oiseaux de Tunisie. Alauda 33(1):1-33.
1957	**Long, C.A.** *1973.* Exponential relations of standard metabolic rates of birds and the weights of eggs. Wilson Bull. 85(3):323-326.
1958	**Long, O.D.** *1968.* Feeding birds of prey (and others) in captivity. Avic. Mag. 74(4):142-144.
1959	**Loppenthin, B.** *1935.* Lille skrigeorn, *Aquila p. pomarina,* paa Samso. Dansk Ornithol. 29(3/4):120-121.
1960	**Lord, E.A.R.** *1956.* The birds of the Murphy's Creek District, southern Queensland. Emu 56(2):100-128
1961	**Louette, M.** *1973.* Ornithological observations near fresh and brackish water in Morocco during summer 1971. Gerfaut 63(1/2):121-132. (In English)
1962	**Love, G.** *1916.* A golden eagle's nest in Decator County, Kansas. Oologist 33(1):12.
1963	**Low, G.P.** *1898.* Electrocuted eagles. Osprey 2(10):124.
1964	**Lowe, W.P.** *1938.* Verreaux's eagle *(Aquila verreauxi).* Ibis (14th Ser.) 2(3):536.
1965	**Lucanus, F., von.** *1919.* Zug und wanderung der vogel Europas nach den ergebnissen des ringversuchs. J. Ornithol. 67(1):1-73.
1966	**Lucas, E.L.** *1938.* The eagle in flight. Yosemite Nat. Notes 17(11):149.
1967	**Ludlow, F.** *1933.* A contribution to the ornithology of Chinese Turkestan. Ibis (13th Ser.) 3(4):658-694.
1968	**Lugovoi, A.E.** *1975.* Ptitsi Mordovii. (Birds of Mordovia.) Gorki. p. 1-300.
1969	**Luhder, W.** *1868. Aquila imperialis* in Pommern. J. Ornithol. 16(5):352-353.
1970	**Lumley, E.D.** *1934.* Additional material on food habits. Hawk and Owl Soc. Bull. p. 22-23.
1971	**Lumley, E.D.** *1939.* The two eagles of North America. Emergency Cons. Comm., Publ. No. 78. 22 pp.
1972	**Lumsden, H.G.** *1964.* Golden eagle nesting in Ontario. Auk 81(1):91.
1973	**Lund, S.** *1978.* Eagles of Jackson's Canyon. Wyoming Wildl. 42(4):29-32.
1974	**Luniak, M.** *1971.* Birds of the middle reaches of the Vistula. Acta Ornithol. 13(2):15-98.
1975	**Luniak, M., W. Kalbarczyk and W. Pawlowski.** *1964.* Ptaki Warszawy. (Birds of Warsaw.) Acta Ornithol. 8(6):175-285. (English Summ.)
1976	**Lups, P.** *1973.* Steinadler mit Schnabelmissbildung. Ornithol. Beob. 70(4):180-182.

Master No.	Citation
1977	**Luscher, W.** *1943.* Steinadler und segelflieger. Ornithol. Beob. 40(9/10):131-132.
1978	**Luther, D.** *1967.* Warum sterben vogel aus? Falke 14(12):408-413.
1979	**Lutken, E.** *1965.* Ynglefuglene pa Nord-Fugloy, Nord-Nord, deres udbredelse og anatl. (The breeding birds of Nord Fugly (North Bird Island) northern Norway, distribution and numbers.) Dansk Ornithol. 58(4):166-192.
1980	**Lynes, H.** *1912.* Bird-notes in two Andalucian Sierras. Ibis 6(23):454-489.
1981	**Lynes, H.** *1920.* Ornithology of the Maroccan "Middle-Atlas.' Ibis (11th Ser.) 2(1):260-300.
1982	**Lynes, H.** *1925.* On the birds of north and central Darfur, with notes on the west-central Kordofan and north Nuba Provinces of British Sudan. (Part IV.) Ibis 1(2):344-416.
1983	**Lysenko, V.I.** *1974.* O redkikh ptitsakh Melitopolstchiny. (Rare birds of Melitopol district.) Ornithologia 11:391-392.
1984	**M.-D., A.** *1921.* Aigle royal *(Aquila fulva,* L.). Ornithol. Beob. 18(12):191.
2002	**Machura, L.** *1950.* Zum vorkommen des steinadlers in Osterreich. Natur Und Land 36(7/8):122-125.
2003	**Mack, D.B.** *1961.* Birds of the upper Murray of South Australia. S. Aust. Ornithol. 23(5):69-79.
2004	**Mack, K.J.** *1970.* Birds of the north-east of South Australia. S. Aust. Ornithol. 25(5):126-141.
2005	**Mackworth-Praed, C.N.** *1917.* A collection of birds from two Districts of British East Africa. Ibis (10th Ser.) 5(3):362-421.
2006	**Mackworth-Praed, C.W. and C.H.B. Grant.** *1970.* Birds of west central and western Africa. Jarroid and Sons, Norwich. 671 pp.
1985	**MacDonald, D.** *1926.* Some notes on the Golden eagle. Brit. Birds 19(9):218-225.
1986	**MacFarlane, R.S.** *1891.* Notes on and list of birds and eggs collected in arctic America, 1861-1866. Proc. U.S. Natl. Mus. 14:413-446.
1987	**MacIntyre, D.** *1933.* In defense of the golden eagle. Field (Apr. 15):809. (Title Only)
1988	**MacIntyre, D.** *1949.* How golden eagles hunt. Field 194:217. (Title Only)
1989	**MacIntyre, D.** *1950.* How eagles mate. Field 196:69. (Title Only)
1990	**MacLean, G.L.** *1957.* A summary of the birds of Westminster O.F.S. and surroundings (1953-1955). Ostrich 28(4):217-232.
1991	**MacLean, G.L.** *1965.* Further ornithological records from southern South West Africa. Ostrich 36(1):9-11.
1992	**MacLean, G.L.** *1970.* The breeding seasons of birds in the south-western Kalahari. Ostrich (Suppl. 8):179-192.
1993	**MacLean, I.M.** *1968.* Golden eagle calling. Scott. Birds 5(3):173.
1994	**MacLeod, J.G.R., E.M. d' C. Murray and C. d' C. Murray.** *1951.* The birds of the Hottentots Holland. (Part 1). Ostrich 22(3):179-189.
1995	**MacLulich, D.A.** *1938.* Birds of Algonquin Provincial Park, Ontario. Contrib. of the Royal Ontario Mus. of Zool., No. 13.
1996	**MacMillan, A.T.** *1970.* Scottish bird report 1969. Scott. Birds 6(2):62-128.
1997	**MacNally, L.** *1964.* Some notes on golden eagles. Scott. Birds 3(1):26-27.
1998	**MacNally, L.** *1977.* The ways of an eagle. Collins and Harvill Press. 143 pp.
1999	**MacNally, L.** *1979.* Peregrine apparently killed by golden eagle. Scott. Birds 10(6):234.
2000	**MacPherson, H.B.** *1909.* The home-life of the golden eagle. H.G. Witherby & Comp., London.
2001	**MacRadish, R.** *1891.* Measurements of a golden eagle's nest. Zoologist 1891:353-354.
2007	**Magnenat, D.** *1960.* Observations de l'aigle royal en dehors du massif alpin. Nos Oiseaux 25(8):228.
2008	**Mailliard, J.** *1911.* Golden eagle and dog. Condor 13(4):138.
2009	**Makatsch, W.** *1949.* Ornithologische beobachtungen aus Macedonien. Vogelwelt 70(2):44-49.
2010	**Makatsch, W.** *1949.* Ornithologische beobachtungen aus Macedonien. 2. Der wald von Kuri. Vogelwelt 70(5):139-144.
2011	**Malbrant, R.** *1954.* Contribution a l'etude des oiseaux du Borku-Ennedi-Tibesti. Oiseau Rev. Fr. Ornithol. 24(1):1-47.
2012	**Malchevski, A.S. and Y.B. Pukinski.** *1975.* Khistchnye ptitsi i nashe otnoshenie k nim. (Birds of prey and man's attitude to them.) Nasha Okhota, Sbornik 5, Leningrad. p. 189-211.
2013	**Malick, D.** *1962.* The eagles of Colorado. Colo. Outdoors 11(1):14-19.
2014	**Mallette, R.D.** *1968.* Status of rare and endangered species, management and special survey investigations. Calif. Dept. of Fish and Game, Upland Game Invest., Job Comp. Rep. Proj. No. W-47-R-6, Job No. 2. 10 pp.
2015	**Mallette, R.D.** *1970.* Raptor survey. Calif. Dept. of Fish and Game, Special Wildl. Invest., Job Prog. Rep. Proj. No. W-54-R-2, Job-III-6. 15 pp.
2016	**Mallette, R.D.** *1973.* Raptor survey. Calif. Dept. of Fish and Game, Special Wildl. Invest., Job Prog. Rep. Proj. No. W-54-R-5, Job-III-6. 17 pp.
2017	**Mallette, R.D.** *1974.* Raptor survey. Calif. Dept. of Fish and Game, Special Wildl. Invest., Job Prog. Rep. Proj. No. W-54-R-6, Job-III-6. 20 pp.
2018	**Mallette, R.D. and G.I. Gould, Jr.** *1976.* Raptors of California. Calif. Dept. of Fish and Game. 85 pp.
2019	**Mallette, R.D. and R.W. Schlorff.** *1978.* A plan for California raptors. Calif. Dept of Fish and Game. 42 pp.
2020	**Malmerendi, D.** *1943.* Catture rare in Romagna. Riv. Ital Ornitol. (2nd Ser.) 13(3):95-96.
2021	**Malone, E.V.** *1964.* Golden eagle - monarch of the heights. Birds Illus. 9(11):332-333,335.
2022	**Malone, E.V.** *1964.* The golden eagle in Ireland. Birds and Country 27(74):129-130.
2023	**Maltitz, F., von.** *1978.* Behaviour of the lesser spotted eagle, *Aquila pomarina,* on a balchatri trap. Safring 7(1):21.
2024	**Mann, C.F.** *1971.* Distributional notes on some Uganda birds. Bull. Brit. Ornithol. Club 91(4):111-113.
2025	**Mannix, D.P.** *1943.* Making "eagle vs. dragon." Am. Falconer 2(1):16-20.
2026	**Mannsberg, A.** *1951.* *Aquila ch. chrysaetos* L. - golden eagle - in Borzsony-woods. Aquila 55-58:281.
2027	**Mannsberg, A., von.** *1911.* Haufiges vorkommen von adlern in Erdely. Aquila 18:391-392.
2028	**Manuel, F.** *1962.* Aigle contre avion. Nos Oiseaux 26(12):319.
2029	**Marcetic, M.** *1955-1956.* Eagles in biotopes of Vojvodina. Larus 9-10:143-148.
2030	**Marchant, S.** *1963.* Migration in Iraq. Ibis 105(3):369-398.

Master No.	Citation
2031	**Marietan, J.** *1946.* L'aigle royal a Zinal. Nos Oiseaux 18:139.
2032	**Marion, W.R. and R.A. Ryder.** *1975.* Perch-site preferences of four diurnal raptors in northeastern Colorado. Condor 77(3):350-352.
2033	**Markgren, G.** *1960.* (On the spotted eagle *(Aquila clanga)* and its occurrence in Sweden). Var Fagelvarld 19:273-285. (Title Only)
2034	**Markgren, M.** *1956.* Kejsarorn *(Aquila heliaca)* sannolikt iakttagenvid Yddingen i syovaster Skane. Var Fagelvarld 15(2):128-130. (English Summ.)
2035	**Markgren, M.** *1960.* Fugitive reactions in avian behaviour. Acta Vertebrata 2(1):1-143.
2036	**Marquiss, M., I. Newton and D.A. Ratcliffe.** *1978.* The decline of the raven, *Corvus corax,* in relation to afforestation in southern Scotland and northern England. J. Appl. Ecol. 15(1):129-144.
2037	**Marsh, A.J.** *1943.* Golden eagle near South Pittsburg. Migrant 14(1):16.
2038	**Marshall, C.H.T.** *1884.* Notes on the birds of Chamba, in the N.W. Himalayas. Ibis (5th Ser.) 2(8):404-425.
2039	**Marshall, G.A.K.** *1900.* Notes on Mashonaland birds. Ibis (7th Ser.) 6(22):221-270.
2040	**Marshall, W.H.** *1940.* An "eagle guard" developed in Idaho. Condor 42(3):166.
2041	**Marston, M.A.** *1971.* The loss of eagles in Wyoming: how one federal agency replied to the protests. Illinois Audubon Bull. 160:13-14.
2042	**Martens, G.H.** *1910.* Ornis fokiensis. J. Ornithol. 58(2):209-275.
2043	**Martin, D.** *1974.* The great feather war. Defenders 49(3):370-372.
2044	**Martin, J.E. and R. Martin.** *1976.* Notes from western Calvinia. Bokmakierie 28(1):20-23.
2045	**Martorelli, G.** *1895.* Gli uccelli d'Italia. Rizzoli & Co., Milano. 752 pp.
2046	**Marzaro, L.** *1936.* Catture di aquila anatraia. Riv. Ital. Ornitol. 6(4):281.
2047	**Masarey, A.** *1924.* Steinadler im Sudtessin. Ornithol. Beob. 21(7):106.
2048	**Masarey, A.** *1929.* Die vogelwelt in der Umgebung von Tunis. Ornithol. Beob. 26(10):174-176.
2049	**Massa, B.** *1977.* II. The situation of the Falconiformes in Sicily. Pages 131-132. *in:* R.D. Chancellor, Ed. Proc. of the World Conf on Birds of Prey. 1-3 October, 1975. Vienna. I.C.B.P.
2050	**Massa, B., S. Bagilieri and G. Cangialosi.** *1979.* Nouvelles donnees pour l'etude des migrations a travers la Mediterranee centrale. Alauda 47(1):17-27. (English Summ.)
2051	**Massey, T.** *1967.* Eagle attacks goose. Ky. Warbler 43(1):38.
2052	**Masters, J.R. and A.L. Milhinch.** *1974.* Birds of the Shire of Northam, about 100 km east of Perth, Wa. Emu 74(4):228-244.
2053	**Matheson, W.E.** *1976.* The birds of the Redcliff Point deveopment area. S. Aust. Ornithol. 27(4):125-130.
2054	**Mathews, G.M.** *1912.* A reference-list to the birds of Australia. Novitates Zool. 18:171-446.
2055	**Mathews, G.M. and T. Iredale.** *1915.* On the ornithology of the Dictionnaire des Sciences Naturelle (Levrault). Aust. Avian Rec. 3(1):5-20.
2056	**Mathey, W.J.** *1966. Isospora buteonis* Henry 1932 in an American kestrel *(Falco sparverius)* and a golden eagle *(Aquila chrysaetos).* Bull. Wildl. Dis. Assoc. 2(2):20-22.
2057	**Mathisen, J.** *1975.* Kings of the avian way. Minn. Vol. (July/Aug.):27-32.
2058	**Mathisen, J.E. and A. Mathisen.** *1968.* Species and abundance of diurnal raptors in the panhandle of Nebraska. Wilson Bull. 80(4):479-486.
2059	**Matschie, P.** *1889.* Die kennzeichen der deutschen raubvogel. Eine anleitung zur sicheren bestimmung unserer deutschen tag- und nachtraubvogel. J. Ornithol. 37(1):67-72.
2060	**Matschie, P.** *1890.* Anleitung zum bestimmen der deutschen tagraubvogel. J. Ornithol. 38(1):89-93.
2061	**Matthes, J. and M. Neubauer.** *1976.* Beitrag zur avifauna der Bialowiezer heide (VR Polen). Falke 23(6):186-193.
2062	**Matthes, J. and M. Neubauer.** *1977.* Der schreiadler im Bezirk Rostock. Falke 24(2):42-47.
2063	**Matvejev, S.D.** *1947.* Ornithological material of Serbia (1836-1947.) Larus 1:77-95.
2064	**Mauersberger, G.** *1960.* Faunistische notizen aus Nordost-Bulgarien (19 Mai Juni 1959). J. Ornithol. 101(1/2):113-125.
2065	**Mauve, L.** *1938.* Der zug der grobvogel uber den Bosporus. J. Ornithol. 86(2):261-301.
2066	**Maxwell, H.** *1919.* The golden eagle in Galloway. Scottish Nat. 87-88:47-48.
2067	**Mayaud, N.** *1934.* Coup d'oeil sur l'avifaune des Causses. Alauda 6(2):222-259.
2068	**Mayaud, N.** *1945.* Commentaires sur l'ornithologie Francaise. Alauda 17-18(2):80-94.
2069	**Mayaud, N.** *1953.* Liste des oiseaux de France. Alauda 21(1):1-63.
2070	**Mayaud, N.** *1960.* Notes d'ornithologie Francaise. Nouveautes et mise au point. Alauda 28(4):287-302.
2071	**Mayaud, N.** *1962.* Notes d'ornithologie Francaise. V. Alauda 30(1):46-64.
2072	**Mayaud, N.** *1963.* Notes d'ornithologie Francaise. VI. Alauda 31(1):36-51.
2073	**Mayaud, N.** *1965.* Contribution a l'ornithologie du nord-ouest de l'Afrique. Alauda 33(1):34-50.
2074	**Mayaud, N.** *1967.* Notes d'ornithologie Francaise. IX. Alauda 35(2):125-139.
2075	**Mayaud, N.** *1970.* Additions et contribution a l'avifaune du nord-ouest de l'Afrique. Alauda 38(1):27-43.
2076	**Mayer, G.T.** *1974.* Die nachweise des kaiseradlers *(Aquila h. heliaca* Sav.) fur Oberosterreich. Egretta 17(1):34-35.
2077	**Maynard, C.J.** *1871.* A catalogue of the birds of Coos Co., N.H., and Oxford Co., Me.; with annotations relative to the breeding habits, migrations, etc. Proc. Boston Soc. Nat. Hist. 14:356-385.
2078	**Mayr, E. and L.L. Short.** *1970.* Species taxa of North American birds. Publ. of the Nuttall Ornithol. Club, No. 9.
2079	**McAtee, W.L.** *1935.* Food habits of common hawks. U.S.D.A. Circular No. 370.
2080	**McBee, C.E.** *1927.* Looking over new collection territory. Oologist 44:119-122.
2081	**McClelland, B.R.** *1973.* Autumn concentrations of bald eagles in Glacier National Park. Condor 75(1):121-123.
2082	**McCrea, R.** *1977.* Black eagle: prolonged use of nest sites. Honeyguide 91:38.
2083	**McCreary, O.** *1939.* Wyoming Bird Life. (Rev. Ed.). Burgess Publ. Comp., Minneapolis. 124 pp.
2084	**McDermond, W.S.** *1883.* Golden eagle. Ornithol. and Ool. 8(2):14.

Master No.	Citation
2085	**McEneaney, T.P.** *1979.* Movements and activities of breeding golden eagles. Paper Presented At Raptor Res. Foundation Ann. Meeting, 8-12 November, 1979. Davis, CA. 15 pp.
2086	**McGahan, J.** *1966.* Ecology of the golden eagle. M.A. Thesis. Univ. of Montana, Missoula, MT. 78 pp.
2087	**McGahan, J.** *1967.* Quantified estimates of predation by a golden eagle population. J. Wildl. Manage. 31(3):496-501.
2088	**McGahan, J.** *1968.* Ecology of the golden eagle. Auk 85(1):1-12.
2089	**McGill, A.R.** *1952.* Branch reports. New South Wales. Emu 52(2):130-139.
2090	**McGilp, J.N.** *1923.* Birds of Lake Frome District, South Australia. Emu 22(4):274-287.
2091	**McGilp, J.N.** *1925.* Wedge-tailed eagles destroying lambs. Emu 25(2):120.
2092	**McGilp, J.N.** *1934.* The hawks of South Australia. S. Aust. Ornithol. 12(7):225-258.
2093	**McGilp, J.N.** *1936.* Wedge-tailed eagles. Emu 36(2):99-102.
2094	**McGregor, P.J.C.** *1906.* Notes on birds observed at Monastir, Turkey in Europe. Ibis 6(22):285-307.
2095	**McGregor, P.J.C.** *1906.* Notes on birds observed at Monastir, Turkey in Europe. Ibis 6(22):285-307.
2096	**McGregor, W.W.** *1956.* Should it be legal to shoot an eagle? Forest. Out. 52(Jan.):12-13,22. (Title Only)
2097	**McKenna, M.G. and R.W. Seabloom.** *1976.* Threatened and unique wildlife of North Dakota: Initial status report. Univ. of North Dakota, Inst. for Ecological Studies, Report No. 13. 101 pp.
2098	**McKinlay, S.L.** *1971.* Feeding behaviour of golden eagle. Scott. Birds 6(5):275-276.
2099	**McLachlan, G.R.** *1969.* Thirteenth ringing report. Ostrich 40(2):37-50.
2100	**McLandress, M.R. and I. McLandress.** *1979.* Blue-phase ross' geese and other blue-phase geese in western North America. Auk 96(3):544-550.
2101	**McLean, D.D.** *1925.* The golden eagle as a deer enemy. Yosemite Nat. Notes 4(4):27.
2102	**McLean, I.M.** *1968.* Golden eagle calling. Scott. Birds 5(3):173.
2103	**McMillan, I.I.** *1962.* Eagles and our ruined grasslands. Audubon 64(2):76-79.
2104	**McMillan, R.P.** *1901.* Wedge-tailed eagle capturing live prey. West. Aust. Nat. 2(6):144. (Title Only)
2105	**Meabe, J. and H. Van Der Vloet.** *1956.* De avifauna van het verdronken land van Saaftinge. Gerfaut 46(3):151-190.
2106	**Mead, C.J.** *1973.* Movements of British raptors. Bird Study 20(4):259-286.
2107	**Meade-Waldo, E.G.B.** *1905.* A trip to the forest of Marmora, Morocco. Ibis 5(18):161-164.
2108	**Mearns, E.A.** *1878.* The golden eagle in the Hudson Highlands. Bull. Nuttall Ornithol. Club 3(2):100-101.
2109	**Meiners, B.** *1971.* Sanctuary for birds of prey. Our Public Lands (Bureau of Land Management). p. 15-18.
2110	**Meinertzhagen, R.** *1920.* Notes on the birds of Quetta. Ibis 2(1):132-194.
2111	**Meinertzhagen, R.** *1920.* Notes on the birds of southern Palestine. Ibis 2(1):195-259.
2112	**Meinertzhagen, R.** *1924.* Notes on a small collection of birds made in Iraq in the winter of 1922-23. Ibis 6(4):601-625.
2113	**Meinertzhagen, R.** *1924.* A contribution towards the birds of the Aden Protectorate. Ibis (11th Ser.) 6(4):625-642.
2114	**Meinertzhagen, R.** *1927.* Systematic results of birds collected at high altitudes in Ladak and Sikkim. (Part II.) Ibis (12th Ser.) 3(4):571-633.
2115	**Meinertzhagen, R.** *1940.* Autumn in central Morrocco. Part II. Ibis 4(2):187-234.
2116	**Meinertzhagen, R.** *1940.* How do larger raptorial birds hunt their prey? Ibis 4(3):530-535.
2117	**Meinertzhagen, R.** *1949.* Notes on Saudi Arabian birds. Ibis 91(3):465-482.
2118	**Meinertzhagen, R.** *1951.* Some relationships between African, oriental, and palearctic genera and species, with a review of the genus *Monticola.* Ibis 93(3):443-459.
2119	**Meinertzhagen, R.** *1954.* Some aspects of spring migration in Palestine. Ibis 96(2):293-298.
2120	**Meinertzhagen, R.** *1954.* Birds of Arabia. Oliver and Boyd, Edinburgh. 624 pp.
2121	**Meinertzhagen, R.** *1959.* Pirates and Predators. Oliver + Boyd, London. 230 pp.
2122	**Melides, M.** *1959.* Caesar - My golden eagle. Birds Illus. 5(8):167-169.
2123	**Menatory, G.** *1974.* A propos de la reproduction de l'aigle imperial dans les Marismas du Guadalquivir. Alauda 42(4):507-508.
2124	**Menatory, G.** *1976.* L'Aigle Royal. Atlas Visuels, Payot Lausanne.
2125	**Mendelssohn, H.** *1972.* The impact of pesticides on bird life in Israel. IX Bull. Int. Council Bird Pres. p. 75-104.
2126	**Mendelssohn, H.** *1975.* Report on the status of some bird species in Israel in 1974. XII Bull. International Council for Bird Preservation. p. 265-270.
2127	**Mendelssohn, H. and U. Marder.** *1970.* Problems of reproduction in birds of prey in captivity. Int. Zoo Yearb. 10:6-11.
2128	**Mendelssohn, H. and U. Paz.** *1977.* Mass mortality of birds of prey caused by azodrin, an organophosphorous insecticide. Biology Cons. (11):163-170.
2129	**Mendelssohn, H., P. Schlueter and Y. Aderet.** *1979.* Report on azodrin poisoning of birds of prey in the Huleh Valley in Israel. III Bull. Int. Council Bird Pres. p. 124-129.
2130	**Menegaux, A.** *1921.* Oiseaux collectes ou observes au Maroc dans l'atlas moyen par le Cap. Lynes. Oiseau Rev. Fr. Ornithol. 7(145):74-78.
2131	**Menzbier, M.** *1884.* On the geographical distribution of birds in European Russia north of the Caucasus. Part I. Rapaces diurnae. Ibis (5th Ser.) 2(7):278-315.
2132	**Meran, F.** *1947.* Bird-observations in the mountain Vertes. Aquila 51-54:186.
2133	**Mercier, A.** *1921.* Valeur economique de quelques rapaces. Gerfaut 11(3/4):93-99.
2134	**Meric, J.-D.** *1972.* A propos d'un mode de chasse de l'aigle royal. Alauda 40(4):400-401.
2135	**Merikallio, E.** *1924.* Kittilan lintujen nimiluettelo. Ornis Fenn. 1(4):69-71.
2136	**Merikallio, E.** *1958.* Finnish birds, their distribution and numbers. Fauna Fenn. 5:1-181.
2137	**Merlet, F.** *1969.* Aigles pomarins dans le centre de la France. Alauda 37(4):350.
2138	**Merlitsch, J.** *1898-1899.* Horsten des steinadlers bei imst in Tirol. Die Schwalbe (New Ser.) 1:140-141.
2139	**Merriam, C.H.** *1881.* Preliminary list of birds ascertained to occur in the Adirondack region, northeast New York. Bull. Nuttall Ornithol. Club 6(4):225-235.

Master No.	Citation
2140	**Merriam, C.H.** *1882.* List of birds ascertained to occur within ten miles from Point de Monts, Province of Quebec, Canada, based chiefly upon the notes of Napoleon A. Comeau. Bull. Nuttall Ornithol. Club 7(4):233-242.
2141	**Merriam, C.H., J.A. Allen, W. Brewster, J. Dwight, Jr., C.W. Richmond, R. Ridgway and W. Stone.** *1902.* Eleventh Supplement to the American Ornithologists' Union Check-list of North American birds. Auk 19(3):315-342.
2142	**Merwe, N.J., van der** *1959.* A preliminary list of the birds on the Percy Fyfe Nature Reserve. Ostrich 30(3):130-133.
2143	**Metcalf, R.** *1981.* Wedge-tailed eagle prey. Aust. Raptor Assoc. News 2(4):12.
2144	**Meyburg, B.-U.** *1969.* Beobachtungen am horst des steinadlers *(Aquila chrysaetos).* (Observations of the aerie of the golden eagle *(Aquila chrysaetos.)* Ornithol. Mitt. 21(6):113-119. (English Summ.)
2145	**Meyburg, B.-U.** *1970.* Zur biologie des schreiadlers *(Aquila pomarina.)* Deutsche Falkenorden 1969:32-66.
2146	**Meyburg, B.-U.** *1971.* Versuche zur kuntstlichen Steigerung der Vermehrungsrate des schreiadlers *(Aquila pomarina)* zu seinem Schutze. Beitr. Vogelkd. 17(3):207-227.
2147	**Meyburg, B.-U.** *1973.* Observations sur l'abondance relative des rapaces (Falconiformes) dans le nord et l'ouest de l'Espagne. Ardeola 19(1):129-150.
2148	**Meyburg, B.-U.** *1973.* Studies of less familiar birds: 172 lesser spotted eagle. Brit. Birds 66:439-447.
2149	**Meyburg, B.-U.** *1974.* Zur brutbiologie und taxonomischen stellung des schreiadlers. Falke 21(4/5):126-134,166-171.
2150	**Meyburg, B.-U.** *1974.* Sibling aggression and mortality among nestling eagles. Ibis 116:224-228.
2151	**Meyburg, B.-U.** *1974.* Quatre poussins dans un nid de l'aigle imperial d'Espagne *Aquila heliaca adalberti.* Alauda 42(1):1-6. (English Summ.)
2152	**Meyburg, B.-U.** *1975.* On the biology of the Spanish imperial eagle *(Aquila heliaca adalberti)* . Ardeola 21(Special):245-283.
2153	**Meyburg, B.-U.** *1976.* Status, bedrohung, und schutz der greifvogel (Falconiformes) in Westspanien. (Status, threat to and protection of birds of prey (Falconiformes) in western Spain.) Angew. Ornithol. 5(1):13-31.
2154	**Meyburg, B.-U.** *1977.* Protective management of eagles by reduction of nestling mortality. Pages 397-398. *in:* R.D. Chancellor, Ed. Proc. of the World Conf. on Birds of Prey. 1-3 October, 1975. Vienna. I.C.B.P.
2155	**Meyburg, B.-U.** *1978.* Sibling aggression and cross-fostering of eagles. Pages 195-201. *in:* S.A. Temple, Ed. Endangered Birds, Management Techniques for Preserving Threatened Species. Univ. of Wisconsin Press, Madison.
2156	**Meyburg, B.-U.** *1978.* Productivity manipulation in wild eagles. Pages 81-92. *in:* T.A. Geer, Ed. Bird of Prey Management Techniques. British Falconers' Club.
2157	**Meyburg, B.-U. and J. Garzon.** *1973.* Sobre la proteccion del aguila imperial *(Aquila heliaca adalberti)* aminorando artificialmente la mortandad juvenil. Ardeola 19(1):107-128.
2158	**Meyburg, B.-U. and J. Garzon.** *1974.* Sobre un nido de *Aquila heliaca adalberti* can 4 pollos. Ardeola 19(2):465-466.
2159	**Meyburg, B.-U. and C. Meyburg.** *1978.* Spanish refuge for Europe's birds of prey. Oryx 19:337-342.
2160	**Meyburg, B.-U. and J. Svehlik.** *1976.* Viererbruten des kaiseradlers *(Aquila heliaca)* . J. Ornithol. 117(4):462-464.
2161	**Meyburg, B.-U., J. Garzon, and J. Svehlik.** *1972.* Greifvogelschutz durch kunstliche herabsetzung der nestlingssterblichkeit. Ber. Deutsche Sekt. Int. Rat. Vogelschutz 12:54-57.
2162	**Meylan, A.** *1960.* Observations de l'aigle royal en dehors du massif alpin. Nos Oiseaux 25(8):227-228.
2163	**Michael, E.** *1929.* The sky battle. Yosemite Nat. Notes 8(5):55-56.
2164	**Michel-Schild, J.** *1924.* Steinadler im Berner - Oberland. Ornithol. Beob. 22(2):31.
2165	**Michie, R.H.** *1972.* Eagles in Northland. Notornis 19(3):272.
2166	**Miczynski, K.** *1962.* Ptaki Dublan (Ulkrainska SSR). (Birds of Dublany (Ukrainian SSR).) Acta Ornithol. 6(10):117-180. (English Summ.)
2167	**Mihaly, H.** *1955.* Second occurence of the tawny eagle in Hungary. Aquila 59-62:432.
2168	**Mikkola, H.** *1976.* Owls killing and killed by other owls and raptors in Europe. Brit. Birds 69(4):144-154.
2169	**Milek, B.** *1969.* The golden eagle... killer or saint? Field & Stream 74(5):10-12,14-16.
2170	**Miles, H.M.** *1950.* Provisional club check list. Rhodesia Bird Club Provisional Club Check List. p. 1-88.
2171	**Milewski, A.V.** *1975.* New distributional data: 6. black eagle, *Aquila verreauxi.* Ostrich 46:167.
2172	**Miller, A.H.** *1932.* The summer distribution of certain birds in central and northern Arizona. Condor 34(2):96-99.
2173	**Miller, L.** *1918.* First flights of a young golden eagle. Condor 20(6):212.
2174	**Miller, L.H.** *1910.* Fossil birds from the quatenary of southern California. Condor 12(1):12-15.
2175	**Miller, L.H.** *1911.* Series of eagle tarsi from the Pleistocene of Rancho La Brea. Univ. of Calif. Publ. Geol. 6(12):305-316.
2176	**Miller, L.H.** *1922.* Fossil birds from the Pleistocene of McKittrick, California. Condor 24(4):122-125.
2177	**Miller, L.H.** *1935.* A second avifauna from the McKittrick Pleistocene. Condor 37(2):72-79.
2178	**Miller, L.H.** *1957.* Bird remains from an Oregon indian midden. Condor 59(1):59-63.
2179	**Miller, V.H.** *1928.* The wedge-tailed eagle. Victorian Nat. 45:48-50.
2180	**Millet-Horsin, D.** *1912.* Notes ornithologiques sur la Tunisie. Oiseau Rev. Fr. Ornithol. 2(36):267-271.
2181	**Millet-Horsin, D.** *1915.* Notes ornithologiques d'A.O.F. (1) les rapaces observes d'Avril 1913 a Mai 1914 dans la presqu'ile de Dakar. Oiseau Rev. Fr. Ornithol. 7(69):1-6.
2182	**Milligan, C.J.** *1931.* Occurrence of the golden eagle *Aquila chrysaetos chrysaetos* (L.) in Co. Antrim, 1926-1930. Irish Nat. J. 3(12):254-255.
2183	**Mills, S.P.** *1976.* Search for the Spanish imperial eagle in Morocco. World Wildl. Fund., Proj. No. 1264. Final Report. 5 pp.
2184	**Millsap, B.A.** *1981.* Distributional status of Falconiformes in westcentral Arizona... with notes on ecology, reproductive success, and management. U.S.D.I. Bureau of Land Management, Tech. Note 355. 102 pp.
2185	**Milon, P.** *1966.* Capture en Bretagne, en 1960, d'un *Aquila rapax* . Alauda 34(3):245.

Master No.	Citation
2186	**Milstein, E. le S.** *1971.* The biology of Barberspan, with special reference to the avifauna. Ostrich (Suppl. 10):1-74.
2187	**Miner, N.R.** *1954.* Golden eagles attacking coyote. Condor 56(4):223.
2188	**Minoranski, V.A.** *1962.* O stenom orle *(Aquila rapax orientalis* L.). (*Aquila rapax orientalis.*) Zool. Zhur. 41(2):295-296. (In Russian, English Summ.)
2189	**Mintus, A.** *1916.* Die aus Nieder-Osterreich als horstvogel verschwundenen raubvogelarten. Ornithol. Jahrbuch. 27(1/2):33-44.
2190	**Mirabelli, P.** *1978.* Distribuzione dei Falconiformi in Calabria. Riv. Ital. Ornitol. (2nd Ser.) 48(2):157-171.
2191	**Mironov, B.S.** *1962.* Vidovoi sostav i chislennost khistchnykh ptits v poime srednego techenia R. Ilek. (Species composition and number of raptors in the valley of River Ilek.) 1st Nauchn. Sovesstchanie Zool. Pedagog. Inst. RSFSR, Moscow. p. 97.
2192	**Misonne, X.** *1956.* Notes sur les oiseaux de la Syrie et de l'Iran. Gerfaut 46(3):191-195.
2193	**Mitchell, G.A.** *1968.* A golden eagle nest in the pine ridge. Nebr. Bird Rev. 36(2):33-35.
2194	**Mitchell, K.D.G.** *1964.* Further observations of birds from aircraft. Brit. Birds 57(8):315-324.
2195	**Mitchell, W.I.** *1898.* The summer birds of San Miguel County, New Mexico. Auk 15(4):306-311.
2196	**Moffat, R.E.** *1972.* Natural pox infection in a golden eagle. J. Wildl. Dis. 8(2):161-162.
2197	**Mojsisovics, A., von.** *1894. Aquila imperialis* in Steiermark. Ornithol. Monatsber. 2(2):23.
2198	**Mojsisovics, A., von.** *1894.* Aquila imperialis (Bechst.) in Steiermark. Ornithol. Jahrbuch 5(1):26-27.
2199	**Moksnes, A.** *1971.* Ornitologiske undersokelser iForradalsomradet i Nord-Trondelag sommeren 1970. (Ornithological investigations in the Forradal area in Norway during the breeding season 1970.) Sterna 10(2):65-89.
2200	**Mollhagen, T.R., R.W. Wiley and R.L. Packard.** *1972.* Prey remains in golden eagle nests: Texas and New Mexico. J. Wildl. Manage. 36(3):784-792.
2201	**Moltoni, E.** *1927.* Moltoni on rare birds in Italy. Ibis (12th Ser.) 3:555.
2202	**Moltoni, E.** *1937.* Osservazioni bromatologiche sugli uccelli rapaci Italiani. Riv. Ital. Ornitol. 7(2):61-119.
2203	**Moltoni, E.** *1939.* A proposito di una recente cattura di aquila della steppa- *Aquila nipalensis orientalis* Cabanis. Riv. Ital. Ornitol. 9(3):208-210.
2204	**Moltoni, E.** *1943.* L'alimentazione dell'aquila anatria *(Aquila clanga,* Pallas). Riv. Ital. Ornitol. (2nd Ser.) 13(3):97-100.
2205	**Moltoni, E.** *1945.* Elenco degli uccelli Italiani con l'attuale nome scientifico e relativa pronuncia in riguardo all'accento. Riv. Ital. Ornitol. 15(2/3):33-78.
2206	**Moltoni, E.** *1945.* Catture de aquila macchiata *(Aquila clanga,* Pallas) in Lombardia. Riv. Ital. Ornitol. (2nd Ser.) 15(1):32.
2207	**Moltoni, E.** *1946.* L'etimologia ed il significato dei nomi volgari e scientifici degli uccelli Italiani. Riv. Ital. Ornitol. (2nd Ser.) 16(2):33-50.
2208	**Moltoni, E.** *1965.* Osservazioni ornitologiche fatte nel 1964. Riv. Ital. Ornitol. (2nd Ser.) 35(1/2):34-118.
2209	**Moltoni, E. and P. Brichetti.** *1978.* Elenco degli uccelli Italiani. Riv. Ital. Ornitol. (2nd Ser.) 48(2):65-142.
2210	**Monastero, S.** *1942.* The capture of a golden eagle in Sicily *(Aquila chrysaetos chrysaetos,* L.). Soc. Di Sci. Natur. Econ. Palermo, Boll., (New Ser.) 24:24-30. (In Italian) (Title Only)
2211	**Mondor, C. and P. Juurand.** *1975.* The Churchill River: a resource for conservation and recreation. Musk Ox 15:44-52.
2212	**Monk, G.** *1976.* The Raptors of Yakima Canyon. U.S.D.I. Bureau of Land Management, Spokane, WA. 38 pp.
2213	**Moodie, R.L.** *1929.* Excess callus in a Pleistocene bird. Am. J. Sci. (5th Ser.) 17(97):81-84.
2214	**Moody, C.S.** *1909.* An aerial gymnast. Bird-Lore 11(6):252-253.
2215	**Mooney, N.** *1981.* Birds of prey in Tasmania. Hawk Trust, Ann. Rep. 11:41-45.
2216	**Moor, C.** *1877.* Golden eagle, rough-legged buzzard, sc., near Woodbridge. Zoologist (3rd Ser.) 1:25.
2217	**Moran, J.** *1967.* Eagles at Antler Creek. Birds Illus. 13(3):78-85.
2218	**More, A.G.** *1865.* On the distribution of birds in Great Britain during the nesting-season. Ibis 1(1):1-27.
2219	**Moreau, R.E.** *1944.* Clutch-size: A comparative study, with special reference to African birds. Ibis 86:286-347.
2220	**Moreau, R.E.** *1966.* The bird faunas of Africa and its islands. Academic Press, New York. 424 pp.
2221	**Moreau, R.E.** *1972.* The Palearctic-African Bird Migration Systems. Academic Press, London and New York. 384 pp.
2222	**Morel, G.** *1962.* La reproduction des oiseaux dans une region semi-aride: La Vallee du Senegal. Alauda 30(3):161-203.
2223	**Morgan, A.M.** *1932.* The spread and weight of the wedge-tailed eagle. S. Aust. Ornithol. 11(6):156-157.
2224	**Moriarty, L.J.** *1966.* Bird's nests in South Dakota. A.O.U. 349 golden eagle. S. Dakota Bird Notes 18(4):82.
2225	**Moriarty, T.K.** *1972.* Birds of Wanjarri, Wa. Emu 72:1-7.
2226	**Morillo, C. and J. Lalanda.** *1973.* Sobre fechas de puestas del aquila real *(Aquila chrysaetos)* en Espana central. Ardeola 19(2):445-446.
2227	**Moriseau, A.** *1959.* Capture d'un aigle criard *(Aquila clanga)* en Haute-Marne. Oiseau Rev. Fr. Ornithol. 29(3):249.
2228	**Moriseau, A.** *1960.* Complement a la note sur l'aigle criard en Haute-Marne. Oiseau Rev. Fr. Ornithol. 30(1):89.
2229	**Morris, F.O.** *1853.* A Natural History of the Nests and Eggs of British Birds. Groombridge and Sons, London. 154 pp.
2230	**Morris, F.T.** *1976.* Growth of plumage from chick to juvenile in the wedge-tailed eagle. Emu 76(2):86.
2231	**Morris, F.T.** *1977.* A species count of birds of prey. Aust. Bird Watcher 7(2):49-51.
2232	**Morris, R.O.** *1903.* Springfield, Mass., Bird notes. Auk 20:439.
2233	**Morris, S.** *1962.* They're killing off the eagle. Am. Forests 68(8):5,51-52.
2234	**Morrison, C.F.** *1889.* Golden eagle in Montana. Ornithol. and Ool. 14(2):25.
2235	**Morrison, C.F.** *1889.* A list of the birds of Colorado. Ornithol. and Ool. 14(1):6-9.
2236	**Morrison, K.D.** *1955.* Bird protection laws show progress. Audubon 57(5):222-225.
2237	**Morzenti, A. and M. Fenn.** *1975.* A new lease on life for birds of prey. Outdoor Calif. (May-June):10-11.
2238	**Mosher, J.A. and C.M. White.** *1976.* Directional exposure of golden eagle nests. Can. Field-Nat. 90(3):356-359.
2239	**Mosher, J.A., C.M. White, J.R. Murphy and M.A. Jenkins.** *1978.* Raptors of the Uinta National Forest, Utah. Great Basin Nat. 38(4):438-446.

Master No.	Citation
2240	**Mountfort, G.** *1957.* First pictures of the Spanish imperial eagle. Sphere 230(2992):163-165.
2241	**Mountfort, G.** *1958.* Portrait of a wilderness. Hutchison + Co., London. 240 pp.
2242	**Mountfort, G. and I.J. Ferguson-Lees.** *1961.* The birds of the Coto-Donana. Ibis 103(1):86-109.
2243	**Mountfort, G. and I.J. Ferguson-Lees.** *1961.* Observations on the birds of Bulgaria. Ibis 103(3):443-471.
2244	**Mouritz, L.B.** *1914.* Notes on birds observed in Katanga, Belgian Congo. Ibis (10th Ser.) 2(1):26-38.
2245	**Mouritz, L.B.** *1915.* Notes on the ornithology of the Matopo District, Southern Rhodesia. Ibis 3(2):185-216.
2246	**Mrugasiewicz, A. and J. Whitkowski.** *1962.* An ornithological sketch of the Barycz valley in Poland. Brit. Birds 55(7):245-272.
2247	**Mueller, H.C. and D.D. Berger.** *1961.* Weather and fall migration of hawks at Cedar Grove, Wisconsin. Wilson Bull. 73(2):171-192.
2248	**Mueller, H.C. and D.D. Berger.** *1973.* The daily rhythm of hawk migration at Cedar Grove, Wisconsin. Auk 90(3):591-596.
2249	**Muir, J.** *1940.* Afrikaans bird-names in Riversdale, C.P. Ostrich 11(1):1-19.
2250	**Muirhead, L.P., Jr.** *1903.* Young eagles. St. Nicholas 30(10):940.
2251	**Muller-Schneider, P.** *1975.* Steinadlerkampfe in Graubunden. Ornithol. Beob. 72(3):115-116.
2252	**Munn, P.W.** *1894.* On the birds of the Calcutta District. Ibis (6th Ser.) 6(21):39-77.
2253	**Munro, J.A.** *1930.* Miscellaneous notes on some British Columbia birds. Condor 32(1):65-68.
2254	**Munson, S.A.** *1882.* Golden eagles. Ornithol. and Ool. 6(12):94.
2255	**Muntyanu, A.I.** *1977.* Gnezdovanie stepnogo orla v vostochnykh raionakh Rostovskoi oblasti. (Nesting of *Aquila nipalensis* in eastern parts of Rostov district.) 7 Vsesoyuzn. Ornithol. Konf., Tezisy Dokl., Chast 2, Kiev. p. 236-237. (In Russian)
2256	**Munyer, E.A. and P.W. Parmalee.** *1966.* Stomach contents of Illinois raptorial birds. Audubon Bull. 138:9-12.
2257	**Murie, A.** *1944.* The wolves of Mount McKinley. U.S.D.I., National Park Service, Fauna Series No. 5. 238 pp.
2258	**Murie, A.** *1946.* Observations on the birds of Mount McKinley National Park, Alaska. Condor 48(6):253-261.
2259	**Murie, O.J.** *1952.* A price on his golden head. Audubon 54(4):232-236.
2260	**Murphy, J.R** *1977.* Status of eagle populations in the western United States. Pages 57-63. *in:* R.D. Chancellor, Ed. Proc. of the World Conf on Birds of Prey. 1-3 October, 1975. Vienna. I.C.B.P.
2261	**Murphy, J.R.** *1975.* Status of a golden eagle population in central Utah, 1967-1973. Pages 91-96. *in:* J.R. Murphy, C.M. White and B.E. Harrell, Eds. Population Status of Raptors, Proc. of the Conf. on Raptor Conserv. Techniques. 22-24 March, 1973. Fort Collins, Co. Raptor Res. Rep. No.3.
2262	**Murphy, J.R.** *1977.* Eagles and livestock - some management considerations. Pages 307-314. *in:* R.D. Chancellor, Ed. Proc. of the World Conf. on Birds of Prey. 1-3 October, 1975. Vienna. I.C.B.P.
2263	**Murphy, J.R.** *1978.* Management considerations for some western hawks. Trans. of the 43rd N. American Wildl. and Nat. Res. Conf., 18-22 March, 1978. Phoenix, Az. p. 241-251.
2264	**Murphy, J.R.** *1978.* Analysis of potential impacts of Central Utah Project features on golden eagles. Part I. Jordanelle Reservoir Area - Municipal and Industrial System, Bonneville Unit. Brigham Young Univ., Dept. of Zool., Final Rep. 9 pp.
2265	**Murphy, J.R.** *1979.* Analysis of potential impact of Central Utah Project features on golden eagles. Part III. Upalco Unit. Brigham Young Univ., Dept. of Zool., Final Rep. 11 pp.
2266	**Murphy, J.R.** *1979.* Analysis of potential impacts of Central Utah Project features on golden eagles. Part IV. Diamond Fork Power System, Bonneville Unit. Brigham Young Univ., Dept. of Zool., Final Rep. 16 pp.
2267	**Murphy, J.R.** *1979.* Analysis of potential impacts of Central Utah Project features on golden eagles. Part II. Uintah Unit. Brigham Young Univ., Dept. of Zool., Final Rep. 9 pp.
2268	**Murphy, J.R.** *1979.* Golden eagle and potential peregrine falcon nesting habitat for the 1979 nesting season in the Uintah, Upalco and Bonneville Units, Central Utah Project. Brigham Young Univ., Dept. of Zool., Final Rep. 15 pp.
2269	**Murphy, J.R., F.J. Camenzind, D.G. Smith, and J.B. Weston.** *1969.* Nesting ecology of raptorial birds in central Utah. Brigham Young Univ. Science Bull., Biology Series 10(4):1-36.
2270	**Murphy, R.** *1965.* The golden eagle. E.P. Dutton & Co., N.Y. 157 pp.
2271	**Murray, D.K.W.** *1931.* North Sea migrants, 1930. Brit. Birds 25(1):6-11.
2272	**Murray, L.T. and Z.E. Murray** *1934.* Notes on the food habits of the golden eagle. Auk 51(3):371.
2273	**Musselman, T.E.** *1942.* Eagles of western Illinois. Auk 59(1):105-107.
2275	**Musselwhite, D.W.** *1919-1920.* Eggs of the golden eagle. Bull. Brit. Ornithol. Club 40:44.
2274	**Musselwhite, D.W.** *1927.* Eggs of golden eagle. Bull. Brit. Ool. Assoc. 1(9):94.
2276	**Myrberget, S.** *1959.* Fugeliv pa Helgelandskysten. Sterna 3(8):321-339.
2277	**Mysterud, I. and H. Dunker.** *1972.* Observasjoner av smagnagere og predatorer i Atnasjofeltet, Sor-Norge i 1963. Fauna 25(4):235-246. (English Summ.)
2278	**Nagy, A.C.** *1977.* Population trend indices based on 40 years of autumn counts at Hawk Mountain Sanctuary in northeastern Pennsylvania. Pages 243-253. *in:* R.D. Chancellor, Ed. Proc. of the World Conf. on Birds of Prey. 1-3 October, 1975. Vienna. I.C.B.P.
2279	**Nagy, L.** *1960-1961.* A volt bihari Sarret jelenlegi Madarvilaga. (The bird-life of the drained marsh-land of "Sarret" in east Hungary.) Aquila 67-68:151-157.
2280	**Nappe, L.** *1973.* Kills of Nevada eagles force change of policy. Foresta Inst. for Ocean and Mountain Studies 1(1):1-2.
2281	**Nash, A.I.** *1903.* Capture of a golden eagle. J. Maine Ornithol. Soc. 5(3):41-42.
2282	**Nass, R.D.** *1977.* Mortality associated with sheep operations in Idaho. J. Range Manage. 30(4):253-258.
2283	**Naurois, R. de and E. Virebayre.** *1955.* L'aigle royal en Lozere. Oiseau Rev. Fr. Ornithol. 25(3):199-205.

Master No.	Citation
2284	**Naurois, R., de.** *1962.* Recherches sur l'avifaune de la Cote Atlantique du Maroc. Du Detroit de Gibraltar aux les de Mogador. Alauda 30(2):81-97.
2285	**Neary, J.** *1977.* Eagle doctor of Tesuque. Audubon 79(4):90-91,92,94,96,98.
2286	**Neary, J.** *1978.* Samaritan for eagles. Quest 2(4):6-11.
2287	**Neave, S.A.** *1910.* On the birds of northern Rhodesia and the Katanga District of Congoland. Ibis (9th Ser.) 4(13):78-155.
2288	**Nedobity, A.** *1890.* Ueber Syrnium uralense, Pall., *Aquila naevia,* Wolf und Nuci feaga caryocatactes, L. im Bohmerwalde. Ornithol. Jahrbuch 1(8):153-155.
2289	**Neese, G., W.M. Longhurst and W.E. Howard.** *1976.* Predation and the sheep industry in California. Univ. of Calif., Div. of Agric. Sciences, Bull. 1878. 63 pp.
2290	**Neilson, J.A.** *1924.* A Saturday's afternoon ride. Oologist 41(6):64-65.
2291	**Nelson, E.W.** *1876.* Birds of north-eastern Illinois. Bull. Essex Inst. 8:90-155.
2292	**Nelson, M.** *1980.* Historic overview of raptor - powerline problems and raptor management priorities. Pages 6-8. *in:* R.P. Howard and J.F. Gore, Eds. Proc. of A Workshop on Raptors and Energy Developments. 25-26 January. Boise, Id. Bonneville Power Admin., U.S. Fish and Wildlife Service, Idaho Power Comp. and Idaho Chapter of the Wildl. Soc.
2293	**Nelson, M.W.** *1957.* Hawks and eagles in conservation. Soil Cons. 22(9):204-208.
2294	**Nelson, M.W.** *1962.* Hunting characteristics of eagles. Pages 96-99. *in:* Trans. 13th Interstate Antelope Conf., 4-6 Dec., Boise, Id.
2295	**Nelson, M.W.** *1969.* Trichomoniasis in the golden eagle. Page 446. *in:* J.J. Hickey, Ed. Peregrine Falcon Populations; Their Biology and Decline. Univ. of Wisconsin Press, Madison.
2296	**Nelson, M.W.** *1969.* Research needs in reestablishing local raptorial bird populations. Pages 403-407. *in:* J.J. Hickey, Ed. Peregrine Falcon Populations; Their Biology and Decline. Univ. of Wisconsin Press, Madison.
2297	**Nelson, M.W.** *1978.* Preventing electrocution deaths and the use of nesting platforms on power lines. Pages 42-46. *in:* T.A. Geer, Ed. Bird of Prey Management Techniques. British Falconers' Club.
2298	**Nelson, M.W. and P. Nelson.** *1977.* Power lines and birds of prey. Pages 228-242. *in:* R.D. Chancellor, Ed. Proc. of the World Conf. on Birds of Prey. 1-3 October, 1975. Vienna. I.C.B.P.
2299	**Nelson, M.W. and P. Nelson.** *1976.* Power lines and birds of prey. Idaho Wildl. Rev. (March-April):3-7.
2300	**Nemeth, F.** *1942.* Contribution to the knowledge of the external morphology of species of the genus *Aquila* in Morrocco. Bull. Soc. Des Sci. Nat. Phys. Du Maroc 20:77-79. (Title Only)
2301	**Netting, M.G.** *1926.* Notes on a golden eagle in captivity. Bird-Lore 28(5):326-328.
2302	**Neumann, O.** *1899.* Beitrage zur vogelfauna von ost- und central Afrika. J. Ornithol. 47(1):33-74.
2303	**Neumann, O.** *1904.* Vogel von Schoa und Sud Athiopien. J. Ornithol. 52(3):321-410.
2304	**Newland, B.C.** *1958.* Notes from the north. S. Aust. Ornithol. 22(5/6):70-71.
2305	**Newman, E.** *1846.* Description of the spotted eagle. Zoologist 4:1207-1209.
2306	**Newman, K.B.** *1977.* Observations on raptor migration. Bokmakierie 28(3):75-77.
2307	**Newton, I.** *1972.* Birds of prey in Scotland: Some conservation problems. Scott. Birds 7(1):5-23
2308	**Newton, I.** *1979.* Population ecology of raptors. Buteo Books, Vermillion, S. Dakota. 339 pp.
2309	**Nice, M.M.** *1953.* Incubation periods of birds of prey. Vogelwarte 16(4):154-157.
2310	**Nichols, A.C.** *1965.* Some of the commoner birds of prey around Bulawayo. Birds Illus. 11(8):241-243.
2311	**Nicholson, E.M.** *1957.* The rarer birds of prey: their present status in the British Isles. Golden eagle. Brit. Birds 50(4):129-155.
2312	**Nicholson, E.M., I.J. Ferguson-Lees and P.A.D. Hollom.** *1957.* The Camarque and the Coto Donana. Brit. Birds 50(12):497-519.
2313	**Nicoll, M.J.** *1909.* Contributions to the ornithology of Egypt. No. II. Birds of the Province of Giza. Part 3. Ibis (9th Ser.) 3(12):623-649.
2314	**Niederwolfsgruber, F.** *1964.* On the golden eagle *(Aquila chrysaetos)* in Tyrol. Z. Jagdwissenschaft 10(2):54-61. (Title Only)
2315	**Niederwolfsgruber, F.** *1964.* The golden eagle *Aquila chrysaetos* in Austria, especially in Tyrol. Pages 124-129. *in:* Proc. of the Working Conf. on Birds of Prey and Owls. 10-12 April, 1964. Caen. I.C.B.P.
2316	**Niederwolfsgruber, F.** *1966.* On the status of the golden eagle, *Aquila chrysaetos,* in the Alps, and the dangers threatening it. Report of the 15th World Conf. 1966. Cambridge. I.C.B.P. 8 pp.
2317	**Niederwolfsgruber, F.** *1967.* Steinadler-forschung. Angew. Ornithol. 2(4):162-163.
2318	**Nielsen, B.P.** *1969.* Further spring observations of the birds of Gilan, northern Iran. Dansk Ornithol. 63(1):50-73.
2319	**Nielsen, B.P. and S. Christensen.** *1969.* On the autumn migration of spotted eagles and buzzards in the Middle East. Ibis 111(4):620-621.
2320	**Nielsen, B.P. and H.-J. Speyer.** *1967.* Some observations of birds in northern Iran. Dansk Ornithol. 61(1):30-39.
2321	**Nielsen, B.P. and S. Christensen.** *1970.* Observations on the autumn migration of raptors in the Lebanon. Ornis Scand. 1(2):65-73.
2322	**Nielsen, D.B. and D.M. Curle.** *1970.* Predator costs to Utah's range sheep industry. Utah State Univ., Dept. Economics. Report. 10 pp.
2323	**Nielsen, S.** *1980.* Eagles. American Taxidermist Mag. 14(4):4-7,10-11,14-17.
2324	**Niemeyer, C.** *1975.* Montana golden eagle removal and translocation project. U.S.D.I. Fish and Wildlife Service. Final Rep. 13 pp.
2325	**Niemeyer, C.** *1976.* Montana golden eagle removal and translocation project. U.S.D.I., Fish and Wildlife Service. Final Report. 19 pp.

Master No. **Citation**

2326 **Niemeyer, C.** *1977.* Montana golden eagle removal and translocation project. U.S.D.I. Fish and Wildlife Service. Final Rep. 20 pp.

2327 **Nilsen, K.A.** *1974.* Kongeorn drept i kamp med rev. (Golden eagle killed in fight with a red fox.) Fauna 27(1):52. (English Summ.)

2328 **Nilsson, S.G.** *1981.* (The size of the breeding populations of diurnal raptors in Sweden.) Var Fagelvarld 40:249-262.

2329 **Ninni, E.** *1923.* La collezione ornitologica G. Bisacco- Palazzi in Chirignago (Mestre). Riv. Ital. Ornitol. 6(1):22-26.

2330 **Nisbet, I. and C. Smout.** *1957.* Eagles over the Bosphorus. Country Life 121(3150):1098-1099.

2331 **Nisbet, I.C.T. and T.C. Smout.** *1957.* Autumn observations on the Bosphorous and Dardanelles. Ibis 99(3):483-499.

2332 **Noakes, D.** *1982.* Raptors and the public. Aust. Raptor Assoc. News 3(1):12.

2333 **Noble, H.** *1902.* Forty-four days' nesting in Andalucia. Ibis (8th Ser.) 2(5):67-89.

2334 **Nordstrom, G.** *1961.* Bird-banding in Finland in the years 1913-1960. Ornis Fenn. 38(3/4):146-154.

2335 **Nordstrom, G.** *1962.* Finnische wiederfunde im Ausland beringter vogel. Ornis Fenn. 39(4):131-151.

2336 **Nordstrom, G.** *1963.* Einige ergebnisse der vogelberingung in Finnland in den jahren 1913-1962. Ornis Fenn. 40(3):81-124.

2337 **Norris, J.P.** *1890.* A series of the eggs of the golden eagle. Ornithol. and Ool. 15(2):17-18.

2338 **Norris, J.P., Jr.** *1926.* A catalogue of sets of Accipitres' eggs in the collection of Joseph Parker Norris, Jr., Philadelphia, Pa., U.S.A. Ool. Rec. 6(2):25-41.

2339 **North, A.J.** *1889.* Descriptive catalogue of the nest and eggs of birds found breeding in Australia and Tasmania. F.W. White, Sydney. 416 pp.

2340 **North, A.J.** *1911.* Nests and eggs of birds found breeding in Australia and Tasmania. Vol. III. Australian Mus., Sydney. Spec. Cat. No. 1. 362 pp.

2341 **North, M.E.W.** *1939.* Field notes on certain raptorials and water-birds in Kenya Colony. Ibis 3(4):617-643.

2342 **North, M.E.W.** *1944.* Some East African birds of prey. Ibis 86:117-138.

2343 **Northeast, C.J.** *1977.* Aerial roll display of golden eagle in winter. Brit. Birds 70(3):119.

2344 **Northeast, C.J.** *1978.* Golden eagle persistently attacking red deer calf. Brit. Birds 71(1):36-37.

2345 **Novelletto, A.** *1980.* Ecologia dell'aquila reale negli Appennini. Riv. Ital. Ornitol. 50(2/3):127-142.

2346 **Nygaard, A. and A. Cleve.** *1971.* Avifaunistiske Undersokelser ved Borsesjo 1954-1970. Sterna 10(2):131-145.

2347 **O'Brien, D. and M. Pulkrabek.** *1974.* An inventory of raptor nesting in Harding County, South Dakota - 1973. South Dakota Dept. of Game, Fish and Parks. Proj. W-95-R-7, Job No. 9-1. 6 pp.

2348 **O'Donnell, J.A., III, R. Garbett and A. Morzenti.** *1978.* Normal fasting plasma glucose levels in some birds of prey. J. Wildl. Dis. 14(4):479-481.

2349 **O'Gara, B.** *1978.* Sheep depredation by golden eagles in Montana. *in:* Proc. of the Vertebrate Pest. Conf. 8:206-213.

2350 **O'Gara, B.W.** *1982.* Predation by golden eagles on domestic lambs in Montana. Pages 345-358. *in:* J.M. Peek and P.D. Dalke, Eds. Wildlife Livestock Relationships Symposium: Proceedings.

2351 **Oates, E.W.** *1882.* A list of the birds of Pegu. Stray Feathers 10(4):175-248.

2352 **Oberholser, H.C.** *1900.* Notes on birds collected by Doctor W.L. Abbott in central Asia. Proc. U.S. Natl. Museum 22:205-228.

2353 **Oberholser, H.C.** *1906.* The North American eagles and their economic relations. U.S.D.A. Biology Survey, Bull. No. 27. 32 pp.

2354 **Oberholser, H.C.** *1923.* Eighth annual list of proposed changes in the A.O.U. check-list of North American birds. Auk 40(4):677-682.

2355 **Oberholser, H.C.** *1945.* Birds of the White-Fuller expedition to Keyna, East Africa. Cleveland Mus. Nat. Sci. Publ. 4(3):43-122.

2356 **Oblinger, H. and J. Konrad.** *1965.* Jagende steinadler *(A.c.)* in den Allgauer Alpen. Natur. Verein. Schwab E. Bericht 69:69-73. (Title Only)

2357 **Oblinger, H. and J. Konrad.** *1966.* Steinadler auf gemsenjagd. Ornithol. Mitt. 18(3):46-48.

2358 **Oddi, A.** *1899.* L'*Aquila rapax* (Tem.) ed il *Buteo desertorum* (Daud) per la prima volta osservati in Itali. Avicula 3:21-22.

2359 **Oddi, E.A.D.** *1898.* Elenco degli uccelli rari o piu diffici li ad aversi conservati nella sua collezione ornitologica Italiana al 31 Dicembre 1898. Ornis 9:199-249.

2360 **Oelke, H.** *1977.* Methoden der bestandserfassung von vogeln: Nestersuche-Revierkartierung. Ornithol. Mitt. 29(8):151-166.

2361 **Oesterreich, R., von and E.F. von Homeyer.** *1879.* Zwolf fruhlingstage an der mittleren Donau. J. Ornithol. 27(1):1-83.

2362 **Oesterreich, R., von.** *1879.* Ornithologische beobachtungen in den Auwaldern der Donau bei Wein. J. Ornithol. 27(2):97-129.

2363 **Ogburn, J.S.** *1976.* To save an eagle. Texas Parks and Wildl. (May):2-5.

2364 **Ogilvie-Grant, W.R. and J.D.D. La Touche.** *1907.* On the birds of the Island of Formosa. Ibis (9th Ser.) 1(2):254-279.

2365 **Ogilvie-Grant, W.R.** *1915.* The greater spotted eagle in Herefordshire. Brit. Birds 9(7):186.

2366 **Okulewicz, J.** *1974.* Birds of Olsztyn and its vicinity. Acta Ornithol. 13(4):99-140. (English Trans.)

2367 **Okulewicz, J. and L. Tomialojc.** *1964.* Jesienne obserwacje na Mierzei Wislanej w 1961 roku. (The 1961 autumn observations at Mierzeja Wislana (Baltic Coast).) Acta Ornithol. 8(7):330-337. (English Summ.)

2368 **Olendorff, R.R.** *1972.* The large birds of prey of the Pawnee National Grassland: nesting habits and productivity, 1969-1971. U.S. Int. Biology Program, Grassland Biome, Tech. Rep. No. 151. 59 pp.

2369 **Olendorff, R.R.** *1972.* Eagles, sheep and power lines. Colorado Out. 21(1):3-11.

2370 **Olendorff, R.R.** *1973.* The ecology of the nesting birds of prey of northeastern Colorado. U.S. Int. Biology Program, Tech. Rep. No. 211. Natural Res. Ecol. Lab., Colorado State Univ., Ft. Collins, CO. 233 pp.

2371 **Olendorff, R.R.** *1975.* Golden Eagle Country. A.A. Knopf, Inc., N.Y. 208 pp.

Master No. **Citation**

2372 **Olendorff, R.R.** *1975.* Population status of large raptors in northeastern Colorado- 1970-1972. Pages 185-205. *in:* J.R. Murphy, C.M. White and B.E. Harrell, Eds. Population Status of Raptors. Proc. of the Conf. on Raptor Conserv. Techniques. 22-24 March, 1973. Fort Collins, Co. Raptor Res. Rep. No. 3.

2373 **Olendorff, R.R.** *1976.* The food habits of North American golden eagles. Am. Midland Nat. 95(1):231-236.

2374 **Olendorff, R.R. and M.N. Kochert.** *1977.* Land management for the conservation of birds of prey. Pages 294-307. *in:* R.D. Chancellor, Ed. Proc. of the World Conf. on Birds of Prey. 1-3 October, 1975. Vienna. I.C.B.P.

2375 **Olendorff, R.R. and S.E. Olendorff.** *1968.* An Extensive Bibliography on Falconry, Eagles, Hawks, Falcons and Other Diurnal Birds of Prey. Privately Printed. Ft. Collins, CO.

2376 **Olendorff, R.R. and J.W. Stoddard, Jr.** *1974.* The potential for management of raptor populations in western grasslands. Pages 44-88. *in:* F.N. Hamerstrom, Jr., B.E. Harrell and R.R. Olendorff, Eds. Management of Raptors. Proc. of the Conf. on Raptor Conserv. Techniques. 22-24 March, 1973. Fort Collins, Co. Raptor Res. Rep. No. 2.

2377 **Olendorff, R.R., A.D. Miller and R.N. Lehman.** *1981.* Suggested practices for raptor protection on powerlines. The state of the art in 1981. Raptor Res. Foundation, Inc. Raptor Res. Report No. 4. 111 pp.

2378 **Oliveira, N.G.** *1978.* European news. Brit. Birds 71(6):254-258.

2379 **Olivier, G.** *1948.* Capture d'un aigle fauve *(Aquila c. chrysaetos* (L.)) en Seine-Inferieure. Oiseau Rev. Fr. Ornithol. 18:172.

2380 **Olivier, G.** *1948.* Capture d'un aigle fauve *(Aquila c. chrysaetos* (L.)). Oiseau Rev. Fr. Ornithol. 18:172.

2381 **Olsen, J.** *1980.* Wedge-tailed eagles raise three young. Aust. Raptor Assoc. News 1(4):13.

2382 **Olsen, J. and P. Olsen.** *1980.* Information derived from captive raptors: a selected review. Corella 4(2):25-29.

2383 **Orde, J.W.P.** *1861.* Letter on the breeding of *Aquila chrysaetos* in a fir-tree in Perthshire. Ibis 3(9):112-113.

2384 **Orford, H.J.L.** *1962.* Rearing a tawny eagle. Afr. Wild Life 16(1):71-75.

2385 **Orlando, C.** *1937.* Catture di aquile in Sicilia. Riv. Ital. Ornitol. 7(2):141-142.

2386 **Orlando, C.** *1938.* Catture di aquile in Sicilia. Riv. Ital. Ornitol. 8(2):96.

2387 **Orlando, C.** *1943.* Catture di aquile in Sicilia. Riv. Ital. Ornitol. 13(2):62.

2388 **Orlando, C.** *1955.* Catture di aquile in Sicilia. Riv. Ital. Ornitol. 25(4):195-197.

2389 **Orlando, C.** *1962.* Aquile in Sicilia. Riv. Ital. Ornitol. 32(4):317.

2390 **Orlova, E.A. and V.Y. Ilyiashenko.** *1978.* Materialy po pitaniyu nekotorykh dnevnykh khistchenykh ptits i sov yugo-vostochno-go Altaya. (Data on food of some diurnal birds of prey and owls in southeastern Altai.) Trudy Zool. Inst. An SSR 76:94-100.

2391 **Orton, D.A.** *1974.* Talon grappling by golden eagles. Bird Study 21(3):219-220.

2392 **Osgood, W.H.** *1892.* After golden eagles. Oologist 9(5):134-135.

2393 **Osmolovskaya, V.I.** *1953.* (Geographic distribution of birds of prey in the lowlands of Kazakhstan and their significance to the destruction of pest rodents.) Trudy Inst. Geograph. 54(1):219-307. (In Russian)

2394 **Ostroumov, A.G.** *1961.* O zimovanii nekotoryh ptits na Kamchatke. (Wintering of some birds in Kamchatka.) Bull. MOIP, Otdel Biology 66(6):151-153. (In Russian, English Summ.)

2395 **Over, W.H. and C.S. Thoms.** *1920.* Birds of South Dakota. South Dakota Geol. and Nat. Hist. Survey, Series 21, Bull. 9. 142 pp.

2396 **Owre, O.T.** *1968.* Records of Falconiformes from the Lake Rudolph area, Kenya. Bull. Brit. Ornithol. Club 88(8):151-152.

2397 **P., G.** *1960.* The golden eagle. Wyoming Wildl. 24(6):7-8.

2398 **Paccaud, O.** *1962.* Observations d'ete dans la region des Plans-sur-Bex, Alpes vaudoises. Nos Oiseaux 26(10/11):257-273.

2399 **Pack, A.N.** *1928.* Blood money for eagles. N. Am. Rev. 226(1):22-26.

2400 **Packard, R.L., E.G. Bolen, M.K. Rylander and F.A. Hudson.** *1969.* Golden eagle - livestock relationships: a survey. Texas Tech. College, Int. Center for Arid and Semi-Arid Land Studies. Spec. Rep. No. 20. 82 pp.

2401 **Page, J.L. and D.J. Seibert.** *1973.* Inventory of golden eagle nests in Elko County, Nevada. California-Nevada Wildl. p. 1-7.

2402 **Paget-Wilkes, A.H.** *1924.* Notes on the birds of the Grahamstown District, Cape Province, South Africa. Ibis (11th Ser.) 6(4):720-763.

2403 **Paget-Wilkes, A.H.** *1925.* Migration in Cameroon. Ibis (12th Ser.) 1(1):525-539.

2404 **Paget-Wilkes, A.H.** *1928.* The birds of the region south of Lake Nyasa. Part I. Non-Passerine birds. Ibis (14th Ser.) 4(4):690-748.

2405 **Paget-Wilkes, A.H.** *1938.* Notes on the breeding of some species in north-eastern Uganda. Ibis 2(1):119-129.

2406 **Paget-Wilkes, A.H. and A.G.L. Sladen.** *1930.* Notes on the breeding habits and eggs of some of the birds of the region south of Lake Nyasa. Ibis 6(3):436-453.

2407 **Palasthy, J. and B.-U. Meyburg.** *1973.* Zur ernahrung des schreiadlers *(Aquila pomarina)* in der Ostslowakei unter atypischen klimatischen Bedingungen. Ornithol. Mitt. 25(4):62-72. (English Summ.)

2408 **Palasthy, J., A. Mosansky and J. Voskar.** *1967.* Geographische verbreitung und populationsdichte des steinadlers *(Aquila chrysaetos)* in der Ostslowakei. Zool. Listy 16(3):253-268.

2409 **Palmar, C.E.** *1949.* Photographing the golden eagle. Country Life 105(2716):246-248.

2410 **Palmar, C.E.** *1952.* The way of an eagle: photographs of the king of birds in the Scottish highlands. Illus. London News 220(5096):1090-1091.

2411 **Palmar, C.E.** *1954.* Scotland's golden eagles at home. Natl. Geog. 105(2):273-286.

2412 **Palmar, C.E.** *1957.* Adventures with golden eagles. Country Life 121(3147):912-915.

2413 **Palmer, E.L.** *1937.* Are they vermin? Cornell Rural School, Leaflet No. 2, Vol. 31. 32 pp.

Master List of Citations (Cont.)

Master No.	Citation
2414	**Palmer, R.S.** *1949.* Notes on birds of the Upper Allagash Drainage and vicinity of Maine, 1948. Bull. Maine Audubon Soc. 5(4):63-76.
2415	**Palmer, R.S.** *1949.* The birds of Maine. Bull. Mus. Comp. Zool. Vol. 102.
2416	**Pankin, N.S.** *1977.* Ekologiya bolshogo podorlika na yuge verkhnego priamuriya. (Ecology of *Aquila clanga* in a southern part of a region of upper Amur River.) Zhivotnyi Mir Dalnego Vostoka, Blagoveshensk. p. 34-43. (In Russian)
2417	**Paris, P.** *1912.* Oiseaux fossiles de France. Oiseau Rev. Fr. Ornithol. 2(37):283-298.
2418	**Parker, J.** *1981.* Raptor vagrants to Britain and north-west Europe. Hawk Trust, Ann. Rep. 11:16-21.
2419	**Parmalee, P.W.** *1958.* Remains of rare and extinct birds from Illinois indian sites. Auk 75(2):169-176.
2420	**Parmalee, P.W.** *1977.* The avifauna from prehistoric Arikara sites in South Dakota. Plains Anthropol. 22:189-222.
2421	**Parrot, C.** *1899.* Zum gegenwartigen stande der schreiadler-frage. J. Ornithol. 47(1):1-32.
2422	**Parslow, J.L.F.** *1967.* Changes in status among breeding birds in Britain and Ireland. Brit. Birds 60(1):2-47.
2423	**Parsons, A.G.** *1959.* First Swedish record of steppe eagle. Var Fagelvarld 18:37-41. (Title Only)
2424	**Passburg, R.E.** *1959.* Bird notes from northern Iran. Ibis 101(2):153-169.
2425	**Paterson, J.** *1908.* Notes on the eagles of Ayrshire. Glasgow Nat. 1(1/2):28-32.
2426	**Patkai, I.** *1951.* Ragadozomadar-kutatasok az 1949. Es 1950. Evekben. Aquila 55-58:75-79. (English Summ.)
2427	**Patkai, I.** *1951.* *Aquila ch. chrysaetos* L. -golden eagle - over Budapest. Aquila 55-58:281.
2428	**Patkai, I.** *1957.* The raptorial-bird collection of M. Hanko in the Ornithological Institute. Aquila 63-64:332.
2429	**Patkai, I.** *1967.* A madartani intezet 1960-1966. Evi madarjelolesei. XXV. Gyuruzesi jelentes. (Bird-banding of the Hungarian Ornithological Institute 1960-1966. 25th report on bird banding). Aquila 73-74:81-107.
2430	**Paton, J.B.** *1975.* Birds of the Gawler Ranges, South Australia. S. Aust. Ornithol. 26(8):180-193.
2431	**Payn, W.H.** *1948.* Notes from Tunisia and eastern Alberia: February 1943 to April 1944. Ibis 90:1-21.
2432	**Peacock, E.** *1980.* Powerline electrocution of raptors. Pages 2-5. *in:* R.P. Howard and J.F. Gore, Eds. Proc. of A Workshop on Raptors and Energy Developments. 25-26 January. Boise, Id. Bonneville Power Admin., U.S. Fish and Wildlife Service, Idaho Power Comp. and Idaho Chapter of the Wildl. Soc.
2433	**Peake, A.** *1929.* Golden eagle attacks sheep. Murrelet 10(2):41.
2434	**Pearson, A.G.** *1946.* Golden eagles. Bird Notes and News 22(1):5.
2435	**Pearson, D.J.** *1978.* East African bird report 1977. Scopus 1(5):114-131.
2436	**Pearson, D.J.** *1979.* East African bird report 1978. Scopus 2(5):105-125
2437	**Pearson, D.J. and B.S. Meadows.** *1979.* Lesser spotted eagles, *Aquila pomarina* in Kenya during 1978/79. With comments on the identification of the species. Scopus 3(2):48-53.
2438	**Pearson, T.G.** *1928.* Eagles and the Alaskan bounty. Bird-Lore 30(1):86-90.
2439	**Pedler, L.P. and G.B. Ragless.** *1978.* Birds observed near Lakes Frome and Callabonna. S. Aust. Ornithol. 27(8):274-276.
2440	**Peiter, W.** *1899.* Das vogelleben in Flur und Wald des deutsch-bohmischen mittelgebirges. J. Ornithol. 47(2):151-207.
2441	**Pelle, I.** *1967.* Ornithological report from East Vojvodina. Aquila 73-74:204.
2442	**Pelle, T., I. Ham, J. Rasajski and T. Gavrilov.** *1977.* Pregled gnezdarica Voivodine. Larus 29-30:171-197.
2443	**Pennide, C.R.** *1903.* Die fange der raubvogel. Vogelwelt 28(7):273-274.
2444	**Pennie, I.D.** *1966.* Scottish ornithologists. 2. Martin Martin. c. 1656-1719. Scott. Birds 4(1):64-73.
2445	**Pennycuick, C.J.** *1972.* Soaring behaviour and performance of some East African birds, observed from a motor glider. Ibis 114(2):178-218.
2446	**Penzhorn, B.L.** *1972.* Bird watching in New York. Bokmakierie 24(2):33-36.
2447	**Perco, F.** *1969.* Sulla cattura di un'Aquila imperiale nei pressi di monfalcone. Riv. Ital. Ornitol. 39(3):236-238.
2448	**Pererva, V. and T. Grebenkina.** *1975.* Stepnoi surok v Kazakhstane. (*Marmota bobak* in Kazakhstan.) Okhota and Okhotn. Khozyaistvo 7:20-21.
2449	**Pererva, V.I.** *1977.* (Quantitative and qualitative estimation of the role of raptors in the biocoenoses of Naurzum Reserve Kazakhstan.) 7th Vsesoyuzn Ornithol. Konf., Tezisy Doklady, Kiev. 1:297-299. (In Russian)
2450	**Perez Chiscano, J.L. and M. Fernandez Cruz.** *1973.* Sobre presencia y conducta reproductora de *Aquila chrysaetos* en Badajoz. Ardeola 19(2):461-464.
2451	**Perfiliev, V.I.** *1976.* K voprosy o pitanii krupnykg khistchnykh ptits srednei Kolymy. (Feeding of large raptors in middle Kolyma River.) Prirodynye Resursy Yakutii, Ikh Ispolzovanie and Okhrana, Yakutsk. p. 57-61. (In Russian)
2452	**Perry, D.H.** *1950.* Wing span of wedge-tailed eagle. W. Aust. Nat. 2(4):89.
2453	**Peters, H.S.** *1936.* A list of external parasites from birds of the eastern part of the United States. Bird-Band. 7(1):9-27.
2454	**Peterson, R.T.** *1961.* Field Guide to Western Birds. Houghton Mifflin Co., Boston. 366 pp.
2455	**Peterson, R.T.** *1963.* A Field Guide to the Birds of Texas. Houghton Mifflin Comp., Boston. 304 pp.
2456	**Peterson, R.T., G. Mountfort and P.A.D. Hollom.** *1954.* A Field Guide to the Birds of Britain and Europe. Houghton Mifflin Co., Boston. 318 pp.
2457	**Petit, E.** *1912.* Capture d'un aigle fauve dans les Pyrenees. Oiseau Rev. Fr. Ornithol. 7(145):82.
2458	**Pettersson, J.** *1977.* En iakttagelse av havsorn *Haliaeetus albicilla* som tar byte fran kungsorn *Aquila chrysaetos.* Var Fagelvarld 36(1):54-55. (English Summ.)
2459	**Pettingill, O.S., Jr. and D. Lancaster. (eds.)** *1971.* Rescue of a golden eagle. Cornell Lab. Ornithol. Newsl. 59(Winter):1.
2460	**Philips, J.R. and D.L. Dindal.** *1977.* Raptor nests as a habitat for invertebrates: a review. Raptor Res. 11(4):86.
2461	**Phillips. A., J. Marshall and G. Monson.** *1964.* The Birds of Arizona. Univ. of Arizona Press, Tucson. 212 pp.
2462	**Phillips, A.R.** *1934.* The present status of hawks and owls in Arizona. Hawk and Owl Society Bull. No. 4. p. 15-17.

Master No. **Citation**

2463 **Phillips, E.L.** *1898.* Narrative of a visit to Somaliland in 1897, with field notes on the birds obtained during the expedition. Ibis 4(15):382-425.

2464 **Phillips, R.L. and A.E. Beske.** *1981.* Golden eagles and coal development in the eastern Powder River Basin of Wyoming. U.S.D.I., Denver Wildlife Res. Center, Ann. Report for 1981. 58 pp.

2465 **Phillips, R.L. and J.M. Lockhart.** *1975.* The effects of coal development on the ecology of birds of prey in southeastern Montana and northern Wyoming. Proj. Report, Montana State Univ. 10 pp.

2466 **Phillips, R.S.** *1949.* A fair deal for our birds of prey? Audubon 51(6):377-381,392-397.

2467 **Pianka, H.D. and E.R. Pianka.** *1970.* Bird censuses from desert localities in western Australia. Emu 70(1):17-22.

2468 **Picchi, C.** *1904.* Elenco degli uccelli conservati nella sua collezione ornitologica Italiana al 29 Febbraio1904 con notizie intorno alla distribozione e nidificazione in Italia. Ornis 12(4):381-562.

2469 **Pichler, A.** *1906.* Beitrage zur kentniss der avifauna der umgebung von-mostar. Vogelwelt 31(7):378-396.

2470 **Picozzi, N. and D. Weir.** *1974.* Breeding biology of the buzzard in Speyside. Brit. Birds 67(5):199-210.

2471 **Pienaar, U. de V.** *1969.* Observations on the nesting habits and predators of breeding colonies of red-billed queleas *Queleas quelea lathami* (A. Smith) in the Kruger National Park. Bokmakierie 21(3,suppl.):XI-XV.

2472 **Pierce, W.M.** *1927.* Examining the nest of a golden eagle in pictures. Ool. Rec. 7(2):34-36.

2473 **Pierce, W.M.** *1927.* Nest of the golden eagle. Ool. Rec. 7(1):11-12.

2474 **Pierce, W.M.** *1927.* Status of the golden eagle in southern California. Calif. Fish and Game 13(1):66.

2475 **Pierce, W.M. and W.L. Finley.** *1939.* The golden eagle. Nature Mag. 32(6):333.

2476 **Pihl, S.** *1977.* Birds in Eilat, spring 1977. Society for the Protection of Nature in Israel, Eilat Field Study Center. 27 pp.

2477 **Pike, O.G.** *1926.* Letter (Lifting power of the golden eagle). Brit. Birds 19(11):291.

2478 **Pike, W.** *1877.* On the habits of the golden eagle. Zoologist (3rd Ser.) 1:103-104.

2479 **Pineau, J.** *1975.* Premiere donnee sur l'aigle criard *Aquila clanga* au Maroc. Alauda 43(2):196.

2480 **Pineau, J. and M. Giraud-Audine.** *1975.* Notes complementaires sur les migrations dans l'extreme nord-ouest du Maroc. (Complementary notes on migration in the extreme northwest of Morrocco.) Alauda 43(2):135-141. (English Summ.)

2481 **Pineau, J. and M. Giraud-Audine.** *1977.* Notes sur les oiseaux nicheurs de l'extreme nord-ouest du Maroc: reproduction et mouvements. (Notes on nesting birds of northwest Morrocco: reproduction and movements.) Alauda 45(1):75-104. (English Summ.)

2482 **Pineau, J. and M. Giraud-Audine.** *1974.* Notes sur les migrateurs traversant l'extreme nord-ouest du Maroc. Alauda 42(2):159-188.

2483 **Pineau, J., E.C. Preston and M. Giraud-Audine.** *1974.* A propos d'une ponte de quarte oeufs de l'aigle imperial d'Espagne *Aquila heliaca adalberti* au Coto Donana. Alauda 42(2):225-226.

2484 **Pinn, G.** *1967.* Die nahrung und jagd des steinadlers *(Aquila chrysaetos).* (Feeding and hunting by the golden eagle.) Anz. Ornithol Ges. Bayern 8(1):59-62.

2485 **Pitman, C.R.S.** *1935.* The eggs of Wahlberg's eagle *Aquila wahlbergi* Sundevall. Ool. Rec. 15(2):33-38.

2486 **Pitman, C.R.S.** *1960.* An unusual case of predation by *Aquila verreauxi.* Bull. Brit. Ornithol. Club 80(4):67.

2487 **Plachetka, K.** *1952.* Orel kriklavy *(Aquila pomarina)* a kane belochvost *(Buteo rufinus)* na rychnovsku. Sylvia 14:121-123. (Russian and German Summ.)

2488 **Platt, J.B.** *1971.* A survey of nesting hawks, eagles, falcons, and owls in Curlew Valley, Utah. Great Basin Nat. 31(2):51-65.

2489 **Platt, J.B. and S.K. Sherrod** *1974.* Interactions between corvids and a golden eagle on a pheasant kill. Auk 91(1):180-181.

2490 **Platt, M.I.** *1951.* Diet of golden eagle. Scott. Nat. 63(1):67.

2491 **Platt, S.W.** *1975.* Population estimates of breeding prairie falcons and golden eagles in New Mexico. New Mexico Dept. of Game and Fish. Endang. Species Proj. FW-17-R. 29 pp.

2492 **Platt, S.W.** *1975.* Status and management recommendations for raptorial birds inhabiting the Challis Planning Unit, Salmon District, BLM, U.S.D.I., Idaho. U.S.D.I., Bureau of Land Management, Salmon, ID. 31 pp.

2493 **Pleske, T.** *1878.* Ornithologische notizen aus Ost-Russland. J. Ornithol. 26(1):89-94.

2494 **Podenzana, G.** *1912.* Aggiunte all'avifauna della Lunigiana. Riv. Ital. Ornitol. 2(1):10.

2495 **Poljakow, G.** *1923.* Verbreitung der vogel im Beckem des oberen Irtysch. Anz. Ornithol. Ges. Bayern 15(4):359-382.

2496 **Pollard, J.R.T.** *1946.* Note on a nest of Verreaux's eagle in Kenya Colony. Ibis 88:521-522.

2497 **Poncy, R.** *1936.* Notes ornithologiques concernant le departement de la Haute-Savoie. Alauda 8(3/4):332-341.

2498 **Poole, E.L.** *1932.* Some recent records from Reading, Pennsylvania. Auk 49(2):233-235.

2499 **Poole, E.L.** *1934.* The hawk migration along the Kittatinny Ridge in Pennsylvania. Auk 51(1):17-20.

2500 **Poole, E.L.** *1938.* Weights and wing areas in North American birds. Auk 55(3):511-517.

2501 **Poole, E.L.** *1964.* Pennsylvania Birds. Livingston Publ. Comp. Narberth, Pa. 94 pp.

2502 **Popov, V.A.** *1977.* Ptitsi Volzhsko-Kamskogo kraya. (Birds of Volga-Kama region. Non-Passeriformes.) Nauka, Moscow. p. 1-126.

2503 **Porter, R. and I. Willis.** *1968.* The autumn migration of soaring birds at the Bosphorous. Ibis 110(4):520-536.

2504 **Porter, R.F.** *1969.* Imperial eagle. Birds of the World, I.P.C. Magazine, Ltd., London.

2505 **Porter, R.F., I. Willis, S. Christensen and B.P. Nielsen.** *1974.* Flight Identification of European Raptors. T. & A.D. Poyser, Berkhamstead, England.

2506 **Portmann, A.** *1947.* Etudes sur la cerebralisation chez les oiseaux. Alauda 15(1):1-15.

2507 **Poslawski, A.N.** *1967.* Der steppenadler. Falke 14(5):156-158.

2508 **Pourtois, A.** *1972.* Nouvelle donnee a propos de l'aigle pomarin. Alauda 40(3):297.

2509 **Pousar, V.** *1905.* Ornithologische beobachtungen aus dem Kirchspial Tammela in Finnland. Ornithol. Jahrbuch 16(5/6):161-185.

Master No.	Citation
2510	**Powys, T.L.** *1860.* Notes on birds observed in the Ionian Islands, and the Provinces of Albania proper, Edipus, Acarnania and Montenegro. Ibis 2(5):1-10.
2511	**Pralle, W.A.E.** *1852.* Ueber den standort des horstes schreiadlers, *Aquila naevia.* Naumannia 2(1):27.
2512	**Prange, H.** *1975.* See- *(Haliaeetus albicilla)* und steinadler *(Aquila chrysaetos)* bei Barhoft nordlich von Straslund. Beitr. Vogelkd. 21(6):487-488.
2513	**Pratt & Son.** *1892.* Spotted eagle in Suffolk. Zoologist (3rd Ser.) 16:25-26.
2514	**Praz, J.-C.** *1973.* A propos des aigles *Aquila chrysaetos* du val de Bagnes. Nos Oiseaux 32:21-22.
2515	**Praz, J.-C. and R. Fellay.** *1974.* Exceptionelle nichee de 3 aiglons *Aquila chrysaetos* menee a bien en Valais. (Exceptional nest of 3 eagle chicks *Aquila chrysaetos* raised successfully in Valais.) Nos Oiseaux 32:237-238.
2516	**Prazak, J.P.** *1897.* Materialien zu einer ornis Ost-Galiziens. J. Ornithol. 45:225-479.
2517	**Prescher, H.** *1934.* Beobachtunger uber den schreiadler *(Aquila pomarina)* 1934. Beitr. Fortpfl.-Biology Vogel 10(6):227-228.
2518	**Preston, F.W.** *1953.* The shapes of birds' eggs. Auk 70(2):160-182.
2519	**Prestt, I.** *1977.* A review of the status of birds of prey in Great Britain. Pages 114-118. *in:* R.D. Chancellor, Ed. Proc. of the World Conf. on Birds of Prey. 1-3 October, 1975. Vienna. I.C.B.P.
2520	**Prestwich, A.A.** *1947.* Retrospect. The Knowsley sale. Avic. Mag. 53(1):16-18.
2521	**Prestwich, A.A.** *1955.* Records of Birds of Prey Bred in Captivity. Arthur A. Prestwich, London.
2522	**Price, B. and R. Kindschy.** *1978.* Raptor and raven inventory of the BLM Vale, Oregon District. U.S.D.I., Bureau of Land Management, Annual Rep., 8 pp.
2523	**Prigogine, A.** *1976.* Additions a l'avifaune du Zaire. Gerfaut 66(3/4):307-308. (English Summ.)
2524	**Priklonski, S.G.** *1958.* (Downy nestlings of *Aquila clanga* by observations in Oka Reserve in 1954-1955.) Trudy Oskogo Gos. Zapovednika 2:165-169. (In Russian)
2525	**Priklonski, S.G.** *1960.* Chislennost i pitanie dnevnykh khistchnykh ptits v Okskom zapovednike. (Number and food of diurnal birds of prey in Oka Reserve.) Trudy 1 Vsesoyuzn Ornithol. Konf. p. 216-222.
2526	**Priklonski, S.G.** *1960.* O pitanii bolshogo podorlika v ustie reki Beloi. (Food of *Aquila clanga* in river Belaya estuary.) Ornitologia 3:174-179. (In Russian)
2527	**Priklonski, S.G.** *1977.* Chislennost redkikh dnevnykh khistchnykh ptits v Okskom zapovednike. (Number of rare birds of prey in Oka Reserve.) 7th Vsesoyuzn Ornithol. Konf., Tezisy Dokl., Kiev. 2:239-240.
2528	**Prill, A.G.** *1922.* Nesting birds of Lake County, Oregon. Wilson Bull. 34(3):131-140.
2529	**Primault, E.** *1962.* Letter concerning aircraft-eagle collison. Nos Osieaux 26(12):319-320.
2530	**Pringle, H.L.** *1951.* Notes on eagles in the Bedford District. Bee-Eater 2(4):2-4.
2531	**Pringle, V. and H. Glen.** *1953.* Observations on a pair of black eagles *(Pteroaetus verreauxi).* Bee-Eater 4(2):3.
2532	**Pringle, V.L.** *1951.* Should eagles be classed as vermin? Bee-Eater 2(3):4.
2533	**Priolo, A.** *1956.* Catture di Aquile in Sicilia. Riv. Ital. Ornitol. (2nd Ser.) 26(1):36-37.
2534	**Proctor, C.A.** *1952.* Golden eagles in New Hampshire. New Hampshire Bird News 5(4):15.
2535	**Prout-Jones, D.V.** *1974.* An Introduction to the Birds of Prey of South Africa. Purnell, Cape Town, South Africa. 127 pp.
2536	**Prozesky, O.P.M.** *1964.* Comprehensive bird concentration at Lake Ngami. Afr. Wild Life 18(2):137-142.
2537	**Pukinski, Y.B.** *1966.* K pitaniyu bolshogo podorlika v gnezdovoi period. (Food of *Aquila clanga* in the nesting period.) 6 Pribalt, Ornithol. Konf., Materialy, Vilnus. p. 125-127.
2538	**Pukinski, Y.B.** *1969.* K metidike aviaucheta khistchnykh ptits. (Methods for aerial survey of raptors.) 5 Vsesoyuzn. Ornithol. Konf., Materialy, Kniga 2, Ashkhabad. p. 517-519.
2539	**Pukinski, Y.B.** *1971.* Redkie khistchnye ptitsi. (Rare birds of prey.) Okhrana Prirody, Moscow. 3:83-90.
2540	**Pulkrabek, M. and D. O'Brien.** *1974.* An inventory of raptor nesting in Harding County, South Dakota - 1974. South Dakota Dept. of Game, Fish and Parks. Proj. W-95-R-8. Job No. 9-1.
2541	**Pumphrey, R.J.** *1948.* The sense organs of birds. Ibis 90(2):171-199.
2542	**Purdy, J.B.** *1899.* The golden eagle and barn owl at Northville, Wayne Co., Mich. Auk 16(1):77.
2543	**Pyman, G.A.** *1960.* Report on rare birds in Great Britain and Ireland in 1958. Brit. Birds 53(4):153-173.
2544	**Quednau, A.** *1930.* Schelladler *(Aquila clanga* Pall.) am Mauersee. Ornithol. Monatsber. 38(3):69-72.
2545	**Radde, G.** *1885.* Zweiter nachtrag zur ornis Causica fur das jahr 1884. J. Ornithol. 33(1):74-81.
2546	**Radetski, V.R.** *1978.* (Seasonal and geographic changes in food of *Aquila clanga* in some parts of its nesting range.) Trudy Oskogo Gos. Zapovednika 14:123-129. (In Russian)
2547	**Ragless, G.B.** *1957.* Some nesting notes. S. Aust. Ornithol. 22(4):52-53.
2548	**Raines, R.J.** *1962.* The distribution of birds in northeast Greece in summer. Ibis 104(4):490-502
2549	**Raitt, R.J. and S.L. Pimm.** *1976.* Dynamics of bird communities in the Chihuahuan Desert, New Mexico. Condor 78(4):427-442.
2550	**Ramsay, E.P.** *1863.* Notes on birds breeding in the neighbourhood of Sydney. Ibis 5(20):445-447.
2551	**Rand, A.L. and R.L. Fleming.** *1957.* Birds from Nepal. Fieldiana: Zool. 41(1):1-218.
2552	**Rand, A.L. and E.T. Gilliard.** *1967.* The Handbook of New Guinea Birds. Weidenfeld and Nicholson, London. 612 pp.
2553	**Randall, D.** *1975.* Shorty. Defenders 50(4):362-364.
2554	**Randall, D.** *1976.* Eye to eye with the kings of birds. Defenders 51(5):295-299.
2555	**Randall, D.** *1978.* Texas eagle killers convicted and fined. Defenders (April):75-79.
2556	**Randla, T.** *1976.* Nekotorye redkie vidy sredi khistchnykh ptits v Estonii, ikh chislennost i okhrana. (Some rare species of raptors in Estonia, their number and protection.) 9 Pribalt. Ornithol. Konf., Materialy, Vilnus. p. 220-221. (In Russian)
2557	**Randla, T.** *1976.* Eesti roolvlinnud kullilised ja kakulised. Valgus, Tallin. p. 1-200. (In Estonian, English Summ.)

Master No.	Citation
2558	**Randla, T.** *1976.* White-tailed sea eagle, *Haliaeetus albicilla,* and golden eagle, *Aquila crhysaetos,* in Estonia. Ornis Fenn. 53(4):125-127.
2559	**Ratcliffe, D.A.** *1951.* Diet of golden eagle. Scottish Nat. 63(3):189.
2560	**Ratcliffe, D.A.** *1960.* Broken eggs in the nests of sparrowhawk and golden eagle. Brit. Birds 53(3):128-130.
2561	**Ratcliffe, D.A.** *1962.* Breeding density in the peregrine *Falco peregrinus* and raven *Corvus corax* . Ibis 104:13-39.
2562	**Ratcliffe, D.A.** *1965.* Organo-chlorine residues in some raptor and corvid eggs from northern Britain. Brit. Birds 58(3):65-81.
2563	**Ratcliffe, D.A.** *1967.* Decrease in eggshell weight in certain birds of prey. Nature 215(5097):208-210.
2564	**Ratcliffe, D.A.** *1969.* Golden eagle. Birds of the World, I.P.C. Magazines, Ltd. London.
2565	**Ratcliffe, D.A.** *1970.* Changes attributable to pesticides in egg breakage frequency and eggshell thickness in some British birds. J. Appl. Ecol. 7(1):67-115.
2566	**Raub, M.W.** *1892.* The golden eagle in Pennsylvania. Auk 9(2):200.
2567	**Rauhe, H.** *1967.* Steinadler *(Aquila chrysaetos chrysaetos)* im Ahlenmoor. Natur. Kult. Jagd. 20:68-69. (Title Only)
2568	**Raveling, D.G., W.E. Crews and W.D. Klimstra.** *1972.* Activity patterns of Canada Geese during winter. Wilson Bull. 84(3):278-295.
2569	**Rawles, T.H. and L.W. Rawles.** *1938.* Notes on a young golden eagle in Colorado. Condor 40(3):124.
2570	**Rawson, C.L.** *1897.* Another golden eagle in Connecticut. Auk 14(2):215-216.
2571	**Ray, M.S.** *1928.* Record set of eggs of the golden eagle. Condor 30(4):250.
2572	**Raynor, H.M.** *1919.* Golden eagle at Moriches, N.Y. Auk 36(3):421.
2573	**Reagan, A.B.** *1908.* The birds of the Rosebud Indian Reservation, South Dakota. Auk 25(4):462-467.
2574	**Realini, G.** *1974.* Elenco aggiornato degli uccelli accerati nella provincia de Pavia con notizie riguardanti anche la loro attuale frequenza. Riv. Ital. Ornitol. 44(1):1-42.
2575	**Reboussin, R.** *1929.* Faune ornithologique caracteristique du Departement de Loir-et-Cher. Oiseau Rev. Fr. Ornithol. 10(11):655-664.
2576	**Redig, P.T.** *1978.* Raptor rehabilitation: Diagnosis, prognosis, and moral issues. Pages 29-41. *in:* T.A. Geer, Ed. Bird of Prey Management Techniques. British Falconers' Club.
2577	**Reeb, F.** *1977.* Contribution a l'etude de l'avifaune et les migrations en Afghanistan. Alauda 45(4):293-333.
2578	**Reed, C.A.** *1965.* North American Birds Eggs. Dover Publ., Inc., New York.
2579	**Reeh, H.** *1938.* Kongeorn, *Aquila chrysaetos,* par Amager. Dansk Ornithol. 32(3/4):159-178.
2580	**Reese, J.G.** *1973.* Bald eagle migration along the upper Mississippi River in Minnesota. Loon 45(1):19-26.
2581	**Reichardt, M.** *2000.* Tourist guide to the birds of Lesotho. Easa (P.T.L.) Ltd., Lesotho.
2582	**Reichel, W.L., E. Cromartie, T.G. Lamont, B.M. Mulhern and R.M. Prouty.** *1969.* Pesticide residues in eagles. Pest. Monit. J. 3(3):142-144.
2583	**Reid, G.** *1881.* The birds of Lucknow Civil Division. Stray Feathers 10(1/3):1-88.
2584	**Reidinger, R.F., Jr. and D.G. Crabtree.** *1974.* Organochlorine residues in golden eagles, United States -March 1964-July 1971. Pest. Monit. J. 8(1):37-43.
2585	**Reiger, G.** *1974.* In the war on predators there is no middle ground. Natl. Wildl. 12(4):34-39.
2586	**Reiser, O.** *1906.* Greift der steinadler menschen an? Ornithol. Beob. 5(11):175-176.
2587	**Reiser, O.** *1906.* Der steinadler in der Schweiz. Ornithol. Beob. 5(7):102-103.
2588	**Renaut, W.E.** *1930.* Eggs of the golden eagle. Bull. Brit. Ool. Assoc. 2(24):143-144.
2589	**Renkhoff, M.** *1972.* Bemerkenswerte beobachtungen 1971 im Balik-Golu-Gebiet (Nordturkei). Ornithol. Mitt. 24(4):63-73.
2590	**Renn, F.B.** *1976.* Birch Creek-Little Lost raptor inventory. Idaho State Univ., Report. 19 pp.
2591	**Renwald, J.D. and M.A. Taylor.** *1980.* Comparative winter golden eagle populations in a cattle area and sheep area in southeast New Mexico. U.S.D.I. Bureau of Land Management, Manuscript. 9 pp.
2592	**Repetto, L.** *1973.* Rinvenimento di *Aquila anatraia minore* in Provincia di Pavia. Riv. Ital. Ornitol. 43(3):485.
2593	**Retfalvi, L.** *1963.* Notes on the birds of San Juan Islands, Washington. Murrelet 44(1):12-13.
2594	**Reynolds, H.V., III.** *1969.* Population status of the golden eagle in south-central Montana. M.S. Thesis. Univ. of Montana, Missoula, MT. 61 pp.
2595	**Rhoads, S.N.** *1895.* Contributions to the zoology of Tennessee, No. 2. Birds. Proc. Acad. Nat. Sci. Phila. 47:463-501.
2596	**Richard, A.** *1915.* Excursions au nid de l'aigle royal. Nos Oiseaux 2:1-15.
2597	**Richard, A.** *1919.* Excursions au nid de l'aigle royal. II. Un vieux nid. Nos Oiseaux 4:197-204.
2598	**Richard, A.** *1919.* Excursions au nid de l'aigle royal. III. Escalade d'une aire. Nos Oiseaux 4:213-221.
2599	**Richard, A.** *1920.* Excursions au nid de l'aigle royal. IV. Une couvee abandonnee. Nos Oiseaux 4:295-302.
2600	**Richard, A.** *1925.* Excursions au nid de l'aigle royal. V. Une aire avec deux aiglons. Nos Oiseaux 8:1-7.
2601	**Richard, A.** *1931.* Excursions au nid de l'aigle royal. VI. Marquage d'un aiglon. Nos Oiseaux 11:83-88.
2602	**Richard, A.** *1936.* Excursions au nid de l'aigle royal. VII. Aire nouvelle. Deux aiglons. Nos Oiseaux 13:81-89.
2603	**Richardson, F.** *1943.* Functional aspects of the inner vane of remiges. Auk 60(1):44-50.
2604	**Richmond, C.W.** *1902. Aquila chrysaetos* . Auk 19(1):79.
2605	**Richmond, K.** *1955.* Attack by an eagle. Field (March 12):821. (Title Only)
2606	**Richmond, W.K.** *1959.* Eagle watching on the high hills. Country Life 125(3247):770-771.
2607	**Rickett, C.B. and J.D.D. La Touche.** *1896.* Additional observations on the birds of the Province of Fohkien. Ibis (7th Ser.) 2(8):489-495.
2608	**Ridgway, R.** *1873.* Catalogue of the ornithological collection of the Boston Society of Natural History. Part II. Falconidae. Proc. Boston Soc. Nat. Hist. 16:43-106.
2609	**Ridgway, R.** *1875.* Outlines of a natural arrangement of the Falconidae. U.S. Department of the Interior. Report.
2610	**Ridiger, W.** *1903.* Der schreiadler, *Aquila naevia* M. et W. (pomarina). Deutsche Jager-Zeitung 41:75-77. (Title Only)
2611	**Ridpath, M.G.** *1969.* The wedge-tailed eagle. Aust. Nat. Hist. 16(7):209-212.

Master No.	Citation
2612	**Ridpath, M.G. and M.G. Brooker.** *1975.* The breeding of the Australian wedge-tailed eagle *Aquila audax* in semi-arid habitats. Emu 74(Suppl.):291. (Abstract)
2613	**Riedal, E.J.** *1931.* Occurence of the black eagle, *Aquila verreauxi* Less., in Tanganyika Territory. Bateleur 3(3):72.
2614	**Riemann, H., D. Behymer, M. Fowler, D. Ley, T. Shultz, R. Ruppanner and J. King.** *1977.* Serological investigation of captive and free living raptors. Raptor Res. 11(4):104-111.
2615	**Riggenbach, H.E. and H. Budich.** *1954.* Bericht der beratungsstelle fur praktischen vogelschutz uber das jahr 1953. Ornithol. Beob. 51(1/2):36-39.
2616	**Riney, T.** *1951.* Relationships between birds and deer. Condor 53(4):178-185.
2617	**Riols, C.** *1969.* Observation d'un aigle presume criard en Brenne. Alauda 37(4):348.
2618	**Ripley, S.D. and G.M. Bond.** *1971.* Systematic notes on a collection of birds from Kenya. Smithsonian Inst. Contrib. Zool. No. 111. p. 1-21.
2619	**Ripley, S.D., II.** *1961.* A Synopsis of the Birds of India and Pakistan. Bombay Nat. Hist. Soc. 703 pp.
2620	**Rippon, G.** *1901.* On the birds of the southern Shan States, Burma. Ibis (8th Ser.) 1(4):525-561.
2621	**Rising, J.D.** *1966.* Mid-winter bird count for 1965. Kansas Ornithol. Soc. Bull. 17(1):1-10.
2622	**Ritter, A.** *1899.* Zei seltene erscheinungen der steierischen ornis *(Aquila clanga* Pall. und *Lestris parasitica* (L.)). Ornithol. Jahrbuch 10(2):72-74.
2623	**Ritter, A.** *1977.* Steinadlerbeobachtung im Kreis Strasburg. Falke 24(12):427.
2624	**Riviere, B.B.** *1935.* Ornithological report for Norfolk for 1934. Brit. Birds 28(12):354-363.
2625	**Rivoire, A.** *1952.* Les oiseaux de la Montagne Sainte-Victoire. Alauda 20(2):85-101.
2626	**Rix, C.E.** *1970.* Birds of the northern Territory. S. Aust. Ornithol. 25(6):147-190.
2627	**Rizzardi, E.** *1977.* Birds of prey in Italy. I. General report. Pages 129-133. *in:* R.D. Chancellor, Ed. Proc. of the World Conf. on Birds of Prey. 1-3 October, 1975. I.C.B.P.
2628	**Roads, K.M.** *1938.* The golden eagle again noted in Ohio. Wilson Bull. 50(3):202.
2629	**Robb, W.H.** *1932.* The death of beauty. Hawk and Owl Society Bull. No. 1. p. 37.
2630	**Robbins, C.S., B. Bruun, H.S. Zim and A. Singer.** *1966.* Birds of North America. Golden Press, New York. 340 pp.
2631	**Robel, D.** *1973.* Zum vorkommen des steinadlers in den bradenburgischen Bezirken. Falke 20(4):131-133.
2632	**Roberts, A.** *1862.* Occurrence of the golden eagle near Driffield. Zoologist 20:7880.
2633	**Roberts, A.** *1906.* Ornithological notes from Wolmaransstad, Transvaal. J. S. African Ornithol. Union 2(1):10-14.
2634	**Roberts, A.** *1913.* Egg-collecting in the Bushveld. J. S. African Ornithol. Union 9(1):1-45.
2635	**Roberts, A.** *1930.* Some recent important records of South African birds. Ostrich 1(2):61-68.
2636	**Roberts, A.** *1935.* Dr. H. Exton and his unpublished notes on South African birds. Ostrich 6(1):1-33.
2637	**Roberts, A.** *1971.* Birds of South Africa. (Rev. Ed.). MacLachan and Liversidge, Cape Town, S. Africa. 542 pp.
2638	**Roberts, A.J.R.** *1916.* Brooding of the golden eagle. Country Life 40(1040):695-699.
2639	**Roberts, A.M.** *1893.* Canadian eagle. Ornithol. and Ool. 18(7):97.
2640	**Roberts, T.S.** *1932.* The birds of Minnesota. Univ. of Minnesota Press, Minn., Mn. 821 pp.
2641	**Robien, P.** *1938.* Schelladler auf der naturwarte Monne. Ornithol. Monatsber. 46(2):53.
2645	**Robinson, P.** *1978.* Illegal management. Pages 149-156. *in:* T.A. Geer, Ed. Bird of Prey Management Techniques. British Falconers' Club.
2642	**Robinson, D.** *1979.* Golden eagle and soaring hawks. Defenders 54(5):281-289.
2643	**Robinson, E.R. and C.T. Stuart.** *1975.* Bird feeding records from South West Africa. Ostrich 46(2):190.
2644	**Robinson, J.** *1972.* Where the eagle can be free. Am. Forests 78(5):24-28.
2646	**Rochlitzer, R.** *1969.* Der schreiadler brutvogel an der Mittelelbe. Apus 1:277-281. (Title Only)
2647	**Rochon-Duvigneaud, A.** *1933.* Les aigles de Suisse, de Carl Stemmler. Alauda 5(4):511-516.
2648	**Rochon-Duvigneaud, A.** *1935.* Lescun, Aubisque, Gavarnie, Ordesa (Pyrenees). Alauda 7(4):506-509.
2649	**Rochon-Duvigneaud, A.** *1952.* Les armes des rapaces. Oiseau Rev. Fr. Ornithol. 22(2):91-106.
2650	**Rochon-Duvigneaud, D.** *1934.* Un aigle criard *(Aquila clanga)* dans les Gorges du Tarn? Alauda 6(1):118-119.
2651	**Rodd, E.H.** *1861.* Specimen of the spotted eagle *(Aquila naevia).* Ibis 3(9):118.
2652	**Rodd, E.H.** *1861.* Occurrence of the spotted eagle *(Falco naevius)* in Cornwall. Zoologist 19:7311-7312.
2653	**Roest, A.I.** *1961.* Interspecific relationships among birds. Auk 78(3):433-435.
2654	**Roever, J.M. and W. Roever.** *1973.* The North American Eagles. Steck-Vaughan., Austin, TX. 30 pp.
2655	**Rogers, M.T.** *1944.* A specimen of the golden eagle from Milledgeville. Oriole 9(1/2):12.
2656	**Roop, L.** *1973.* Eagles today. Wyoming Wildl. 37(6):14-19.
2657	**Roos, G.** *1964.* Competition between a wintering golden eagle *(Aquila chrysaetos)* and a white-tailed eagle *(Haliaetus albicilla)* in south Sweden. Var Fagelvarld 23(4):416-418. (Title Only)
2658	**Rosen, M.N., K. D'Amico and E.J. O'Neill.** *1973.* First record of a golden eagle death due to avian cholera. Calif. Fish and Game 59(3):209-210.
2659	**Ross, W.M.** *1941.* Aerial display by a pair of golden eagles. Brit. Birds 35(4):82-83.
2660	**Rothschild, E.V.** *1959.* Sovmestnoe gnezdovanie vorobiev i khistchnykh ptits v pustyne. (Joint nesting of sparrows and birds of prey in the desert.) Zool. Zhur. 38(6):935-938. (In Russian, English Summ.)
2661	**Rothschild, L.** *1900.* Exhibit of *Aquila fluvescens.* Ibis (7th Ser.) 6(23):530-531.
2662	**Rothschild, L.** *1921.* On a collection of birds from west-central and north-western Yunnan. Novitates Zool. 28:14-67.
2663	**Rothschild, L.** *1926.* On the avifauna of Yunnan, with critical notes. Novitates Zool. 33(3):189-343.
2664	**Rothschild, L.** *1927.* Exhibition and remarks on the eggs of eagles and vultures, and of the smaller Falconidae. Bull. Brit. Ool. Assoc. 1(9):89-90.
2665	**Rothschild, L.** *1930.* Exhibition of golden eagle eggs. Bull. Brit. Ool. Assoc. 2(12):139-141.
2666	**Rothschild, W.** *1900.* Untitled. Bull. Brit. Ornithol. Club 10:51-52.

Master No.	Citation
2667	**Rothschild, W. and E. Hartert.** *1915.* Notes on Papuan birds. Novitates Zool. 22:46-60.
2668	**Rouse, G.H.** *1888.* How we shot a golden eagle. Oologist 5(1):4-5.
2669	**Rowe, E.G.** *1947.* The breeding biology of *Aquila verreauxi* Lesson. Ibis 89(Oct.):576-606.
2670	**Rowe, E.G.** *1947.* The breeding biology of *Aquila verreauxi* Lesson. Ibis 89:387-410.
2671	**Rowley, I.** *1961.* Birds of Dookie Agricultural College, Victoria. Emu 61(1):7-17.
2672	**Rowley, I.** *1970.* Lamb predation in Australia: incidence, predisposing conditions, and the identification of wounds. CSIRO Wildl. Res. 15(1):79-123.
2673	**Rowley, I.** *1974.* Bird Life. Taplinger Publ. Co., Inc., NY. 284 pp.
2674	**Rubatto, F.** *1945.* Cattura di *Aquila macchiata* od anatraia maggiore nel Chierese (Torino). Riv. Ital. Ornitol. (2nd Ser.) 15(2/3):84-85.
2675	**Rucner, D.** *1952.* Birds of the Neretva Valley. Larus 6-7:41-111. (English Trans.)
2676	**Rucner, D.** *1968.* Further data on the birds of the Neretva Valley. Larus 21-22:95-106. (English Trans.)
2677	**Rucner, D. and R. Rucner.** *1970.* Contribution to the study of bird populations in the biotopes of Baranja. Larus 24:35-63. (English Trans.)
2678	**Rucner, R.** *1954.* Bird banding in 1953. Larus 8:5-26.
2679	**Rucner, R.** *1959.* Bird banding in 1957 and 1958. Larus 12-13:7-36. (English Trans.)
2680	**Rucner, R.** *1968.* The avifauna of Lonjsko Polje. Larus 21-22:33-64. (English Trans.)
2681	**Rudebeck, G.** *1950.* The migration of birds of prey in southern Sweden, particularly at Falsterbo. Pages 317-319. *in:* S. Horstadius, Ed. Proc. 10th Int. Ornithol. Cong., Upsala.
2682	**Rudebeck, G.** *1950.* Studies on bird-migration based on field studies in southern Sweden. Var Fagelvarld (Suppl. 1):1-48.
2683	**Rudebeck, G.** *1950.* Studies on bird migration. Berlingsha Boktrycheriet, Lund. 148 pp.
2684	**Rudeen, S. and L. R. Powers.** *1978.* Body temperature of a nestling golden eagle. Condor 80(4):447-449.
2685	**Rudiger, W.** *1915.* Zur vogelfauna der umgebung von eberswalde. Vogelwelt 36(2):107-110.
2686	**Rudiger, W.** *1920.* Beobachtungen an schreiadlern. Z. Ool. Ornithol. 25:20-25. (Title Only)
2687	**Rudy, C.** *1970.* The 1969 foray bird list. Redstart 37(1):2-12.
2688	**Ruhl, H.D. and G.W. Brandt.** *1946.* Sheep-killing eagles? Jack Pine Warb. 24(3):105-107.
2689	**Rukovsky, N.N. and A.G. Kupriyanov.** *1968.* Pine marten *(Martes martes* L.) as prey of the golden eagle *(Aquila chrysaetos* L). Zool. Zhur. 47(3):476.
2690	**Ruthenberg, H.** *1965.* Tod eines schreiadlers durch herbizide. Falke 12(12):421.
2691	**Ruthke, P.** *1933.* Steinadler in Pommern. Ornithol. Monatsber. 41(4):122.
2692	**Ruthke, P.** *1938.* Beobachtungen am Harpartrask, Tenala. Ornis Fenn. 15(2/3):70-73.
2693	**Ruthke, P.** *1939.* Wieder ein schelladler in Pommern. Ornithol. Monatsber. 47(2):58-59.
2694	**Ryabov, V.F.** *1969.* Biologiya pitanita stepnogo orla i kurgannika v severo-zapadnom Kazakhstane. (Feeding biology of *Aquila nipalensis* and *Buteo rufinus* in north-western Kazakhstan.) 5 Vsesoyuzn. Ornithol. Konf., Materialy. 2 Kniga, Ashkabad. p. 551-554. (In Russian)
2695	**Rydzewski, W.** *1949.* Sprawozdanie z dzialalnosci Stac j i Badania wedrowek ptakow za rok 1938. Acta Ornithol. 4(1):1-113. (In Polish, French Summ.)
2696	**Rydzewski, W.** *1951.* A historical review of bird marking. Dansk Ornithol. 45(2):61-95.
2697	**Rydzewski, W.** *1976.* Longevity records. Ring 86-87:17-18.
2698	**Rydzewski, W.** *1978.* The longevity of ringed birds. Ring 8(96/97):218-262.
2699	**Safriel, U.** *1968.* Bird migration at Elat, Israel. Ibis 110(3):283-320.
2700	**Sage, B.L.** *1962.* Albinism and melanism in birds. Brit. Birds 55(6):201-222.
2701	**Sage, B.L.** *1974.* Ecological distribution of birds in the Atigun and Sagavanirktok River Valleys, arctic Alaska. Can. Field-Nat. 88(3):281-291.
2702	**Sage, B.L.** *1975.* Recent observations in the Wrangell Mountains, Alaska. Condor 77(2):206-207.
2703	**Saghy, A.** *1955.* Adatok a Gerecse-Hegyseg es a Kozepso-Duna Madarvilagahoz. Aquila 59-62:191-200.
2704	**Saghy, A.** *1965.* Excerpts from the letter of Dr. M. Vasvari. Aquila 71-72:246.
2705	**Sailler, E.** *1979.* Observations sur le comportement charognard de quelques oiseaux en Corse. Nos Oiseaux 35(3):101-112.
2706	**Salvadori, T.** *1880.* Ornitologia della Papuasia e delle Molucche. Stamperia Reale Della Ditta G.B. Paravia E Comp., Torino. 573 pp.
2707	**Salvan, J.** *1968.* Contribution a l'etude des oiseaux de Tchad. Oiseau Rev. Fr. Ornithol. 38(1):53-85.
2708	**Salvan, J.** *1972.* Notes ornithologiques du Congo-Brazzaville. Oiseau Rev. Fr. Ornithol. 42(4):240-252.
2709	**Salvin, O.** *1859.* Five months' birds'-nesting in the eastern Atlas. Ibis 1(2):174-191.
2710	**Sampson, R.N.** *1971.* The mighty eagles of Snake River. All Outdoors 25(7):28-32.
2711	**Sanborn, H.C.** *1932.* Notes on a captive golden eagle. Migrant 3(4):39-40.
2712	**Sandeman, P.W.** *1957.* The breeding success of golden eagles in the southern Grampians. Scott. Nat. 69(3):148-152.
2713	**Sanders, B.A.** *1976.* Prescribed burning for golden eagle habitat management. Pages 109-110. *in:* Proc. of the Fire by Prescription Symposium. 13-15 October. Atlanta.
2714	**Sass, H.R.** *1925.* Golden wanderer. Sat. Even. Post 197(31):43-44,46,48.
2715	**Saunders, A.A.** *1921.* A distributional list of the birds of Montana. Pac. Coast Avifauna, No. 14. 194 pp.
2716	**Saunders, H.** *1869.* Notes on the ornithology of Italy and Spain. Ibis 5(20):391-403.
2717	**Saunders, H.** *1869.* Ornithological rambles in Spain. Ibis 5(18):170-186.
2718	**Saunders, H.** *1871.* A list of the birds of southern Spain. Ibis 1(1):54-68.
2719	**Saunders, H.** *1884.* Notes on the birds of the Pyrenees. Ibis (5th Ser.) 2(8):365-392.
2720	**Saunders, H.** *1891.* Notes on birds observed in Switzerland; chiefly in the Cantons of Vaud and Neuchatel. Ibis (6th Ser.) 3(10):157-188.

Master No. Citation

2721 **Saunders, W.E.** *1883.* Golden eagle. Ornithol. and Ool. 8(2):14.

2722 **Saunders, W.E.** *1937.* Golden eagle and Richardson's owl in Brant County, Ontario. Can. Field-Nat. 51(3):44.

2723 **Saurola, P.** *1976.* Finnish raptor censuses. Ornis Fenn. 53(4):135-139.

2724 **Saurola, P.** *1978.* Artificial nest construction in Europe. Pages 72-80. *in:* T.A. Geer, Ed. Bird of Prey Management Techniques. British Falconers' Club.

2725 **Saurola, P.** *1979.* Rengastettujen petolintujemme loytymistavat. Lintumies 14:15-21.

2726 **Saurola, P.** *1981.* Maa- ja merikotkan muutosta rengasloytojen valossa. (Recoveries of the golden eagle and the white-tailed eagle ringed in Finland.) Lintumies 16:173-177.

2727 **Savitski, B.P., E.E. Padutov and G.M. Savitskaya.** *1974.* (Changes in number of main groups of nesting birds in Kalmykia under influence of man-made factors.) Biology Issledovania Cheloveka I Zhivotnykh, Elista. p. 6-14. (In Russian)

2728 **Savory, A.** *1953.* Black or Verreaux's eagle. Afr. Wild Life 7:323-324. (Title Only)

2729 **Sayers, B.C.** *1976.* Records of breedings under controlled conditions in Britain: birds of prey. Avic. Mag. 82(2):106-107.

2730 **Sayre, R. (ed.)** *1974.* Antique dealer arrested for selling indian artifacts made of eagle feathers. Audubon 76(2):102.

2731 **Sayre, R. (ed.).** *1974.* A Spanish imperial eagle in California. Audubon 76(1):122.

2732 **Schafer, E.** *1938.* Ornithologische ergebnisse zweier forschungsreisen nach Tibet. J. Ornithol. 86(Special):1-349.

2733 **Schafer, E. and R.M. De Schauensee.** *1939.* Zoological results of the second Dolan expedition to western China and eastern Tibet, 1934-1936. Proc. Acad. Nat. Sci. Philadelphia 90:185-260.

2734 **Schalow, H.** *1908.* Beitrage zur vogelfauna centralasi ens. J. Ornithol. 56(1):72-121.

2735 **Schauensee, R.M., de.** *1957.* On some avian types, principally Gould's, in the collection of the Academy. Proc. of the Acad. Nat. Sci. Phila. 109:123-246.

2736 **Schenk, H.** *1977.* III. Status and conservation of birds of prey in Sardinia. Pages 132-136. *in:* R.D. Chancellor, Ed. Proc. of the World Conf. on Birds of Prey. 1-3 October, 1975. Vienna. I.C.B.P.

2737 **Schenk, J.** *1908.* Der vogelzug in Ungarn im Fruhjahre 1907. Aquila 15:1-141.

2738 **Schenk, J.** *1909.* Der vogelzug in ungarn im Fruhjahre 1908. Aquila 16:1-128.

2739 **Schenk, J.** *1929-1930.* Edelreiherschutz - naturschutz. Bericht 1929-31. Aquila 36-37:361-363.

2740 **Schenk, J.** *1931-1934.* Edelreiherschutz - naturschutz. Bericht 1932-34. Aquila 38-41:334-338.

2741 **Schettino, M.** *1918.* Catture accidentali di *Aquila chrysaetos* (Linneo) nella Campania. Riv. Ital. Ornitol. 4:36-37.

2742 **Schettino, M.** *1923.* Catture di due aquile reali - *Aquila chrysaetos* (L.) - nella Campania. Riv. Ital. Ornitol. 6(1):30.

2743 **Schiemenz, H.** *1964.* Tagung des arbeitskreises greifvogelschutz beim BFA ornithologie und vogelschutz Dresden. Falke 11(2):63-64.

2744 **Schiemenz, H.** *1972.* Die situation der vom Aussterben bedrohten vogel in der DDR. Falke 19(2):42-47.

2745 **Schifferli, A.** *1922.* Steinadler im Weisstannental. Ornithol. Beob. 19(11):175.

2746 **Schifferli, A.** *1936.* 11. und 12. Bericht der Schweiz. Vogelwarte sempach (1934 und 1935). Ornithol. Beob. 33(7):113-142.

2747 **Schifferli, A.** *1955.* Schweizerische ringfundmeldung. Ornithol. Beob. 52(6):184-202.

2748 **Schifferli, A.** *1961.* Schweizerische ringfundmeldung fur 1959 und 1960. Ornithol. Beob. 58(5/6):166-196.

2749 **Schifferli, A. and C. Imboden.** *1972.* Schweizerische ringfundmeldung fur 1969 und 1970. Ornithol. Beob. 69(2):70-109.

2750 **Schifferli, A. and B. Jacquat.** *1973.* Schweizerische ringfundmeldung fur 1971 und 1972. Ornithol. Beob. 70(5/6):227-265.

2751 **Schlegel, R.** *1912.* Beitrage zur avifauna des rachelgebietes (bayerischer wald). Ornithol. Jahrbuch 23(3/4):92-112.

2752 **Schlenker, R.** *1975.* Schreiadler, *Aquila pomarina,* wurde 17 jahre alt. Vogelwarte 28(2):159.

2753 **Schmalzried, J.T.** *1976.* Nesting and food habits of the golden eagle on the Laramie plains. M.S. Thesis. Univ. of Wyoming, Laramie, WY. 86 pp.

2754 **Schmidt, E.** *1958.* Ornithological notes from the surroundings of Pecel. Aquila 65:362.

2755 **Schodde, R.** *1959.* Numbers of native birds handled by commerce. S. Aust. Ornithol. 23(1):26-29.

2756 **Schorger, A.W.** *1945.* The golden eagle in Wisconsin. Pass. Pigeon 7(2):48-51.

2757 **Schroot, M.** *1938.* Beitrag zur brutbiologie des schreiadlers. Dohrniana 17:91-96. (Title Only)

2758 **Schueler, D.G.** *1978.* Incident at Eagle Ranch. Audubon 80(3):41-72.

2759 **Schueler, D.G.** *1980.* Incident at Eagle Ranch. Sierra Club Books, San Francisco. 297 pp.

2760 **Schuhmacher, E.** *1934.* Beobachtungen an einem steinadlerhorst im Karwendelgebirge. Anz. Ornithol. Ges. Bayern 2(8):320-322.

2761 **Schuhmacher, E.** *1944.* Ein beitrag zur kenntnis der brutpflege des steinadlers *(Aquila chrysaetos fulva* (L.)). Beitr. Fortpfl.-Biology Vogel. 20(5/6):113-117.

2762 **Schuhmacher, E.** *1949.* Ein deutches steinadler-gelege. Vogelwelt 70(2):54.

2763 **Schuhmacher, E.** *1949.* Uber die lebensweise des steinadlers *(Aquila chrysaetos* L.). Vogelwelt 70(4):97-101.

2764 **Schuster, L.** *1913.* Oologisches aus deutschostafrika. J. Ornithol. 61(3):540-546.

2765 **Schuster, L.** *1926.* Beitrage zur verbreitung und biologie der vogel deutsch-Ostafrikas. J. Ornithol. 74(3):521-541.

2766 **Schuster, L.** *1928.* Aus der Brutzeit 1927. Beitr. Fortpfl.-Biology Vogel 4(1):9-12.

2767 **Schuster, W.** *1906.* Zum schutze des steinadlers. Ornithol. Beob. 5(12):188.

2768 **Schuz, E.** *1930.* Krabbentaucher, *Alle alle* L., und steinadler, *Aquila chrysaetos* L., als ostpreubische Gaste. Ornithol. Monatsber. 38(3):87.

2769 **Schuz, E.** *1957.* Das verschlingen eigener junger (Kronismus) bei vogeln und seine bedeutung. Vogelwarte 19(1):1-15.

2770 **Sclater, P.L.** *1865.* Remarks on "Notes on the ornithology of Spain." Ibis 1(3):359-360.

2771 **Sclater, P.L.** *1890.* Remarks on the fifth cubital remex of the wing in the Carinatae. Ibis 2(5):77-83.

2772 **Sclater, W.L.** *1912.* On the birds collected by Mr. Claude H.B. Grant at various localities in South Africa. Ibis (9th Ser.) 6(21):1-63.

Master No. **Citation**

2773 **Sclater, W.L. and C. Mackworth-Praed.** *1919.* A list of the birds of the Anglo-Egyptian Sudan, based on the collections of Mr. A.L. Butler,Mr. A. Chapman and Capt. H. Lynes,R.N., and Major Cuthbert Christy. Ibis (11th Ser.) 1(4):628-707.

2774 **Sclater, W.L. and R.E. Moreau.** *1933.* Taxonomic and field notes on some birds of north-eastern Tanganyika Territory. Ibis (13th Ser.) 3(3):399-440.

2775 **Scott, J.W.** *1942.* Mating behavior of the sage grouse. Auk 59(4):477-498.

2776 **Scully, J.** *1876.* A contribution to the ornithology of eastern Turkestan. Stray Feathers 4(1/3):41-205.

2777 **Scully, J.** *1879.* A contribution to the ornithology of Nepal. Stray Feathers 8(2/5):204-368.

2778 **Scully, J.** *1881.* A contribution to the ornithology of Gilgit. Stray Feathers 10(1/3):88-146.

2779 **Sedalistchev, V.T. and V.N. Vinocurov.** *1979.* (Ecology of *Aquila chrysaetos* in Yakutia.) Tezisy Vsesotuzn. Konf. Molodykg Uchenykh, Ecologiya Gnezdovania Ptits I Metody Eje Isuchenia, Samarkand. p. 195-197. (In Russian)

2780 **Sedgwick, E.H.** *1952.* Bird life at Leonora, western Australia. Emu 52(4):285-296.

2781 **Seebohm, H.** *1882.* Notes on the birds of Archangel. Ibis 6(23):371-386.

2782 **Seebohm, H.** *1882.* Notes on the birds of Astrakhan. Ibis 6(22):204-232.

2783 **Seebohm, H.** *1883.* Nomenclature concerning *Aquila naevia.* Ibis 1(1):121-124.

2784 **Seebohm, H.** *1883.* Notes on the birds of the Caucasus. Ibis 1(1):1-37.

2785 **Seebohm, H.** *1884.* Notes on a collection of birds from Lankoran. Ibis (5th Ser.) 2(8):425-429.

2786 **Segnestam, M. and B. Helander.** *1976.* Rovfaglarna i Sverige. (Birds of prey in Sweden). Sverige Natur 67(1):8-15.

2787 **Segnestam, M. and B. Helander.** *1977.* Birds of prey in Sweden. Pages 170-178. *in:* R.D. Chancellor, Ed. Proc. of the World Conf. on Birds of Prey. 1-3 October, 1975. Vienna. I.C.B.P.

2788 **Seibert, D.J., R. J. Oakleaf, J. M. Laughlin and J. L. Page.** *1976.* Nesting ecology of golden eagles in Elko County Nevada. U.S.D.I. Bureau of Land Management, Tech. Note TN-281. 17 pp.

2789 **Seidensticker, J.C., IV.** *1968.* Response of juvenile raptors to DDT in the diet. M.S. Thesis. Univ. of Montana, Missoula, MT. 82 pp.

2790 **Seidensticker, J.C., IV and H.V. Reynolds III.** *1965.* Preliminary studies on the use of a general anaesthetic in Falconiform birds. J. Am. Vet. Med. Assoc. 155(7):1044-1045.

2791 **Selby, P.J.** *1823.* Some observations on the *Falco chrysaetos* and *F. fulvus* authors, proving the identity of the two supposed species. Mem. Wernerian Nat. Hist. Soc. 4(2):428-433.

2792 **Semago, L.L.** *1978.* Migratsii i zimovka berkuta i belokhvosta v verkhnem Pridonie. (Migration and wintering of *Aquila chrysaetos* and *Haliaeetus albicilla* near upper Don River.) 2 Vsesoyuzn. Konf. Po Migratsiyam Ptits, Tezisy, Chast 2, Alma-Ata. p. 136-137. (In Russian)

2793 **Semenov, N.M., A.G. Agafonov, D.S. Rezinko and A.A. Rozhkov.** *1962.* (Dependence of distribution and number of *Aquila nipalensis* on density of sousliks in the steppe near Sarpinski Lakes.) Voprosy Ecol. 6:132-133. (In Russian)

2794 **Sennett, G.B.** *1882.* Capture of the golden eagle in Crawford County, Pennsylvania. Bull. Nuttall Ornithol. Club 7(1):58.

2795 **Serle, W.** *1943.* Further field observations on northern Nigerian birds. Ibis 85:264-300.

2796 **Serle, W., G.J. Morel and W. Hartwig.** *1977.* A Field Guide to the Birds of West Africa. Collins, London. 351 pp.

2797 **Serventy, D.L.** *1952.* New South Wales. Emu 52(2):130-139.

2798 **Serventy, V.N.** *1953.* Branch reports. Western Australia. Emu 53(1):20-33.

2799 **Serventy, V.N.** *1954.* Birds as vermin. Emu 54(1):47-50.

2800 **Servheen, C.** *1976.* Deck-feather molt in bald and golden eagles in relation to feather mounting of radio transmitters. Raptor Res. 10(2):58-60.

2801 **Servheen, C.W.** *1978.* Mountain beaver as a prey species of the golden eagle. Murrelet 59(2):77.

2802 **Seth-Smith, L.M.** *1913.* Notes on birds around Mpumu, Uganda. Ibis (10th Ser.) 1(4):485-508.

2803 **Shamburger, P.** *1952.* Cub vs. eagle. Flying 51(6):32.

2804 **Sharland, M., W.R.B. Oliver and G.M. Storr.** *1956.* Branch reports. Emu 56(1):52-61.

2805 **Sharp, C.S.** *1904.* A set of abnormally large eggs of the golden eagle. Condor 6(6):165-168.

2806 **Sharp, W.M.** *1951.* Observations on predator-prey relations between wild ducks, trumpeter swans, and golden eagles. J. Wildl. Manage. 15(2):224-226.

2807 **Sharpe, H.B.** *1930.* The birds of Marsabil Mountain, Kenya Colony. Bateleur 3(1):9-14.

2808 **Sharpe, R.B.** *1891.* Notes on a second collection of birds made by Mr. W.D. Cumming at Fao, in the Persian Gulf. Ibis (6th Ser.) 3(9):103-116.

2809 **Sharpe, R.B.** *1892.* On the birds collected by Mr. F.J. Jackson, F.Z.S., during his recent expedition to Uganda through the Territory of the Imperial British East-African Company. Ibis (6th Ser.) 4(16):534-555.

2810 **Sharpe, R.B.** *1904.* On a collection of birds from the District of Deelfontein in Cape Colony. Ibis (8th Ser.) 4(13):1-29.

2811 **Sharrock, J.T.R.** *1976.* The Atlas of Breeding Birds in Britain and Ireland. T. & A.D. Poyser, Berkhamsted, England. 479 pp.

2812 **Shaw, T.-H.** *1938.* Double ovaries in some Chinese birds of prey. Nature 142(3607):1079.

2813 **Sheldon, C.** *1909.* List of birds observed on the upper Toklat River near Mt. McKinley, Alaska 1907-1908. Auk 26(1):66-70.

2814 **Sheldon, H.H.** *1909.* Notes on some birds of Kern County. Condor 11(5):168-172.

2815 **Shelley, G.E.** *1871.* Contributions to the ornithology of Egypt. Ibis 1(1):38-54.

2816 **Shelley, G.E.** *1882.* On a collection of birds made by Mr. J.S. Jameson in south-eastern Africa, with notes by Mr. T. Ayres. Ibis (4th Ser.) 22:236-265.

2817 **Sherihai, H.** *1981.* Autumn migration of steppe eagles *(Aquila nipalensis)* over Eilat, 1980. Torgos 1(2):17-20.

2818 **Shevchenko, V.L.** *1976.* Gibel stepnykh orlov. (Destruction of steppe eagles *(Aquila nipalensis*).) Priroda 8:144.

Master No.	Citation
2819	**Shubin, I.G.** *1962.* K pitaniyu khistchnykh zverei ptits Kazakhskogo nagoriya. (Feeding of predatory mammals and birds in the Kazakhstan plateau). Trudy Inst. Zool. An Kazakh. SSR. 17:183-191. (In Russian)
2820	**Shufeldt, R.W.** *1891.* Tertiary fossils of North American birds. Auk 8(4):365-368.
2821	**Shufeldt, R.W.** *1892.* A study of the fossil avifauna of the Equus Beds of the Oregon Desert. J. Acad. Nat. Sci. Philadelphia (2nd Ser.) 9:389-425.
2822	**Shufeldt, R.W.** *1913.* Further studies of fossil birds with descriptions of new and extinct species. Bull. Am. Mus. Nat. Hist. 32(16):285-306.
2823	**Shufeldt, R.W.** *1913.* Review of the fossil fuana of the desert region of Oregon, with a description of additional material collected there. Am. Mus. Nat. Hist. Bull. 32(4):123-178.
2824	**Shufeldt, R.W.** *1913.* Contributions to avian Paleontology. Auk 30:29-39.
2825	**Shufeldt, R.W.** *1922.* On the skeleton of the wedge-tailed eagle (*Uroaetus audax,* Latham). Emu 21(4):294-306.
2826	**Shultz, K.A. and A.W. Shultz.** *1977.* A preliminary analysis of roadside census data. New Mexico Dept. of Game and Fish. Report. 10 pp.
2827	**Shultz, P.I.** *1971.* Birds of prey. New Mexico Wildl. (Sept.-Oct.):16-19.
2828	**Sieber, H.** *1965.* Junger steinadler im Mecklenburg gefangen. Falke 12(7):244.
2829	**Siegfried, W.B.** *1968.* Breeding season, clutch and brood sizes in Verreaux's eagle. Ostrich 39(2):139-145.
2830	**Siewert, H.** *1932.* Der schreiadler. J. Ornithol. 80(1):1-40.
2831	**Silvestri, A.** *1975.* L'aquila in Romagna Riv. Ital. Ornitol. 45(3):302-304.
2832	**Simison, R.L.** *1973.* Some natural enemies join forces to curb electrocution of eagles. Defenders 48(4):419-420.
2833	**Simmons, K.E.L.** *1951.* Raptor migration in the Suez Area: autumn 1949-spring 1950. Ibis 93(3):402-406.
2834	**Simpson, N.H.** *1861.* A fortnight in the Dobrudscha. Ibis 3(12):361-374.
2835	**Simpson, R.B.** *1910.* Trapping a golden eagle. Oologist 27(1):2-3.
2836	**Simpson, W.H.** *1860.* On the nesting of *Aquila imperialis* and *Falco sacer.* Ibis 2(8):375-378.
2837	**Simpson, W.H.** *1860.* Ornithological notes from Mesolonghi and southern Aetola. Ibis 2(7):279-296.
2838	**Simson, F.B.** *1882.* Notes on birds found near Dacca and in the surrounding District of Eastern Bengal. Ibis 6(21):84-95.
2839	**Singer, F.J.** *1974.* Status of the osprey, bald eagle and golden eagle in the Adirondacks. N.Y. Fish and Game J. 21(1):18-31.
2840	**Singh, G.** *1961.* The eastern steppe eagle *(Aquila nipalensis nipalensis* (Hodgson) on the South Col of Everest. J. Bombay Nat. Hist. Soc. 58(1):270.
2841	**Sjostedt, K.** *1954.* Storre skrikorn, *Aquila clanga,* Pall., sedd i Liljendal. (The spotted eagle observed in the commune of Liljendal.) Ornis Fenn. 30-31:95. (English Summ.)
2842	**Skead, C.J.** *1967.* Ecology of birds in the eastern Cape Province. Ostrich (Suppl. 7):1-103.
2843	**Skinner, M.P.** *1924.* The Yellowstone Nature Book. A.C. McClurg and Comp., Chicago. 228 pp.
2844	**Skinner, M.P.** *1925.* The birds of the Yellowstone National Park. Roosevelt Wildl. Bull. 3(1):11-183.
2845	**Skinner, M.P.** *1928.* Yellowstone's winter birds. Condor 30(4):237-242.
2846	**Sladek, J.** *1955.* K ochrane orla krikl'aveho *(Aquila pomarina* Brehm.) na Slovensku. (Der schutz des schreiadlers in der Slowakei.) Ochrana Prirody 10(6):176-181. (German Summ.)
2847	**Sladek, J.** *1957.* Beitrag zur kenntnis der nidobiologie und der postembryonalen entwicklung des schreiadlers *(Aquila pomarina)* . Acta Rer. Nat. Mus. Sloven. 3:1-9. (Title Only)
2848	**Sladek, J.** *1959.* Zum problem des vorzeitigen absterbens des zweiten jungen beim schreiadler *(Aquila pomarina)* . Biologia Bratislava 14:448-454. (Title Only)
2849	**Sladek, J.** *1959.* Prispevok k poznaniu orla Kriklaveho. Zool. Listy 8(2):105-1139 (In Slovakian, German Summ.)
2850	**Sladek, J.** *1959.* Priciny rozsirovania arealu orla kralovskeho *(Aquila heliaca)* v strednej Europe a jeho hniezdenie na Slovensku. Sylvia 16:79-95.
2851	**Slevin, J.R.** *1929.* A contribution to our knowledge of the nesting habits of the golden eagle. Proc. Calif. Acad. of Sciences 18(3):45-71.
2852	**Smeenk, C.** *1974.* Comparative-ecological studies of some East African birds of prey. Ardea 62(1/2):1-97.
2853	**Smith, A.C.** *1868.* A sketch of the birds of Portugal. Ibis (New Ser.) 4(16):428-460.
2854	**Smith, D.G.** *1971.* Population dynamics, habitat selection, and partitioning of breeding raptors in the eastern Great Basin of Utah. Ph.D. Dissertation. Brigham Young Univ., Provo, UT. 260 pp.
2855	**Smith, D.G. and J.R. Murphy.** *1972.* Unusual causes of raptor mortality. Raptor Res. 6(1):4-5.
2856	**Smith, D.G. and J.R. Murphy.** *1973.* Breeding ecology of raptors in the eastern Great Basin of Utah. Brigham Young Univ., Sci. Bull. Biology Ser. 18(3):1-76.
2857	**Smith, E.C.** *1974.* Some additional infomation on birds in Afganistan. Ardea 62(3/4):226-235.
2858	**Smith, E.E. and M. Bush.** *1978.* Haematologic parameters on various species of Strigiformes and Falconiformes. J. Wildl. Dis. 14(4):447-450.
2859	**Smith, G.A.** *1973.* The 1972 spring migration at Derby Hill with remarks on the period 1963-1971. Kingbird 23(1):13-27.
2860	**Smith, K.D** *1968.* The identification of raptors. Brit. Birds 61(6):273-275.
2861	**Smith, K.D.** *1938.* Notes on Corsican birds. Ibis (14th Ser.) 2(2):345-346.
2862	**Smith, K.D.** *1941.* Notes on the birds of Mashonaland, Southern Rhodesia. Ibis (14th Ser.) 5(2):296-301.
2863	**Smith, K.D.** *1955.* Recent records from Eritrea. Ibis 97(1):65-80.
2864	**Smith, K.D.** *1957.* An annotated check list of the birds of Eritrea. Ibis 99(1):1-26.
2865	**Smith, K.D.** *1960.* The passage of Palearctic migrants through Eritrea. Ibis 102(4):356-544.
2866	**Smith, K.D.** *1965.* On the birds of Morrocco. Ibis 107(4):493-526.
2867	**Smith, K.D. and G.B. Popov.** *1953.* On birds attacking desert locust swarms in Eritrea. Entomologist 86(1):3-7.
2868	**Smith, K.V.** *1935.* The birds of "Chowilla" Station and district. S. Aust. Ornithol. 13(2):42-47.
2869	**Smithwick, J.W.P.** *1891.* The eagles of North America. Oologist 8(4/5):97-99.

Master No. **Citation**

2870 **Snelling, J.C.** *1969.* A raptor study in the Kruger National Park. Bokmakierie 21(3,suppl.):7-11.

2871 **Snelling, J.C.** *1970.* Some information obtained from marking large raptors in the Kruger National Park, Republic of South Africa. Ostrich (8 Suppl.):417-427.

2872 **Snelling, J.C.** *1975.* Raptor rehabilitation at the Oklahoma City Zoo. Raptor Res. 9(3/4):33-45.

2873 **Snow, C.** *1973.* Golden eagle *(Aquila chrysaetos)* . Habitat management series for unique or endangered species. U.S.D.I. Bureau of Land Management. Tech. Note TN-239. 52 pp.

2874 **Snow, C.** *1976.* Bennett Hills Planning Unit raptor survey May/June, 1976. U.S.D.I. Bureau of Land Management, Report. 17 pp.

2875 **Snow, D.W.** *1952.* A contribution to the ornithology of north-west Africa. Ibis 94(3):473-498.

2876 **Snyder, C.K.** *1910.* Golden eagles. Oologist 27(4):43-44.

2877 **Snyder, L.L.** *1948.* Additional instances of paired ovaries in raptorial birds. Auk 65(4):602.

2878 **Snyder, L.L.** *1949.* On the distribution of the golden eagle in eastern Canada. Can. Field-Nat. 63(1):39-41.

2879 **Snyder, L.L.** *1957.* Arctic Birds of Canada. Univ. of Toronto Press, Toronto.

2880 **Snyder, N.F.R. and J.W. Wiley.** *1976.* Sexual size dimorphism in hawks and owls of North America. Am. Ornithol. Union, Ornithol. Monogr. No. 20. 96 pp.

2881 **Snyder, R.S.** *1972.* Peregrinations of a golden eagle. Raven 43(3):45-46.

2882 **Solomatin, A.O.** *1970.* Feeding of imperial eagles in the Naurzum Preserve. Ekologiya 1(3):247-249. (English Trans.)

2883 **Solomatin, A.O.** *1974.* (*Aquila heliaca* and *Falco cherrug* in Turgai plateau and their environment.) Bull. MOIP, Otdel Biology 79(3):40-51. (In Russian, English Summ.)

2884 **Sondevall, C.J.** *1886.* On the wings of birds. Ibis 4(16):389-457.

2885 **Sooter, C.A.** *1944.* Methods of grasping and carrying prey. Condor 46(2):88.

2886 **Sopyev, O.** *1962.* Nekotorye dannye po ekologii berkuta i domovogo sycha v vostochnykh Kara-Kumakh. (Some data on the ecology of *Aquila chrysaetos* and *Athene noctua* in east Kara-Kum.) Izvestia An Turkmen SSR, Seria Biology Nauk 2:79-83. (In Russian)

2887 **Sopyev, O.** *1967.* (On the breeding of some birds in the Kara Kum desert.) Ornitologia 8:221-251. (Title Only)

2888 **Sorci, G., B. Massa and G. Cangialosi.** *1973.* Avifauna delle Isole Egadi con notizie riguardanti quella della provincia di Trapani (Sicilia). Riv. Ital. Ornitol. 43(1):1-119.

2889 **Sorensen, L.H.** *1972.* Identification des rapaces en vol. (2 partie). (Identification of raptors in flight.) Aves 9(6):201-225.

2890 **Sorensen, N.V.** *1957.* Albinistisk ung kongeorn *(Aquila chrysaetos* L.) observeret i Sydsjaelland. Dansk Ornithol. 51(3):131-136.

2891 **Soucy, L.J., Jr.** *1975.* Kittatinny Mountains raptor banding station. Station Report, 1975 Results. 14 pp.

2892 **Soucy, L.J., Jr.** *1976.* Raptor report - Kittatinny Mountains. N. Am. Bird Bander 1(3):108-113.

2893 **Sovago, M.** *1958.* Hortobagyi jegyzetek. (Notes from the Hortobagy.) Aquila 65:314-316,364. (English Summ.)

2894 **Sowerby, J.L.** *1898.* On a collection of birds from Fort Chiquaqua, Mashonaland. Ibis (7th Ser.) 4(16):567-575.

2895 **Spangenberg, E. and G. Dementiev.** *1935.* Contribution a l'avifaune du Thian-Chan Occidental (Monts Alexandrowski). Alauda 7(3):336-366.

2896 **Spangenberg, E.P.** *1951.* (Ornithological observations along the state shelterbelt in the steppe of Stavropol District and along the river Manych.) Okhrana Prirody 13:57-65. (In Russian)

2897 **Spangenberg, E.P.** *1972.* Redkie i maloizuchennye ptitsi Darvinskogo Zapovednika. (Rare and less-known birds of Darwin Reserve.) Ornitologia 10:139-150. (In Russian)

2898 **Spencer, C.C.** *1943.* Notes on the life history of Rocky Mountain bighorn sheep in the Tarryall Mountains of Colorado. J. Mammal. 24(1):1-11.

2899 **Spencer, R.** *1958.* Report on bird ringing for 1957. Brit. Birds 51(Suppl.):449-487.

2900 **Spencer, R.** *1960.* Report on bird ringing for 1959. Brit. Birds 53(Suppl.):457-502.

2901 **Spencer, R.** *1969.* Report on bird-ringing for 1968. Brit. Birds 62(10):393-442.

2902 **Spencer, R. and R. Hudson.** *1978.* Report on bird ringing for 1977. Ringing and Migration 2(2):57-104.

2903 **Spencer, R. and R. Hudson.** *1979.* Report on bird-ringing for 1978. Ringing and Migration 2(4):161-208.

2904 **Sperry, C.C.** *1957.* Golden eagle attacks decoy duck. Wilson Bull. 69(1):107-108.

2905 **Spivey, M.** *1973.* A school opens for hunting birds to learn to live. Smithsonian (April):32-38.

2906 **Spofford, S.** *1967.* Plight of golden eagle. Florida Nat. 40(3):113.

2907 **Spofford, W.R.** *1945.* Wintering of golden eagles in Tennessee. Migrant 16(1):6-8.

2908 **Spofford, W.R.** *1946.* Observations on two golden eagles. Auk 63(1):85-87.

2909 **Spofford, W.R.** *1953.* Eagles in New York. Conservationist (Oct.-Nov.):26-27.

2910 **Spofford, W.R.** *1959.* Nesting of the golden eagle at Sugluk Inlet, Ungava, Quebec. Can. Field-Nat. 73(3):177.

2911 **Spofford, W.R.** *1961.* White epaulettes in some Appalachian golden eagles. Prothonotary 27(9):99.

2912 **Spofford, W.R.** *1962.* The status of the golden eagle in North America. Audubon 64(3):149.

2913 **Spofford, W.R.** *1964.* Golden eagle in the Trans-Pecos and Edwards Plateau of Texas. Audubon Cons. Report No. 1. 47 pp.

2914 **Spofford, W.R.** *1964.* Recommendations for conservation and control of the golden eagle. Audubon 66(1):46-47.

2915 **Spofford, W.R.** *1964.* Golden eagle 509-50214. Bird-Band. 35(2):123-124.

2916 **Spofford, W.R.** *1969.* Problems of the golden eagle in North America. Pages 345-347. *in:* J.J. Hickey, Ed. Peregrine Falcon Populations; Their Biology and Decline. Univ. of Wisconsin Press, Madison.

2917 **Spofford, W.R.** *1969.* Hawk Mountain counts as population indices in northeastern North America. Pages 323-333. *in:* J.J. Hickey, Ed. Peregrine Falcon Populations; Their Biology and Decline. Univ. of Wisconsin Press, Madison.

2918 **Spofford, W.R.** *1971.* The breeding status of the golden eagle in the Appalachians. Am. Birds 25(1):3-7.

2919 **Spofford, W.R.** *1971.* The golden eagle - rediscovered. Conservationist 26(1):6-8.

Master No.	Citation
2920	**Spofford, W.R.** *1975.* The problems of golden eagles in the west. Pages 127-128. *in:* J.R. Murphy, C.M. White and B.E. Harrell, Eds. Population Status of Raptors. Proc. of the Conf. on Raptor Conserv. Techniques. 22-24 March, 1973. Fort Collins, Co. Raptor Res. Rep. No. 3.
2921	**Sprunt, A., Jr.** *1935.* Do eagles steal children? Am. Forests 41(6):263-265.
2922	**Sprunt, A., Jr.** *1955.* North American Birds of Prey. Harper and Brothers, New York. 227 pp.
2923	**St. John, O.B.** *1889.* On the birds of southern Afghanistan and Kelat. Ibis (6th Ser.) 1(2):145-180.
2924	**Stabler, R.M.** *1941.* Further studies on trichomoniasis in birds. Auk 58(4):558-562.
2925	**Stabler, R.M. and P.A. Holt.** *1965.* Hematoza from Colorado birds. II: Falconiformes and Strigiformes. J. Parasitol. 51(6):927-928.
2926	**Stabler, R.M. and N.D. Hoy.** *1942.* Measurements of tarsal circumferences from living raptorial birds. Bird-Band. 13(1):9-12.
2927	**Staerk, O.J.** *1954.* The golden eagle in the blackforest. Badischen Landes. Natur. Mitt. 6:110-111. (Title Only)
2928	**Stahlecker, D.W.** *1975.* Impacts of a 230 KV transmission line on Great Plains wildlife. M.S. Thesis. Colorado State Univ., Ft. Collins, CO. 67 pp.
2929	**Stahlecker, D.W.** *1975.* Trends in wintering diurnal raptor populations from central Colorado Christmas Bird Counts. Am. Birds 29(5):936-940.
2930	**Stahlecker, D.W.** *1976.* The impact of an underground nuclear fracturing experiment on cliff-nesting raptors. Wilson Bull. 88(1):151-154.
2931	**Stahlecker, D.W.** *1978.* Effect of a new transmission line on wintering prairie raptors. Condor 80(4):444-446.
2932	**Stahlecker, D.W. and T.E. Behlke.** *1974.* Winter diurnal raptor populations of the three habitat types in northeastern Colorado. Colo. Field Ornithol. 20:6-17.
2933	**Stanley, A.J.** *1937.* Sexual dimorphism in North American hawks. I. Sex organs. J. Morphol. 6(2):321-349.
2934	**Stanley, A.J. and E. Witschi.** *1940.* Germ cell migration in relation to asymmetry in the sex glands of hawks. Anat. Rec. 76(3):329-342.
2935	**Stansell, S.S.S.** *1909.* Birds of central Alberta. Auk 26(4):390-400.
2936	**Staraj, I.** *1956.* Phenological field report of the Opatija Hunting Club. Larus 9-10:197-198. (English Trans.)
2937	**Stauber, E.** *1973.* Suspected riboflavin deficiency in a golden eagle. J. Am. Vet. Med. Assoc. 163(15):645-646.
2938	**Stearns, M.** *1977.* Golden eagles nesting in northwest Kansas. Kansas Ornithol. Soc. Bull. 28(3).
2939	**Stebbins, R.** *1971.* Eagles are not for killing. Am. Forests 77(10):12-14,56-57.
2940	**Stegmann, B.** *1930.* Die vogel des dauro-mandschurischen uebergangsgebietes. J. Ornithol. 78(4):389-471.
2941	**Steinbacher, G.** *1975.* Zum vorkommen kleiner adler im Norden des bayerischen Regierungsbezirks Schwaben. Ornithol. Mitt. 27(12):248-249.
2942	**Steinbacher, J.** *1951.* Der schreiadler. Natur und Volk 81(7):167-170.
2943	**Steinbeck, W.** *1884.* The golden eagle *(Aquila chrysaetos.)* Ornithol. and Ool. 9(5):58.
2944	**Steinfatt, O.** *1932.* Der Bosporous als Landbrucke fur den vogelzug zwischen Europa und Kleinasien. J. Ornithol. 80(3):354-383.
2945	**Steinfatt, O.** *1955.* Vogelkundliche beobachtungen in Attika. J. Ornithol. 96(1):92-101.
2946	**Stelter, L.S. and J.G. Nagy.** *1979.* Raptor food habits in the oil shale region of northwestern Colorado. Colo. Field Ornithol. 13(2):7-12.
2947	**Stembeck, W.** *1883.* Golden eagles's nest and eggs. Ornithol. and Ool. 8(5):36.
2948	**Stemmler, C.** *1932.* Die adler der Schweiz. Grethlein & Co., A.-G., Zurich and Leipzig. 248 pp.
2949	**Stemmler, C.** *1933.* Vom steinadler im schweizerischen Hochgebirge. Beitr. Fortpfl.-Biology Vogel 9(3):73-77.
2950	**Stemmler, C.** *1934.* The golden eagle in the high mountains of Switzerland. Terre Et Vie 4:477-488. (In French)
2951	**Stemmler, C.** *1937.* Une aire d'aigle royal sur un arbre. Alauda 9(3/4):300-302.
2952	**Stemmler, C.** *1943.* Steinadlerhorste in den Schweizer Alpen. (Golden eagle nest in the Swiss Alps.) Beitr. Fortpfl.-Biology Vogel 19(3):97-98.
2953	**Stemmler, C.** *1954.* Die steinadler-brutpaare der Schweiz. (Golden eagle breeding pairs in Switzerland.) Vogelwelt 75(2):65.
2954	**Stemmler, C.** *1955.* Der Steinadler in der Schweiger Alpen. Druck: Verlagsdruckerie Holfinger and Comp., Stuttgart. 338 pp.
2955	**Stemmler, C.** *1955.* Der steinadler in der Schweizer alpen. Published by Author, Schaffhausen. 338 pp. (Title Only)
2956	**Sten, I.** *1974.* Bird ringing activities in Finland 1913-1972. Ornis Fenn. 51(3):195-208.
2957	**Stepanyan, L.S.** *1978.* Materialy k ornithologicheskoi faune yugo-zapodnoi chasti Zangezurskogo khrebta. (Data on the avifauna of south-western Zangezur Ridge.) Sbornik Trudov Zool. Museya, Mosk. Univ. 17:25-51. (In Russian)
2958	**Stephen, D.** *1950.* Portrait of a golden eagle. Field 196:288-289. (Title Only)
2959	**Stephen, D.** *1954.* The golden eagle. Scott. Field (Sept.):13-15. (Title Only)
2960	**Stephen, D.** *1957.* The eagle's amazingly varied diet. Scotsman (Sept.):9. (Title Only)
2961	**Stephens, H.A.** *1967.* Observations on eagles in Kansas. Kansas Ornithol. Soc. Bull. 18(1):23-25.
2962	**Stevens, W.F.** *1975.* The biology of the European rabbit, *Oryctolagus cuniculus,* on San Juan Island, Washington. M.S. Thesis, Univ. of Washington, Seattle, Wa. 151 pp.
2963	**Stevenson, B.H.** *1936.* Golden eagle captured in Burke County, Georgia. Oriole 1(2):18-19.
2964	**Stevenson, H.** *1869.* Golden eagle in Norfolk. Zoologist (2nd Ser.) 4:1863.
2965	**Stevenson, H.M. and A. Stupka.** *1948.* The altitudinal limits of certain birds in the mountains of the southeastern states. Migrant 19(3):33-60.
2966	**Stevin, P.** *1980.* Further observations on the tawny eagle. Ostrich 51(1):54-55.
2967	**Stewart, P.A. and R.W. Skinner.** *1967.* Weights of birds from Alabama and North Carolina. Wilson Bull. 79(1):37-42.
2968	**Stewart, R.E. and C.S. Robbins.** *1958.* Birds of Maryland and the District of Columbia. U.S.D.I., N. American Fauna. No. 62. 401 pp.

Master No.	Citation
2969	**Steyn, P.** *1962.* Observations on Wahlberg's eagle. Bokmakierie 14(1):7-14.
2970	**Steyn, P.** *1968.* Notes on the nesting of Wahlberg's eagle *Aquila wahlbergi.* Ostrich 39(3):192-193.
2971	**Steyn, P.** *1968.* Photographs. Bokmakierie 20(4):98-99.
2972	**Steyn, P.** *1973.* Eagle Days. Purnell, New York. 158 pp.
2973	**Steyn, P.** *1973.* Observations on the tawny eagle. Ostrich 44(1):1-22.
2974	**Steyn, P.** *1975.* Observations on the African hawk-eagle. Ostrich 46:87-105.
2975	**Steyn, P.** *1980.* Further observations on the tawny eagle. Ostrich 51(1):54-55.
2976	**Steyn, P.** *1980.* Observations on the prey and breeding success of Wahlberg's eagle. Ostrich 51(1):56-59.
2977	**Stickel, L.F.** *1973.* Pesticide residues in birds and mammals. Pages 254-312. *in:* C.A. Edwards, Ed. Environmental Pollution by Pesticides. Plenum Press, London and New York.
2978	**Stickel, L.F., V.A. Adomaitis, G.E. Bagley, N.J. Chura, L.N. Locke, C.M. Menzie, R.M. Prouty, W.L. Reichel and P.A. Stewart.** *1965.* Bald eagle. Pages 6-7. *in:* Effects of Pesticides on Fish and Wildlife. U.S.D.I. Fish and Wildlife Service, Circ. No. 226.
2979	**Stinson, C.H.** *1979.* On the selective advantage of fratricide in raptors. Evolution 33(4):1219-1225.
2980	**Stoddard, H.L.** *1917.* Notes on a few of the rarer birds of Sauk and Dane Counties, Wisconsin. Auk 34(1):63-67.
2981	**Stokes, A.W.** *1961.* Voice and social behavior of the chukar partridge. Condor 63(2):111-127.
2982	**Stone, W.** *1896.* Golden eagles in Virginia. Auk 13(3):256-257.
2983	**Stoneham, H.F.** *1926.* Field notes on a collection of birds from Uganda. Ibis (12th Ser.) 2(1):26-91.
2984	**Stoneham, H.F.** *1944.* Steppe-eagle in Trans-Nzoia. J. E. African Nat. Hist. Soc. 17(5/6):395.
2985	**Stoner, D.** *1939.* The golden eagle in eastern New York. Univ. of the State of N.Y., Bull. to Schools 25(13):114-117.
2986	**Stoner, E.A.** *1921.* Newspaper ornithology. Oologist 38(11):150.
2987	**Stoner, E.A.** *1934.* Golden eagle electrocuted. Oologist 51(1):12.
2988	**Stoney, C.V.** *1926.* Status of the golden eagle in Ireland. Brit. Birds 19(10):258-259.
2989	**Stoole, E.W.L.** *1971.* A case of allopreening by *Aquila wahlbergi* . Ostrich 42(4):301.
2990	**Stophlet, J.J.** *1962.* The golden eagle; king of the bird world. Natl. Parks Mag. 36(173):4-7.
2991	**Stott, K., Jr.** *1943.* Eagles in captivity. Aviculture 13(3):46-47.
2992	**Stott, K., Jr.** *1948.* Notes on the longevity of captive birds. Auk 65(3):402-405.
2993	**Strandtmann, R.W.** *1956.* A case of infertility in a pair of golden eagles, *(Aquila chrysaetos,* Aves). Southwest. Nat. 1(2):89.
2994	**Strandtmann, R.W.** *1962.* Notes on nest building and mating in the golden eagle. Southwest. Nat. 7(3/4):267-268.
2995	**Stresemann, E.** *1937.* Beitrage zur ornithologie von nordwest-Kansu asch den forschungen von Walter Bieck in den Jahren 1926-1933. J. Ornithol. 85(3):375-576.
2996	**Stresemann, E.** *1943.* Ueberlick uber die vogel kretas und den vogelzug in der Aegaeis. J. Ornithol. 91(4):448-514.
2997	**Stresemann, E.** *1955.* Bemerkungen zu den verbeitungskarten in: Peterson-Mountfort-Hollom, die vogel Europas. J. Ornithol. 96(1):107-114.
2998	**Stresemann, E.** *1956.* Bausteine zu einer ornithologie von Kreta. J. Ornithol. 97(1):44-72.
2999	**Stresemann, E.** *1957.* Eine ornithologische studienfahrt in den Parco Nazionale d'Abrusso. J. Ornithol. 98(1):1-21.
3000	**Stresemann, E., M. Schoenwetter and E. Stresemann.** *1938.* (Study of ornithology of northwest Kansu, research of Walter Bieck in the years of 1926-1933.) J. Ornithol. 86(2):171-221.
3001	**Stresemann, V. and E. Stresemann.** *1960.* Die handschwingenmauser der Tagraubvogel. J. Ornithol. 101(4):373-403.
3002	**Strode, W.S.** *1894.* Golden eagles. Nidiologist 1(6):90.
3003	**Stromar, L.** *1975.* Prstenovanje ptica u godinama 1971. (Bird banding in 1971.) Larus 26-28:5-43. (English Summ.)
3004	**Stromar, L.** *1977.* Prstenovanje ptica u godin 1973. (Bird banding in 1973.) Larus 29-30:295-321. (English Summ.)
3005	**Strong, W.A.** *1922.* The geese and the eagle. Oologist 39(6):96.
3006	**Stubbs, F.J.** *1931.* Former nesting of the eagle in Yorkshire. Naturalist 892:135-137.
3007	**Stupka, A.** *1936.* Eagles released in the Smokies. Migrant 7(1):25.
3008	**Styan, F.W.** *1887.* On a collection of birds from Foochow. Ibis 5(18):215-234.
3009	**Styan, F.W.** *1891.* On the birds of the Lower Yangtse Basin. Ibis (6th Ser.) 3(11):316-359.
3010	**Styan, F.W.** *1899.* On birds from West China. Ibis (7th Ser.) 5(18):289-300.
3011	**Suetens, W. and P. van Groenendael.** *1971.* Note succinte sur la nidification d'un couple d'aigles imperiaux *(Aquila heliaca adalberti)* dans la Province de Caceres. Ardeola (Special):575-580.
3012	**Sukhinin, A.N.** *1958.* O razmnozhenii i pitanii nekotorykh khistchnykh ptits i sov Badkhyza. (Breeding and feeding of some birds of prey in Badkhyz.) Trudy Inst. Zool. An Turkmen SSR 3:47-118. (In Russian)
3013	**Sukhinin, A.N.** *1971.* Ekologiya sov i khistchnykh ptits Badkhyza (yufo-vostochnaya Turkmeniya). (Ecology of birds of prey in Badkbyz (south-eastern Turkmenia).) Ylym. Ashkhabad. p. 1-102. (In Russian)
3014	**Sulkava, S.** *1959.* Kotkan ravinnosta poronhoitoalueen Eletapuolella. (On the food habits of the golden eagle south of the Reindeer Management Area.) Suomen Riista 13:69-72.
3015	**Sulkava, S. and P. Rajala.** *1966.* Kotkan pesimisaikaisesta ravinnosta poronhoitoalueella. (Diet of the golden eagle *(Aquila chrysaetos)* during the nesting period in the Finnish reindeer husbandry area.) Suomen Riista 19:7-19.
3016	**Sullivan, C.** *1931.* Notes from north-western New South Wales. Emu 31(2):124-135.
3017	**Sultana, J. and C. Gauci.** *1977.* The situation of birds of prey in Malta 1975. Pages 136-139. *in:* R.D. Chancellor, Ed. Proc. of the World Conf. on Birds of Prey. 1-3 October, 1975. Vienna. I.C.B.P.
3018	**Summers, G.** *1976.* Owned by an Eagle. E.P. Dutton, N.Y. 223 pp.
3019	**Sumner, E.L.** *1929.* Comparative studies in the growth of young raptors. Condor 31(3):85-111.
3020	**Sumner, E.L., Jr.** *1929.* Notes on the growth and behavior of young golden eagles. Auk 46(2):161-169.
3021	**Sumner, E.L., Jr.** *1933.* The growth of some young raptorial birds. Univ. of Calif. Publ. in Zool. 40(4):277-308.

Master No.	Citation
3022	**Sumner, E.L., Jr.** *1934.* The behavior of some young raptorial birds. Univ. of Calif. Publ. in Zool. 40(8):331-362.
3023	**Sumner, E.L., Jr.** *1940.* Longevity in raptorial birds as indicated by banding records. Condor 42(1):39-40.
3024	**Sumner, F.B.** *1931.* Golden eagle kills a cat. Condor 33(6):248.
3025	**Sundevall, C.J.** *1886.* On the wings of birds. Ibis (5th Ser.) 4(16):389-457.
3026	**Surber, T.** *1894.* Capture of the golden eagle at Covington, Virginia. Auk 11(1):77.
3027	**Survillo, A.V., V.B. Sandzhiev, O.T. Ulumdzhiev, G.I. Chervichenko, V.D. Kolesnikov and M.O. Mastyukov.** *1977.* O chislennosti iekologii stepnoge orla v tsentralnykh raionakh Kalmykii. (Number and ecology of *Aquila nipalensis* in central Kalmykia.) 7 Vsesoyuzn. Ornithol. Konf., Tezisy Dokl., Chast 2, Kiev. p. 247-248. (In Russian)
3028	**Suschkin, P.** *1900.* On the species of *Aquila* . Bull. Brit. Ornithol. Club 11:6-10.
3029	**Suschkin, P.P.** *1914.* Die vogel der mittleren kirgisensteppe. J. Ornithol. 62(4):557-607.
3030	**Suschkin, P.P.** *1914.* Die vogel der mittleren kirgisensteppe. J. Ornithol. 62(3):297-333.
3031	**Suter, G.W., II and J.L. Joness.** *1981.* Criteria for golden eagle, ferruginous hawk, and prairie falcon nest site protection. Raptor Res. 15(1):12-18.
3032	**Sutter, E.** *1959.* Adler und Gemse. Ornithol. Beob. 56(3):98-100.
3033	**Sutter, E.** *1975.* Zum geschlecht kampfender steinadler. Ornithol. Beob. 72(3):116-117.
3034	**Sutton, G.M.** *1928.* Abundance of the golden eagle in Pennsylvania in 1927-28. Auk 45(3):375.
3035	**Sutton, G.M.** *1928.* An Introduction to the Birds of Pennsylvania. J. Horace MacFarland Comp., Harrisburg. 161 pp.
3037	**Sutton, G.M.** *1934-1935.* Notes on the birds of the western panhandle of Oklahoma. Ann. Carnegie Mus. Pitts. 24:1-50
3036	**Sutton, G.M.** *1967.* Oklahoma Birds, Their Ecology and Distribution. Univ. of Oklahoma Press, Norman. 674 pp.
3038	**Sutton, J.** *1933.* Birds of Salt Creek District and some island-sanctuaries in the Coorong. S. Aust. Ornithol. 12(1):5-18.
3039	**Svehlik, J. and B.-U. Meyburg.** *1979.* Gelegrobe und bruterfolg des schreiadlers *(Aquila pomarina)* und des kaiseradlers *(Aquila heliaca)* in den ostslowakischen Karpaten 1960-1978. J. Ornithol. 120(4):406-415.
3040	**Svensson, L.** *1975.* Storre skrikorn *Aquila clanga* och mindre skrikorn *A. pomarina* - problemet att artbestamma dem. (Spotted eagle and lesser spotted eagle - the problem of identification.) Var Fagelvarld 34(1):1-26. (English Summ.)
3041	**Svensson, L.** *1976.* Ett fynd av stapporn *Aquila rapax* och nagra reflexioner om svarigheten att artbestamma ornar samt att bedoma raritetsrapporter. (A record of a steppe eagle and lesson taught.) Var Fagelvarld 35(4):305-306. (English Summ.)
3042	**Svensson, L.** *1976.* Sallsynta faglar i Sverige 1975. Var Fagelvarld 35(4):328-334.
3043	**Svensson, L.** *1978.* Sallsynta faglar i Sverige 1976- rapport fran SOF:S raritetskommitte. (Report on rare birds in Sweden in 1976.) Var Fagelvarld 37(1):73-809. (English Summ.)
3044	**Swann, H.K.** *1925.* Remarks on races of the golden eagle. Bull. Brit. Ornithol Club 45:64-73.
3045	**Swann, H.K.** *1925.* Remarks on *Aquila rapax culleni* . Bull. Brit. Ornithol. Club 45:110-111.
3046	**Swann, H.K. and A. Wetmore (Eds.).** *1936.* A Monograph of the Birds of Prey (Order Accipitres). Wheldon and Wesley, Ltd. London.
3047	**Swann, R.L. and A.D.K. Ramsay** *1978.* Avian prey of large raptors on Canna. Brit. Birds 71(1):46.
3048	**Swanquist, R.** *1981.* Raptor migration summary, Eilat, spring 1978-1979. Torgos 1(2):14-16.
3049	**Swarth, H.S.** *1914.* A distributional list of the birds of Arizona. Pac. Coast Avifauna No. 10. 133 pp.
3050	**Swartz, L.G., W. Walker II, D.G. Roseneau and A.M. Springer.** *1975.* Populations of gyrfalcons on the Seward Peninsula, Alaska, 1968-1972. Pages 71-75. *in:* J.R. Murphy, C.M. White and B.E. Harrell, Eds. Population Status of Raptors. Proc. of the Conf. on Raptor Conserv. Techniques. 22-24 March, 1973. Fort Collins, Co. Raptor Res. Rep. No. 3.
3051	**Swiesle, D.** *1921.* Steinadler, in Wurttemberg geschossen. Ornithol. Beob. 18(10):158-159.
3052	**Swinhoe, C.** *1882.* On the birds of southern Afghanistan. Ibis (4th Ser.) 6(21):95-126.
3053	**Swinhoe, C. and H. Barnes.** *1885.* On the birds of central India. Part 1. Ibis (5th Ser.) 3(9):52-69.
3054	**Swynnerton, C.F.M.** *1907.* On the birds of Gazaland, Southern Rhodesia. Part II. Ibis (9th Ser.) 1(2):279-311.
3055	**Swynnerton, C.F.M.** *1908.* Further notes on the birds of Gazaland. Ibis 2(7):391-443.
3056	**Sylven, M.** *1975.* Forekomst av storreskrikorn *Aquila clanga* och mindre skrikorn *A. pomarina* i Sverige. (Occurence of the spotted eagle and lesser spotted eagle in Sweden.) Var Fagelvarld 34(1):27-36. (English Summ.)
3057	**Sylven, M. and G. Gustafson.** *1976.* Breeding attempt of the golden eagle *Aquila chrysaetos* in Skane, South Sweden. Anser 15(2):141-144. (Title Only)
3058	**Symons, R.E.** *1920.* Bird life in the Drakensberg Mountains. The black eagle *(Aquila verreauxi)* . S. African J. Nat. Hist. 2(2):253-258.
3059	**Szczepski, J.B.** *1970.* Sprawozdanie z dzialalnoscí Stacji ornitologicznej za lata 1955-1957. (Report of the Ornithological Station for 1955-1957.) Acta Ornithol. 12(4):59-128.
3060	**Szczepski, J.B. and M.W. Szczepska.** *1959.* Sprawozdanie z dzialalnosci stacji ornitologicznejza rok 1952. (Report of the ornithological station for 1952.) Acta Ornithol. 5(8):233-282. (In Polish, English Summ.)
3061	**Szemere, L.** *1912.* Am horst von *Aquila melanaetus* Gm. und *Circaetus gallicus* Gm. Aquila 19:441-451.
3062	**Szemere, L.** *1920.* Uber den schreiadler. Aquila 27:59-60.
3063	**Szielasko, A.** *1895.* Einiges uber *Aquila pomarina* Br. wahrend des Brutgeschafts. Ornithol. Monatsber. 3:94-95.
3064	**Szielasko, A.** *1895.* Einiges uber *Aquila pomarina* Br. wahrend des brutgeschaftes. Ornithol. Jahrbuch 6(3):163-164.
3065	**Szielasko, A.** *1896.* Aquila pomarina Br. am Brutplatze. Ornithol. Jahrbuch 7(2):75-77.
3066	**Szielasko, A.** *1905.* Die gestalt der vogeleier. J. Ornithol. 53(2):273-297.
3067	**Szielasko, A.** *1913.* Die bedeutung der eischalenstruktur der vogel fur die systematik. J. Ornithol. 61(1):52-117.

Master No.	Citation
3068	**Szielasko, A.** *1913.* Die bedeutung der eischalenstruktur der vogel fur die systematik. J. Ornithol. 61(2):229-361.
3069	**Szlivka, L.** *1957.* On the birdlife of Vojvodina. Larus 11:31-35. (English Trans.)
3070	**Szlivka, L.** *1961.* A study of the avifauna of Obekska Bara and its environs. Larus 15:77-101. (English Trans.)
3071	**Taczanowski, L.** *1872.* Bericht uber die ornithologischen untersuchungen des Dr. Dybowski in Ost-Sibirien. J. Ornithol. 20(5):340-366.
3072	**Taczanowski, L.** *1873.* Bericht uber die ornithologischen untersuchungen des Dr. Dybowski in Ost-Sibiriens. J. Ornithol. 21(2):81-119.
3073	**Taczanowski, L.** *1874.* Zweiter nachtrag zum bericht uber die ornithologischen untersuchungen des Dr. Dybowski in Ost-Sibirien. J. Ornithol. 22(3):315-337.
3074	**Taczanowski, L.** *1888.* Liste des oiseaux observes depuis cinquante ans dans le Royaume de Pologne. Ornis 4(1):441-516.
3075	**Tait, W.C.** *1887.* A list of the birds of Portugal. Ibis (5th Ser.) 5(19):302-314.
3076	**Takagi, N. and M. Sasaki.** *1974.* A phylogenetic study of bird karyotypes. Chromosoma 46(1):91-120.
3077	**Talsky, J.** *1890.* Ornithologischer bericht aus Mahren. Ornithol. Jahrbuch 1(2):35-37.
3078	**Tapfer, D.** *1973.* Der kaiseradler in Pannonien. Falke 20(12):402-407.
3079	**Tarboton, W.** *1977.* Nesting, territoriality and food habits of Wahlberg's eagle. Bokmakierie 29(2):46-50.
3080	**Tarboton, W.** *1979.* Eagles of the Transvaal. Transvaal Nature Cons. Div. 34 pp.
3081	**Tate, R.C.** *1923.* Some birds of the Oklahoma panhandle. Proc. Oklahoma Acad. Sci. 3:41-51.
3082	**Taverner, P.A.** *1934.* Birds of Canada. National Mus. Canada, Bull. No. 72. 445 pp.
3083	**Taverner, P.A.** *1939.* Canadian Water Birds, Game Birds, Birds of Prey. David McKay Comp., Philadelphia. 291 pp.
3084	**Taylor, C.H. and J.A. Bucknill.** *1907.* Notes on a collection of birds made in the Amsterdam District upon the Transvaal - Swazieland border between the months of June and October, 1906. J. S. African Ornithol. Union 3(1):12-42.
3085	**Taylor, D.C.** *1968.* Black eagle. Lammergeyer 8(March):52.
3086	**Taylor, E.C.** *1859.* Ornithological reminiscences of Egypt. Ibis 1(1):41-55.
3087	**Taylor, E.C.** *1867.* Egypt revisited. Ibis 3(9):48-73.
3088	**Taylor, E.C.** *1878.* A few additional notes on birds of Egypt. Ibis (4th Ser.) 2(7):368-374.
3089	**Taylor, E.C.** *1880.* Remarks on new species *Aquila culleni.* Ibis 4(13):143-144.
3090	**Taylor, E.C.** *1896.* A few notes on birds of Egypt, from observations made at Cairo in the months of January and February 1896. Ibis (7th Ser.) 2(8):477-482.
3091	**Taylor, H.R.** *1888.* Nesting of the golden eagle. Ornithol. and Ool. 13(11):172-173.
3092	**Taylor, H.R.** *1890.* Nesting habits of the golden eagle. Zoe 1(2):42-44.
3093	**Taylor, H.R.** *1890.* Nesting habits of the golden eagle in California. Zoologist (3rd Ser.) 14:377-379.
3094	**Taylor, H.R.** *1894.* Notes on the golden eagle. Nidologist 1(7):98.
3095	**Taylor, L.E.** *1909.* Notes from Cape Colony. J. S. African Ornithol. Union 5(2):81-87.
3096	**Taylor, W.E. and A.H. Van Vleet.** *1889.* Notes on Nebraska birds. Ornithol. and Ool. 14(11):163-165.
3097	**Taylor, W.P.** *1923.* Note on the sense of smell in the golden eagle and certain other birds. Condor 25(1):28.
3098	**Tekke, M.J.** *1977.* Ornithologie van Nederland 1974 en 1975. Limosa 50(1/2):34-60.
3099	**Temperley, G.W.** *1951.* Ornithological report for Northumberland and Durham. Naturalist 838(July-Sept.):115-138.
3100	**Ten Kate, C.G.B.** *1933.* Steenarend, *Aquila chr. chrysaetus* (L.), bij Barneveld gevangen. Limosa 5(4):169.
3101	**Tener, J.S.** *1954.* Three observations of predators attacking prey. No. 3. Golden eagle killing a Canada goose. Can. Field-Nat. 68(4):181-182.
3102	**Terrasse, J.-F.** *1965.* La dimunution recentes des effectifs de rapaces en France et ses causes. Terre Et Vie 112(3):273-291.
3103	**Terrasse, M. and J.F. Terrasse.** *1978.* Je decouvre les rapaces, 1978. Andre Leson, Paris. 153 pp.
3104	**Terres, J.K.** *1957.* Economic study proposed for the golden eagle in Texas. Audubon 59(1):35.
3105	**Terres, J.K.** *1967.* Eagles over Hawk Mountain. Audubon 69(4):28-32.
3106	**Terrill, S.E. and C.E. Rix.** *1950.* The birds of South Australia their distribution and habitat. S. Aust. Ornithol. 19(6/8):53-100.
3107	**Test, C.D.** *1946.* Exploring nests of golden eagles. Audubon Bull. 59:1-8.
3108	**Thacker, R.** *1971.* Estimations relative to birds of prey in captivity in the United States of America. Raptor Res. News 5(4):108-122.
3109	**Thake, M.A.** *1976.* Visible migration of raptors over Buskett - autumn 1975. Il-Merill 17:21-24.
3110	**Thelander, C.G.** *1974.* Nesting territory utilization by golden eagles *(Aquila chrysaetos)* in California during 1974. Calif. Dept. of Fish and Game. Spec. Wildlife Investigations, Proj. W-054-R-06, Wk. Pl. 02, Job 5.8. 22 pp.
3111	**Thevenot, M., P. Bergier and P. Beaubrun.** *1982.* Repartition actuelle et statut des rapaces au Maroc. *in:* Workshop on the Conservation and Management of Birds of Prey in the Mediterranean. World Conf. on Birds of Prey. 26-29 April, 1982. Thessaloniki, Greece. I.C.B.P. (English Summ.)
3112	**Thielcke, G.** *1977.* Birds of prey in the German Federal Rebublic. Pages 107-114. *in:* R.D. Chancellor, Ed. Proc. of the World Conf. on Birds of Prey. 1-3 October, 1975. Vienna. I.C.B.P.
3113	**Thienemann, J.** *1909.* Jahresbericht (1908) der vogelwarte rossitten der deutschen ornithologischen gesellschaft. J. Ornithol. 57(3):384-502.
3114	**Thienemann, J.** *1931.* Bericht uber den vogelberingungsversuch in den jahren 1925 und 1926 auf der vogelwarte Rossitten. J. Ornithol. 79(2):255-268.
3115	**Thiollay, J.-M.** *1966.* La migration d'automne des rapaces diurnes aux cols de Cou et Bretolet. Nos Oiseaux 28(9/10):229-251.
3116	**Thiollay, J.-M.** *1967.* La migration d'automne des rapaces diurnes aux cols de Cou et Bretolet (II). Nos Oiseaux 29(4):69-97.

Master No.	Citation
3117	**Thiollay, J.-M.** *1967.* La migration d'automne des rapaces diurnes aux cols de Cou et Bretolet (III). Nos Oiseaux 29(5):105-126.
3118	**Thiollay, J.-M.** *1967.* Essai sur les rapaces du midi de la France, distribution - ecologie tentative de denombrement. Alauda 35(2):140-150.
3119	**Thiollay, J.-M.** *1971.* L'exploitation des feux de brousse par les oiseaux en Afrique occidental. Alauda 39(1):54-72. (English Summ.)
3120	**Thiollay, J.-M.** *1973.* Place des oiseaux dans les Chaines Tropiques d'une zone preforestiere en Cote-d'Ivoire. Alauda 41(3):273-300. (English Summ.)
3121	**Thiollay, J.-M.** *1975.* Les rapaces d'une zone de contact Savane-foret en Cote-d'Ivoire: densite, dynamique et structure du peuplement. Alauda 43(4):387-416.
3122	**Thiollay, J.-M.** *1975.* Les rapaces des Parcs Nationaux du Tchad Meridional. Oiseau Rev. Fr. Ornithol. 45(1):27-40.
3123	**Thiollay, J.-M.** *1975.* Les rapaces d'une zone de contact Savane-Foret en Cote-d'Ivoire: presentation du peuplement. Alauda 43(1):75-102.
3124	**Thiollay, J.-M.** *1975.* Migrations de rapaces Africains en Ouganda et au Rwanda. Oiseau Rev. Fr. Ornithol. 45(2):192-194.
3125	**Thiollay, J.-M.** *1975.* Les rapaces des parcs nationaux de Cote D'Ivoire; analyse du peuplement. Oiseau Rev. Fr. Ornithol. 45(3):241-257. (English Summ.)
3126	**Thiollay, J.-M.** *1977.* L'importance des hivernants palearctiques dans le peuplement de rapaces d'Afrique tropicale. Nos Oiseaux 34(2):59-64.
3127	**Thiollay, J.-M.** *1977.* Les rapaces d'une zone de contact Savane-foret en Cote-d'Ivoire: modes d'exploitation du milieu. Alauda 45(2/3):197-218.
3128	**Thiollay, J.-M.** *1977.* Distribution saisonnaiere des rapaces diurnes en Afrique Occidental. Oiseau Rev. Fr. Ornithol. 47(3):253-285. (English Summ.)
3129	**Thiollay, J.-M.** *1977.* Importance des populations de rapaces migrateurs en Mediterranee Occidentale. Alauda 45(1):115-121.
3130	**Thiollay, J.-M.** *1977.* Le peuplement de Falconiformes d'une Savane Ougandaise: structure et fluctuations a court terme. Oiseau Rev. Fr. Ornithol. 47(2):193-205. (English Summ.)
3131	**Thiollay, J.-M.** *1978.* Comparaisons entre les peuplements de Falconiformes des Plaines Cotieres du Mexique et de Cote-d'Ivore. Gerfaut 68(2):139-162. (English Summ.)
3132	**Thiollay, J.-M.** *1978.* Precisions nouvelles sur le statut de certains rapaces en Afghanistan. Alauda 46(1):98-100.
3133	**Thiollay, J.-M.** *1978.* Les plaines du Nord Cameroun, centre d'hivernage de rapaces Palearctiques. Alauda 46(4):319-326. (English Summ.)
3134	**Thiollay, J.-M.** *1978.* Population structure and seasonal fluctuations of the Falconiformes in Uganda National Parks. E. African Wildl. J. 16:145-151.
3135	**Thiollay, J.-M.** *1979.* La migration des grues a travers l'himalaya et la predation par les aigles royaux. Alauda 47(2):83-92. (English Summ.)
3136	**Thiollay, J.-M. and A.R. Dupuy.** *1970.* Les rapaces du Parc National du Niokolo-Koba. Donnees preliminaire. Oiseau Rev. Fr. Ornithol. 40(2):115-130.
3137	**Thiollay, J.-M. and L. Duhautois.** *1976.* Notes sur les oiseaux du Nord Yemen. Oiseau Rev. Fr. Ornithol. 46(3):261-271.
3138	**Thiollay, J.-M. and A. Perthuis.** *1975.* La migration d'automme a Gibraltar (1er au 20 Octobre 1974): analyse et interpretation. Ardeola 21(Special):595-614.
3139	**Thomas, J.W., C.V. Hoozer and R.G. Marburger.** *1964.* Wild turkey behavior affected by the presence of golden eagles Wilson Bull. 76(4):384-385.
3140	**Thompson, A.V.** *1891.* More about the Iowa eagles. Oologist 8(8):159-160.
3141	**Thompson, F.** *1961.* Scream of eagles. Sports Afield 144(7):51,85-86.
3142	**Thompson, H.N. and W.H. Craddock.** *1902.* Notes on the occurence of certain birds in the southern Shan states of Burma. J. Bombay Nat. Hist. Soc. 14(3):600.
3143	**Thompson, W.** *1838.* Golden and sea eagle; *Aquila chrysaetos* and *A. (Haliaeetus) albicilla* in Ireland. Charles W. Mag. Nat. Hist. 2:164. (Title Only)
3144	**Thompson, W.K.** *1949.* Predation on antelope. J. Wildl. Manage. 13(3):313-314.
3145	**Thorpe, W.H., P.T. Cotton and P.F. Holmes.** *1936.* Notes on the birds of Lakes Ochrid, Malik, and Prespa and adjacent parts of Yugoslavia, Albania, and Greece. Ibis (13th Ser.) 6(3):557-580.
3146	**Thurow, T.L., C.M. White, R.P. Howard and J.F. Sullivan.** *1980.* Raptor ecology of Raft River Valley, Idaho. E.G. & G., Rep. Prepared for U.S. Dept. of Energy, Idaho. 45 pp.
3147	**Ticehurst, C.B.** *1923.* The birds of Sind. Ibis (11th Ser.) 5(2):235-275.
3148	**Ticehurst, C.B.** *1931.* New name for the Himalayan golden eagle. Bull. Brit. Ornithol. Club 52(353):24-25.
3149	**Ticehurst, C.B.** *1932.* The correct name of the Himalayan golden eagle. Bull. Brit. Ornithol. Club 53(362):21-22.
3150	**Ticehurst, C.B. and H. Whistler.** *1932.* On the ornithology of Albania. Ibis (13th Ser.) 2(1):40-93.
3151	**Ticehurst, C.B. and H. Whistler.** *1933.* Some notes on the birds of Portugal. Ibis (13th Ser.) 3(1):97-112.
3152	**Tigner, J.R.** *1973.* Golden eagle predation on a pronghorn fawn. Southwest. Nat. 18(3):346-348.
3153	**Tigner, J.R. and G.E. Larson.** *1982.* Golden eagles. Scavengers and predators on domestic lambs. Pages 359-361. *in:* J.M. Peek and P.D. Dalke, Eds. Wildlife and Livestock Relationships Symposium: Proceedings.
3154	**Tigner, J.R. and G.E. Larson.** *1977.* Sheep losses on selected ranches in southern Wyoming. J. Range Manage. 30(4):244-252.
3155	**Tilden, P.M.** *1962.* Golden eagle protection. Nat. Hist. 71(6):68.
3156	**Tilly, F.** *1972.* Raccoon Ridge daily hawk counts - autumn 1971. N.J. State Museum, Sci. Notes No. 7. 4 pp.
3157	**Tischler, F.** *1912.* *Aquila maculata* (Gm.) in Ostpreufsen erlegt. Ornithol. Monatsber. 20(12):187-188.
3158	**Tischler, F.** *1919.* Das angebliche vorkommen des steppenadlers *(Aquila orientalis* Cab.) in Pommern. Ornithol. Monatsber. 27(1/2):3-5.

Master No. **Citation**

3159 **Tischler, F.** *1943.* Zur vogelfauna des urwaldes von Bialowies. Ornithol. Monatsber. 51(3/4):80-83.

3160 **Tjernberg, M.** *1977.* (Individual recognition of golden eagles, *Aquila chrysaetos* in the field and results of winter censuses in southwest Uppland, central Sweden.) Var Fagelvarld 36(1):21-32. (English Summ.)

3161 **Todd, R.L.** *1978.* Winter bald eagle in Arizona. Ariz. Game and Fish Dept., Proj. W-53-R, Wk. Pl. 5, Job 01. 36 pp.

3162 **Tohme, G. and J. Neuschwander.** *1974.* Nouvelles donnees sur l'avifaune de la Republique Libanaise. Alauda 42(3):243-258.

3163 **Tolbert, F.X.** *1956.* He hunts Texas eagles in plane. Audubon 58(1):31.

3164 **Tomek, A. and M.P. Krzemien.** *1974.* Orzel przedni na 'Czerwonym bagnie. " (The golden eagle in the Czerwone Bagno Nature Reserve.) Chronmy Przyrode Ojczysta 26(2):39-42.

3165 **Tomialojc, L.** *1963.* Ptaki okolic Dobergo Miasta w powiecie lidzbarksim. Acta Ornithol. 7(13):427-463.

3166 **Tomialojc, L.** *1976.* Birds of Poland, A list of species and their distribution. Published for U.S. Fish and Wildlife Service and Natl. Science Foundation by Foreign Scientific Publ. Dept. of National Center for Scientific, Tech. and Econ. Information. 256 pp.

3167 **Tornielli, A.** *1963.* La raccolta ornitologica del Parmense. Riv. Ital. Ornitol. 33(1/2):14-28.

3168 **Townsend, C.W.** *1925.* The winking of the water ouzel. Condor 27(4):143-144.

3169 **Tratz, E.P.** *1950.* Das steinadlervorkommen in Osterreich. Columba 2(3):54-55.

3170 **Tratz, E.P.** *1950.* Vom steppenadler. Columba 2(1):11.

3171 **Tratz, E.P.** *1953.* Der steinadler in Osterreich. Vogelwelt 74(5):161-162.

3172 **Traylor, M.A.** *1965.* A collection of birds from Barotseland and Bechuanaland. Ibis 107(2):137-172.

3173 **Tree, A.J.** *1961.* Observations on migrating movements of some migratory birds in northern Rhodesia. Ostrich 32(2):86-89.

3174 **Tree, A.J.** *1962.* The birds of the Leopardshill Area of the Zambesi Escarpment. Ostrich 33(4):3-23.

3175 **Tree, A.J.** *1963.* Records of fighting and display in eagles in northern Rhodesia. Ostrich 34(3):180.

3176 **Tree, A.J.** *1966.* Notes on the Palearctic migrants in the North Kaufe Basin, Zambia. Ostrich 37(3):184-190.

3177 **Tree, A.J.** *1975.* New distributional data: 6. lesser spotted eagle, *Aquila pomarina* and Wahlberg's eagle, *Aquila wahlbergi.* Ostrich 46:167.

3178 **Trischitta, A.** *1923.* Catture di uccelli rari a Messina. Riv. Ital. Ornitol. 6(2):45-45.

3179 **Tristram, H.B.** *1859.* On the ornithology of northern Africa. Ibis 1(3):277-301.

3180 **Tristram, H.B.** *1860.* A few leaves from a naturalist's note-book in eastern Algeria. Ibis 2(8):361-375.

3181 **Tristram, H.B.** *1865.* On the ornithology of Palestine. Ibis 1(3):241-263.

3182 **True, D.** *1981.* A family of eagles. Everest House, New York, NY.

3183 **Truman, H.V.** *1946.* What a golden eaglet eats. Audubon 48(1):21-26.

3184 **Truslow, F.K.** *1965.* Water, Prey and Game Birds of North America. Natl. Geographic Soc., Washington. 464 pp.

3185 **Tucker, J.** *1959.* Golden eagle at Chattanooga. Migrant 30(4):57.

3186 **Tucker, R.K. and D.G. Crabtree.** *1970.* Handbook of toxicity of pesticides to wildlife. U.S.D.I. Fish and Wildlife Service Resource Publ. No. 84. 131 pp.

3187 **Tuer, V.** *1973.* Notes on the African hawk eagle and Wahlberg's eagle in the Matopos. Honeyguide 75:18-21.

3188 **Tuer, V.V.** *1965.* Observation on the black eagle. Honeyguide 47:29-33. (Title Only)

3189 **Turbiville, V.W.** *1954.* Alta, my friend, the eagle. Audubon 56(2):66-67,84-85.

3190 **Turcek, F.J.** *1956.* On the bird population of the spruce forest community in Slovakia. Ibis 98(1):24-33.

3191 **Turner, J.** *1971.* Eagles: Vanishing Americans? Sierra Club Bull. 56(9):14-19.

3192 **Turner, J.F.** *1971.* Eagles over Wyoming. Wyoming Wildl. 35(9):8-13,34.

3193 **Turner, L.M.** *1885.* List of the birds of Labrador, including Ungava, East Main, Moose, and Gulf Districts of the Hudson Bay Company, together with the islands of Anticosti. Proc. U.S. Natl. Mus. 8:233-254.

3194 **Tyler, B.P. and R.B. Lyle.** *1933.* Winter birds of northeastern Tennessee. Migrant 4(3):25-29.

3195 **Tyler, C.** *1966.* A study of the egg shells of the Falconiformes. J. Zool. 150:413-425.

3196 **Tyler, J.G.** *1933.* Items from an oologist's notebook. Condor 35(5):186-188.

3197 **Tyler, S.J.** *1978.* Some observations of birds in Fah, northeast Eritrea. Bull. Brit. Ornithol. Club 98(3):80-87.

3198 **Tyurekhodzhaev, Z.M.** *1973.* (Influence of *Aquila nipalensis* on population structure and number fluctuations of *Citellus pygmaeus* . Trudy Kazakh. Inst. Zastchity Rastenii. 12:69-76. (In Russian)

3199 **Tyurekhodzhaev, Z.M.** *1977.* Stepnoi orel i baloban v Uralskoi oblasti. (*Aquila nipalensis* and *Falco cherrug* in Uralsk district.) Redkie I Ischezayustchie Zveri I Ptitsy Kazakstana, Alma Ata. p. 240-243. (In Russian)

3200 **U.S. Congress.** *1972.* Bald Eagle Protection Act. Comm. on Commerce, U.S. Senate 92nd Congress, Second Session, S. 2547, H.R. 12186 and H.R. 14731, June 29, 1972, Ser. No. 92-63. 78 pp.

3201 **U.S. Department of Agriculture.** *1972.* Powerline contacts by eagles and other large birds. U.S.D.A. Rural Electrification Admin., REA Bull. 61-10. 7 pp.

3202 **U.S. Department of Agriculture.** *1976.* Arroyo Seco/Indians Station Complex. U.S.D.A. Forest Service, Los Padres National Forest. Environmental Analysis Report. 63 pp.

3203 **U.S. Department of the Interior.** *1979.* Snake River Birds of Prey Study Area, Summary Report. U.S.D.I. Bureau of Land Management.

3204 **U.S. Department of the Interior.** *1979.* Snake River birds of Prey Environmental Statement. U.S.D.I. Bureau of Land Management, Final Environmental Impact Statement.

3205 **U.S. Department of the Interior.** *1978.* Golden eagle management, a conceptual plan. U.S.D.I. Fish and Wildlife Service, Report. 17 pp.

3206 **U.S. Department of the Interior.** *1979.* Whooping crane falls to eagle. U.S.D.I. Fish and Wildlife Service, Endang. Spec. Tech. Bull. 4(11):3.

3207 **U.S. Department of the Interior.** *1979.* Snake River Birds of Prey Special Research Report. U.S.D.I. Bureau of Land Management, Boise Dist. Spec. Res. Rep. of the Secretary of the Interior. 142 pp.

Master No.	Citation
3208	**U.S. Department of the Interior.** *[?].* America's eagle heritage. U.S.D.I. Fish and Wildlife Service. Report. 10 pp.
3209	**Udvardy, M.** *1959.* Supplement to the bird fauna of the Hortobagy. Aquila 66:217-223.
3210	**Ulfstrand, S.** *1956.* Fagelstracket vid Falsterbo 1949-1950. Var Fagelvarld 15(3):187-199. (English Summ.)
3211	**Ulfstrand, S.** *1960.* (Migration of African kites, *Milvus migrans* (Bodd.) - *M. aegyptius* Sharpe and Bouvier, and Wahlberg's eagles, *Aquila wahlbergi* Sund., at Lake Tanganyika.) Kungl. Fysiografiska Sallsk. I Lund Forhandlingar 30:31-39.
3212	**Ulfstrand, S.** *1962.* Supplementary notes on the imperial eagles at Hackeberga, Scania, and a comparison between imperial and tawny eagles. Var Fagelvarld 21:250-252. (Title Only)
3213	**Undeland, R.E.** *1964.* Verreaux's eagle *Aquila verreauxii* in Sinai. Ibis 106(2):258.
3214	**Unwin, W.H.** *1874.* On the breeding of the golden eagle *(Aquila chrysaetos)* in northwestern India. Proc. Zool. Soc. London 42:208-212.
3215	**Urndt, E.** *1890.* Goldadlerhorst in Bosnien. Vogelwelt 15(10):267-268.
3216	**Uttendorfer, O.** *1932.* Beobachtungen uber die ernahrung unserer tagraubvogel und eulen im jahr 1931. J. Ornithol. 80(3):284-293.
3217	**Uttendorfer, O.** *1934.* Beobachtungen uber die ernahrung unserer tagraubvogel und eulen im jahre 1932. J. Ornithol. 82(2):210-221.
3218	**Uttendorfer, O.** *1952.* Neue ergebnisse uber die ernahrung der greifvogel und eulen. Eugen Ulmer, Stuttgart. 230 pp.
3219	**Uys, C.J. and J.G.R. Macleod.** *1967.* The birds of the De Hoop Lei region, Bredasdorp, and the effect of the 1957 inundation over a 10-year period (1957-1966) on the distribution of species, numbers and breeding. Ostrich 38(4):233-254.
3220	**Vagliano, C.** *1977.* The status of birds of prey in Greece. Pages 118-125. *in:* R.D. Chancellor, Ed. Proc. of the World Conf. on Birds of Prey. 1-3 October, 1975. Vienna. I.C.B.P. (In French, English Summ.)
3221	**Vaillant, G.** *1981.* Observation d'un aigle des steppes en Tunisie. Alauda 49(3):229-230.
3222	**Valet, G.** *1970.* A propos de l'aigle pomarin *(Aquila pomarina).* Alauda 38(2):161-162.
3223	**Vallon, G.** *1914.* Besprechung von zwei arbeiten des Professors Martorelli, Welche den Falco feldeggi behandeln. Falco 10(1):1-163.
3224	**Vallon, G.** *1918.* Uccelli rari comparsi nella Provincia del Friuli durante l'anno 1914. Riv. Ital. Ornitol. 4:16-18.
3225	**Valverde, J.A.** *1958.* An ecological sketch of the Coto Donana. Brit. Birds 51(1):1-23.
3226	**Valverde, J.A.** *1960.* La population d'aigles imperiaux *(Aquila heliaca adalberti)* des Marismas du Guadalquivir: son evolution depuis un siecle. Alauda 28(1):20-26.
3227	**Valyus, M., E. Kalindrene, A. Pyatraitis and V. Skoudis.** *1976.* Istoricheskie i sovremennye materialy po ornithologii Litovskoi SSR. (Historical and recent data on the ornithology of Lithuanian SSR.) Ecologia Ptits Litovsk. SSR. Vilnus. p. 5-32. (In Russian)
3228	**Valyus, V.** *1977.* Materialy po ekologii nekotorykh vidov khistchnykh ptits v Litve. (Data on ecology of some species of birds of prey in Lithuania.) Ecologia Ptits Litovsk. SSR, Vilnus. p. 121-128. (In Russian)
3229	**Van der Akker, J.B.** *1954.* A wintering concentration of eagles in Oklahoma. Wilson Bull. 66(2):136.
3230	**Van Dobay, L.** *1934.* Adatok a bekaszo sasok biologiajahoz. Kocsag 7:31-39. (In Hungarian, German Summ.)
3231	**Van Horn, D. and D.J. Kelly.** *1975.* Fall and winter raptors of Turnbull National Wildlife Refuge, 1972 to 1973. Murrelet 56(3):12-16.
3232	**Van Impe, J.** *1977.* L'avifaune estivale du complexe lagunaire Razelm-Sinoie (Roumaine). Alauda 45(1):17-52.
3233	**Van Rossem, A.J.** *1945.* A distributional survey of the birds of Sonora, Mexico. Louisiana St. Univ., Mus. Zool., Occas. Pap. No. 21. 379 pp.
3234	**Van Someren, V.G.L.** *1932.* Birds of Kenya and Uganda, being addenda and corrigenda to my previous paper in "Novitates Zoologicae" XXIX, 1922. Novitates Zool. 37:252-380.
3235	**Van Someren, V.G.L.** *1956.* Days with birds. Studies of habits of some East African species. Fieldiana: Zoology 38:1-520.
3236	**Vance, J.M.** *1973.* Lord of the skies. Missouri Cons. 34(10):16-19.
3237	**Vande Weghe, J.-P.** *1978.* Les rapaces Palearctiques au Rwanda. Gerfaut 68(4):493-519. (English Summ.)
3238	**Vande Weghe, J.P.** *1973.* Les periodes de nidification des oiseaux du Parc National de l'Akagera au Rwanda. Gerfaut 63(3/4):235-255. (English Summ.)
3239	**Varney, J.R. and D.H. Ellis.** *1974.* Telemetering egg for use in incubation and nesting studies. J. Wildl. Manage 38(1):142-148.
3240	**Varshavski, B.S.** *1973.* (Some landscape-ecological characteristics of feeding of *Buteo rufinus, Aquila heliaca* and *Aquila nipalensis* north of the Aral Sea.) Bull. MOIP Otdel. Biology 78(3):30-37. (In Russian, English Summ.)
3241	**Varshavski, S.N.** *1952.* (Results of the ornithological survey of birds of prey in the steppe.) Metody Ucheta Chislennosti I Geographicheskogo Raspredelenia Nazemnykh Pozvonochnykh. p. 97-104. (In Russian)
3242	**Varshavski, S.N.** *1959.* (Some results of analysis of raptor pellets to study distribution and number of small mammals north of the Aral Sea.) 2 Vsesoyuzn. Ornithol. Konf., Tezisy Doklady., Chast 3. Moskow. p. 7-8. (In Russian)
3243	**Varshavski, S.N.** *1962.* (Hunting methods of raptors in open landscapes in relation to ecological characteristics and feeding specialization of various species.) 3 Vsesoyuzn. Ornithol. Konf., Materialy, Kniga 1, Lvov. p. 57-59. (In Russian)
3244	**Varshavski, S.N.** *1968.* O pitanii berkuta na yugozapadnom Ustyurte. (Feeding of *Aquila chrysaetos* in southwestern Ust-urt.) Ornitologia 9:146-149. (In Russian)

Master No.	Citation
3245	**Varshavski, S.N.** *1974.* (Biocenotic relations, feeding specialization and distribution of diurnal birds of prey in zonal semi-desert and desert landscapes.) 6 Vsesoyuzn. Ornithol. Konf., Materialy, Chast 2, Moskow. p. 36-38. (In Russian)
3246	**Vasic, V.F.** *1974.* Observations ornithologiques en Afghanistan. Alauda 42(3):259-280.
3247	**Vasse, R. and S. Vasse.** *1974.* A golden eagle vs. the geese. Ill. Audubon Bull. 169:12.
3248	**Vasvari, M.** *1955.* Magyarorszagi madarak meretei. (Measurements of Hungarian birds.) Aquila 59-62:167-184. (English Summ.)
3250	**Vasvari, N.** *1931-1934.* Uber einige neve oder seltene vogel unserer ornithofauna. Aquila 38-41:298-307.
3249	**Vasvari, N.** *1939.* Die verbreitung und okologie des kaiseradlers. Festschr. 60. Geburtstag Prof. E. Strand. 5:290-317. (Title Only)
3251	**Vaucher, A.** *1915.* Liste des oiseaux observes au Maroc de 1884 a 1914. Oiseau Rev. Fr. Ornithol. 4(74):94-96.
3252	**Vaucher, C.** *1945.* Encoure une histoire d'aigle. Nos Oiseaux 18:92-93.
3253	**Vaucher, C.A.** *1971.* Notes sur l'ethologie de l'aigle de Bonnelli. Nos Oiseaux 31(5/6):101-111.
3254	**Vaughan, M.** *1879.* Reported nesting of the golden eagle in Shetland. Zoologist (3rd Ser.) 3:461.
3255	**Vaughan, R.E. and K.H. Jones.** *1913.* The birds of Hong Kong, Macao, and the West River or Si Kiang in south-eastern China: with special reference to their nidification and seasonal movements. Ibis (10th Ser.) 1(2):163-201.
3256	**Vaurie, C.** *1964.* Physiography and phytogeography of Mongolia. Bull. Am. Mus. Nat. Hist. 127(3):103-144.
3257	**Vaurie, C.** *1965.* The Birds of the Palearctic Fauna. A Systematic Reference. Non Passeriformes. H.F. & G. Witherby, Ltd., London. 763 pp.
3258	**Verheyen, R.** *1950. Aquila pomarina* (Brehm). Gerfaut 40(3):174.
3259	**Vernon, C.J.** *1965.* The 1964 black eagle survey in the Matopos, Rhodesia. Arnoldia 2(6):1-9.
3260	**Vernon, C.J.** *1967.* Some observations from the journals of K.W. Greenhow. Ostrich 38(1):48-49.
3261	**Vernon, C.J.** *1972.* On the life-expectancy of Matopos black eagles. Ostrich 43(4):206-210.
3262	**Vernon, C.J.** *1977.* Breeding seasons of birds in deciduous woodland at Zimbabwe, Rhodesia, from 1970 to 1974. Ostrich 49(3):102-115.
3263	**Vernon, C.J.** *1978.* A review of the status of raptors in the eastern Cape. Pages 87-95. *in:* Proc. of the Symposium on African Predatory Birds. 29 August - 1 September, 1977. Pretoria. N. Transvaal Ornithol. Soc.
3264	**Verrill, A.E.** *1897.* Golden eagle *(Aquila chrysaetos)* taken near New Haven, Conn. Auk 14(1):89.
3265	**Vertse, A.** *1958.* Madarvedelmi beszamolo 1952-57. Aquila 65:39-49.
3266	**Vielliard, J.** *1969.* Etude des possibilites de reproduction de l'aigle pomarin *Aquila pomarina* en France. Alauda 37(4):348-350.
3267	**Vielliard, J.** *1969.* Donnees biogeographiques sur l'avifaune d'Asie Occidentale. I. Afghanistan (Premiere partie.) Alauda 37(4):273-300.
3268	**Vielliard, J.** *1971.* Donnees biogeographiques sur l'avifaune d'Afrique centrale. Alauda 39(3):227-248. (English Summ.)
3269	**Vielliard, J.** *1972.* Donnees biogeographiques sur l'avifaune d'Afrique centrale. Alauda 40(1):63-92.
3270	**Vielliard, J.** *1974.* Donnees inedites sur la reproduction de l'aigle imperial dans les Marismas du Guadalquivir. Alauda 42(1):7-9.
3271	**Viksne, J.** *1975.* Berkut *(Aquila chrysaetos* (L.)) vse zhe ne perestal gnezditsya v Latvii. (*Aquila chrysaetos* has not ceased nesting in Latvia.) Trudy Mus. Zool. Latvia Univ. 12:69-78. (In Russian)
3272	**Vincent, A.W.** *1945.* On the breeding habits of some African birds. Ibis 87(2):203-216.
3273	**Vinogradov, V.V., G.M. Rusanov, G.A. Krivonosov and D.V. Bondarev.** *1977.* O chislennosit i razmnozhenii stepnogo orla v Kalmykii i Dagestane. (Number and breeding of *Aquila nipalensis* in Kalmykia and Dagestan.) 7th Vsesoyuzn. Ornithol. Konf., Tezisy Doklady, Chast 2, Kiev. p. 196-197. (In Russian)
3274	**Vinokurov, V.N.** *1979.* Zametki po ekologii gnezdovaniya khistchnykg ptits tsentralnoi Yakutii. (Notes on nesting ecology of birds of prey in central Yakutia.) Migratsii I Ecologia Ptits Sibiri, Tezisy Doklady, Yakutsk. p. 129-130. (In Russian)
3275	**Visher, S.S.** *1909.* A list of the birds of western South Dakota. Auk 26(2):144-153.
3276	**Visher, S.S.** *1913.* An annotated list of the birds of Sanborn County. Auk 30(4):561-573.
3277	**Visser, J.** *1963.* Black eagle of Zuurhoek, Jansenville. Afr. Wild Life 17(3):191-194.
3278	**Vittery, A.** *1978.* Wahlberg's eagle nesting in Ethiopia. Scopus 2(1):18-19.
3279	**Vogue, G., de.** *1948.* Notes sur l'avifaune de la region de Megeve (haute-savoie). Alauda 16:128-142.
3280	**Voisin, R.** *1968.* Neuf jours au nid chocard a bee jaune. Nos Oiseaux 29(11):286-292.
3281	**Volkov, N.I.** *1975.* Chislennosti nekotorykh aistoobraznykh, tsaplevykh i dnevnykh khistchnykh ptits v Zapovednike. (Numbers of some storks, herons, and diurnal birds of prey in the Reserve.) Berezinsk. Zapovadnik, Minsk. 4:194-200. (In Russian)
3282	**Voous, K.H.** *1973.* Steppenarend in Nederland. (Tawny eagle in the Netherlands.) Limosa 46(3/4):233-238. (English Summ.)
3283	**Voskar, J., A. Mosansky and J. Palasthy.** *1969.* (Breeding success and ecological distribution of golden eagles in Czechoslovakia.) Zool. Listy 18(1):39-54.
3284	**Vreeland, F.K.** *1926.* A battle among the crags. Am. Forests 32(387):136-138.
3285	**Vrijdagh, J.M.** *1949.* Observations ornithologiques en region occidentale du Lac Albert et principalement de la Plaine d'Ishwa. Gerfaut 39(1/2):1-115.
3286	**Wackernagel, H.** *1964.* Hochstalter europaischer vogel im Zoologischen Garten Basel. Ornithol. Beob. 61(4):139-140.
3287	**Wagner, A.** *1932.* The golden eagle as a pet. Iowa Bird Life 3(4):51.
3288	**Walker, A.** *1911.* Golden eagle victim of peculiar accident. Oologist 28(12):184.
3289	**Walker, A.** *1916.* Some raptores of Douglas County, South Dakota. Condor 18(3):130.

Master No.	Citation
3290	**Walker, L.W. and M. Walker.** *1939.* From egg to eaglehood. Nat. Hist. 43(5):284-289.
3291	**Walker, L.W. and M. Walker.** *1940.* Headlines on eagles. Nature 33(6):320-323.
3292	**Wallace, A.R.** *1868.* On the raptorial birds of the Malay Archipelago. Ibis (New Ser.) 4(13):1-27.
3293	**Wallace, G.J.** *1937.* Notes from Vermont. Auk 54(4):547-548.
3294	**Wallis, H.M.** *1895.* Notes on the birds of the central Pyrenees. Ibis (7th Ser.) 1(1):64-85.
3295	**Wallis, H.M.** *1896.* Eggs of golden eagle. Ibis 2(6):261.
3296	**Wallis, H.M.** *1896. Aquila chrysaetos* nesting in confinement. Bull. Brit. Ornithol. Club 7:21-22.
3297	**Wallis, H.M.** *1912-1913.* Account of the birds of the Balkans. Bull. Brit. Ornithol. Club 31:92-96.
3298	**Wampole, J.H.** *1947.* Hawk populations in Nebraska, 1945-1946. Nebr. Game. Forestation and Parks Comm. Proj. W-015-R-02, Job Sp 1. 5 pp.
3299	**Ward, A.L., P.L. Hegdal, V.B. Richens and H.P. Tietjen.** *1967.* Gophacide, a new pocket gopher control agent. J. Wildl. Manage. 31(2):332-338.
3300	**Ware, R.** *1930.* Egg of golden eagle in Scotland. Bull. Brit. Ool. Assoc. 2(24):144.
3301	**Warschawski, S.N. and M.N. Schilov.** *1979.* (Zonal and geographic specialization of food of the steppe eagle in the western part of its range.) 7th Vses. Zoogeogr. Konf., Moscow, 1980. Tez Doklady. p. 104-107.
3302	**Waterston, G.** *1959.* Golden eagle with tuberculosis and aspergillosis. Brit. Birds 52(6):197-198.
3303	**Watson, A.** *1957.* Golden eagle display in late June. Scott. Nat. 69(3):184.
3304	**Watson, A.** *1957.* The breeding success of golden eagles in the north-east highlands. Scott. Nat. 69(3):153-169.
3305	**Watson, A.** *1966.* Hill birds of the Cairngorms. Scott. Birds 4(2):179-203.
3306	**Watson, A. and N.C. Morgan.** *1964.* Residues of organo-chlorine insecticides in a golden eagle. Brit. Birds 57(9):341-344.
3307	**Watson, D.M.** *1970.* Notes on some birds of prey at Abilene Zoo. Int. Zoo Yearb. 10:12-13.
3308	**Wauer, R.H.** *1964.* Ecological distribution of the birds of the Panamint Mountains, California. Condor 66(4):287-301.
3309	**Weaver, R.A. and J.L. Mensch.** *1970.* Observed interaction between desert bighorn sheep, *Ovis canadensis,* and reported predator species. Calif. Fish and Game 56(3):206-207.
3310	**Webber, A.C.** *1971.* The case of the slaughtered eagles, how '750 were killed from the air.' Am. Rifleman 119(10):24-26.
3311	**Weber, J.W. and E.J. Larrison.** *1977.* Birds of Southeastern Washington. Univ. of Idaho Press, Moscow, Idaho.
3312	**Weddle, F.** *1973.* The eagles - victims once again. Defenders (Oct.):554-556.
3313	**Wedemeyer, K.O.** *1936.* Ueber das vorkommen des schreiadlers in der Provinz Hannover und ihren Crenzgebieten. Ornithol. Monatsber. 44(3):71-73.
3314	**Weir, D.N.** *1968.* Recoveries of raptors ringed on Speyside. Scott. Birds 5(3):172-173.
3315	**Weir, D.N.** *1972.* Mortality of hawks and owls in Speyside. Bird Study 18(3):147-154.
3316	**Weir, D.N.** *1973.* A case of lamb-killing by golden eagles. Scott. Birds 7(6):293-301.
3317	**Weir, D.N.** *1978.* Effects of poisoning on ravens, buzzards and golden eagles in Scotland. Brit. Birds 71(5):227-228.
3318	**Weir, D.N.** *1978.* Effects of poisoning on raven and raptor populations. Scott. Birds 10(1):31.
3319	**Weir, T.** *1964.* Days with golden eagles. Country Life 135(3505):1102-1103.
3320	**Weise, W.** *1964.* Ein steinadler, *Aquila ch. chrysaetos* , in Sachsen. Beitr. Vogelkd. 9(6):446-448.
3321	**Wellein, E.G. and T. Ray.** *1964.* Eagle investigation - New Mexico and western Texas. Unpublished Rep. 26 pp.
3322	**Wendland, V.** *1932.* Zur biologie des schreiadlers *(Aquila pomarina* A.). Brutbiologie. Beitr. Fortpfl.-Biology Vogel 8(1):1-9.
3323	**Wendland, V.** *1932.* Zur biologie des schreiadlers *(Aquila pomarina).* Beitr. Fortpfl.-Biology Vogel. 8(2):47-53.
3324	**Wendland, V.** *1934.* Funf- und sechsjahrige beobachtungen uber die raubvogel zweier norddeutscher waldgebiete (mit berucksichtigung ihrer Siedlungsdichte). Beitr. Fortpfl.-Biology Vogel 10(4):130-138.
3325	**Wendland, V.** *1951.* Zwanzigjahre beobachtungen uber den schreiadler *Aquila pomarina.* Vogelwelt 72(1):4-11.
3326	**Wendland, V.** *1958.* Zum problem des vorzeitigen sterbens von jungen greifvogeln und eulen. Vogelwarte 19(3):186-191.
3327	**Wendland, V.** *1959.* Schreiadler und Schelladler. Die Neue Brehm-Bucherei, A. Ziemsen Verlag, Wittenberg Lutherstadt. 71 pp.
3328	**Wendland, V.** *1972.* Uber de vogelwelt z weier Sudtiroler Alpentaler. Egretta 15(2):49-54.
3329	**Wendland, V.** *1972.* Die vogel des rauristales (Hohe Tauern). Egretta 15(2):41-48.
3330	**West, C.** *1976.* Wildlife of the Okanogan National Forest: Habitat requirements and management guidelines. U.S.D.A., Forest Service, Okanogan Natl. Forest, Washington. Report.
3331	**Westernhagen, W., von.** *1962.* Greifvogel in Andalusien. Ornithol. Mitt. 14(3):41-48.
3332	**Weston, F.M., Jr.** *1925.* Golden eagle near Pensacola, Florida. Auk 42(3):443.
3333	**Wet, J.R., de.** *1945.* Flight of Verreaux's eagle. Ostrich 16(3):235-236.
3334	**Wetmore, A.** *1923.* Avian fossils from the Miocene and Pliocene of Nebraska. Bull. Am. Mus. Nat. Hist. 48(12):483-507.
3335	**Wetmore, A.** *1933.* The eagle, king of birds, and his kin. Natl. Geog. 64(1):43-63.
3336	**Wetmore, A.** *1933.* Oligocene eagle from Wyoming. Smith. Inst. Misc. Coll. 87(19):1-9.
3337	**Wetmore, A.** *1933.* Status of *Minerva antiqua, Aquila ferox,* and *Aquila lydekkeri* as fossil birds. Am. Mus. Novitates 680:1-4.
3338	**Whatmough, R.J.** *1978.* Birds of the Torrens River, Adelaide. S. Aust. Ornithol. 28(1):1-15.
3339	**Wheat, M.C., Jr.** *1960.* Move to protect golden eagle. Kingbird 10(4):168.
3340	**Whistler, H.** *1916.* Notes on the birds of the Jhelum District of the Punjab. Ibis (10th Ser.) 4(1):35-118.
3341	**Whistler, H.** *1922.* The birds of Jhang District, S.W Punjab. Part II. Non-Passerine birds. Ibis (11th Ser.) 4(3):401-437.
3342	**Whistler, H.** *1926.* The birds of Kangra District, Punjab. Part II. Ibis (12th Ser.) 2(4):724-783.
3343	**Whistler, H.** *1930.* The birds of Rawal Pindi District, N.W. India. Ibis (12th Ser.) 6(2):247-279.
3344	**Whistler, H.** *1936.* Further observations from Albania. Ibis (13th Ser.) 6(2):335-356.
3345	**Whistler, H.** *1940.* How do large raptorial birds hunt their prey? Ibis 4(4):732-735.
3346	**Whitaker, J.J.S.** *1894.* Notes on some Tunisian birds. Ibis (6th Ser.) 6(21):78-100.
3347	**White, C.A.** *1893.* The raptores of Omaha and vicinity. Oologist 10(5):138-140.
3348	**White, C.M.** *1965.* Roadside raptor count through Utah, Colorado, and Kansas. Kansas Ornithol. Soc. Bull. 16(3):18-19.

Master No. **Citation**

3349 **White, C.M. and S.K. Sherrod.** *1973.* Advantages and disadvantages of the use of rotor-winged aircraft in raptor surveys. Raptor Res. 7(3/4):97-104.

3350 **White, C.M. and J.H. Streater.** *1971.* Survey of raptorial birds along the proposed Trans Alaska Pipeline system. Raptor Res. News 5(3):96-99.

3351 **White, C.M., T.D. Ray and L.W. Sowl.** *1977.* The 1970-1972-1974 raptor surveys along the Trans Alaska Oil Pipeline. Pages 222-227. *in:* R.D. Chancellor, Ed. Proc. of the World Conf. on Birds of Prey. 1-3 October, 1975. Vienna. I.C.B.P.

3352 **White, C.M.N.** *1945.* The ornithology of the Kaonde-Lunda Province, northern Rhodesia. - Part II. Systematic list. Ibis 87:185-202.

3353 **White, C.M.N.** *1951.* Systematic notes on African birds. Ostrich 22(1):25-26.

3354 **White, E.J.B.** *1968.* Encounter with a golden eagle. Blue Jay 26(1):59-60.

3355 **White, H.L.** *1918.* Notes upon the eggs of the wedge-tailed eagle *(Uroaetus audax)* . Emu 17(3):149-150.

3356 **White, S.E.** *1890.* Caged eagles. Oologist 7(4):65-66.

3357 **Whitehead, C.H.T.** *1909.* On the birds of Kohat and Kurram, Northern India. Ibis (9th Ser.) 3(10):214-284.

3358 **Whitehead, C.H.T.** *1909.* Additions and corrections to the "Birds of Kohat." Ibis (9th Ser.) 3(12):620-623.

3359 **Whitehead, R.** *1908.* The First Book of Eagles. Franklin Watts, Inc., New York. 45 pp.

3360 **Whitfield, D.W.A., D.W. Davis, J.M. Gerrard and W.J. Maher.** *1969.* Golden eagles in central Saskatchewan. Blue Jay 27(2):74-79.

3361 **Whitherby, H.F.** *1921.* Spotted eagle in Essex. Brit. Birds 14(10):237.

3362 **Whitlock, F.L.** *1909.* Notes on birds observed on the Pilbarra Goldfield, north-western Australia. Emu 8(4):173-194.

3363 **Whitlock, F.L.** *1928.* Wedge-tailed eagle attacks child. Emu 27(4):287-289.

3364 **Whitten, L.** *1968.* Pinion, The Golden Eagle. D. Van Nostrand Comp., Inc. Princeton, N.J. 117 pp.

3365 **Wikstrom, D.** *1948.* Paivapetolintujen tunteminen. (The knowledge of birds of prey.) Suomen Riista 3:156-174. (English Summ.)

3366 **Wildburg, B.** *1915.* Uber den steppenadler *(Aquila nipalensis orientalis* Cab.) Ornithol. Jahrbuch 26(1/2):54-56.

3367 **Wiley, R.W. and E.G. Bolen.** *1971.* Eagle-livestock relationships: livestock carcass cenus and wound characteristics. Southwest. Nat. 16(2):151-169.

3368 **Wilhelmsen, S.** *1961.* Noen notater om kongeorn. (New notes on the golden eagle.) Sterna 4(7):265-268.

3369 **Willard, F.C.** *1916.* The golden eagle in Cochise County, Arizona. Oologist 33(1):2-10.

3370 **Willard, F.C.** *1916.* Notes on the golden eagle in Arizona. Condor 18(5):200-201.

3371 **Willett, G.** *1912.* Birds of the Pacific Slope of southern California. Pac. Coast Avifauna No. 7. p. 1-113.

3372 **Willgohs, J.F.** *1968.* Bedre vern av vare rovfugler. (Better protection for our birds of prey.) Sterna 8(1):1-14.

3373 **Willgohs, J.F.** *1969.* Status and prospects of eagles in Norway. Biology Cons. 2(1):71-72.

3374 **Willgohs, J.F.** *1977.* Birds of prey in Norway. Pages 143-152. *in:* R.D. Chancellor, Ed. Proc. of the World Conf. on Birds of Prey. 1-3 October, 1975. I.C.B.P.

3375 **Williams, C.H. and W. Neubrech.** *1976.* Eagles, modern passenger pigeon? *in:* C.H. Williams and W. Neubrech, Eds. Indian Treaties, American Nightmare. Outdoor Empire Publ. Comp., Seattle, Wa.

3376 **Williams, R.B.** *1947.* Infestations of raptorials by *Ornithodoros aquilae.* Auk 64(2):185-188.

3377 **Williams, W.J.** *1926.* Golden eagle and marsh-harrier in Ireland. Brit. Birds 20(4):107-108.

3378 **Wilson, C.** *1973.* How the woolgrowers fleece the public. Audubon 75(3):119-121.

3379 **Wilson, G.L.** *1928.* Hidatsa eagle trapping. Am. Mus. Nat. Hist., Anthropol. Papers 30(4):99-245.

3380 **Wilson, J.** *1823.* Remarks on the different opinions entertained regarding the specific distinction, or identity, of the ring-tailed and golden eagles. Mem. Wernerian Nat. Hist. Soc. 4(2):434-448.

3381 **Wilson, M.** *1974.* Birds of the Simpson Desert. Emu 74(3):169-176.

3382 **Wilson, S.B.** *1887.* Notes on some Swiss birds. Ibis 5(18):130-150.

3383 **Winkler, A. and P.W. Bonnot.** *1958.* Man meets bird in Swiss alpine adventure: Ringing an eaglet in his eyrie on an inaccessible rockface. Illus. London News 233(6233):908-909.

3384 **Winterbottom, J.M.** *1936.* Distributional and other notes on some northern Rhodesian birds. Ibis (13th Ser.) 6(4):763-791.

3385 **Winterbottom, J.M.** *1939.* Miscellaneous notes on some birds of northern Rhodesia. Ibis (14th Ser.) 3(4):712-734.

3386 **Winterbottom, J.M.** *1959.* Notes on the status of some birds in northern Rhodesia. Ostrich 30(1):1-12.

3387 **Winterbottom, J.M.** *1962.* Some manuscript notes of S.F. Townsend for the period 1878-1925. Ostrich 33(2):66-71.

3388 **Winterbottom, J.M.** *1971.* Priest's Eggs of Southern African Birds. Winchester Press, Johannesburg.

3389 **Wirth, H.** *1964.* Ein steinadler vor dem erfrierungstod gerettet. Falke 5:165-166.

3390 **Wisecarver, J. and G. Bogue.** *1974.* Raptor rehabilitation at Alexander Lindsay Junior Museum. Raptor Res. 8(1/2):6-10.

3391 **Witherby, H.F.** *1921.* Spotted eagle in Essex. Brit. Birds 14(10):237.

3392 **Witherby, H.F.** *1921.* Notes on British records of the spotted eagle and steppe buzzard and on the British tawny owl. Brit. Birds 14(8):180-183.

3393 **Witherby, H.F.** *1921.* British records of the spotted eagle. Brit. Birds 14(9):209.

3394 **Witherby, H.F.** *1921.* The St. Columb spotted eagle. Brit. Birds 15(3):69.

3395 **Witherby, H.F.** *1928.* On the birds of central Spain, with some notes on those of south-east Spain. Ibis (12th Ser.) 4(4):587-663.

3396 **Witherby, H.F.** *1931.* Primary feather of a golden eagle from Ireland. Bull. Brit. Ornithol. Club 52:20.

3397 **Witherby, H.F., F.C.R. Jourdain, N.F. Ticehurst and B.W. Tucker.** *1939.* The Handbook of British Birds. Vol. III. H.F. & G. Witherby Ltd., London. 384 pp.

3398 **Witherington, G.** *1921.* Lifting and weight-carrying power of golden eagles. Brit. Birds 14(12):284.

3399 **Withers, A.** *1981.* Wedge-tailed eagle kills a large goanna. Aust. Raptor Assoc. News 2(1):10.

3400 **Witowski, J.** *1965.* Birds of "Pond Nowokuznicki Nature Reserve" in 1963-1964. Acta Ornithol. 9(3):170-180.

Master No. **Citation**

3401 **Wodzicki, G.C.** *1857.* Notiz uber *Aquila chrysaetos* und *Strix bubo* in Galicien. J. Ornithol. 5(4):192-293.

3402 **Woffinden, N.D. and J.R. Murphy.** *1977.* A roadside raptor census in the eastern Great Basin. Raptor Res. 11(3):62-66.

3403 **Wolfe, L.R.** *1928.* The breeding Accipitres of Utah. Ool. Rec. 8(4):90-102.

3404 **Wolfe, L.R.** *1938.* Eggs of the Falconiformes. Ool. Rec. 18(3):49-60.

3405 **Wolfe, L.R.** *1938.* A synopsis of North American birds of prey and their related forms in other countries. Bull. Chicago Acad. Sci. 5(8):167-208.

3406 **Wolfe, L.R.** *1950.* Notes on the birds of Korea. Auk 67(4):433-455.

3407 **Wolk, K.** *1973.* Zbiory zoologiczne Muzeum Lasu i Drewna w Rogowie. (Zoological collection in the Forest and Timber Museum in Rogow.) Przeglad Zool. 17(1):138-141.

3408 **Wood, D.T.** *1946.* Eye-witness account of golden eagle killing calf. Condor 48(3):143.

3409 **Wood, K.A.** *1941.* Aerial display by a pair of golden eagles. Brit. Birds 35(7):158-159.

3410 **Wood, N.A.** *1951.* The Birds of Michigan. Univ. of Michigan, Museum Zool. Misc. Publ. No. 75. 559 pp.

3411 **Wood, R.V.** *1945.* Wahlberg's eagle - *Aquila wahlgergi* (sic) on Natal south coast. Ostrich 16(3):233.

3412 **Woodbury, A.M. and H.N. Russell, Jr.** *1945.* Birds of the Navajo country. Univ. of Utah, Bull., Biology Ser. 9(1):1-157.

3413 **Woodgerd, W.** *1952.* Food habits of the golden eagle. J. Wildl. Manage. 16(4):457-459.

3414 **Work, T.H. and A.J. Wool.** *1947.* At home with the golden eagle. Nat. Hist. 56(9):412-417,428.

3415 **Workman, W.H.** *1902.* Golden eagle in Co. Donegal. Zoologist (4th Ser.) 6:150.

3416 **Wormell, P.** *1965.* Manx shearwaters and other sea-birds as prey of peregrines and golden eagles. Brit. Birds 58(4):149.

3417 **Worthington, W.W.** *1891.* Golden eagle at Shelter Island, New York. Auk 8(1):113.

3418 **Wrakestraw. G.F.** *1973.* The 1973 Wyoming bald and golden eagle survey. Am. Birds 27(4):716-718.

3419 **Wrakestraw, G.F.** *1972.* 1972 Wyoming bald and golden eagle survey. Wyoming Dept. of Fish and Game. Proj. No. W-50-R-21. Wk. Pl. 7, Job. No. 31. 7 pp.

3420 **Wray, R.S.** *1887.* On some points in the morphology of the wings of birds. Proc. Zool. Soc. London 1887:343-357.

3421 **Wright, C.A.** *1864.* List of the birds observed in the Islands of Malta and Gozo. Ibis 6(21):42-73.

3422 **Wright, C.A.** *1869.* Third appendix to a list of birds observed in Malta and Gozo. Ibis 5(19):245-256.

3423 **Wright, C.A.** *1874.* Fifth appendix to a list of birds observed in Malta and Gozo. Ibis 4(15):223-241.

3424 **Wright, G.M.** *1928.* Fare of the golden eagle. Yos. Nature Notes 7(4):29.

3425 **Wright, G.M.** *1933.* Men and birds in joint occupation of National parks. Condor 35(6):213-218.

3426 **Wright, W.C.** *1902.* Golden eagle in Co. Donegal. Zoologist (4th Ser.) 6:150.

3427 **Wright, W.L., Jr.** *1977.* Ecology of the Cascade mountain goat, Mount Baker-Snoqualmie National Forest, Washington. M.S. Thesis. Western Wash. State Coll., Bellingham, WA.

3428 **Wust, W.** *1936.* Steinadler - paar im Walchenseegebiet. Ornithol. Monatsber. 44(3):86.

3429 **Wust, W.** *1961.* Ein schreiadler *(Aquila pomarina)* am Chimsee. Ornithol. Mitt. 13(2):28.

3430 **Wust, W.** *1962.* Prodromos einer "avifauna bayerns.' Anz. Ornithol. Ges. Bayern 6(4):305-358.

3431 **Wust, W.** *1963.* Steinadler, *(A. c.)* wieder in Gefahr. Ber. Deutsche Sekt. Int. Rat Vogelschutz 3:29-32. (Title Only)

3432 **Wustnei, C.** *1899.* Beitrage zur ornis Mecklenburgs. J. Ornithol. 47(2):125-151.

3433 **Wustnei, C.** *1900.* Beobachtungen aus der ornis Mecklenburgs im jahre 1899. J. Ornithol. 48(3):314-349.

3434 **Wustnei, C.** *1901.* Beobachtungen aus der ornis Mecklenburgs im jahre 1900. J. Ornithol. 49(2):246-264.

3435 **Wyss, J.** *1924.* Eine steinadlererbeutung im Tessin. Ornithol. Beob. 21(7):106-107.

3436 **Xampeny Baro, J.** *1974.* Presencia en Cataluna de un *Aquila heliaca* joven. Ardeola 20:345-347.

3437 **Yamashina, Y.** *1941.* A Natural History of Japanese Birds. Published in Japan. 1080 pp.

3438 **Yoakum, J.D.** *1959.* Eagle predation of wild pigeon. Murrelet 40(3):34.

3439 **Young, H.** *1973.* Breeding of a pair of wedge-tailed eagles. Aust. Bird Watcher 5(3):99-101.

3440 **Youngworth, W.** *1933.* Migration records of eagles and snowy owls in the upper Missouri Valley. Wilson Bull. 45(1):32-33.

3441 **Yrsaliev, D.** *1962.* Materialy po biologii berkuta. (Data on the biology of *Aquila chrysaetos.*) Izvestia An Kirgiz SSR, Seria Bio. Nauk. 4(1):83-87. (In Russian)

3442 **Zangheri, P.** *1936.* Elenco ragionato di tutte le specie fino ad oggi osservate in Romagna e notizie di indole generale. Riv. Ital. Ornitol. 6(2):73-82.

3443 **Zarkhidze, V.A.** *1969.* (Raptors of northwestern Turkmenia and their significance in the destruction of desert rodents.) 5 Vsesoyuzn. Ornithol. Konf., Materialy, Kniga 2, Ashkabad. p. 233-235.

3444 **Zastrov, M.** *1946.* Om kungsornens *(Aquila chr. chrysaetos* L.) utbredning och biologi i Estland. (On the distribution and biology of the golden eagle in Estonia.) Var Fagelvarld 5(2):64-80.

3445 **Zebe, V.** *1942.* Der schreiadler als brutvogel in der Bartschniederung. Ber. Ver. Schles. Ornithol. 27:27-30. (Title Only)

3446 **Zedlitz, O.G.** *1908.* Hat die im winter niedergegangene regenmenge einen einflufs auf das brutgeschaft einzelner vogel im Chott-Gebiete sudlich des Atlas? J. Ornithol. 56(3):480-487.

3447 **Zedlitz, O.G.** *1909.* Ornithologische beobachtungen aus Tunesien, speziell dem Chott-Gebiete. J. Ornithol. 57(3):241-322.

3448 **Zedlitz, O.G.** *1910.* Meine ornithologische ausbeute in Nordost-Afrika. J. Ornithol. 58(2):290-394.

3449 **Zedlitz, O.G.** *1914.* Das Sud-Somaliland als zoogeographisches gebiet. J. Ornithol. 62(4):608-678.

3450 **Zedlitz, O.G.** *1920.* Die avifauna des westlichen Pripjet-Sumpfes im Lichte der forschung deutscher ornithologen in den jahren 1915-1918. J. Ornithol. 68(3/4):350-388.

3451 **Zedlitz, O.G.** *1926.* Einige notizen uber den beginn der brutzeiten in Nord-Afrika. Beitr. Fortpfl.-Biology Vogel 2(3):58-64.

3452 **Zhezherin, V.P.** *1969.* On taxonomic relations of *Aquila clanga* Pall. and *Aquila pomarina* Brehm. Zbirn. Prats. Zool. Mus. 33:91-97. (In Ukrainian, English Summ.) (Title Only)

3453 **Zhordaniya, R.G. and G.S. Gogilashvili.** *1976.* Birds of Svanetia (Georgian SSR). Acta Ornithol. 15(6):243-257. (English Trans.)

3454 **Zubarovski, V.M.** *1977.* (Birds of prey, fauna of Ukraine. Vol. 5. Birds 2.) Naukova Dumka, Kiev. 328 pp. (In Ukrainian)

Master No.	Citation
3455	**Zubtsovski, N.E. and V.K. Ryabitsev.** *1976.* Novye dannue o ptitsah poluostrova Kanin. (New data on birds of Kanin peninsula.) Ornitologia 12:228-231. (In Russian)
3456	**Zunino, F.** *1969.* Brevi note sull'avifauna della Valle de Rhemes. Riv. Ital. Ornitol. 39(3):169-197.
3457	**Zwickel, F.C.** *1972.* Grazing and predation on blue grouse. Murrelet 53:52-53.
3458	**Zwiesele, D.** *1905.* Steinadler *(Aquila chrysaetos* L.). Ornithol. Beob. 4(11):171.
3459	**Zwinger, A.** *1977.* The eagle's fate and mine are one. Audubon 79(4):50-88.

Chapter 4

Permuted List of Keywords

Master No.	Keywords
324	**Abandonment** • Reproductive Success • Food Requirements • Territory • Mortality • Cain/Abel • Density • Dispersal • Breeding Biology • Nest Survey • A. chrysaetos • Idaho
465	**Abandonment** • Behavior • Nest • Egg • Raptors • A. audax • Australia
1102	**Abandonment** • Disturbance • Egg • Mortality • Nestling • Photography • Banding • Raptors • Tech. • A. chrysaetos • North America
1192	**Abandonment** • A. chrysaetos • Switzerland
1778	**Abandonment** • Nest Survey • Nest Chronology • Prey • Reproductive Success • Marking • Roadside Survey • Territory • Population Dynamics • Raptors • SRBPNA • A. chrysaetos • Idaho
1780	**Abandonment** • Nest Survey • Nest Chronology • Prey • Reproductive Success • Marking • Roadside Survey • Territory • Population Dynamics • Raptors • SRBPNA • A. chrysaetos • Idaho
1950	**Abandonment** • Animal Damage • Nest Survey • Prey • Management • A. chrysaetos • Scotland
2488	**Abandonment** • Nest • Prey • Raptors • A. chrysaetos • Utah
2599	**Abandonment** • Egg • Illus. • A. chrysaetos • Switzerland
2753	**Abandonment** • Monograph • Habitat • Nest Survey • Behavior • Prey • Pellet • A. chrysaetos • Wyoming
2854	**Abandonment** • Population Dynamics • Nest • Density • Prey • Breeding Biology • Distribution • Habitat • Directional Exposure • Reproductive Success • Mortality • Disturbance • Interspecific Behavior • Territory • A. chrysaetos • Utah
3102	**Abandonment** • Population Decline • Mortality • Raptors • A. chrysaetos • France
60	**Abnormal Behavior** • Interspecific Behavior • Flight • A. chrysaetos • New Mexico
186	**Abnormal Behavior** • Falconry • A. chrysaetos • Siberia
325	**Abnormal Behavior** • Monograph • Breeding Biology • Density • Mortality • Cain/Abel • Prey • Egg Measure • Marking • A. chrysaetos • Idaho
542	**Abnormal Behavior** • Nest • Cain/Abel • Habitat • A. verreauxi • Rhodesia
768	**Abnormal Behavior** • Toxicology • Raptors • A. chrysaetos • Great Britain
954	**Abnormal Behavior** • Disturbance • Interspecific Behavior • A. chrysaetos • Colorado
1283	**Abnormal Behavior** • A. chrysaetos • Texas
1291	**Abnormal Behavior** • A. chrysaetos • Scotland
1315	**Abnormal Behavior** • A. chrysaetos • Washington
1588	**Abnormal Behavior** • A. chrysaetos • California
1944	**Abnormal Behavior** • Marking • Courtship • Nest Failure • Telemetry • Raptors • A. chrysaetos • Wyoming • Idaho
1976	**Abnormal Behavior** • Anatomy • Collect • Photo Account • A. chrysaetos • Switzerland
1997	**Abnormal Behavior** • Measure • Prey • A. chrysaetos • Scotland
2486	**Abnormal Behavior** • A. verreauxi • South Africa
2563	**Abnormal Behavior** • Egg Measure • Nest Failure • Pesticides • Raptors • A. chrysaetos • Great Britain
3354	**Abnormal Behavior** • A. chrysaetos • Manitoba
233	**Aerial Survey** • Attitude • Reproductive Success • Density • Wintering Population • Food Requirements • Habitat • Management • Nest Survey • Raptors • A. chrysaetos • Montana
422	**Aerial Survey** • Nest Survey • Aircraft • A. chrysaetos • New Mexico • Wyoming • Colorado
423	**Aerial Survey** • Wintering Population • Nest Survey • Banding • Roadside Survey • Electrocution • A. chrysaetos • North America
424	**Aerial Survey** • Age Structure • Nest Survey • A. chrysaetos • United States
425	**Aerial Survey** • Management • Animal Damage • Wintering Population • A. chrysaetos • Texas • New Mexico
427	**Aerial Survey** • Age Structure • Nest Survey • Prey • Banding • Wintering Population • A. chrysaetos • North America
428	**Aerial Survey** • Wintering Population • Reproductive Success • Status • Proj. Out. • A. chrysaetos • Colorado • Idaho • Montana • Utah • New Mexico • Wyoming
757	**Aerial Survey** • Nest Survey • Reproductive Success • Density • Raptors • A. chrysaetos • Colorado
758	**Aerial Survey** • Nest Survey • Density • Reproductive Success • Winter Survey • Wintering Population • Status • A. chrysaetos • Colorado
969	**Aerial Survey** • Wintering Population • Eagles • A. chrysaetos • Colorado
1500	**Aerial Survey** • A. chrysaetos • Idaho • Oregon
1501	**Aerial Survey** • Winter Survey • Age Structure • Eagles • A. chrysaetos • Wyoming
1777	**Aerial Survey** • Predator/Prey • Electrocution • A. chrysaetos • North America
1779	**Aerial Survey** • Reproductive Success • Prey • Banding • Marking • Population Dynamics • Raptors • SRBPNA • A. chrysaetos • Idaho
1782	**Aerial Survey** • Marking • Reproductive Success • Prey • Banding • Population Dynamics • SRBPNA • Raptors • A. chrysaetos • Idaho
1889	**Aerial Survey** • Artificial Nest • Management • Raptors • A. chrysaetos • North America
1943	**Aerial Survey** • Time Lapse • Animal Damage • Nest Survey • Prey • Habitat • Winter Survey • Capture • Age Structure • Food Requirements • Shooting • A. chrysaetos • Texas

Master No.	Keywords
2184	**Aerial Survey** • Reproductive Success • Management • Survey • Clutch Size • Distribution • Habitat • Nest Survey • Prey • Raptors • A. chrysaetos • Arizona
2400	**Aerial Survey** • Animal Damage • Predator/Prey • Habitat • Status • Density • Population Dynamics • Food Habits • Nest • Interspecific Competition • Literature Review • A. chrysaetos • Texas • New Mexico
2538	**Aerial Survey** • Raptors • Tech. • A. chrysaetos • A. clanga • Russia
2788	**Aerial Survey** • Nest Survey • Habitat • Density • Photo Account • A. chrysaetos • Nevada
3161	**Aerial Survey** • Wintering Population • Eagles • A. chrysaetos • Arizona
3321	**Aerial Survey** • Capture • Hunting Behavior • Animal Damage • A. chrysaetos • Texas • New Mexico
3349	**Aerial Survey** • Aircraft • Disturbance • Raptors • A. chrysaetos • Alaska
3351	**Aerial Survey** • Nest Survey • Behavior • Raptors • A. chrysaetos • Alaska
3419	**Aerial Survey** • Winter Survey • Population Estimate • Eagles • A. chrysaetos • Wyoming
424	**Age Structure** • Aerial Survey • Nest Survey • A. chrysaetos • United States
427	**Age Structure** • Aerial Survey • Nest Survey • Prey • Banding • Wintering Population • A. chrysaetos • North America
495	**Age Structure** • Wintering Population • Prey • Distribution • Hunting Behavior • A. nipalensis • Africa
1434	**Age Structure** • Migration • Raptors • A. chrysaetos • Pennsylvania
1439	**Age Structure** • Migration • Raptors • Bake Oven • A. chrysaetos • Pennsylvania
1441	**Age Structure** • Migration • Bake Oven • Raptors • A. chrysaetos • Pennsylvania
1442	**Age Structure** • Migration • Bake Oven • Raptors • A. chrysaetos • Pennsylvania
1443	**Age Structure** • Migration • Raptors • Bake Oven • A. chrysaetos • Pennsylvania
1444	**Age Structure** • Migration • Raptors • Bake Oven • A. chrysaetos • Pennsylvania
1501	**Age Structure** • Aerial Survey • Winter Survey • Eagles • A. chrysaetos • Wyoming
1773	**Age Structure** • Winter Survey • Distribution • Population Dynamics • Prey • Wintering Population • Eagles • A. chrysaetos • Washington
1943	**Age Structure** • Aerial Survey • Time Lapse • Animal Damage • Nest Survey • Prey • Habitat • Winter Survey • Capture • Food Requirements • Shooting • A. chrysaetos • Texas
2591	**Age Structure** • Roadside Survey • Wintering Population • Animal Damage • Prey • Raptors • A. chrysaetos • New Mexico
2920	**Age Structure** • Animal Damage • Prey • A. chrysaetos • North America
3160	**Age Structure** • Wintering Population • Survey • Territory • Field Id • Supplemental Feeding • Illus. • A. chrysaetos • Sweden
304	**Aging** • Plumage • Intraspecific Behavior • Behavior • A. heliaca • A. nipalensis • A. chrysaetos • A. pomarina • A. clanga • Europe
384	**Aging** • Taxonomy • Capture • Field Id • Photo Account • A. nipalensis • A. rapax • Southwest Africa
490	**Aging** • Flight Id • Wintering Population • Eagles • A. nipalensis • A. pomarina • A. rapax • A. wahlbergi • A. verreauxi • Africa
1056	**Aging** • Growth • Falconry • Parental Behavior • Measure • Plumage • Dimorphism • Prey • Photo Account • A. audax • Australia
1688	**Aging** • Molt • Plumage • A. chrysaetos • North America
2325	**Aging** • Capture • Banding • Marking • Animal Damage • Sexing • Management • Translocation • Prey • Survey • Measure • Mortality • Captive Care • A. chrysaetos • Montana
2326	**Aging** • Capture • Banding • Marking • Animal Damage • Sexing • Management • Translocation • Prey • Measure • Mortality • Captive Care • Food Requirements • A. chrysaetos • Montana
2085	**Agonistic Behavior** • Telemetry • Territory • Display • Behavior • Nest • A. chrysaetos • Wyoming • Montana
2251	**Agonistic Behavior** • A. chrysaetos • Switzerland
2445	**Agonistic Behavior** • Flight • Aircraft • Prey • Raptors • A. rapax • Kenya • Tanzania
3033	**Agonistic Behavior** • Sexing • Museum Specimen • A. chrysaetos • Switzerland
104	**Aircraft** • Mortality • A. chrysaetos • Nevada
149	**Aircraft** • Shooting • Attitude • Eagles • Photo Account • Popular Monograph • A. chrysaetos • Wyoming • Colorado
153	**Aircraft** • Shooting • Conservation • Protection • Reward • Attitude • Economics • A. chrysaetos • United States
279	**Aircraft** • Shooting • A. chrysaetos • California
319	**Aircraft** • Folklore • Shooting • Animal Damage • Attitude • Popular Monograph • A. chrysaetos • Texas
422	**Aircraft** • Nest Survey • Aerial Survey • A. chrysaetos • New Mexico • Wyoming • Colorado
634	**Aircraft** • Shooting • Attitude • Animal Damage • A. audax • Australia
956	**Aircraft** • Capture • Measure • Tech. • A. chrysaetos • Montana
2194	**Aircraft** • Flight • Altitude • Raptors • A. heliaca • Turkey
2445	**Aircraft** • Flight • Agonistic Behavior • Prey • Raptors • A. rapax • Kenya • Tanzania
2555	**Aircraft** • Shooting • Punishment • Attitude • Eagle • Popular Monograph • A. chrysaetos • Texas
2803	**Aircraft** • Attitude • Shooting • Animal Damage • Economics • Popular Monograph • A. chrysaetos • Texas
3163	**Aircraft** • Shooting • Attitude • A. chrysaetos • Texas
3200	**Aircraft** • Protection • Attitude • Punishment • Reward • Raptors • A. chrysaetos • United States
3202	**Aircraft** • Disturbance • Protection • Mitigation • Environmental Impact • A. chrysaetos • California
3349	**Aircraft** • Aerial Survey • Disturbance • Raptors • A. chrysaetos • Alaska

Permuted List of Keywords (Cont.)

Master No.	Keywords
560	**Aircraft Collision** • Behavior • A. chrysaetos • Switzerland
1188	**Aircraft Collision** • A. chrysaetos • Switzerland
1606	**Aircraft Collision** • A. chrysaetos • Switzerland
2028	**Aircraft Collision** • A. chrysaetos • Switzerland
2529	**Aircraft Collision** • A. chrysaetos • Switzerland
69	**Albinism** • Museum Specimen • A. nipalensis • Russia
839	**Albinism** • A. chrysaetos • North America
1304	**Albinism** • Egg • Abnormality • A. chrysaetos • North America
1883	**Albinism** • Measure • Trapping • A. audax • Australia
2700	**Albinism** • Raptors • A. chrysaetos • Great Britain
2890	**Albinism** • A. chrysaetos • Denmark
221	**Altitude** • Attack • Flight • Lift • Prey Behavior • A. chrysaetos • A. pomarina
615	**Altitude** • Flight • A. audax • Australia
2194	**Altitude** • Aircraft • Flight • Raptors • A. heliaca • Turkey
2840	**Altitude** • Measure • Mortality • A. nipalensis • India
734	**Anaesthesia** • Raptors • A. rapax • Kenya
735	**Anaesthesia** • Sexing • Raptors • A. rapax • Kenya
1077	**Anaesthesia** • Physiology • Raptors • A. chrysaetos
2576	**Anaesthesia** • Rehabilitation • Shooting • Trapping • Disease • Blood Sample • Raptors • A. chrysaetos • United States
2790	**Anaesthesia** • A. chrysaetos • Montana
97	**Anatomy** • Measure • A. chrysaetos • United States
202	**Anatomy** • Paleontology • A. chrysaetos • Cuba
305	**Anatomy** • Breeding Biology • Cain/Abel • Distribution • A. pomarina • A. clanga • South Africa
406	**Anatomy** • Museum Specimen • A. chrysaetos • A. pomarina • A. heliaca • A. rapax • A. audax • A. verreauxi • A. wahlbergi • A. clanga • Great Britain
546	**Anatomy** • Hunting Behavior • Status • A. chrysaetos • A. heliaca • A. audax • A. gurneyi • A. rapax • A. nipalensis • A. clanga • A. verreauxi • A. wahlbergi • A. pomarina
604	**Anatomy** • Physiology • Captivity • Raptors • A. chrysaetos • A. verreauxi • A. rapax • A. wahlbergi
655	**Anatomy** • Field Id • Raptors • A. chrysaetos • North America
785	**Anatomy** • Raptors • A. chrysaetos • North America
787	**Anatomy** • Measure • Physiology • Raptors • A. chrysaetos • Yugoslavia
972	**Anatomy** • Flight • Raptors • A. chrysaetos • North America
1170	**Anatomy** • Behavior • Prey • Raptors • A. chrysaetos • United Soviet Socialists Republic
1217	**Anatomy** • Raptors • A. chrysaetos • Ohio
1277	**Anatomy** • Raptors • A. chrysaetos
1325	**Anatomy** • Taxonomy • A. heliaca • A. nipalensis
1334	**Anatomy** • Senses • A. wahlbergi
1347	**Anatomy** • Plumage • Behavior • Taxonomy • Raptors • Theory • A. chrysaetos
1575	**Anatomy** • Paleontology • Measure • A. chrysaetos • North America
1587	**Anatomy** • Raptors • A. chrysaetos
1683	**Anatomy** • Taxonomy • Osteology • Museum Specimen • Raptors • A. chrysaetos • A. rapax
1684	**Anatomy** • Taxonomy • Osteology • Museum Specimen • Raptors • A. chrysaetos • A. rapax
1685	**Anatomy** • Taxonomy • Osteology • Museum Specimen • Raptors • A. chrysaetos • A. rapax
1686	**Anatomy** • Taxonomy • Osteology • Museum Specimen • Raptors • A. chrysaetos • A. rapax
1689	**Anatomy** • Raptors • A. chrysaetos
1740	**Anatomy** • A. chrysaetos
1766	**Anatomy** • Plumage • A. clanga • A. pomarina • A. nipalensis • A. heliaca
1957	**Anatomy** • Physiology • Raptors • A. chrysaetos • North America
1976	**Anatomy** • Collect • Abnormal Behavior • Photo Account • A. chrysaetos • Switzerland
2300	**Anatomy** • (Title Only) • A. chrysaetos • A. heliaca • Morocco
2443	**Anatomy** • Talons • Photo Account • A. pomarina • A. clanga
2500	**Anatomy** • Wing Loading • Measure • Raptors • A. chrysaetos • North America
2506	**Anatomy** • Measure • Raptors • A. chrysaetos
2541	**Anatomy** • Senses • Raptors • A. chrysaetos
2603	**Anatomy** • Flight • A. chrysaetos
2609	**Anatomy** • Raptors • A. chrysaetos • North America
2649	**Anatomy** • Measure • Raptors • A. chrysaetos • A. rapax
2771	**Anatomy** • A. chrysaetos
2812	**Anatomy** • Paired Ovaries • Raptors • A. clanga • China
2825	**Anatomy** • A. audax • Australia
2877	**Anatomy** • Paired Ovaries • Raptors • A. chrysaetos • Canada
2884	**Anatomy** • Raptors • A. chrysaetos
2933	**Anatomy** • Paired Ovaries • Dimorphism • Raptors • Gonads • A. chrysaetos • North America
2934	**Anatomy** • Physiology • Measure • Raptors • Ovary • Testes • A. chrysaetos
3025	**Anatomy** • Raptors • A. chrysaetos
3067	**Anatomy** • Taxonomy • Egg • Egg Measure • Raptors • A. chrysaetos • A. heliaca • A. pomarina • A. clanga • A. nipalensis
3068	**Anatomy** • Taxonomy • Egg • Egg Measure • Raptors • A. chrysaetos • A. heliaca • A. pomarina • A. clanga • A. nipalensis
3365	**Anatomy** • Conservation • Bounty • Protection • Raptors • Illus. • A. chrysaetos • Finland
3420	**Anatomy** • Raptors • A. chrysaetos
22	**Animal Damage** • Hunting Behavior • Prey • Talon Wounds • A. chrysaetos • Texas
41	**Animal Damage** • Prey • A. chrysaetos • Nevada
44	**Animal Damage** • Collect • Raptors • A. chrysaetos • Iowa
82	**Animal Damage** • Prey • A. chrysaetos • Great Britain
136	**Animal Damage** • Conservation • Popular Monograph • A. chrysaetos • North America
154	**Animal Damage** • Banding • Captivity • Rehabilitation • Popular Monograph • A. chrysaetos • North America
198	**Animal Damage** • Prey • Monograph • Nest Survey • Habitat • Reproductive Success • Literature Review • Disturbance • Pellet • A. chrysaetos • Utah
201	**Animal Damage** • Economics • A. chrysaetos • North America

Permuted List of Keywords (Cont.)

Master No.	Keywords
231	**Animal Damage** • Economics • (Title Only) • A. chrysaetos • Texas • New Mexico
294	**Animal Damage** • Poisoning • Shooting • Captivity • Lift • A. audax • Australia
319	**Animal Damage** • Folklore • Shooting • Attitude • Aircraft • Popular Monograph • A. chrysaetos • Texas
425	**Animal Damage** • Aerial Survey • Management • Wintering Population • A. chrysaetos • Texas • New Mexico
434	**Animal Damage** • Economics • Attitude • Management • Survey • A. chrysaetos • North America
441	**Animal Damage** • Prey • A. chrysaetos • Sweden
444	**Animal Damage** • Attitude • Bounty • Conservation • Raptors • A. verreauxi • A. rapax • South Africa
461	**Animal Damage** • A. chrysaetos • Rhodesia
464	**Animal Damage** • Nest • Habitat • Egg Collection • Behavior • Vocalization • A. audax • Australia
470	**Animal Damage** • A. chrysaetos • Montana
501	**Animal Damage** • Prey • Pellet • Attack • Hunting Behavior • Habitat • Cache • A. audax • Australia
552	**Animal Damage** • Nest Survey • Prey • Habitat • Behavior • Status • Reproductive Success • Food Requirements • Intraspecific Behavior • Predator/Prey • A. chrysaetos • Scotland
578	**Animal Damage** • Hunting Behavior • Shooting • Attitude • A. chrysaetos • Texas
605	**Animal Damage** • Hunting Behavior • Poisoning • A. audax • Australia
606	**Animal Damage** • Economics • Attitude • Prey • Conservation • Raptors • A. chrysaetos • North America
609	**Animal Damage** • Wintering Population • Banding • Capture • Eagles • A. chrysaetos • Tennessee
634	**Animal Damage** • Shooting • Aircraft • Attitude • A. audax • Australia
704	**Animal Damage** • Trapping • Measure • A. chrysaetos • North Carolina
783	**Animal Damage** • Protection • Conservation • Prey • Hunting Behavior • Prey Behavior • Raptors • A. chrysaetos • Sweden
784	**Animal Damage** • Conservation • Attitude • Prey • Raptors • A. heliaca • A. audax
794	**Animal Damage** • Shooting • Bounty • Prey • Eagles • A. chrysaetos • California
802	**Animal Damage** • Protection • Beneficial Effect • Raptors • A. clanga • Siberia
819	**Animal Damage** • Mortality • Shooting • Poisoning • Eagles • A. chrysaetos • Ohio
820	**Animal Damage** • Shooting • A. chrysaetos • Ohio
842	**Animal Damage** • Courtship • Hunting Behavior • Prey Beh. • A. audax • Australia
921	**Animal Damage** • Capture • Banding • A. chrysaetos • North Carolina
949	**Animal Damage** • Conservation • Prey • Raptors • A. chrysaetos • Scotland
1042	**Animal Damage** • Descriptive Summary • Measure • Hunting Behavior • Prey • Raptors • A. chrysaetos • North America
1044	**Animal Damage** • Poisoning • A. chrysaetos • Alabama
1079	**Animal Damage** • Prey • Eagles • Popular Monograph • A. audax • Australia
1089	**Animal Damage** • Poisoning • A. audax • Australia
1090	**Animal Damage** • Economics • Attitude • Conservation • Mortality • Popular Monograph • A. chrysaetos • Wyoming
1131	**Animal Damage** • Folklore • Hunting Behavior • Trapping • Protection • Eagles • A. chrysaetos • Tennessee
1168	**Animal Damage** • Prey • Hunting Behavior • A. audax • Australia
1204	**Animal Damage** • Lift • Prey • A. chrysaetos • Great Britain
1222	**Animal Damage** • Prey • Raptors • A. chrysaetos • Kansas
1365	**Animal Damage** • Prey • A. chrysaetos • North America
1454	**Animal Damage** • Hunting Behavior • Raptors • A. chrysaetos • A. nipalensis
1473	**Animal Damage** • Protection • Prey • A. chrysaetos • Switzerland
1482	**Animal Damage** • A. chrysaetos • Sweden
1583	**Animal Damage** • A. chrysaetos • New Mexico
1659	**Animal Damage** • Predator/Prey • A. chrysaetos • Scotland
1717	**Animal Damage** • Monograph • Prey • Hunting Behavior • Eagles • A. chrysaetos • North America
1764	**Animal Damage** • A. chrysaetos • Nevada
1783	**Animal Damage** • Status • Raptors • A. chrysaetos • India
1867	**Animal Damage** • Hunting Behavior • Prey • A. audax • Australia
1900	**Animal Damage** • Mortality • Prey • Nest • Dispersal • Pellet • Habitat • Bounty • Protection • Photo Account • A. audax • Australia
1942	**Animal Damage** • Food Habits • Nest • Time Lapse • Behavior • Economics • Reproductive Success • Status • A. chrysaetos • Texas
1943	**Animal Damage** • Aerial Survey • Time Lapse • Nest Survey • Prey • Habitat • Winter Survey • Capture • Age Structure • Food Requirements • Shooting • A. chrysaetos • Texas
1950	**Animal Damage** • Nest Survey • Abandonment • Prey • Management • A. chrysaetos • Scotland
1971	**Animal Damage** • Conservation • Protection • Eagles • A. chrysaetos • North America
2037	**Animal Damage** • Capture • A. chrysaetos • Tennessee
2091	**Animal Damage** • Hunting Behavior • Poisoning • A. audax • Australia
2092	**Animal Damage** • Field Id • Descriptive Summary • Hunting Behavior • Nest • Egg • Attitude • Piracy • Raptors • A. audax • Australia
2093	**Animal Damage** • Hunting Behavior • A. audax • Australia
2096	**Animal Damage** • Protection • (Title Only) • A. chrysaetos • North America
2103	**Animal Damage** • Conservation • Folklore • Habitat • Shooting • A. chrysaetos • United States
2124	**Animal Damage** • Monograph • Breeding Biology • Migration • Prey • Hunting Behavior • Attack • Folklore • Falconry • Popular Monograph • A. chrysaetos

Master No.	Keywords
2179	**Animal Damage** • Museum Display • Attitude • Folklore • Popular Monograph • A. audax • Australia
2200	**Animal Damage** • Prey • Economics • Nest Survey • Trapping • A. chrysaetos • New Mexico • Texas
2261	**Animal Damage** • Nest Survey • Disturbance • Prey • Population Dynamics • Management • Status • Artificial Nest • Predator/Prey • A. chrysaetos • Utah
2262	**Animal Damage** • Management • Conservation • Raptors • A. chrysaetos • United States
2273	**Animal Damage** • Shooting • Wintering Population • Measure • Eagles • A. chrysaetos • Illinois
2282	**Animal Damage** • A. chrysaetos • Idaho
2289	**Animal Damage** • Economics • A. chrysaetos • California
2311	**Animal Damage** • Attitude • Status • Population Estimate • Mortality • Protection • Prey • Disturbance • Raptors • A. chrysaetos • Great Britain
2315	**Animal Damage** • Protection • Status • A. chrysaetos • Austria
2322	**Animal Damage** • Economics • Questionnaire • A. chrysaetos • Utah
2324	**Animal Damage** • Capture • Banding • Management • Prey • Sexing • Translocation • Marking • Measure • Captive Care • Tech. • Eagles • A. chrysaetos • Montana
2325	**Animal Damage** • Capture • Banding • Marking • Sexing • Management • Translocation • Prey • Survey • Measure • Mortality • Captive Care • Aging • A. chrysaetos • Montana
2326	**Animal Damage** • Capture • Banding • Marking • Sexing • Management • Translocation • Prey • Measure • Mortality • Captive Care • Aging • Food Requirements • A. chrysaetos • Montana
2349	**Animal Damage** • Feeding Behavior • Capture • Predator/Prey • Protection • Management • Economics • A. chrysaetos • Montana
2350	**Animal Damage** • Economics • A. chrysaetos • Montana
2369	**Animal Damage** • Prey • Folklore • Lift • Conservation • Electrocution • Eagles • Popular Monograph • A. chrysaetos • North America
2373	**Animal Damage** • Prey • Literature Review • Food Requirements • A. chrysaetos • Colorado
2400	**Animal Damage** • Predator/Prey • Aerial Survey • Habitat • Status • Density • Population Dynamics • Food Habits • Nest • Interspecific Competition • Literature Review • A. chrysaetos • Texas • New Mexico
2532	**Animal Damage** • Prey • Raptors • A. verreauxi • South Africa
2554	**Animal Damage** • Conservation • Nest • Behavior • Interspecific Behavior • Prey • Trapping • Rehabilitation • Photo Account • Popular Monograph • A. chrysaetos • Wyoming
2572	**Animal Damage** • Prey • A. chrysaetos • New York
2585	**Animal Damage** • Bounty • Attitude • Photo Account • Raptors • Popular Monograph • A. chrysaetos • North America
2591	**Animal Damage** • Roadside Survey • Wintering Population • Age Structure • Prey • Raptors • A. chrysaetos • New Mexico
2628	**Animal Damage** • Lift • A. chrysaetos • Ohio
2672	**Animal Damage** • Prey • Hunting Behavior • Feeding Behavior • A. audax • Australia
2688	**Animal Damage** • Hunting Behavior • Prey • Eagles • A. chrysaetos • Michigan
2759	**Animal Damage** • Shooting • Conservation • Punishment • Attitude • Reward • Translocation • Popular Monograph • A. chrysaetos • Texas
2802	**Animal Damage** • Status • Raptors • A. wahlbergi • Uganda
2803	**Animal Damage** • Attitude • Shooting • Aircraft • Economics • Popular Monograph • A. chrysaetos • Texas
2906	**Animal Damage** • Attitude • Mortality • Prey • Raptors • A. chrysaetos • Texas
2913	**Animal Damage** • Shooting • Survey • Prey • Plumage • A. chrysaetos • Texas
2920	**Animal Damage** • Age Structure • Prey • A. chrysaetos • North America
2963	**Animal Damage** • Rarity • Measure • Captivity • Behavior • A. chrysaetos • Georgia
3024	**Animal Damage** • Prey Behavior • A. chrysaetos • California
3058	**Animal Damage** • Habitat • Prey • Prey Behavior • Nest • Shooting • Egg • Behavior • A. verreauxi • South Africa
3095	**Animal Damage** • Nest • Prey • Behavior • A. verreauxi • South Africa
3099	**Animal Damage** • A. chrysaetos • Great Britain
3104	**Animal Damage** • A. chrysaetos • Texas
3153	**Animal Damage** • Economics • A. chrysaetos • Wyoming
3154	**Animal Damage** • A. chrysaetos • Utah
3194	**Animal Damage** • Status • A. chrysaetos • Tennessee
3205	**Animal Damage** • Management • Status • Distribution • Wintering Population • A. chrysaetos • United States
3263	**Animal Damage** • Status • Museum Specimen • Shooting • Raptors • A. verreauxi • A. nipalensis • A. rapax • South Africa
3316	**Animal Damage** • Prey • Food Requirements • Management • A. chrysaetos • Scotland
3321	**Animal Damage** • Aerial Survey • Capture • Hunting Behavior • A. chrysaetos • Texas • New Mexico
3332	**Animal Damage** • A. chrysaetos • Florida
3367	**Animal Damage** • Attitude • Feeding Behavior • A. chrysaetos • Texas
3372	**Animal Damage** • Protection • Population Decline • Mortality • Conservation • Bounty • A. chrysaetos • Norway
3378	**Animal Damage** • Economics • Protection • Popular Monograph • A. chrysaetos • North America
3408	**Animal Damage** • Hunting Behavior • Measure • A. chrysaetos • California
3435	**Animal Damage** • Capture • A. chrysaetos • Switzerland
3440	**Animal Damage** • Shooting • A. chrysaetos • South Dakota • Iowa • Minnesota • Nebraska
834	**Anthropology** • Raptors • A. chrysaetos • California
1305	**Anthropology** • Monograph • Descriptive Summary • A. chrysaetos • A. clanga • A. rapax • A. audax • A. verreauxi • A. gurneyi • A. wahlbergi • A. pomarina • A. clanga • A. heliaca
1383	**Anthropology** • A. chrysaetos • Arizona • Utah

Master No.	Keywords
1385	**Anthropology** • Photo Account • A. audax • A. chrysaetos
2178	**Anthropology** • Symbolism • A. chrysaetos • Oregon
2420	**Anthropology** • Symbolism • Capture • Historical Account • Raptors • Photo Account • A. chrysaetos • South Dakota
3379	**Anthropology** • Capture • Captivity • Symbolism • Mortality • A. chrysaetos • North America
503	**Art** • Raptors • A. chrysaetos • North America
705	**Art** • Conservation • Population Decline • Refuge • A. heliaca • Spain
706	**Art** • A. chrysaetos • North America
840	**Art** • A. chrysaetos • United States
1	**Artificial Incubation** • Disturbance • A. chrysaetos • California
1019	**Artificial Incubation** • Captive Breeding • Measure • Raptors • A. chrysaetos • A. rapax • A. nipalensis • West Germany
1293	**Artificial Incubation** • Captive Breeding • Behavior • Egg Measure • Artificial Insemination • Photography Account • A. chrysaetos • United States
1292	**Artificial Insemination** • Photo Account • Popular Monograph • A. chrysaetos • New York
1293	**Artificial Insemination** • Captive Breeding • Behavior • Egg Measure • Artificial Incubation • Photography Account • A. chrysaetos • United States
1361	**Artificial Insemination** • Story • Falconry • Capt. Breed. • Photography Account • Popular Monograph • A. chrysaetos • United States
146	**Artificial Nest** • Falconry • Raptors • A. chrysaetos • California
244	**Artificial Nest** • Nest Survey • Prey • Habitat • Raptors • A. chrysaetos • California
358	**Artificial Nest** • Status • Management • Population Estimate • Protection • Raptors • A. chrysaetos • A. clanga • Finland
610	**Artificial Nest** • Habitat • Management • Photo Account • Raptors • A. chrysaetos • North America
837	**Artificial Nest** • Eagles • A. rapax • South Africa
960	**Artificial Nest** • Electrocution • Disturbance • Raptors • A. chrysaetos • North America
1278	**Artificial Nest** • Popular Monograph • A. chrysaetos • Idaho
1889	**Artificial Nest** • Aerial Survey • Management • Raptors • A. chrysaetos • North America
2261	**Artificial Nest** • Nest Survey • Disturbance • Prey • Population Dynamics • Management • Status • Animal Damage • Predator/Prey • A. chrysaetos • Utah
2297	**Artificial Nest** • Electrocution • Raptors • A. chrysaetos • United States
2299	**Artificial Nest** • Electrocution • Conservation • Management • Shooting • Eagles • Popular Monograph • A. chrysaetos • Idaho
2376	**Artificial Nest** • Habitat • Management • Nest Chronology • A. chrysaetos • North America
2724	**Artificial Nest** • Raptors • A. chrysaetos • Europe
3080	**Artificial Nest** • Cain/Abel • Distribution • Educational Material • Field Id • Habitat • Hunting Behavior • Prey • Nest • Incubation • Wintering Population • Conservation • Photo Account • Eagles • A. chrysaetos • A. wahlbergi • A. verreauxi • A. nipalensis • A. pomarina • South Africa
2298	**Artificial Nest** • Electrocution • Mortality • Shooting • Eagles • A. chrysaetos • Idaho
221	**Attack** • Flight • Lift • Altitude • Prey Behavior • A. chrysaetos • A. pomarina
501	**Attack** • Prey • Pellet • Hunting Behavior • Animal Damage • Habitat • Cache • A. audax • Australia
700	**Attack** • A. audax • Australia
1246	**Attack** • Behavior • Hunting Behavior • Lift • Trapping • Photography • Behavior • Photo Account • Popular Monograph • A. chrysaetos • Scotland
1698	**Attack** • A. chrysaetos • Scotland
2124	**Attack** • Monograph • Breeding Biology • Migration • Prey • Hunting Behavior • Folklore • Falconry • Animal Damage • Popular Monograph • A. chrysaetos
2586	**Attack** • Nest • A. chrysaetos • Yugoslavia
2605	**Attack** • Popular Monograph • (Title Only) • A. chrysaetos • Great Britain
3363	**Attack** • Hunting Behavior • Capture • A. audax • Australia
86	**Attitude** • Historical Account • Flight • Status • Prey • Photo Account • A. chrysaetos • Great Britain
149	**Attitude** • Shooting • Aircraft • Eagles • Photo Account • Popular Monograph • A. chrysaetos • Wyoming • Colorado
151	**Attitude** • Poisoning • Mortality • Wintering Population • Conservation • Punishment • A. chrysaetos • Wyoming
153	**Attitude** • Shooting • Conservation • Protection • Aircraft • Reward • Economics • A. chrysaetos • United States
170	**Attitude** • Shooting • Punishment • Eagles • A. chrysaetos • Mississippi
195	**Attitude** • Nest Survey • Wintering Population • Habitat • Distribution • Raptors • A. chrysaetos • Arizona
196	**Attitude** • Nest Survey • Habitat • Wintering Population • Falconry • Raptors • A. chrysaetos • Arizona
205	**Attitude** • Rarity • Rehabilitation • A. clanga • Great Britain
211	**Attitude** • Photo Account • A. chrysaetos • Great Britain
233	**Attitude** • Reproductive Success • Density • Wintering Population • Food Requirements • Habitat • Management • Aerial Survey • Nest Survey • Raptors • A. chrysaetos • Montana
246	**Attitude** • Prey • Protection • A. chrysaetos • Ohio
265	**Attitude** • Prey • Nest • Raptors • A. pomarina • United Soviet Socialists Republic
267	**Attitude** • Shooting • Conservation • Raptors • A. chrysaetos • Scotland
319	**Attitude** • Folklore • Shooting • Animal Damage • Aircraft • Popular Monograph • A. chrysaetos • Texas

Permuted List of Keywords (Cont.)

Master No.	Keywords
332	**Attitude** • Field Id • Prey • Raptors • Popular Monograph • A. audax • Australia
434	**Attitude** • Animal Damage • Economics • Management • Survey • A. chrysaetos • North America
444	**Attitude** • Animal Damage • Bounty • Conservation • Raptors • A. verreauxi • A. rapax • South Africa
457	**Attitude** • Conservation • Shooting • Protection • Status • A. chrysaetos • Austria
462	**Attitude** • Habitat • Conservation • Disturbance • Flight Speed • Eagles • Photo Account • Popular Monograph • A. chrysaetos • North America
466	**Attitude** • Distribution • Egg Collection • Prey • Raptors • A. audax • Australia
471	**Attitude** • Prey • Raptors • A. chrysaetos • United States
557	**Attitude** • Shooting • Photo Account • Popular Monograph • A. chrysaetos • Montana
578	**Attitude** • Hunting Behavior • Shooting • Animal Damage • A. chrysaetos • Texas
606	**Attitude** • Animal Damage • Economics • Prey • Conservation • Raptors • A. chrysaetos • North America
618	**Attitude** • Prey • Nest • Behavior • Breeding Biology • Flight • Poisoning • Molt • Hunting Behavior • Vocalization • Folklore • Lift • Photo Account • A. chrysaetos • Montana
634	**Attitude** • Shooting • Aircraft • Animal Damage • A. audax • Australia
784	**Attitude** • Conservation • Animal Damage • Prey • Raptors • A. heliaca • A. audax
791	**Attitude** • Prey • Raptors • A. chrysaetos • A. pomarina • France
813	**Attitude** • Folklore • A. chrysaetos • Switzerland
1090	**Attitude** • Economics • Conservation • Animal Damage • Mortality • Popular Monograph • A. chrysaetos • Wyoming
1259	**Attitude** • Poetry • Captivity • Protection • Flight • Folklore • Raptors • A. chrysaetos • Great Britain
1339	**Attitude** • Economics • Descriptive Summary • Raptors • A. verreauxi • A. rapax • South Africa
1359	**Attitude** • Conservation • A. chrysaetos • Texas
1372	**Attitude** • Shooting • A. chrysaetos • Switzerland
1487	**Attitude** • Shooting • A. chrysaetos • Switzerland
1508	**Attitude** • Literature Review • Prey • Pellet • Shooting • Predator/Prey • Nest Survey • Reproductive Success • A. chrysaetos • North America
1678	**Attitude** • Bounty • Raptors • A. chrysaetos • North Dakota
1812	**Attitude** • Prey • Behavior • A. chrysaetos • South Dakota
1897	**Attitude** • Conservation • Mortality • Popular Monograph • A. chrysaetos • Wyoming
2004	**Attitude** • Status • Raptors • A. audax • Australia
2012	**Attitude** • Population Decline • Status • Raptors • A. chrysaetos • A. clanga • Russia
2079	**Attitude** • Prey • Distribution • Field Id • Raptors • A. chrysaetos • United States
2092	**Attitude** • Field Id • Descriptive Summary • Hunting Behavior • Animal Damage • Nest • Egg • Piracy • Raptors • A. audax • Australia
2133	**Attitude** • Economics • A. chrysaetos • A. clanga • Belgium
2179	**Attitude** • Museum Display • Folklore • Animal Damage • Popular Monograph • A. audax • Australia
2217	**Attitude** • Nest • Prey • Photo Account • Popular Monograph • A. chrysaetos • Canada
2233	**Attitude** • Shooting • Protection • Punishment • Eagles • Popular Monograph • A. chrysaetos • North America
2304	**Attitude** • Prey • Trapping • Raptors • A. audax • Australia
2311	**Attitude** • Status • Population Estimate • Mortality • Protection • Prey • Animal Damage • Disturbance • Raptors • A. chrysaetos • Great Britain
2316	**Attitude** • Status • Population Decline • Banding • Nest Survey • Disturbance • A. chrysaetos • Europe
2413	**Attitude** • Field Id • Folklore • Descriptive Summary • Raptors • A. chrysaetos • United States
2438	**Attitude** • Bounty • Eagles • A. chrysaetos • Alaska
2555	**Attitude** • Aircraft • Shooting • Punishment • Eagle • Popular Monograph • A. chrysaetos • Texas
2585	**Attitude** • Bounty • Animal Damage • Photo Account • Raptors • Popular Monograph • A. chrysaetos • North America
2758	**Attitude** • Shooting • Conservation • Story • Punishment • Popular Monograph • A. chrysaetos • Texas
2759	**Attitude** • Animal Damage • Shooting • Conservation • Punishment • Reward • Translocation • Popular Monograph • A. chrysaetos • Texas
2803	**Attitude** • Shooting • Aircraft • Animal Damage • Economics • Popular Monograph • A. chrysaetos • Texas
2804	**Attitude** • Bounty • Shooting • Protection • Economics • A. audax • Australia
2835	**Attitude** • Shooting • A. chrysaetos • Kansas
2906	**Attitude** • Mortality • Animal Damage • Prey • Raptors • A. chrysaetos • Texas
2990	**Attitude** • Mortality • Conservation • Protection • Photo Account • Popular Monograph • A. chrysaetos • North America
3002	**Attitude** • Capture • Mortality • A. chrysaetos • Illinois
3163	**Attitude** • Shooting • Aircraft • A. chrysaetos • Texas
3189	**Attitude** • Shooting • Popular Monograph • A. chrysaetos • Texas
3191	**Attitude** • Conservation • Shooting • Poisoning • Electrocution • Eagles • A. chrysaetos • United States
3192	**Attitude** • Protection • Poisoning • Shooting • Electrocution • Popular Monograph • A. chrysaetos • Wyoming
3200	**Attitude** • Protection • Punishment • Reward • Aircraft • Raptors • A. chrysaetos • United States
3367	**Attitude** • Animal Damage • Feeding Behavior • A. chrysaetos • Texas
3375	**Attitude** • Protection • Symbolism • Shooting • Economics • Conservation • A. chrysaetos • United States

Master No.	Keywords
313	**Autopsy** • Poisoning • Toxicology • Captivity • Residues • Thal. Sulfate • A. chrysaetos
836	**Autopsy** • Pesticides • Mortality • Toxicology • Eagles • Popular Monograph • A. chrysaetos • North America
1748	**Autopsy** • Disease • Raptors • A. rapax • A. verreauxi • A. chrysaetos • A. audax • A. heliaca
3302	**Autopsy** • Disease • Nest • A. chrysaetos • Great Britain
3306	**Autopsy** • Pesticides • Mortality • Disease • A. chrysaetos • Great Britain
94	**Banding** • Folklore • Flight • Capture • A. chrysaetos • A. pomarina • Europe
128	**Banding** • A. pomarina • Tanzania • Estonia
131	**Banding** • Photo Account • Popular Monograph • A. chrysaetos • Switzerland
139	**Banding** • A. pomarina • Rhodesia • Latvia
150	**Banding** • Trapping • Raptors • A. audax • Australia
154	**Banding** • Captivity • Rehabilitation • Animal Damage • Popular Monograph • A. chrysaetos • North America
155	**Banding** • Poisoning • Raptors • A. audax • Australia
162	**Banding** • Photo Account • A. audax • Australia
163	**Banding** • Mortality • Raptors • A. audax • Australia
175	**Banding** • Raptors • A. audax • Australia
182	**Banding** • Raptors • A. audax • Australia
383	**Banding** • Photography • Measure • Raptors • A. nipalensis • A. wahlbergi • A. rapax • South Africa
423	**Banding** • Aerial Survey • Wintering Population • Nest Survey • Roadside Survey • Electrocution • A. chrysaetos • North America
427	**Banding** • Aerial Survey • Age Structure • Nest Survey • Prey • Wintering Population • A. chrysaetos • North America
488	**Banding** • Capture • A. verreauxi • South Africa
522	**Banding** • Migration • Raptors • A. chrysaetos • Pennsylvania
584	**Banding** • Shooting • A. chrysaetos • Switzerland
609	**Banding** • Wintering Population • Capture • Animal Damage • Eagles • A. chrysaetos • Tennessee
728	**Banding** • Raptors • A. chrysaetos • California • Tennessee • Virginia • Mexico
753	**Banding** • Behavior • Nest • Shooting • Photography • Story • Photo Account • Raptors • A. verreauxi • South Africa
766	**Banding** • Nest • Raptors • Popular Monograph • A. chrysaetos • Wyoming
876	**Banding** • Migration • A. chrysaetos • Europe
905	**Banding** • A. chrysaetos • Poland • Latvia
906	**Banding** • A. pomarina • East Germany • Poland
921	**Banding** • Animal Damage • Capture • A. chrysaetos • North Carolina
953	**Banding** • Raptors • A. verreauxi • A. wahlbergi • A. rapax • A. pomarina • South Africa
1021	**Banding** • Raptors • A. heliaca • Spain
1022	**Banding** • Raptors • A. heliaca • Spain
1023	**Banding** • Raptors • A. chrysaetos • A. heliaca • Spain
1028	**Banding** • Migration • Measure • Raptors • Photo Account • A. chrysaetos • Ontario
1029	**Banding** • Migration • Raptors • A. chrysaetos • Ontario
1030	**Banding** • Migration • Raptors • A. chrysaetos • Ontario
1098	**Banding** • Survey • Disturbance • Nest • A. chrysaetos • Canada
1102	**Banding** • Abandonment • Disturbance • Egg • Mortality • Nestling • Photography • Raptors • Tech. • A. chrysaetos • North America
1144	**Banding** • Nest Survey • Status • Behavior • Fledging • Mortality • Marking • Cain/Abel • Reproductive Success • Breeding Biology • Interspecific Behavior • Hunting Behavior • Display • Courtship • Raptors • A. verreauxi • Rhodesia
1148	**Banding** • Marking • Fledging • A. verreauxi • Rhodesia
1224	**Banding** • Longevity • A. chrysaetos • Switzerland
1374	**Banding** • A. chrysaetos • Switzerland
1407	**Banding** • Raptors • A. nipalensis • Israel • United Soviet Socialists Republic
1499	**Banding** • Prey • A. chrysaetos • Oregon
1539	**Banding** • Raptors • A. chrysaetos • Norway
1540	**Banding** • Raptors • A. chrysaetos • Norway
1541	**Banding** • Raptors • A. chrysaetos • Norway
1542	**Banding** • Raptors • A. chrysaetos • Norway
1543	**Banding** • Raptors • A. chrysaetos • Norway
1544	**Banding** • Raptors • A. chrysaetos • Norway
1545	**Banding** • Raptors • A. chrysaetos • Norway
1546	**Banding** • Raptors • A. chrysaetos • Norway
1551	**Banding** • Migration • Raptors • A. chrysaetos • Pennsylvania
1565	**Banding** • Nest • Raptors • A. chrysaetos • Saskatchewan
1566	**Banding** • Raptors • A. chrysaetos • Saskatchewan
1647	**Banding** • Longevity • A. chrysaetos • Switzerland
1648	**Banding** • Longevity • A. chrysaetos • Switzerland
1746	**Banding** • Raptors • A. pomarina • Hungary
1747	**Banding** • A. heliaca • Hungary
1775	**Banding** • Monograph • Status • Nest Survey • Growth • Reproductive Success • Cain/Abel • Marking • Mortality • Pesticides • Prey • Residues • A. chrysaetos • Idaho
1779	**Banding** • Reproductive Success • Prey • Aerial Survey • Marking • Population Dynamics • Raptors • SRBPNA • A. chrysaetos • Idaho
1781	**Banding** • Nest Survey • Reproductive Success • Marking • Habitat • Nest Chronology • Behavior • Prey • Territory • Population Dynamics • Raptors • SRBPNA • A. chrysaetos • Idaho
1782	**Banding** • Marking • Reproductive Success • Prey • Aerial Survey • Population Dynamics • SRBPNA • Raptors • A. chrysaetos • Idaho
1838	**Banding** • A. chrysaetos • Northwest Territory • Montana
1919	**Banding** • Migration • Raptors • A. pomarina • Europe
1941	**Banding** • Conservation • Photo Account • Raptors • A. chrysaetos • Nebraska
1965	**Banding** • Migration • A. pomarina • Bulgaria • West Germany
2023	**Banding** • Capture • Behavior • Tech. • A. pomarina • South Africa
2086	**Banding** • Monograph • Mortality • Migration • Breeding Biology • Prey • Food Requirements • A. chrysaetos • Montana

Master No.	Keywords
2088	**Banding** • Breeding Biology • Prey • Migration • Directional Exposure • Mortality • A. chrysaetos • Montana
2099	**Banding** • Raptors • A. verreauxi • A. wahlbergi • South Africa
2106	**Banding** • Raptors • A. chrysaetos • Great Britain
2186	**Banding** • Descriptive Summary • Prey • Raptors • A. wahlbergi • A. rapax • South Africa
2224	**Banding** • Nest • Prey • A. chrysaetos • South Dakota
2316	**Banding** • Status • Population Decline • Nest Survey • Disturbance • Attitude • A. chrysaetos • Europe
2324	**Banding** • Capture • Animal Damage • Management • Prey • Sexing • Translocation • Marking • Measure • Captive Care • Tech. • Eagles • A. chrysaetos • Montana
2325	**Banding** • Capture • Marking • Animal Damage • Sexing • Management • Translocation • Prey • Survey • Measure • Mortality • Captive Care • Aging • A. chrysaetos • Montana
2326	**Banding** • Capture • Marking • Animal Damage • Sexing • Management • Translocation • Prey • Measure • Mortality • Captive Care • Aging • Food Requirements • A. chrysaetos • Montana
2334	**Banding** • Raptors • A. chrysaetos • Finland
2335	**Banding** • A. chrysaetos • Sweden • Finland
2336	**Banding** • Raptors • A. chrysaetos • Finland
2429	**Banding** • A. heliaca • Hungary
2601	**Banding** • Prey • Nest • Photo Account • A. chrysaetos • Switzerland
2611	**Banding** • Prey • Conservation • Nest • Photo Account • Popular Monograph • A. audax • Australia
2676	**Banding** • Raptors • A. heliaca • Yugoslavia
2678	**Banding** • Raptors • A. heliaca • Yugoslavia
2679	**Banding** • Raptors • A. heliaca • Yugoslavia
2695	**Banding** • Raptors • A. clanga • Poland
2696	**Banding** • Raptors • A. nipalensis • A. chrysaetos
2704	**Banding** • A. heliaca • Hungary
2711	**Banding** • Trapping • Shooting • Rehabilitation • A. chrysaetos • Tennessee
2723	**Banding** • Nest Survey • Mortality • Status • Reproductive Success • Raptors • A. chrysaetos • Finland
2725	**Banding** • Raptors • A. chrysaetos • Finland
2726	**Banding** • Protection • Eagles • A. chrysaetos • Finland
2746	**Banding** • A. chrysaetos • Switzerland
2747	**Banding** • A. chrysaetos • Switzerland
2748	**Banding** • A. chrysaetos • Switzerland
2750	**Banding** • A. chrysaetos • Switzerland
2749	**Banding** • A. chrysaetos • Switzerland
2752	**Banding** • Longevity • A. pomarina • West Germany
2870	**Banding** • Survey • Marking • Nest • Capture • Prey • Food Requirements • Raptors • A. chrysaetos • A. wahlbergi • South Africa
2891	**Banding** • Migration • Raptors • A. chrysaetos • New Jersey
2892	**Banding** • Migration • Raptors • A. chrysaetos • New Jersey
2899	**Banding** • Raptors • A. chrysaetos • Great Britain
2900	**Banding** • Raptors • A. chrysaetos • Great Britain
2901	**Banding** • Raptors • A. chrysaetos • Great Britain
2903	**Banding** • Raptors • A. chrysaetos • Great Britain
2902	**Banding** • Raptors • A. chrysaetos • Great Britain
2915	**Banding** • Migration • Shooting • A. chrysaetos • North America
2926	**Banding** • Measure • Raptors • A. chrysaetos
2956	**Banding** • Raptors • A. chrysaetos • Finland
2973	**Banding** • Life History • Prey • Plumage • Cain/Abel • Reproductive Success • Interspecific Behavior • Piracy • Photo Account • Raptors • A. chrysaetos • Rhodesia
3003	**Banding** • Raptors • A. chrysaetos • A. heliaca • Yugoslavia
3004	**Banding** • Raptors • A. chrysaetos • A. heliaca • Yugoslavia
3020	**Banding** • Nest • Growth • Behavior • Clutch Size • Photo Account • A. chrysaetos • California
3023	**Banding** • Longevity • Raptors • A. chrysaetos • California
3059	**Banding** • Raptors • A. chrysaetos • A. clanga • A. pomarina • Poland
3060	**Banding** • Raptors • A. chrysaetos • Poland
3114	**Banding** • Nest • A. pomarina • West Germany
3187	**Banding** • Nest Survey • Prey • Raptors • A. wahlbergi • Rhodesia
3383	**Banding** • Photo Account • A. chrysaetos • Switzerland
3414	**Banding** • Nest • Photo Account • Popular Monograph • A. chrysaetos • California
50	**Behavior** • Capture • Captivity • A. chrysaetos • Colorado
57	**Behavior** • Collect • Prey • A. heliaca • India
89	**Behavior** • Nest • Hunting Behavior • Popular Monograph • A. chrysaetos • California
197	**Behavior** • Nest • Prey • Captivity • A. chrysaetos • Algeria
238	**Behavior** • Nest • A. chrysaetos • New Mexico
262	**Behavior** • Descriptive Summary • Etymology • Status • Nest • Plumage • A. heliaca • A. nipalensis • A. clanga • A. chrysaetos • Cyprus
271	**Behavior** • Nest • Egg Collection • Vocalization • Behavior • Photo Account • Popular Monograph • A. chrysaetos • California
271	**Behavior** • Behavior • Nest • Egg Collection • Vocalization • Photo Account • Popular Monograph • A. chrysaetos • California
274	**Behavior** • Capture • A. chrysaetos • Illinois
287	**Behavior** • Prey • Nest • Photo Account • Raptors • A. chrysaetos • Norway
304	**Behavior** • Plumage • Aging • Intraspecific Behavior • A. heliaca • A. nipalensis • A. chrysaetos • A. pomarina • A. clanga • Europe
306	**Behavior** • Falconry • Raptors • Photo Account • A. chrysaetos • West Germany
349	**Behavior** • Electrocution • Raptors • A. chrysaetos • North America
351	**Behavior** • Hunting Behavior • Measure • Life History • Courtship • Distribution • Vocalization • Plumage • Raptors • Photo Account • A. chrysaetos • North America
388	**Behavior** • Mobbing • Photography Account • A. chrysaetos • France
412	**Behavior** • Nest • Egg Collection • Cain/Abel • A. verreauxi • South Africa
464	**Behavior** • Nest • Habitat • Egg Collection • Vocalization • Animal Damage • A. audax • Australia

Master No.	Keywords
465	**Behavior** • Nest • Egg • Abandonment • Raptors • A. audax • Australia
498	**Behavior** • Taxonomy • Field Id • Molt • Measure • Museum Specimen • Wintering Population • Interspecific Competition • Conservation • Eagles • Photo Account • A. nipalensis • A. pomarina • Africa
544	**Behavior** • Taxonomy • Status • Field Id • Flight • Habitat • Terminology • Distribution • Mortality • Life History • Longevity • Territory • Photo Account • Raptors • A. chrysaetos • A. clanga • Great Britain • Ireland • Scotland
547	**Behavior** • Breeding Biology • Conservation • Knowledge Status • A. chrysaetos • A. heliaca • A. audax • A. rapax • A. nipalensis • A. pomarina • A. clanga • A. gurneyi • A. verreauxi • A. wahlbergi
552	**Behavior** • Nest Survey • Prey • Habitat • Status • Reproductive Success • Food Requirements • Animal Damage • Intraspecific Behavior • Predator/Prey • A. chrysaetos • Scotland
553	**Behavior** • Cain/Abel • Reproductive Success • Eagles • A. verreauxi • A. rapax • A. wahlbergi • Africa
560	**Behavior** • Aircraft Collision • A. chrysaetos • Switzerland
569	**Behavior** • Prey • A. chrysaetos • California
579	**Behavior** • Nest • Photo Account • A. chrysaetos • Wyoming
587	**Behavior** • Nest • Prey • Growth • Vocalization • Nest Adornment • Photo Account • Illus. • A. chrysaetos • Switzerland
588	**Behavior** • Growth • Raptors • A. chrysaetos • North America
595	**Behavior** • Status • Collect • Raptors • A. rapax • A. clanga • India
611	**Behavior** • Habitat • Nest • Field Id • Nest Survey • Photo Account • Raptors • A. chrysaetos • North America
614	**Behavior** • Monograph • Nest Survey • Reproductive Success • Habitat • Directional Exposure • A. chrysaetos • Utah
616	**Behavior** • Nest • Habitat • Egg • Prey • Vocalization • Growth • Trapping • Poisoning • Lift • Photo Account • A. chrysaetos • Montana
617	**Behavior** • Mortality • Egg • Raptors • A. chrysaetos • Montana
618	**Behavior** • Prey • Nest • Attitude • Breeding Biology • Flight • Poisoning • Molt • Hunting Behavior • Vocalization • Folklore • Lift • Photo Account • A. chrysaetos • Montana
619	**Behavior** • Nest • Fledgling • Photo Account • Popular Monograph • A. chrysaetos • Montana
707	**Behavior** • Captivity • A. chrysaetos • California
714	**Behavior** • Food Requirements • Breeding Biology • Predator/Prey • Cain/Abel • Directional Exposure • Growth • Sex Ratio • A. chrysaetos • Idaho
722	**Behavior** • Nest • Egg • Raptors • A. chrysaetos • A. pomarina • Rumania
730	**Behavior** • Nest • Raptors • A. chrysaetos • California
731	**Behavior** • Flight • A. chrysaetos • Great Britain
753	**Behavior** • Nest • Shooting • Photography • Story • Banding • Photo Account • Raptors • A. verreauxi • South Africa
767	**Behavior** • Telemetry • Protection • Nest • Falconry • Pesticides • Photo Account • Popular Monograph • A. chrysaetos • United States
800	**Behavior** • Display • A. chrysaetos • Sweden
883	**Behavior** • Descriptive Summary • Etymology • Prey • Display • A. heliaca • A. rapax • A. pomarina • A. clanga • India
887	**Behavior** • Disturbance • A. verreauxi • Rhodesia
899	**Behavior** • Prey Behavior • A. chrysaetos • California
944	**Behavior** • Nest • Nest Adornment • Prey • Growth • A. audax • Australia
955	**Behavior** • Pesticides • Parental Behavior • Vocalization • Disturbance • Measure • Breeding Biology • Parasite • Disease • Terminology • Growth • A. chrysaetos • Montana
957	**Behavior** • Monograph • Growth • Pesticides • Parental Behavior • Vocalization • Disturbance • Measure • Breeding Biology • Photography • Captivity • Parasite • Disease • Terminology • Photo Account • A. chrysaetos • Montana
958	**Behavior** • Marking • Dye • Tech. • A. chrysaetos
963	**Behavior** • Growth • Nest • Plumage • Prey • Vocalization • Photo Account • A. audax • Australia
974	**Behavior** • Rehabilitation • Shooting • Marking • Eagles • A. chrysaetos • United States
992	**Behavior** • Nest • A. chrysaetos • California
1038	**Behavior** • Nest • A. verreauxi • Kenya
1041	**Behavior** • Status • Prey • Nest • Habitat • Photo Account • A. chrysaetos • Bulgaria
1043	**Behavior** • Status • Raptors • A. audax • Australia
1054	**Behavior** • Captive Breeding • Nest • Plumage • Measure • Photo Account • Popular Monograph • A. audax • Australia
1058	**Behavior** • Taxonomy • Museum Specimen • Collect • Raptors • A. pomarina • A. clanga • A. nipalensis • Nepal
1092	**Behavior** • Nest • Pellet • A. chrysaetos • Switzerland
1144	**Behavior** • Nest Survey • Status • Fledging • Mortality • Banding • Marking • Cain/Abel • Reproductive Success • Breeding Biology • Interspecific Behavior • Hunting Behavior • Display • Courtship • Raptors • A. verreauxi • Rhodesia
1153	**Behavior** • Cain/Abel • Parental Behavior • Clutch Size • Growth • Eagles • A. verreauxi • A. pomarina • Rhodesia
1155	**Behavior** • Physiology • Nest • Temperature • A. verreauxi • Rhodesia
1163	**Behavior** • Nest • Prey • Clutch Size • A. heliaca • Spain
1167	**Behavior** • Nest • Photography • Prey • Photo Account • A. audax • Australia
1170	**Behavior** • Anatomy • Prey • Raptors • A. chrysaetos • United Soviet Socialists Republic
1174	**Behavior** • Vocalization • Nest • Prey • Photo Account • Popular Monograph • A. pomarina • West Germany
1179	**Behavior** • Nest • Raptors • A. audax • Australia
1209	**Behavior** • Plumage • Taxonomy • Captivity • Eagles • A. chrysaetos • Nova Scotia

Master No.	Keywords
1246	**Behavior** • Hunting Behavior • Lift • Trapping • Attack • Photography • Behavior • Photo Account • Popular Monograph • A. chrysaetos • Scotland
1246	**Behavior** • Behavior • Hunting Behavior • Lift • Trapping • Attack • Photography • Photo Account • Popular Monograph • A. chrysaetos • Scotland
1247	**Behavior** • Nest Adornment • A. chrysaetos • Scotland
1248	**Behavior** • Nest • A. chrysaetos • Scotland
1251	**Behavior** • Basking • Popular Monograph • A. chrysaetos • Scotland
1261	**Behavior** • Prey • Nest • Photo Account • Popular Monograph • A. chrysaetos • Scotland
1264	**Behavior** • Popular Monograph • (Title Only) • A. chrysaetos • Scotland
1266	**Behavior** • A. chrysaetos • Scotland
1267	**Behavior** • Fledging • Flight • A. chrysaetos • Great Britain
1270	**Behavior** • A. chrysaetos • Great Britain
1271	**Behavior** • Incubation • Flight • A. chrysaetos • Scotland
1272	**Behavior** • A. chrysaetos • Scotland
1281	**Behavior** • Folklore • Prey • A. chrysaetos • Texas
1282	**Behavior** • Capture • A. chrysaetos • Texas
1293	**Behavior** • Captive Breeding • Egg Measure • Artificial Insemination • Artificial Incubation • Photography Account • A. chrysaetos • United States
1313	**Behavior** • Flight • A. chrysaetos • Switzerland
1343	**Behavior** • Descriptive Summary • (Title Only) • A. clanga • West Germany
1344	**Behavior** • Vocalization • A. chrysaetos • West Germany
1347	**Behavior** • Plumage • Anatomy • Taxonomy • Raptors • Theory • A. chrysaetos
1370	**Behavior** • Nest • A. chrysaetos • California
1427	**Behavior** • Prey • Shooting • Measure • A. chrysaetos • Georgia
1432	**Behavior** • Biology • (Title Only) • A. pomarina • West Germany
1453	**Behavior** • Capture • A. chrysaetos • Alberta
1497	**Behavior** • Fledging • Flight • A. chrysaetos • Idaho • Oregon
1513	**Behavior** • Nest Failure • Prey • Habitat • A. chrysaetos • Alaska
1521	**Behavior** • Nest • A. chrysaetos • California
1564	**Behavior** • Nest • Vocalization • Hunting Behavior • Raptors • A. verreauxi • South Africa
1597	**Behavior** • Prey • Collect • Habitat • Raptors • A. nipalensis • A. clanga • A. pomarina • India
1602	**Behavior** • Nest • A. chrysaetos • California
1639	**Behavior** • Taxonomy • Dimorphism • Measure • Plumage • Distribution • Prey • Management • Habitat • Territory • Mortality • Life History • Economics • Literature Review • A. chrysaetos • Oregon • Washington
1726	**Behavior** • Captive Breeding • Photo Account • A. chrysaetos • North America
1763	**Behavior** • Nest • Prey • Photo Account • Popular Monograph • A. chrysaetos • West Germany
1770	**Behavior** • Falconry • Story • Eagles • Photography Account • A. chrysaetos • Great Britain
1781	**Behavior** • Nest Survey • Reproductive Success • Marking • Habitat • Nest Chronology • Prey • Banding • Territory • Population Dynamics • Raptors • SRBPNA • A. chrysaetos • Idaho
1812	**Behavior** • Prey • Attitude • A. chrysaetos • South Dakota
1829	**Behavior** • A. chrysaetos • Wisconsin
1891	**Behavior** • Egg Collection • Nest Chronology • Nest • Raptors • A. wahlbergi • Rhodesia
1902	**Behavior** • Status • Conservation • Prey • Nest • A. chrysaetos • Israel
1906	**Behavior** • Nest • Prey • A. heliaca • Spain
1908	**Behavior** • Nest • Breeding Biology • Clutch Size • Prey • A. nipalensis • Bulgaria
1942	**Behavior** • Food Habits • Nest • Time Lapse • Economics • Animal Damage • Reproductive Success • Status • A. chrysaetos • Texas
1954	**Behavior** • Prey • Photo Account • Raptors • Popular Monograph • A. chrysaetos • Europe
1985	**Behavior** • Habitat • Nest • Prey • Flight • Photo Account • A. chrysaetos • Scotland
1998	**Behavior** • Reproductive Success • Growth • Prey • Photography Account • Popular Monograph • A. chrysaetos • Scotland
2008	**Behavior** • Folklore • Mortality • A. chrysaetos • California
2023	**Behavior** • Capture • Banding • Tech. • A. pomarina • South Africa
2025	**Behavior** • Falconry • Flight • Eagles • Popular Monograph • A. chrysaetos • North America
2066	**Behavior** • Rarity • Prey • A. chrysaetos • Ireland
2085	**Behavior** • Telemetry • Territory • Display • Agonistic Behavior • Nest • A. chrysaetos • Wyoming • Montana
2138	**Behavior** • Nest • Prey • A. chrysaetos • Switzerland
2145	**Behavior** • Breeding Biology • Habitat • Cain/Abel • Density • Prey • Nest Survey • Photo Account • A. pomarina • Czechoslovakia
2172	**Behavior** • Nest • Raptors • A. chrysaetos • Arizona
2240	**Behavior** • Nest • Prey • Photography • Photo Account • A. heliaca • Spain
2241	**Behavior** • Refuge • Photography • Nest Survey • Reproductive Success • Mortality • Photo Account • Raptors • Popular Monograph • A. heliaca • Spain
2291	**Behavior** • Status • A. chrysaetos • Wisconsin
2294	**Behavior** • Lift • Hunting Behavior • Conservation • A. chrysaetos • North America
2332	**Behavior** • Prey • Conservation • A. audax • Australia
2371	**Behavior** • Popular Monograph • Story • Breeding Biology • Habitat • Nest Survey • Hunting Behavior • Prey • Raptors • Illus. • A. chrysaetos • Colorado
2377	**Behavior** • Conservation • Electrocution • Management • Status • Literature Review • Eagles • A. chrysaetos • North America
2411	**Behavior** • Interspecific Behavior • Nest • Photography • Prey • Nest Adornment • Talon Grappling • Cain/Abel • Photo Account • Popular Monograph • A. chrysaetos • Scotland
2450	**Behavior** • Nest • Measure • A. chrysaetos • Spain
2463	**Behavior** • Collect • Status • Vocalization • Raptors • A. rapax • Somalia

Master No.	Keywords
2531	**Behavior** • Nest • Prey • A. verreauxi • South Africa
2535	**Behavior** • Field Id • Status • Distribution • Raptors • A. verreauxi • A. rapax • A. nipalensis • A. pomarina • A. wahlbergi • South Africa
2542	**Behavior** • Captivity • A. chrysaetos • Michigan
2544	**Behavior** • Collect • Habitat • Prey • A. clanga • West Germany
2554	**Behavior** • Conservation • Nest • Interspecific Behavior • Prey • Trapping • Animal Damage • Rehabilitation • Photo Account • Popular Monograph • A. chrysaetos • Wyoming
2568	**Behavior** • Prey • A. chrysaetos • North America
2569	**Behavior** • Nest • Disturbance • Prey • A. chrysaetos • Colorado
2593	**Behavior** • Talon Grappling • Interspecific Behavior • Raptors • A. chrysaetos • Washington
2625	**Behavior** • Nest • Raptors • Illus. • A. chrysaetos • France
2705	**Behavior** • Supplemental Feeding • Raptors • Photo Account • A. chrysaetos • Corsica
2720	**Behavior** • Status • Museum Specimen • Raptors • A. chrysaetos • A. clanga • Switzerland
2753	**Behavior** • Monograph • Habitat • Nest Survey • Prey • Pellet • Abandonment • A. chrysaetos • Wyoming
2760	**Behavior** • Nest • Prey • Vocalization • A. chrysaetos • West Germany
2763	**Behavior** • Prey • Nest Chronology • A. chrysaetos • West Germany
2772	**Behavior** • Collect • Raptors • A. wahlbergi • South Africa
2880	**Behavior** • Dimorphism • Food Requirements • Theory • Raptors • A. chrysaetos
2886	**Behavior** • Nest • Prey • Raptors • A. chrysaetos • Turkmenia
2947	**Behavior** • Egg Collection • Nest • A. chrysaetos • California
2949	**Behavior** • Nest • Status • Vocalization • Prey • Photo Account • A. chrysaetos • Switzerland
2963	**Behavior** • Rarity • Animal Damage • Measure • Captivity • A. chrysaetos • Georgia
2966	**Behavior** • Prey • Incubation • Cain/Abel • Reproductive Success • A. rapax • South Africa
2969	**Behavior** • Nest Survey • Habitat • Prey • Breeding Biology • Reproductive Success • Growth • Field Id • Vocalization • Migration • Eagles • Photo Account • A. wahlbergi • A. rapax • Rhodesia
3018	**Behavior** • Falconry • Story • Illus. • A. chrysaetos • Great Britain
3022	**Behavior** • Cain/Abel • Growth • Vocalization • Pellet • Feeding Behavior • Raptors • A. chrysaetos • California
3020	**Behavior** • Nest • Growth • Clutch Size • Banding • Photo Account • A. chrysaetos • California
3037	**Behavior** • Talon Grappling • A. chrysaetos • Oklahoma
3058	**Behavior** • Habitat • Prey • Prey Behavior • Nest • Animal Damage • Shooting • Egg • A. verreauxi • South Africa
3084	**Behavior** • Nest • Collect • Raptors • A. wahlbergi • South Africa
3089	**Behavior** • Taxonomy • Captivity • Plumage • A. rapax
3094	**Behavior** • Nest • Egg Collection • A. chrysaetos • California
3095	**Behavior** • Animal Damage • Nest • Prey • A. verreauxi • South Africa
3168	**Behavior** • A. chrysaetos • California
3188	**Behavior** • (Title Only) • A. verreauxi • Rhodesia
3214	**Behavior** • Collect • Nest • Habitat • Growth • Captivity • A. chrysaetos • India
3235	**Behavior** • Growth • Nest • Photography • Prey • Hunting Behavior • Vocalization • Nest Adornment • Egg • Raptors • Popular Monograph • A. rapax • Kenya
3351	**Behavior** • Nest Survey • Aerial Survey • Raptors • A. chrysaetos • Alaska
3369	**Behavior** • Nest Survey • Egg • Vocalization • Habitat • Hunting Behavior • Photo Account • A. chrysaetos • Arizona
3397	**Behavior** • Descriptive Summary • Pellet • Distribution • Plumage • Measure • Raptors • Illus. • A. chrysaetos • A. clanga • Great Britain
3424	**Behavior** • Prey • A. chrysaetos • California
3436	**Behavior** • Supplemental Feeding • Rarity • Plumage • Photo Account • A. heliaca • Spain
802	**Beneficial Effect** • Protection • Animal Damage • Raptors • A. clanga • Siberia
1126	**Beneficial Effect** • Raptors • A. nipalensis • A. heliaca • United Soviet Socialists Republic
2393	**Beneficial Effect** • Distribution • Prey • Raptors • A. heliaca • A. nipalensis • Kazakhstan
34	**Bibliography** • Literature Review • Eagles • A. chrysaetos • Washington
603	**Bibliography** • Raptors • (Title Only) • A. chrysaetos • Alaska
2375	**Bibliography** • Literature Review • Raptors • A. chrysaetos
1221	**Biology** • Breeding Biology • Nest • (Title Only) • A. chrysaetos • Texas
1432	**Biology** • Behavior • (Title Only) • A. pomarina • West Germany
1509	**Biology** • Descriptive Summary • (Title Only) • A. rapax
2848	**Biology** • Fledgling • (Title Only) • A. pomarina • West Germany
243	**Blood Sample** • Raptors • A. chrysaetos • A. heliaca
733	**Blood Sample** • Raptors • A. rapax • Africa
2348	**Blood Sample** • Physiology • A. chrysaetos • North America
2576	**Blood Sample** • Rehabilitation • Shooting • Trapping • Disease • Anaesthesia • Raptors • A. chrysaetos • United States
2614	**Blood Sample** • Disease • Raptors • A. chrysaetos • California
2858	**Blood Sample** • A. chrysaetos • A. rapax • A. heliaca
2925	**Blood Sample** • Parasite • Raptors • A. chrysaetos • Colorado
280	**Bounty** • A. chrysaetos • Alabama
386	**Bounty** • Disturbance • Habitat Loss • Protection • Shooting • Status • Pesticides • Migration • Distribution • Population Decline • Raptors • A. chrysaetos • A. heliaca • A. pomarina • A. clanga • Europe
444	**Bounty** • Attitude • Animal Damage • Conservation • Raptors • A. verreauxi • A. rapax • South Africa

Master No.	Keywords
794	**Bounty** • Animal Damage • Shooting • Prey • Eagles • A. chrysaetos • California
1421	**Bounty** • Prey • Popular Monograph • A. audax • Australia
1678	**Bounty** • Attitude • Raptors • A. chrysaetos • North Dakota
1681	**Bounty** • Shooting • Mortality • A. chrysaetos • Texas
1900	**Bounty** • Mortality • Prey • Nest • Dispersal • Pellet • Habitat • Animal Damage • Protection • Photo Account • A. audax • Australia
2038	**Bounty** • Status • Prey • Raptors • A. chrysaetos • A. heliaca • India
2089	**Bounty** • Protection • Raptors • A. audax • Australia
2399	**Bounty** • Economics • Conservation • Mortality • Eagles • Popular Monograph • A. chrysaetos • North America
2438	**Bounty** • Attitude • Eagles • A. chrysaetos • Alaska
2585	**Bounty** • Animal Damage • Attitude • Photo Account • Raptors • Popular Monograph • A. chrysaetos • North America
2797	**Bounty** • Protection • Raptors • A. audax • Australia
2799	**Bounty** • A. audax • Australia
2804	**Bounty** • Shooting • Attitude • Protection • Economics • A. audax • Australia
3310	**Bounty** • Shooting • Poisoning • Conservation • Punishment • Story • Eagles • Popular Monograph • A. chrysaetos • Wyoming
3365	**Bounty** • Conservation • Anatomy • Protection • Raptors • Illus. • A. chrysaetos • Finland
3372	**Bounty** • Protection • Population Decline • Mortality • Conservation • Animal Damage • A. chrysaetos • Norway
3373	**Bounty** • Status • Conservation • Mortality • Eagles • A. chrysaetos • Norway
3413	**Bounty** • Prey • Lift • Hunting Behavior • A. chrysaetos • Montana
305	**Breeding Biology** • Anatomy • Cain/Abel • Distribution • A. pomarina • A. clanga • South Africa
324	**Breeding Biology** • Reproductive Success • Food Requirements • Territory • Abandonment • Mortality • Cain/Abel • Density • Dispersal • Nest Survey • A. chrysaetos • Idaho
325	**Breeding Biology** • Monograph • Density • Mortality • Cain/Abel • Prey • Abnormal Behavior • Egg Measure • Marking • A. chrysaetos • Idaho
534	**Breeding Biology** • Habitat • Density • Nest Survey • Interspecific Competition • Prey • Reproductive Success • Etymology • Plumage • Vocalization • Nest Adornment • Raptors • A. verreauxi • A. wahlbergi • Kenya
535	**Breeding Biology** • Nest Survey • Etymology • Prey • Reproductive Success • Interspecific Competition • Raptors • Photo Account • A. wahlbergi • A. verreauxi • Kenya
540	**Breeding Biology** • Nest Failure • Food Requirements • Reproductive Success • Home Range • Parental Behavior • Eagles • Theory • A. wahlbergi • A. verreauxi • Africa
547	**Breeding Biology** • Conservation • Knowledge Status • Behavior • A. chrysaetos • A. heliaca • A. audax • A. rapax • A. nipalensis • A. pomarina • A. clanga • A. gurneyi • A. verreauxi • A. wahlbergi
618	**Breeding Biology** • Prey • Nest • Attitude • Behavior • Flight • Poisoning • Molt • Hunting Behavior • Vocalization • Folklore • Lift • Photo Account • A. chrysaetos • Montana
629	**Breeding Biology** • Nest • (Title Only) • A. chrysaetos • West Germany
714	**Breeding Biology** • Food Requirements • Predator/Prey • Cain/Abel • Directional Exposure • Growth • Sex Ratio • Behavior • A. chrysaetos • Idaho
955	**Breeding Biology** • Behavior • Pesticides • Parental Behavior • Vocalization • Disturbance • Measure • Parasite • Disease • Terminology • Growth • A. chrysaetos • Montana
957	**Breeding Biology** • Monograph • Behavior • Growth • Pesticides • Parental Behavior • Vocalization • Disturbance • Measure • Photography • Captivity • Parasite • Disease • Terminology • Photo Account • A. chrysaetos • Montana
1040	**Breeding Biology** • Monograph • Prey • Vocalization • Territory • Taxonomy • Migration • Mortality • Molt • Falconry • A. chrysaetos • A. verreauxi • A. audax
1144	**Breeding Biology** • Nest Survey • Status • Behavior • Fledging • Mortality • Banding • Marking • Cain/Abel • Reproductive Success • Interspecific Behavior • Hunting Behavior • Display • Courtship • Raptors • A. verreauxi • Rhodesia
1145	**Breeding Biology** • Nest • Parental Behavior • Interspecific Behavior • Intraspecific Behavior • Feeding Behavior • Hunting Behavior • Fledging • Prey • Display • Raptors • A. verreauxi • Rhodesia
1156	**Breeding Biology** • Habitat • Prey • Density • Nest Survey • A. verreauxi • Rhodesia
1176	**Breeding Biology** • Nest • Courtship • Food Habits • A. pomarina • West Germany
1220	**Breeding Biology** • Prey • A. clanga • United Soviet Socialists Republic • Siberia
1221	**Breeding Biology** • Biology • Nest • (Title Only) • A. chrysaetos • Texas
1235	**Breeding Biology** • Prey • A. pomarina • Belorussia
1236	**Breeding Biology** • Prey • A. pomarina • Belorussia
1533	**Breeding Biology** • Nest • (Title Only) • A. pomarina • Europe
1687	**Breeding Biology** • Monograph • Molt • Field Id • Literature Review • A. chrysaetos • Colorado
1818	**Breeding Biology** • Hunting Behavior • Prey • Raptors • A. clanga • Siberia
1859	**Breeding Biology** • A. chrysaetos • Uzbek
1908	**Breeding Biology** • Nest • Behavior • Clutch Size • Prey • A. nipalensis • Bulgaria
1909	**Breeding Biology** • Nest • A. clanga • Kazakhstan
1921	**Breeding Biology** • Density • Predator/Prey • A. nipalensis • Russia
1933	**Breeding Biology** • Raptors • A. heliaca • Kazakhstan
1934	**Breeding Biology** • Nest • Density • Territory • A. heliaca • Kazakhstan

Master No.	Keywords
1947	**Breeding Biology** • Prey • Density • (Title Only) • A. chrysaetos • Scotland
2086	**Breeding Biology** • Monograph • Mortality • Banding • Migration • Prey • Food Requirements • A. chrysaetos • Montana
2088	**Breeding Biology** • Prey • Banding • Migration • Directional Exposure • Mortality • A. chrysaetos • Montana
2124	**Breeding Biology** • Monograph • Migration • Prey • Hunting Behavior • Attack • Folklore • Falconry • Animal Damage • Popular Monograph • A. chrysaetos
2145	**Breeding Biology** • Habitat • Cain/Abel • Density • Behavior • Prey • Nest Survey • Photo Account • A. pomarina • Czechoslovakia
2148	**Breeding Biology** • Distribution • Field Id • Management • Taxonomy • Habitat • Nest Survey • Prey • Photo Account • Eagles • A. pomarina • Europe
2149	**Breeding Biology** • Cain/Abel • Nest • Taxonomy • Prey • Photo Account • Popular Monograph • A. pomarina • West Germany
2152	**Breeding Biology** • Habitat • Growth • Parental Behavior • Prey • Interspecific Behavior • Courtship • Display • Photo Account • Raptors • A. heliaca • Spain
2188	**Breeding Biology** • Density • Prey • A. nipalensis • Caucasus
2193	**Breeding Biology** • A. chrysaetos • Nebraska
2370	**Breeding Biology** • Status • Nest • Nest Survey • Monograph • Habitat • Mortality • Nest Chronology • Courtship • Density • Prey • Management • A. chrysaetos • Colorado
2371	**Breeding Biology** • Popular Monograph • Story • Behavior • Habitat • Nest Survey • Hunting Behavior • Prey • Raptors • Illus. • A. chrysaetos • Colorado
2416	**Breeding Biology** • Prey • A. clanga • Siberia
2612	**Breeding Biology** • Prey • Habitat • Dispersal • A. audax • Australia
2669	**Breeding Biology** • Nest • Fledging • Parental Behavior • Food Habits • A. chrysaetos • Tanzania
2670	**Breeding Biology** • Cain/Abel • Prey • Parental Behavior • Growth • A. verreauxi • Africa
2757	**Breeding Biology** • Nest • (Title Only) • A. pomarina • West Germany
2761	**Breeding Biology** • Prey • Directional Exposure • Vocalization • A. chrysaetos • West Germany
2829	**Breeding Biology** • Theory • A. verreauxi • Africa
2830	**Breeding Biology** • Habitat • Cain/Abel • Nest Adornment • Growth • Vocalization • Plumage • Prey • Hunting Behavior • Illus. • A. pomarina • West Germany
2854	**Breeding Biology** • Population Dynamics • Nest • Density • Prey • Distribution • Habitat • Directional Exposure • Reproductive Success • Mortality • Abandonment • Disturbance • Interspecific Behavior • Territory • A. chrysaetos • Utah
2856	**Breeding Biology** • Nest Survey • Habitat • Hunting Behavior • Territory • Interspecific Behavior • Prey • Raptors • A. chrysaetos • Utah
2887	**Breeding Biology** • (Title Only) • AQ • United Soviet Socialists Republic
2969	**Breeding Biology** • Nest Survey • Habitat • Prey • Reproductive Success • Growth • Field Id • Vocalization • Migration • Behavior • Eagles • Photo Account • A. wahlbergi • A. rapax • Rhodesia
2972	**Breeding Biology** • Plumage • Photography • Story • Prey • Photo Account • Popular Monograph • A. verreauxi • A. wahlbergi • A. rapax • Rhodesia • South Africa
3012	**Breeding Biology** • Prey • Raptors • A. nipalensis • Turkmenia
3230	**Breeding Biology** • A. clanga • A. pomarina • Hungary
3283	**Breeding Biology** • Prey • Conservation • Mortality • Distribution • Photo Account • A. chrysaetos • Czechoslovakia
3322	**Breeding Biology** • Nest Survey • Nest Chronology • Cain/Abel • A. pomarina • West Germany
3325	**Breeding Biology** • Status • A. pomarina • Europe
3327	**Breeding Biology** • Monograph • Prey • A. pomarina • A. clanga
3441	**Breeding Biology** • Falconry • Nest • Prey • A. chrysaetos • Kirgizia
3444	**Breeding Biology** • Nest Survey • Habitat • Protection • Prey • Distribution • Reproductive Success • Photo Account • A. chrysaetos • Estonia
501	**Cache** • Prey • Pellet • Attack • Hunting Behavior • Animal Damage • Habitat • A. audax • Australia
781	**Cache** • Captivity • Raptors • A. chrysaetos
305	**Cain/Abel** • Anatomy • Breeding Biology • Distribution • A. pomarina • A. clanga • South Africa
324	**Cain/Abel** • Reproductive Success • Food Requirements • Territory • Abandonment • Mortality • Density • Dispersal • Breeding Biology • Nest Survey • A. chrysaetos • Idaho
325	**Cain/Abel** • Monograph • Breeding Biology • Density • Mortality • Prey • Abnormal Behavior • Egg Measure • Marking • A. chrysaetos • Idaho
412	**Cain/Abel** • Nest • Egg Collection • Behavior • A. verreauxi • South Africa
542	**Cain/Abel** • Nest • Abnormal Behavior • Habitat • A. verreauxi • Rhodesia
553	**Cain/Abel** • Reproductive Success • Behavior • Eagles • A. verreauxi • A. rapax • A. wahlbergi • Africa
714	**Cain/Abel** • Food Requirements • Breeding Biology • Predator/Prey • Directional Exposure • Growth • Sex Ratio • Behavior • A. chrysaetos • Idaho
852	**Cain/Abel** • Clutch Size • Mortality • Nestling • Prey • Pesticides • A. chrysaetos • Spain
1139	**Cain/Abel** • Growth • Photo Account • A. verreauxi • Rhodesia
1140	**Cain/Abel** • Photo Account • A. verreauxi • Rhodesia
1142	**Cain/Abel** • Photo Account • A. verreauxi • Rhodesia
1143	**Cain/Abel** • Photo Account • A. verreauxi • Rhodesia

Master No.	Keywords
1144	**Cain/Abel** • Nest Survey • Status • Behavior • Fledging • Mortality • Banding • Marking • Reproductive Success • Breeding Biology • Interspecific Behavior • Hunting Behavior • Display • Courtship • Raptors • A. verreauxi • Rhodesia
1153	**Cain/Abel** • Parental Behavior • Clutch Size • Behavior • Growth • Eagles • A. verreauxi • A. pomarina • Rhodesia
1158	**Cain/Abel** • Nest Survey • Reproductive Success • Raptors • A. verreauxi • Rhodesia
1181	**Cain/Abel** • A. clanga • West Germany
1306	**Cain/Abel** • A. pomarina • A. clanga • West Germany
1308	**Cain/Abel** • A. clanga • West Germany
1416	**Cain/Abel** • A. chrysaetos • Switzerland
1618	**Cain/Abel** • Mortality • Food Requirements • Raptors • A. chrysaetos • A. verreauxi • A. pomarina
1769	**Cain/Abel** • Descriptive Summary • Nest • Photo Account • Popular Monograph • A. chrysaetos • Scotland
1775	**Cain/Abel** • Monograph • Status • Nest Survey • Growth • Reproductive Success • Banding • Marking • Mortality • Pesticides • Prey • Residues • A. chrysaetos • Idaho
2145	**Cain/Abel** • Breeding Biology • Habitat • Density • Behavior • Prey • Nest Survey • Photo Account • A. pomarina • Czechoslovakia
2146	**Cain/Abel** • Management • A. pomarina • Czechoslovakia
2149	**Cain/Abel** • Breeding Biology • Nest • Taxonomy • Prey • Photo Account • Popular Monograph • A. pomarina • West Germany
2150	**Cain/Abel** • Mortality • Theory • Photo Account • A. chrysaetos • A. pomarina • A. verreauxi • A. heliaca • Czechoslovakia • South Africa • Spain
2154	**Cain/Abel** • Mortality • Reproductive Success • Management • Experiment • A. chrysaetos • A. heliaca
2155	**Cain/Abel** • Management • Experiment • Theory • A. pomarina • A. heliaca • A. verreauxi • A. chrysaetos
2156	**Cain/Abel** • Management • Clutch Size • Translocation • Eagles • Tech. • A. pomarina • A. verreauxi • A. rapax • A. chrysaetos • A. heliaca
2157	**Cain/Abel** • Conservation • Fostering • Reproductive Success • Photo Account • A. heliaca • Spain
2161	**Cain/Abel** • Management • A. pomarina • A. verreauxi • A. heliaca • East Germany
2384	**Cain/Abel** • Captive Care • Photo Account • Popular Monograph • A. chrysaetos • South Africa
2411	**Cain/Abel** • Interspecific Behavior • Nest • Behavior • Photography • Prey • Nest Adornment • Talon Grappling • Photo Account • Popular Monograph • A. chrysaetos • Scotland
2670	**Cain/Abel** • Prey • Parental Behavior • Growth • Breeding Biology • A. verreauxi • Africa
2769	**Cain/Abel** • A. pomarina • West Germany
2830	**Cain/Abel** • Breeding Biology • Habitat • Nest Adornment • Growth • Vocalization • Plumage • Prey • Hunting Behavior • Illus. • A. pomarina • West Germany
2966	**Cain/Abel** • Prey • Behavior • Incubation • Reproductive Success • A. rapax • South Africa
2973	**Cain/Abel** • Life History • Prey • Plumage • Banding • Reproductive Success • Interspecific Behavior • Piracy • Photo Account • Raptors • A. chrysaetos • Rhodesia
2975	**Cain/Abel** • Prey Behavior • Incubation • Reproductive Success • A. rapax • South Africa
2979	**Cain/Abel** • Mortality • Raptors • Theory • A. chrysaetos • A. clanga • A. pomarina • A. verreauxi • A. heliaca • A. rapax
3022	**Cain/Abel** • Growth • Behavior • Vocalization • Pellet • Feeding Behavior • Raptors • A. chrysaetos • California
3080	**Cain/Abel** • Distribution • Educational Material • Field Id • Habitat • Hunting Behavior • Prey • Artificial Nest • Nest • Incubation • Wintering Population • Conservation • Photo Account • Eagles • A. chrysaetos • A. wahlbergi • A. verreauxi • A. nipalensis • A. pomarina • South Africa
3322	**Cain/Abel** • Nest Survey • Nest Chronology • Breeding Biology • A. pomarina • West Germany
3323	**Cain/Abel** • Territory • Incubation • A. pomarina • West Germany
3326	**Cain/Abel** • A. pomarina • West Germany
2521	**Capt. Breed** • Raptors • A. chrysaetos
983	**Captive Breeding** • A. chrysaetos • United States
1019	**Captive Breeding** • Measure • Artificial Incubation • Raptors • A. chrysaetos • A. rapax • A. nipalensis • West Germany
1054	**Captive Breeding** • Nest • Behavior • Plumage • Measure • Photo Account • Popular Monograph • A. audax • Australia
1108	**Captive Breeding** • A. chrysaetos • North America
1293	**Captive Breeding** • Behavior • Egg Measure • Artificial Insemination • Artificial Incubation • Photography Account • A. chrysaetos • United States
1360	**Captive Breeding** • Raptors • Popular Monograph • A. chrysaetos • North America
1361	**Captive Breeding** • Story • Falconry • Artificial Insemination • Photography Account • Popular Monograph • A. chrysaetos • United States
1362	**Captive Breeding** • A. chrysaetos • North America
1725	**Captive Breeding** • A. chrysaetos • Kansas
1726	**Captive Breeding** • Behavior • Photo Account • A. chrysaetos • North America
1757	**Captive Breeding** • A. chrysaetos • Kansas
1758	**Captive Breeding** • Lift • A. chrysaetos • North America
2127	**Captive Breeding** • Raptors • A. nipalensis • Israel
2729	**Captive Breeding** • Raptors • A. chrysaetos • Great Britain
3296	**Captive Breeding** • A. chrysaetos • Great Britain
1364	**Captive Care** • Hacking • Fostering • Raptors • A. chrysaetos • A. pomarina
1958	**Captive Care** • Captivity • Raptors • A. wahlbergi
2122	**Captive Care** • Captivity • A. chrysaetos
2324	**Captive Care** • Capture • Animal Damage • Banding • Management • Prey • Sexing • Translocation • Marking • Measure • Tech. • Eagles • A. chrysaetos • Montana

Master No.	Keywords
2326	**Captive Care** • Capture • Banding • Marking • Animal Damage • Sexing • Management • Translocation • Prey • Measure • Mortality • Aging • Food Requirements • A. chrysaetos • Montana
2384	**Captive Care** • Cain/Abel • Photo Account • Popular Monograph • A. chrysaetos • South Africa
3307	**Captive Care** • A. chrysaetos • Texas
2325	**Captive Care** • Capture • Banding • Marking • Animal Damage • Sexing • Management • Translocation • Prey • Survey • Measure • Mortality • Aging • A. chrysaetos • Montana
14	**Captivity** • Melanism • Capture • A. chrysaetos • Colorado
50	**Captivity** • Capture • Behavior • A. chrysaetos • Colorado
83	**Captivity** • A. chrysaetos • Great Britain
123	**Captivity** • Trapping • Flight • Photo Account • A. chrysaetos • North America
147	**Captivity** • Longevity • A. audax • A. chrysaetos • A. heliaca
154	**Captivity** • Banding • Rehabilitation • Animal Damage • Popular Monograph • A. chrysaetos • North America
197	**Captivity** • Nest • Prey • Behavior • A. chrysaetos • Algeria
240	**Captivity** • Nest • Distribution • A. chrysaetos • Quebec
292	**Captivity** • Molt • Raptors • A. chrysaetos • Hungary
294	**Captivity** • Animal Damage • Poisoning • Shooting • Lift • A. audax • Australia
313	**Captivity** • Poisoning • Toxicology • Autopsy • Residues • Thal. Sulfate • A. chrysaetos
335	**Captivity** • Distribution • Nest • Vocalization • Prey • Raptors • A. chrysaetos • Oregon
394	**Captivity** • Museum Specimen • Mortality • A. chrysaetos • Illinois
604	**Captivity** • Anatomy • Physiology • Raptors • A. chrysaetos • A. verreauxi • A. rapax • A. wahlbergi
631	**Captivity** • Conservation • Story • Trapping • Falconry • Rehabilitation • Trade • Wintering Population • Raptors • Popular Monograph • A. chrysaetos • New York
707	**Captivity** • Behavior • A. chrysaetos • California
726	**Captivity** • Measure • A. verreauxi • United States
736	**Captivity** • Disease • Vaccination • Raptors • A. rapax • Africa
781	**Captivity** • Cache • Raptors • A. chrysaetos
796	**Captivity** • A. chrysaetos • Poland
799	**Captivity** • Status • Plumage • Raptors • A. chrysaetos • A. heliaca • A. clanga • Rumania
909	**Captivity** • Shooting • A. chrysaetos • Great Britain
957	**Captivity** • Monograph • Behavior • Growth • Pesticides • Parental Behavior • Vocalization • Disturbance • Measure • Breeding Biology • Photography • Parasite • Disease • Terminology • Photo Account • A. chrysaetos • Montana
973	**Captivity** • Trade • Raptors • A. rapax • A. clanga • A. pomarina • Great Britain
1013	**Captivity** • Rarity • Poisoning • Measure • Plumage • Photo Account • A. heliaca • Italy
1047	**Captivity** • Longevity • Egg • A. chrysaetos • Great Britain
1048	**Captivity** • Longevity • A. chrysaetos • Great Britain
1063	**Captivity** • Longevity • A. chrysaetos • A. audax • A. verreauxi • A. heliaca • A. rapax • A. pomarina • Europe • Australia
1132	**Captivity** • Raptors • A. chrysaetos • Tennessee
1209	**Captivity** • Plumage • Taxonomy • Behavior • Eagles • A. chrysaetos • Nova Scotia
1259	**Captivity** • Poetry • Attitude • Protection • Flight • Folklore • Raptors • A. chrysaetos • Great Britain
1317	**Captivity** • Longevity • A. heliaca • Spain
1320	**Captivity** • Egg • Raptors • A. audax • Australia
1321	**Captivity** • Egg • Raptors • A. audax • Australia
1328	**Captivity** • Museum Specimen • Plumage • Taxonomy • Eagles • A. chrysaetos • A. heliaca • A. rapax • A. clanga • A. nipalensis
1332	**Captivity** • Longevity • Raptors • A. chrysaetos • A. heliaca
1425	**Captivity** • Prey • A. chrysaetos • Scotland
1503	**Captivity** • Longevity • Egg • Plumage • A. rapax • Tunisia
1660	**Captivity** • Electrocution • A. chrysaetos • California
1736	**Captivity** • Marking • Raptors • A. wahlbergi • South Africa
1831	**Captivity** • Photo Account • A. chrysaetos • A. pomarina • A. verreauxi • West Germany
1843	**Captivity** • Collect • Raptors • A. heliaca • A. pomarina • China
1854	**Captivity** • Nest • Shooting • A. chrysaetos • Texas
1958	**Captivity** • Captive Care • Raptors • A. wahlbergi
1970	**Captivity** • Shooting • A. chrysaetos • Montana
1986	**Captivity** • Nest • Clutch Size • Prey • Nest Survey • Collect • A. chrysaetos • Canada
2047	**Captivity** • Nest • A. chrysaetos • Switzerland
2122	**Captivity** • Captive Care • A. chrysaetos
2125	**Captivity** • Poisoning • Pesticides • Status • Density • Food Requirements • Habitat • Protection • Predator/Prey • Raptors • Photography Account • PCB • A. clanga • A. heliaca • A. nipalensis • A. verreauxi • A. pomarina • Israel
2201	**Captivity** • Rarity • A. pomarina • Italy
2232	**Captivity** • Trapping • A. chrysaetos • Massachusetts
2250	**Captivity** • Nest • A. chrysaetos • United States
2379	**Captivity** • A. chrysaetos • France
2382	**Captivity** • Senses • Raptors • A. audax • Australia
2542	**Captivity** • Behavior • A. chrysaetos • Michigan
2800	**Captivity** • Molt • Telemetry • Eagles • A. chrysaetos • Washington
2963	**Captivity** • Rarity • Animal Damage • Measure • Behavior • A. chrysaetos • Georgia
2983	**Captivity** • Raptors • A. wahlbergi • Uganda
2991	**Captivity** • A. chrysaetos • California
3021	**Captivity** • Growth • Food Requirements • Measure • Molt • Pellet • Temperature • Photo Account • Raptors • A. chrysaetos • California
3089	**Captivity** • Taxonomy • Behavior • Plumage • A. rapax
3108	**Captivity** • Raptors • A. chrysaetos • A. wahlbergi • A. rapax • A. audax • A. verreauxi • A. heliaca • United States
3140	**Captivity** • Plumage • A. chrysaetos • Iowa
3214	**Captivity** • Collect • Nest • Habitat • Behavior • Growth • A. chrysaetos • India
3286	**Captivity** • Longevity • A. chrysaetos
3287	**Captivity** • Shooting • A. chrysaetos • Iowa

Master No.	Keywords
3290	**Captivity** • Lift • Nest • Photo Account • Popular Monograph • A. chrysaetos • California
3295	**Captivity** • Clutch Size • A. chrysaetos • Great Britain
3356	**Captivity** • Capture • Eagles • A. chrysaetos • Michigan
3379	**Captivity** • Capture • Symbolism • Mortality • Anthropology • A. chrysaetos • North America
14	**Capture** • Melanism • Captivity • A. chrysaetos • Colorado
50	**Capture** • Behavior • Captivity • A. chrysaetos • Colorado
76	**Capture** • A. chrysaetos • Scotland
94	**Capture** • Folklore • Flight • Banding • A. chrysaetos • A. pomarina • Europe
159	**Capture** • Trapping • Mortality • Raptors • A. chrysaetos • Nevada
228	**Capture** • A. chrysaetos • Colorado
274	**Capture** • Behavior • A. chrysaetos • Illinois
384	**Capture** • Taxonomy • Aging • Field Id • Photo Account • A. nipalensis • A. rapax • Southwest Africa
488	**Capture** • Banding • A. verreauxi • South Africa
609	**Capture** • Wintering Population • Banding • Animal Damage • Eagles • A. chrysaetos • Tennessee
626	**Capture** • Plumage • Measure • A. pomarina • Tunisia
651	**Capture** • Nest • Economics • Folklore • A. chrysaetos • Tennessee
921	**Capture** • Animal Damage • Banding • A. chrysaetos • North Carolina
934	**Capture** • A. chrysaetos • New York
947	**Capture** • Roadside Survey • Roost • Nest Survey • Prey • Hunting Behavior • Marking • Interspecific Behavior • Shooting • Electrocution • A. chrysaetos • Utah
956	**Capture** • Measure • Aircraft • Tech. • A. chrysaetos • Montana
1282	**Capture** • Behavior • A. chrysaetos • Texas
1453	**Capture** • Behavior • A. chrysaetos • Alberta
1455	**Capture** • Talon Grappling • A. chrysaetos • Great Britain
1609	**Capture** • Trade • A. chrysaetos • Tennessee
1661	**Capture** • Nest • Symbolism • A. chrysaetos • New Mexico
1789	**Capture** • Mortality • A. chrysaetos • Hungary
1876	**Capture** • Rarity • A. clanga • Great Britain
1930	**Capture** • Raptors • A. audax • Australia
1943	**Capture** • Aerial Survey • Time Lapse • Animal Damage • Nest Survey • Prey • Habitat • Winter Survey • Age Structure • Food Requirements • Shooting • A. chrysaetos • Texas
2023	**Capture** • Behavior • Banding • Tech. • A. pomarina • South Africa
2037	**Capture** • Animal Damage • A. chrysaetos • Tennessee
2210	**Capture** • (Title Only) • A. chrysaetos • Italy
2313	**Capture** • Status • Raptors • A. clanga • A. heliaca • Egypt
2324	**Capture** • Animal Damage • Banding • Management • Prey • Sexing • Translocation • Marking • Measure • Captive Care • Tech. • Eagles • A. chrysaetos • Montana
2325	**Capture** • Banding • Marking • Animal Damage • Sexing • Management • Translocation • Prey • Survey • Measure • Mortality • Captive Care • Aging • A. chrysaetos • Montana
2326	**Capture** • Banding • Marking • Animal Damage • Sexing • Management • Translocation • Prey • Measure • Mortality • Captive Care • Aging • Food Requirements • A. chrysaetos • Montana
2349	**Capture** • Feeding Behavior • Animal Damage • Predator/Prey • Protection • Management • Economics • A. chrysaetos • Montana
2380	**Capture** • A. chrysaetos • France
2420	**Capture** • Symbolism • Historical Account • Anthropology • Raptors • Photo Account • A. chrysaetos • South Dakota
2425	**Capture** • Prey • Lift • Historical Account • A. chrysaetos • Scotland
2828	**Capture** • Rarity • A. chrysaetos • West Germany
2870	**Capture** • Banding • Survey • Marking • Nest • Prey • Food Requirements • Raptors • A. chrysaetos • A. wahlbergi • South Africa
3002	**Capture** • Mortality • Attitude • A. chrysaetos • Illinois
3321	**Capture** • Aerial Survey • Hunting Behavior • Animal Damage • A. chrysaetos • Texas • New Mexico
3356	**Capture** • Captivity • Eagles • A. chrysaetos • Michigan
3363	**Capture** • Hunting Behavior • Attack • A. audax • Australia
3379	**Capture** • Captivity • Symbolism • Mortality • Anthropology • A. chrysaetos • North America
3435	**Capture** • Animal Damage • A. chrysaetos • Switzerland
486	**Children's** • Conservation • Popular Monograph • A. chrysaetos • Norway
1311	**Children's** • Story • Conservation • Photo Account • SRBPNA • Popular Monograph • A. chrysaetos • Idaho
1931	**Children's** • A. chrysaetos • A. rapax • A. clanga • A. pomarina • A. nipalensis • A. heliaca • A. audax • A. gurneyi • A. wahlbergi • A. verreauxi
2270	**Children's** • Story • A. chrysaetos • Colorado
1151	**Climatology** • Reproductive Success • Density • Population Dynamics • Prey • Nest • Territory • Interspecific Competition • Mortality • Clutch Size • Pesticides • Eagles • A. verreauxi • Rhodesia
56	**Clutch Size** • Etymology • Descriptive Summary • Plumage • Habitat • Museum Specimen • Measure • A. rapax • A. pomarina • India
58	**Clutch Size** • Nest • Prey • Habitat • Collect • A. pomarina • India
108	**Clutch Size** • Plumage • Nest • Raptors • A. wahlbergi • A. rapax • A. verreauxi • Kenya • Uganda
250	**Clutch Size** • Museum Specimen • Raptors • A. chrysaetos • A. heliaca • A. rapax • Tunisia • Morocco
345	**Clutch Size** • Nest • A. wahlbergi • Malawi
363	**Clutch Size** • Prey • Nest • A. audax • Australia
400	**Clutch Size** • Raptors • A. heliaca • Spain
521	**Clutch Size** • Reproductive Success • Raptors • A. chrysaetos • Morocco

Master No.	Keywords
832	**Clutch Size** • A. chrysaetos • California
851	**Clutch Size** • Prey • Nest • A. chrysaetos • Spain
852	**Clutch Size** • Mortality • Nestling • Prey • Pesticides • Cain/Abel • A. chrysaetos • Spain
1141	**Clutch Size** • Nest Survey • Habitat • Density • Disturbance • Reproductive Success • Raptors • A. verreauxi • Rhodesia
1151	**Clutch Size** • Reproductive Success • Density • Population Dynamics • Prey • Nest • Climatology • Territory • Interspecific Competition • Mortality • Pesticides • Eagles • A. verreauxi • Rhodesia
1153	**Clutch Size** • Cain/Abel • Parental Behavior • Behavior • Growth • Eagles • A. verreauxi • A. pomarina • Rhodesia
1163	**Clutch Size** • Nest • Behavior • Prey • A. heliaca • Spain
1232	**Clutch Size** • Status • Nest • Raptors • A. nipalensis • Ukraine
1279	**Clutch Size** • A. chrysaetos • Scotland
1295	**Clutch Size** • A. chrysaetos • Oregon
1309	**Clutch Size** • Prey • Nest • A. nipalensis • Russia
1310	**Clutch Size** • A. heliaca • A. chrysaetos • West Germany
1391	**Clutch Size** • (Title Only) • A. pomarina • West Germany
1612	**Clutch Size** • Nest • Etymology • A. chrysaetos • Israel
1680	**Clutch Size** • Nest • A. chrysaetos • Washington
1706	**Clutch Size** • Egg • A. chrysaetos • Scotland
1716	**Clutch Size** • Nest • Plumage • Egg • A. pomarina • West Germany
1806	**Clutch Size** • Nest • Egg • Measure • Raptors • A. wahlbergi • Rhodesia
1892	**Clutch Size** • Nest Survey • Egg • Measure • A. wahlbergi • Rhodesia
1908	**Clutch Size** • Nest • Breeding Biology • Behavior • Prey • A. nipalensis • Bulgaria
1914	**Clutch Size** • Distribution • Habitat • Nest • Status • Descriptive Summary • Raptors • A. chrysaetos • New Mexico
1986	**Clutch Size** • Nest • Prey • Captivity • Nest Survey • Collect • A. chrysaetos • Canada
2123	**Clutch Size** • A. heliaca • Spain
2151	**Clutch Size** • Fostering • Nestling • A. heliaca • Spain
2156	**Clutch Size** • Cain/Abel • Management • Translocation • Eagles • Tech. • A. pomarina • A. verreauxi • A. rapax • A. chrysaetos • A. heliaca
2158	**Clutch Size** • Nest • A. heliaca • Spain
2160	**Clutch Size** • A. heliaca • Spain
2184	**Clutch Size** • Aerial Survey • Reproductive Success • Management • Survey • Distribution • Habitat • Nest Survey • Prey • Raptors • A. chrysaetos • Arizona
2219	**Clutch Size** • Raptors • A. verreauxi • A. wahlbergi • Africa
2368	**Clutch Size** • Nest Survey • Density • Habitat • Raptors • A. chrysaetos • Colorado
2381	**Clutch Size** • A. audax • Australia
2383	**Clutch Size** • Nest • A. chrysaetos • Great Britain
2403	**Clutch Size** • A. wahlbergi • Malawi
2404	**Clutch Size** • Status • Nest • Nest Chronology • Raptors • A. wahlbergi • A. rapax • Malawi
2483	**Clutch Size** • A. heliaca • Spain
2507	**Clutch Size** • Nest • Egg • Photo Account • Popular Monograph • A. nipalensis • United Soviet Socialists Republic
2558	**Clutch Size** • Nest Survey • Reproductive Success • Habitat • Protection • Eagles • A. chrysaetos • Estonia
2571	**Clutch Size** • Raptors • A. chrysaetos • California
2970	**Clutch Size** • Nest Survey • Egg • Measure • A. wahlbergi • Rhodesia
3019	**Clutch Size** • Growth • Photo Account • Raptors • A. chrysaetos • California
3020	**Clutch Size** • Nest • Growth • Behavior • Banding • Photo Account • A. chrysaetos • California
3065	**Clutch Size** • Nest • Vocalization • A. pomarina • Austria
3079	**Clutch Size** • Prey • Nest Survey • Reproductive Success • Display • Hunting Behavior • Density • Migration • A. wahlbergi • South Africa
3270	**Clutch Size** • Reproductive Success • A. heliaca • Spain
3295	**Clutch Size** • Captivity • A. chrysaetos • Great Britain
3439	**Clutch Size** • Nest • Prey • Nest Chronology • A. audax • Australia
28	**Collect** • Prey • A. chrysaetos • France
29	**Collect** • Prey • A. chrysaetos • France
36	**Collect** • Status • Raptors • A. chrysaetos • Massachusetts
37	**Collect** • Status • Raptors • A. chrysaetos • Siberia
44	**Collect** • Animal Damage • Raptors • A. chrysaetos • Iowa
57	**Collect** • Prey • Behavior • A. heliaca • India
58	**Collect** • Nest • Clutch Size • Prey • Habitat • A. pomarina • India
79	**Collect** • Nest • Egg • A. pomarina • West Germany
81	**Collect** • A. clanga • A. chrysaetos • West Germany
87	**Collect** • A. chrysaetos • California
98	**Collect** • A. chrysaetos • Switzerland
113	**Collect** • Wintering Population • A. pomarina • Southwest Africa
222	**Collect** • Prey • Raptors • A. rapax • South Africa
224	**Collect** • Prey • Museum Specimen • Raptors • A. wahlbergi • A. verreauxi • South Africa
225	**Collect** • Prey • Raptors • A. rapax • South Africa
247	**Collect** • Raptors • A. pomarina • India
255	**Collect** • Raptors • A. chrysaetos • A. nipalensis • Turkmenia
256	**Collect** • Parasite • Raptors • A. rapax • Tanzania
257	**Collect** • Raptors • A. chrysaetos • Tibet
260	**Collect** • Museum Specimen • Raptors • A. rapax • Somalia
266	**Collect** • Rarity • A. clanga • West Germany
307	**Collect** • Rarity • Raptors • A. clanga • Cyprus
341	**Collect** • Nest • Raptors • A. verreauxi • A. wahlbergi • Malawi
343	**Collect** • Raptors • A. rapax • Malawi
344	**Collect** • Raptors • A. verreauxi • A. rapax • Ethiopia
379	**Collect** • Status • Raptors • A. chrysaetos • A. clanga • Italy
431	**Collect** • Prey • Raptors • A. chrysaetos • Italy
448	**Collect** • Rarity • A. clanga • Czechoslovakia
450	**Collect** • Status • A. chrysaetos • France
452	**Collect** • A. chrysaetos • Ireland
463	**Collect** • Raptors • A. clanga • Italy

Master No.	Keywords
472	**Collect** • A. chrysaetos • West Germany
477	**Collect** • Status • Habitat • Raptors • A. chrysaetos • A. clanga • Italy
484	**Collect** • Raptors • A. wahlbergi • Tanzania
575	**Collect** • Prey • Raptors • A. heliaca • Cyprus
585	**Collect** • Sub-Adult • Nest • A. heliaca • Spain
593	**Collect** • Status • Raptors • A. rapax • Sudan
595	**Collect** • Status • Behavior • Raptors • A. rapax • A. clanga • India
692	**Collect** • Nest • Raptors • A. wahlbergi • A. rapax • Rhodesia
773	**Collect** • Prey • Nest • Raptors • A. heliaca • A. pomarina • India
780	**Collect** • Prey • Plumage • Raptors • A. heliaca • A. clanga • Hungary
793	**Collect** • Raptors • A. chrysaetos • Italy
810	**Collect** • A. chrysaetos • Switzerland
815	**Collect** • Status • Egg Collection • Raptors • A. clanga • A. rapax • A. heliaca • India
845	**Collect** • Rarity • A. chrysaetos • A. clanga • Thailand
865	**Collect** • Raptors • A. chrysaetos • Siberia
877	**Collect** • A. clanga • France
881	**Collect** • Raptors • A. chrysaetos • A. pomarina • A. clanga • A. heliaca • West Germany
908	**Collect** • Flight Id • Photo Account • A. chrysaetos • A. nipalensis • A. heliaca • A. rapax • India
927	**Collect** • A. chrysaetos • Belgium
928	**Collect** • A. chrysaetos • Belgium
929	**Collect** • A. chrysaetos • Belgium
936	**Collect** • Prey • A. chrysaetos • Kansas
970	**Collect** • Status • Prey • A. rapax • India
1001	**Collect** • Status • Egg • Raptors • A. audax • Australia
1012	**Collect** • A. clanga • Italy
1053	**Collect** • Rarity • A. heliaca • Czechoslovakia
1058	**Collect** • Taxonomy • Museum Specimen • Behavior • Raptors • A. pomarina • A. clanga • A. nipalensis • Nepal
1080	**Collect** • Nest • Etymology • Literature Review • Distribution • A. chrysaetos • A. rapax • Tunisia
1081	**Collect** • Taxonomy • Literature Review • A. rapax • Africa
1082	**Collect** • Status • A. chrysaetos • West Germany
1087	**Collect** • Plumage • Museum Specimen • Raptors • A. rapax • Ethiopia • Kenya
1182	**Collect** • Museum Specimen • Prey • Raptors • A. wahlbergi • Cambodia
1198	**Collect** • Nest • Measure • A. clanga • A. chrysaetos • Italy
1210	**Collect** • Measure • A. clanga • Italy
1211	**Collect** • Rarity • Measure • Plumage • Photo Account • A. pomarina • Italy
1229	**Collect** • Rarity • A. clanga • Great Britain
1285	**Collect** • Parasite • Fat • Molt • Museum Specimen • A. chrysaetos • North Carolina
1286	**Collect** • Raptors • A. rapax • Uganda
1318	**Collect** • Raptors • A. clanga • Syria
1337	**Collect** • Synonomy • A. wahlbergi • South Africa
1341	**Collect** • A. pomarina • Austria
1349	**Collect** • Raptors • A. chrysaetos • A. clanga • A. heliaca • West Germany
1392	**Collect** • Distribution • A. gurneyi • Molluca Islands
1393	**Collect** • Plumage • Raptors • A. rapax • A. wahlbergi • Nigeria
1430	**Collect** • Rarity • A. clanga • A. pomarina • Switzerland
1450	**Collect** • Raptors • A. rapax • Rhodesia
1475	**Collect** • Rarity • A. heliaca • Switzerland
1476	**Collect** • Rarity • A. pomarina • Switzerland
1493	**Collect** • A. clanga • West Germany
1511	**Collect** • Raptors • A. chrysaetos • Yugoslavia
1597	**Collect** • Prey • Behavior • Habitat • Raptors • A. nipalensis • A. clanga • A. pomarina • India
1601	**Collect** • A. chrysaetos • Switzerland
1636	**Collect** • Prey • A. chrysaetos • West Germany
1640	**Collect** • Raptors • A. wahlbergi • Uganda
1644	**Collect** • A. chrysaetos • West Germany
1784	**Collect** • Taxonomy • A. chrysaetos • Tunisia
1785	**Collect** • Descriptive Summary • Etymology • Status • Raptors • A. chrysaetos • A. heliaca • A. pomarina • A. clanga • Egypt
1840	**Collect** • Status • Raptors • A. chrysaetos • A. clanga • France
1841	**Collect** • Rarity • Measure • Plumage • A. heliaca • France
1843	**Collect** • Captivity • Raptors • A. heliaca • A. pomarina • China
1844	**Collect** • Egg Collection • Nest • Raptors • A. chrysaetos • China
1845	**Collect** • Trade • Raptors • A. chrysaetos • China
1846	**Collect** • Raptors • A. chrysaetos • China
1873	**Collect** • Prey • Raptors • A. clanga • France
1918	**Collect** • Plumage • A. heliaca • Spain
1964	**Collect** • Habitat • A. verreauxi • Uganda
1976	**Collect** • Anatomy • Abnormal Behavior • Photo Account • A. chrysaetos • Switzerland
1981	**Collect** • Nest • Raptors • A. chrysaetos • Morocco
1986	**Collect** • Nest • Clutch Size • Prey • Captivity • Nest Survey • A. chrysaetos • Canada
2005	**Collect** • Raptors • A. wahlbergi • A. pomarina • A. rapax • Kenya
2039	**Collect** • Nest • Status • Prey • Raptors • A. wahlbergi • Rhodesia
2046	**Collect** • A. clanga • Italy
2067	**Collect** • Status • Raptors • A. chrysaetos • France
2068	**Collect** • Raptors • A. chrysaetos • France
2071	**Collect** • Raptors • A. chrysaetos • France
2076	**Collect** • Rarity • A. heliaca • Austria
2084	**Collect** • Prey • Trade • A. chrysaetos • California
2110	**Collect** • Status • Hunting Behavior • Raptors • A. chrysaetos • A. nipalensis • Pakistan
2112	**Collect** • Status • Raptors • A. nipalensis • A. heliaca • Iraq
2113	**Collect** • Raptors • A. nipalensis • Aden
2115	**Collect** • Raptors • A. chrysaetos • A. rapax • Morocco
2189	**Collect** • Nest • Raptors • A. chrysaetos • A. heliaca • A. pomarina • A. clanga • Australia
2203	**Collect** • Rarity • Photo Account • A. nipalensis • Italy
2206	**Collect** • Prey • Measure • A. clanga • Italy
2227	**Collect** • Rarity • A. clanga • France
2228	**Collect** • A. clanga • France
2244	**Collect** • A. rapax • Congo
2256	**Collect** • Prey • Raptors • A. chrysaetos • Illinois
2287	**Collect** • Raptors • A. wahlbergi • Rhodesia
2302	**Collect** • A. rapax • Kenya
2303	**Collect** • A. rapax • Ethiopia

Master No.	Keywords
2329	**Collect** • Raptors • A. chrysaetos • Italy
2342	**Collect** • Etymology • Status • Habitat • Interspecific Behavior • Nest • Raptors • A. verreauxi • A. nipalensis • A. rapax • Africa
2351	**Collect** • Raptors • A. nipalensis • A. clanga • India
2352	**Collect** • Prey • Plumage • A. nipalensis • A. heliaca • India
2355	**Collect** • Plumage • Taxonomy • Raptors • A. rapax • Kenya
2361	**Collect** • Raptors • A. chrysaetos • A. heliaca • A. pomarina • Austria
2385	**Collect** • Prey • Eagles • A. chrysaetos • Italy
2387	**Collect** • Eagles • A. chrysaetos • Italy
2388	**Collect** • Status • Eagles • A. chrysaetos • A. clanga • A. heliaca • Italy
2396	**Collect** • Prey • Raptors • A. clanga • Kenya
2463	**Collect** • Status • Vocalization • Behavior • Raptors • A. rapax • Somalia
2497	**Collect** • Raptors • A. chrysaetos • France
2498	**Collect** • A. chrysaetos • Pennsylvania
2509	**Collect** • A. chrysaetos • A. clanga • Finland
2516	**Collect** • Taxonomy • Raptors • A. pomarina • A. clanga • A. chrysaetos • A. nipalensis • A. heliaca • Poland
2523	**Collect** • Raptors • A. nipalensis • Zaire
2544	**Collect** • Habitat • Behavior • Prey • A. clanga • West Germany
2545	**Collect** • A. clanga • Caucasus
2551	**Collect** • Raptors • A. clanga • Nepal
2570	**Collect** • A. chrysaetos • Connecticut
2575	**Collect** • Status • A. chrysaetos • France
2583	**Collect** • Prey • Habitat • Status • Raptors • A. heliaca • A. nipalensis • A. clanga • A. rapax • A. pomarina • India
2592	**Collect** • Museum Specimen • A. pomarina • A. clanga • Italy
2622	**Collect** • Rarity • Plumage • A. clanga • West Germany
2636	**Collect** • Egg Collection • Raptors • A. wahlbergi • A. rapax • South Africa
2641	**Collect** • A. clanga • West Germany
2651	**Collect** • Rarity • A. clanga • Great Britain
2662	**Collect** • Raptors • A. nipalensis • China
2663	**Collect** • Raptors • A. chrysaetos • A. nipalensis • China
2667	**Collect** • Raptors • A. gurneyi • New Guinea
2674	**Collect** • A. chrysaetos • Italy
2675	**Collect** • Prey • Wintering Population • Etymology • Status • Raptors • A. chrysaetos • A. clanga • Yugoslavia
2709	**Collect** • Nest • Egg • Etymology • Raptors • A. chrysaetos • A. clanga • A. rapax • Algeria
2733	**Collect** • Raptors • A. chrysaetos • A. nipalensis • China
2734	**Collect** • Raptors • A. chrysaetos • A. nipalensis • China
2741	**Collect** • Prey • A. chrysaetos • Italy
2742	**Collect** • A. chrysaetos • Italy
2751	**Collect** • Nest • Raptors • A. pomarina • West Germany
2756	**Collect** • Status • Shooting • Trapping • Photo Account • A. chrysaetos • Wisconsin
2765	**Collect** • Nest • Egg • A. rapax • Southwest Africa
2768	**Collect** • A. chrysaetos • West Germany
2772	**Collect** • Behavior • Raptors • A. wahlbergi • South Africa
2773	**Collect** • Plumage • Raptors • A. wahlbergi • A. heliaca • A. rapax • A. nipalensis • Sudan
2774	**Collect** • Raptors • A. wahlbergi • Tanzania
2776	**Collect** • Falconry • Etymology • Raptors • A. chrysaetos • China
2778	**Collect** • Museum Specimen • Raptors • A. chrysaetos • India
2808	**Collect** • Raptors • A. clanga • Saudi Arabia
2809	**Collect** • Status • Raptors • A. rapax • Uganda
2816	**Collect** • Nest • Prey • Status • Raptors • A. wahlbergi • South Africa
2894	**Collect** • Plumage • Raptors • A. wahlbergi • Rhodesia
2923	**Collect** • Raptors • A. heliaca • A. clanga • Ethiopia
3010	**Collect** • A. chrysaetos • China
3051	**Collect** • A. chrysaetos • West Germany
3052	**Collect** • Status • Raptors • A. heliaca • Afghanistan
3053	**Collect** • Status • Raptors • A. clanga • A. rapax • India
3061	**Collect** • Nest • Vocalization • Eagles • A. heliaca • Hungary
3062	**Collect** • Nest • Photo Account • A. pomarina • Hungary
3071	**Collect** • Raptors • A. chrysaetos • A. heliaca • A. clanga • Siberia
3072	**Collect** • Taxonomy • Raptors • A. clanga • Siberia
3073	**Collect** • Raptors • A. chrysaetos • A. nipalensis • Siberia
3084	**Collect** • Nest • Behavior • Raptors • A. wahlbergi • South Africa
3088	**Collect** • Museum Specimen • Raptors • A. clanga • Egypt
3090	**Collect** • Raptors • A. clanga • Egypt
3113	**Collect** • Raptors • A. pomarina • West Germany
3142	**Collect** • A. heliaca • Burma
3157	**Collect** • A. clanga • West Germany
3172	**Collect** • Status • Raptors • A. wahlbergi • Rhodesia
3214	**Collect** • Nest • Habitat • Behavior • Growth • Captivity • A. chrysaetos • India
3250	**Collect** • Taxonomy • Raptors • A. clanga • Hungary
3258	**Collect** • Prey • Rarity • A. pomarina • France
3279	**Collect** • Raptors • A. chrysaetos • France
3285	**Collect** • Raptors • A. wahlbergi • A. nipalensis • Congo
3340	**Collect** • Fat • Nest • Raptors • A. nipalensis • A. rapax • India
3346	**Collect** • Raptors • A. chrysaetos • Tunisia
3357	**Collect** • Prey • Status • Piracy • Raptors • A. heliaca • A. nipalensis • A. clanga • A. rapax • India
3392	**Collect** • Rarity • A. clanga • Great Britain
3410	**Collect** • Status • Raptors • A. chrysaetos • Michigan
3432	**Collect** • Raptors • A. clanga • West Germany
3433	**Collect** • Raptors • A. chrysaetos • A. pomarina • West Germany
3434	**Collect** • A. clanga • A. chrysaetos • West Germany
3447	**Collect** • Etymology • Raptors • A. rapax • A. chrysaetos • Tunisia
3448	**Collect** • Plumage • Raptors • A. chrysaetos • A. rapax • Ethiopia
3458	**Collect** • A. chrysaetos • West Germany
102	**Conservation** • Egg Collection • A. chrysaetos • California

Master No.	Keywords
136	**Conservation** • Animal Damage • Popular Monograph • A. chrysaetos • North America
143	**Conservation** • Falconry • Raptors • A. rapax • A. chrysaetos • New York
148	**Conservation** • Protection • A. chrysaetos • North America
151	**Conservation** • Poisoning • Mortality • Wintering Population • Attitude • Punishment • A. chrysaetos • Wyoming
153	**Conservation** • Shooting • Protection • Aircraft • Reward • Attitude • Economics • A. chrysaetos • United States
267	**Conservation** • Attitude • Shooting • Raptors • A. chrysaetos • Scotland
290	**Conservation** • Shooting • Status • Nest • A. chrysaetos • France
421	**Conservation** • Protection • A. audax • Australia
444	**Conservation** • Attitude • Animal Damage • Bounty • Raptors • A. verreauxi • A. rapax • South Africa
457	**Conservation** • Shooting • Attitude • Protection • Status • A. chrysaetos • Austria
462	**Conservation** • Habitat • Attitude • Disturbance • Flight Speed • Eagles • Photo Account • Popular Monograph • A. chrysaetos • North America
474	**Conservation** • Migration • Field Id • Refuge • Raptors • A. chrysaetos • Pennsylvania
479	**Conservation** • Migration • Shooting • Protection • Refuge • Raptors • Hawk Mt. • A. chrysaetos • Pennsylvania
486	**Conservation** • Children's • Popular Monograph • A. chrysaetos • Norway
498	**Conservation** • Taxonomy • Field Id • Molt • Measure • Museum Specimen • Behavior • Wintering Population • Interspecific Competition • Eagles • Photo Account • A. nipalensis • A. pomarina • Africa
527	**Conservation** • Migration • Shooting • Population Decline • Historical Account • Hawk Mt. • Raptors • Popular Monograph • A. chrysaetos • Pennsylvania
547	**Conservation** • Breeding Biology • Knowledge Status • Behavior • A. chrysaetos • A. heliaca • A. audax • A. rapax • A. nipalensis • A. pomarina • A. clanga • A. gurneyi • A. verreauxi • A. wahlbergi
548	**Conservation** • Status • Poisoning • Habitat Loss • A. pomarina • A. clanga • A. heliaca • East Africa • Ethiopia
551	**Conservation** • Monograph • Taxonomy • Courtship • Territory • A. chrysaetos • A. heliaca • A. clanga • A. rapax • A. verreauxi • A. wahlbergi • A. audax • A. gurneyi • A. nipalensis • A. pomarina
568	**Conservation** • Nest • Prey • Photo Account • Popular Monograph • A. chrysaetos • South Dakota
573	**Conservation** • Shooting • Protection • Eagles • A. chrysaetos • United States
606	**Conservation** • Animal Damage • Economics • Attitude • Prey • Raptors • A. chrysaetos • North America
631	**Conservation** • Story • Trapping • Falconry • Rehabilitation • Trade • Wintering Population • Captivity • Raptors • Popular Monograph • A. chrysaetos • New York
705	**Conservation** • Art • Population Decline • Refuge • A. heliaca • Spain
720	**Conservation** • Status • Mortality • Raptors • A. audax • Australia
761	**Conservation** • Status • Field Id • Raptors • A. chrysaetos • Colorado
765	**Conservation** • Nest • History • Management • Raptors • Popular Monograph • A. chrysaetos • United States
769	**Conservation** • Status • A. heliaca • A. pomarina • Rumania
783	**Conservation** • Protection • Prey • Animal Damage • Hunting Behavior • Prey Behavior • Raptors • A. chrysaetos • Sweden
784	**Conservation** • Animal Damage • Attitude • Prey • Raptors • A. heliaca • A. audax
806	**Conservation** • A. chrysaetos • Europe
807	**Conservation** • A. chrysaetos • Switzerland
808	**Conservation** • Protection • A. chrysaetos • Switzerland
858	**Conservation** • Nest Survey • Status • A. chrysaetos • Austria
903	**Conservation** • Status • Protection • Raptors • A. chrysaetos • A. clanga • Belorussia
942	**Conservation** • Educational Material • Eagles • A. chrysaetos
949	**Conservation** • Animal Damage • Prey • Raptors • A. chrysaetos • Scotland
999	**Conservation** • Refuge • Management • Wintering Population • Raptors • A. chrysaetos • Great Britain
1035	**Conservation** • Management • Raptors • A. chrysaetos • United States
1060	**Conservation** • Field Id • Status • Raptors • A. chrysaetos • Colorado
1090	**Conservation** • Economics • Attitude • Animal Damage • Mortality • Popular Monograph • A. chrysaetos • Wyoming
1127	**Conservation** • Raptors • A. chrysaetos • United Soviet Socialists Republic
1160	**Conservation** • Environmental Impact • (Title Only) • A. heliaca • Spain
1162	**Conservation** • Status • Nest Survey • Protection • Shooting • Habitat • Raptors • A. chrysaetos • A. heliaca • Spain
1164	**Conservation** • Status • Pesticides • Disturbance • Population Estimate • Raptors • A. heliaca • A. chrysaetos • Spain
1311	**Conservation** • Story • Children's • Photo Account • SRBPNA • Popular Monograph • A. chrysaetos • Idaho
1359	**Conservation** • Attitude • A. chrysaetos • Texas
1507	**Conservation** • Protection • Economics • Mortality • Reward • Pesticides • Poisoning • Eagles • A. chrysaetos • Wyoming • Colorado
1589	**Conservation** • Population Decline • Status • Distribution • Raptors • Popular Monograph • A. clanga • A. heliaca • Europe
1608	**Conservation** • Poisoning • A. chrysaetos • Scotland
1635	**Conservation** • Density • Nest • Prey • Intraspecific Behavior • A. chrysaetos • Belorussia

Master No.	Keywords
1653	**Conservation** • Punishment • Disturbance • Nest Robbing • Popular Monograph • A. chrysaetos • Scotland
1709	**Conservation** • Habitat • Nest • Raptors • A. chrysaetos • A. pomarina • A. clanga • Hungary
1729	**Conservation** • Prey • Habitat • Mortality • A. chrysaetos • Japan
1820	**Conservation** • Population Decline • Protection • Raptors • A. chrysaetos • A. clanga • A. pomarina • West Germany
1825	**Conservation** • Status • Raptors • A. chrysaetos • A. nipalensis • A. clanga • A. pomarina • A. heliaca • Turkey
1897	**Conservation** • Mortality • Attitude • Popular Monograph • A. chrysaetos • Wyoming
1902	**Conservation** • Status • Prey • Nest • Behavior • A. chrysaetos • Israel
1903	**Conservation** • Nest • Prey • Habitat • Disturbance • Mortality • Hunting Behavior • A. chrysaetos • Israel
1904	**Conservation** • Poisoning • Status • A. chrysaetos • Israel
1922	**Conservation** • Shooting • Status • Raptors • A. chrysaetos • California
1941	**Conservation** • Banding • Photo Account • Raptors • A. chrysaetos • Nebraska
1971	**Conservation** • Protection • Animal Damage • Eagles • A. chrysaetos • North America
1987	**Conservation** • Popular Monograph • (Title Only) • A. chrysaetos • Great Britain
2013	**Conservation** • Symbolism • Field Id • Popular Monograph • A. chrysaetos • Colorado
2021	**Conservation** • Flight Speed • Popular Monograph • A. chrysaetos • Ireland
2103	**Conservation** • Folklore • Habitat • Shooting • Animal Damage • A. chrysaetos • United States
2109	**Conservation** • Educational Material • Refuge • Raptors • A. chrysaetos • Idaho
2157	**Conservation** • Fostering • Reproductive Success • Cain/Abel • Photo Account • A. heliaca • Spain
2159	**Conservation** • Habitat Loss • Forestry • Habitat Protection • Status • Raptors • A. heliaca • Spain
2237	**Conservation** • Rehabilitation • Shooting • Disease • Photo Account • Raptors • A. chrysaetos • California
2259	**Conservation** • Nest • Photo Account • Popular Monograph • A. chrysaetos • United States
2262	**Conservation** • Management • Animal Damage • Raptors • A. chrysaetos • United States
2278	**Conservation** • Migration • Refuge • Survey • Shooting • Hawk Mt. • A. chrysaetos • Pennsylvania
2286	**Conservation** • Rehabilitation • Photo Account • Raptors • Popular Monograph • A. chrysaetos • New Mexico
2293	**Conservation** • Falconry • Prey • Hunting Behavior • Raptors • Popular Monograph • A. chrysaetos • United States
2294	**Conservation** • Lift • Hunting Behavior • Behavior • A. chrysaetos • North America
2299	**Conservation** • Electrocution • Artificial Nest • Management • Shooting • Eagles • Popular Monograph • A. chrysaetos • Idaho
2307	**Conservation** • Status • Density • Mortality • Pesticides • Raptors • A. chrysaetos • Scotland
2332	**Conservation** • Behavior • Prey • A. audax • Australia
2369	**Conservation** • Prey • Folklore • Lift • Animal Damage • Electrocution • Eagles • Popular Monograph • A. chrysaetos • North America
2374	**Conservation** • Management • Habitat Protection • Refuge • Theory • Tech. • Raptors • A. chrysaetos • United States
2377	**Conservation** • Electrocution • Management • Status • Literature Review • Behavior • Eagles • A. chrysaetos • North America
2399	**Conservation** • Economics • Bounty • Mortality • Eagles • Popular Monograph • A. chrysaetos • North America
2426	**Conservation** • Nest Survey • Raptors • A. heliaca • A. pomarina • Hungary
2553	**Conservation** • Story • Shooting • Rehabilitation • Nest • Popular Monograph • A. chrysaetos • Wyoming
2554	**Conservation** • Nest • Behavior • Interspecific Behavior • Prey • Trapping • Animal Damage • Rehabilitation • Photo Account • Popular Monograph • A. chrysaetos • Wyoming
2611	**Conservation** • Prey • Banding • Nest • Photo Account • Popular Monograph • A. audax • Australia
2642	**Conservation** • Mortality • Photo Account • Raptors • A. chrysaetos • United States
2644	**Conservation** • Refuge • Nest • Measure • SRBPNA • Photo Account • Popular Monograph • A. chrysaetos • Idaho
2645	**Conservation** • Shooting • Poisoning • Egg Collection • Nest Robbing • Protection • Trade • Trapping • Raptors • A. chrysaetos • Great Britain
2656	**Conservation** • Winter Survey • Shooting • Poisoning • Protection • Reward • Photo Account • Eagles • Popular Monograph • A. chrysaetos • Wyoming
2736	**Conservation** • Status • Density • Protection • Survey • Raptors • A. chrysaetos • Sardinia
2758	**Conservation** • Attitude • Shooting • Story • Punishment • Popular Monograph • A. chrysaetos • Texas
2759	**Conservation** • Animal Damage • Shooting • Punishment • Attitude • Reward • Translocation • Popular Monograph • A. chrysaetos • Texas
2787	**Conservation** • Status • Protection • Habitat Loss • Mortality • Raptors • A. chrysaetos • A. clanga • Sweden
2827	**Conservation** • Protection • Raptors • A. chrysaetos • New Mexico
2914	**Conservation** • Management • A. chrysaetos • United States
2921	**Conservation** • Folklore • Eagles • A. chrysaetos • United States
2990	**Conservation** • Mortality • Attitude • Protection • Photo Account • Popular Monograph • A. chrysaetos • North America
3017	**Conservation** • Status • Shooting • Migration • Raptors • A. pomarina • Malawi

Master No.	Keywords
3080	**Conservation** • Cain/Abel • Distribution • Educational Material • Field Id • Habitat • Hunting Behavior • Prey • Artificial Nest • Nest • Incubation • Wintering Population • Photo Account • Eagles • A. chrysaetos • A. wahlbergi • A. verreauxi • A. nipalensis • A. pomarina • South Africa
3103	**Conservation** • Distribution • Status • Field Id • Photo Account • Raptors • A. chrysaetos • France
3155	**Conservation** • Protection • Mortality • A. chrysaetos • North America
3191	**Conservation** • Shooting • Poisoning • Electrocution • Attitude • Eagles • A. chrysaetos • United States
3203	**Conservation** • Descriptive Summary • Raptors • SRBPNA • A. chrysaetos • Idaho
3204	**Conservation** • Refuge • Habitat • Raptors • SRBPNA • A. chrysaetos • Idaho
3283	**Conservation** • Breeding Biology • Prey • Mortality • Distribution • Photo Account • A. chrysaetos • Czechoslovakia
3310	**Conservation** • Shooting • Poisoning • Bounty • Punishment • Story • Eagles • Popular Monograph • A. chrysaetos • Wyoming
3312	**Conservation** • Trapping • Raptors • A. chrysaetos • United States
3339	**Conservation** • Protection • A. chrysaetos • North America
3364	**Conservation** • A. chrysaetos • United States
3365	**Conservation** • Anatomy • Bounty • Protection • Raptors • Illus. • A. chrysaetos • Finland
3372	**Conservation** • Protection • Population Decline • Mortality • Animal Damage • Bounty • A. chrysaetos • Norway
3373	**Conservation** • Status • Bounty • Mortality • Eagles • A. chrysaetos • Norway
3374	**Conservation** • Status • Distribution • Protection • Raptors • A. chrysaetos • Norway
3375	**Conservation** • Protection • Symbolism • Shooting • Economics • Attitude • A. chrysaetos • United States
3459	**Conservation** • Photo Account • Raptors • SRBPNA • Popular Monograph • A. chrysaetos • Idaho
296	**Courtship** • Descriptive Summary • Popular Monograph • A. chrysaetos • Montana
351	**Courtship** • Hunting Behavior • Measure • Life History • Behavior • Distribution • Vocalization • Plumage • Raptors • Photo Account • A. chrysaetos • North America
551	**Courtship** • Monograph • Conservation • Taxonomy • Territory • A. chrysaetos • A. heliaca • A. clanga • A. rapax • A. verreauxi • A. wahlbergi • A. audax • A. gurneyi • A. nipalensis • A. pomarina
630	**Courtship** • Longevity • A. chrysaetos • California
790	**Courtship** • A. audax • Australia
842	**Courtship** • Animal Damage • Hunting Behavior • Prey Beh. • A. audax • Australia
895	**Courtship** • Nest • Territory • Egg • Measure • Intraspecific Behavior • Interspecific Behavior • Hunting Behavior • A. chrysaetos • California
1144	**Courtship** • Nest Survey • Status • Behavior • Fledging • Mortality • Banding • Marking • Cain/Abel • Reproductive Success • Breeding Biology • Interspecific Behavior • Hunting Behavior • Display • Raptors • A. verreauxi • Rhodesia
1176	**Courtship** • Breeding Biology • Nest • Food Habits • A. pomarina • West Germany
1254	**Courtship** • Flight • Intraspecific Behavior • Interspecific Behavior • A. chrysaetos • Great Britain
1268	**Courtship** • Intraspecific Behavior • Flight • Courtship • A. chrysaetos • Scotland
1268	**Courtship** • Intraspecific Behavior • Flight • Courtship • A. chrysaetos • Scotland
1275	**Courtship** • A. chrysaetos • Scotland
1944	**Courtship** • Marking • Abnormal Behavior • Nest Failure • Telemetry • Raptors • A. chrysaetos • Wyoming • Idaho
1989	**Courtship** • Popular Monograph • (Title Only) • A. chrysaetos • Great Britain
2152	**Courtship** • Habitat • Growth • Parental Behavior • Prey • Breeding Biology • Interspecific Behavior • Display • Photo Account • Raptors • A. heliaca • Spain
2370	**Courtship** • Status • Nest • Nest Survey • Monograph • Habitat • Mortality • Breeding Biology • Nest Chronology • Density • Prey • Management • A. chrysaetos • Colorado
2994	**Courtship** • Nest Chronology • Incubation • Nest Adornment • Nest Construction • A. chrysaetos • Texas
3175	**Courtship** • Eagles • A. rapax • Rhodesia
8	**Density** • Prey • Nest Failure • Reproductive Success • A. nipalensis • Russia
233	**Density** • Attitude • Reproductive Success • Wintering Population • Food Requirements • Habitat • Management • Aerial Survey • Nest Survey • Raptors • A. chrysaetos • Montana
324	**Density** • Reproductive Success • Food Requirements • Territory • Abandonment • Mortality • Cain/Abel • Dispersal • Breeding Biology • Nest Survey • A. chrysaetos • Idaho
325	**Density** • Monograph • Breeding Biology • Mortality • Cain/Abel • Prey • Abnormal Behavior • Egg Measure • Marking • A. chrysaetos • Idaho
534	**Density** • Habitat • Nest Survey • Breeding Biology • Interspecific Competition • Prey • Reproductive Success • Etymology • Plumage • Vocalization • Nest Adornment • Raptors • A. verreauxi • A. wahlbergi • Kenya
757	**Density** • Nest Survey • Reproductive Success • Aerial Survey • Raptors • A. chrysaetos • Colorado
758	**Density** • Nest Survey • Reproductive Success • Winter Survey • Wintering Population • Status • Aerial Survey • A. chrysaetos • Colorado
1025	**Density** • Hunting Behavior • Nest • Prey • Habitat • Photo Account • Popular Monograph • A. heliaca • Spain
1117	**Density** • Predator/Prey • Prey • A. clanga • Russia
1119	**Density** • Status • Raptors • A. clanga • Russia
1123	**Density** • Protection • A. clanga • A. chrysaetos • United Soviet Socialists Republic
1124	**Density** • Habitat • Raptors • A. clanga • Russia

Master No.	Keywords
1125	**Density** • Protection • Popular Monograph • A. chrysaetos • United Soviet Socialists Republic
1141	**Density** • Nest Survey • Habitat • Disturbance • Reproductive Success • Clutch Size • Raptors • A. verreauxi • Rhodesia
1150	**Density** • Territory • Dispersal • Nest • Survey • Intraspecific Behavior • Marking • Reproductive Success • Display • A. verreauxi • Rhodesia
1151	**Density** • Reproductive Success • Population Dynamics • Prey • Nest • Climatology • Territory • Interspecific Competition • Mortality • Clutch Size • Pesticides • Eagles • A. verreauxi • Rhodesia
1154	**Density** • Nest • Prey • Status • Habitat Protection • Habitat • Reproductive Success • A. verreauxi • Rhodesia
1156	**Density** • Breeding Biology • Habitat • Prey • Nest Survey • A. verreauxi • Rhodesia
1635	**Density** • Nest • Conservation • Prey • Intraspecific Behavior • A. chrysaetos • Belorussia
1638	**Density** • Raptors • A. clanga • A. pomarina • Poland
1724	**Density** • Status • Raptors • A. pomarina • A. clanga • A. chrysaetos • Latvia
1752	**Density** • Distribution • Roadside Survey • Raptors • A. chrysaetos • California
1776	**Density** • Nest Survey • Refuge • Winter Survey • Electrocution • Shooting • Mortality • Prey • Predator/Prey • Raptors • SRBPNA • A. chrysaetos • Idaho
1815	**Density** • Nest • Raptors • A. chrysaetos • Kirgizia
1850	**Density** • Raptors • A. chrysaetos • Siberia
1851	**Density** • Prey • A. chrysaetos • Siberia
1852	**Density** • Distribution • Raptors • A. chrysaetos • Siberia
1868	**Density** • Nest Survey • Raptors • A. chrysaetos • Oregon
1915	**Density** • Nest • Raptors • A. clanga • A. pomarina • Russia
1921	**Density** • Breeding Biology • Predator/Prey • A. nipalensis • Russia
1934	**Density** • Breeding Biology • Nest • Territory • A. heliaca • Kazakhstan
1947	**Density** • Prey • Breeding Biology • (Title Only) • A. chrysaetos • Scotland
2062	**Density** • Nest • Field Id • Photo Account • Popular Monograph • A. pomarina • East Germany
2125	**Density** • Poisoning • Pesticides • Status • Captivity • Food Requirements • Habitat • Protection • Predator/Prey • Raptors • Photography Account • PCB • A. clanga • A. heliaca • A. nipalensis • A. verreauxi • A. pomarina • Israel
2145	**Density** • Breeding Biology • Habitat • Cain/Abel • Behavior • Prey • Nest Survey • Photo Account • A. pomarina • Czechoslovakia
2188	**Density** • Prey • Breeding Biology • A. nipalensis • Caucasus
2191	**Density** • Raptors • A. heliaca • A. nipalensis • Kazakhstan
2212	**Density** • Nest • Reproductive Success • Management • Shooting • Prey • Raptors • A. chrysaetos • Washington
2255	**Density** • Electrocution • Predator/Prey • Nest • Prey • A. nipalensis • Russia
2269	**Density** • Nest Survey • Directional Exposure • Reproductive Success • Raptors • A. chrysaetos • Utah
2307	**Density** • Status • Conservation • Mortality • Pesticides • Raptors • A. chrysaetos • Scotland
2345	**Density** • Prey • Shooting • Nest • Reproductive Success • Eagles • A. chrysaetos • Italy
2368	**Density** • Nest Survey • Habitat • Clutch Size • Raptors • A. chrysaetos • Colorado
2370	**Density** • Status • Nest • Nest Survey • Monograph • Habitat • Mortality • Breeding Biology • Nest Chronology • Courtship • Prey • Management • A. chrysaetos • Colorado
2372	**Density** • Nest Survey • Status • Habitat • Raptors • A. chrysaetos • Colorado
2400	**Density** • Animal Damage • Predator/Prey • Aerial Survey • Habitat • Status • Population Dynamics • Food Habits • Nest • Interspecific Competition • Literature Review • A. chrysaetos • Texas • New Mexico
2464	**Density** • Management • Habitat • Directional Exposure • Disturbance • Mortality • Nest Survey • Nest Chronology • Prey • Reproductive Success • Prog. Rep. • A. chrysaetos • Wyoming
2491	**Density** • Distribution • Habitat • Status • Survey • Raptors • A. chrysaetos • New Mexico
2492	**Density** • Nest Survey • Raptors • A. chrysaetos • Idaho
2594	**Density** • Monograph • Pesticides • Prey • Nest Survey • Mortality • Reproductive Success • A. chrysaetos • Montana
2736	**Density** • Status • Conservation • Protection • Survey • Raptors • A. chrysaetos • Sardinia
2788	**Density** • Nest Survey • Habitat • Aerial Survey • Photo Account • A. chrysaetos • Nevada
2852	**Density** • Roadside Survey • Prey • Monograph • Habitat • Interspecific Competition • Taxonomy • Vocalization • Hunting Behavior • Nest Chronology • Photo Account • Raptors • A. wahlbergi • A. rapax • A. nipalensis • A. heliaca • Kenya
2854	**Density** • Population Dynamics • Nest • Prey • Breeding Biology • Distribution • Habitat • Directional Exposure • Reproductive Success • Mortality • Abandonment • Disturbance • Interspecific Behavior • Territory • A. chrysaetos • Utah
2962	**Density** • Prey • Piracy • Raptors • A. chrysaetos • Washington
3079	**Density** • Prey • Nest Survey • Clutch Size • Reproductive Success • Display • Hunting Behavior • Migration • A. wahlbergi • South Africa
3110	**Density** • Nest Survey • Distribution • Status • Habitat Loss • Density • Reproductive Success • Raptors • Photo Account • A. chrysaetos • California
3110	**Density** • Nest Survey • Distribution • Status • Density • Habitat Loss • Reproductive Success • Raptors • Photo Account • A. chrysaetos • California
3118	**Density** • Distribution • Habitat • Prey • Descriptive Summary • Interspecific Competition • Eagles • A. chrysaetos • France

Master No.	Keywords
3120	**Density** • Raptors • A. wahlbergi • Ivory Coast
3121	**Density** • Roadside Survey • Raptors • A. wahlbergi • A. rapax • Ivory Coast
3122	**Density** • Habitat • Status • Raptors • A. rapax • A. wahlbergi • A. nipalensis • A. pomarina • Chad
3133	**Density** • Wintering Population • Habitat • Prey • Raptors • A. wahlbergi • A. rapax • A. nipalensis • A. pomarina • Cambodia
3171	**Density** • Status • A. chrysaetos • Austria
3190	**Density** • Habitat • Raptors • A. pomarina • A. clanga • Czechoslovakia
3198	**Density** • Predator/Prey • A. nipalensis • Kazakhstan
3207	**Density** • Refuge • Habitat • Nest Survey • Prey • Predator/Prey • Disturbance • Telemetry • Territory • Reproductive Success • Raptors • SRBPNA • Photo Account • A. chrysaetos • Idaho
3226	**Density** • Reproductive Success • Sub-Adult • Plumage • Museum Specimen • Population Decline • Illus. • A. heliaca • Spain
3259	**Density** • Nest Survey • Habitat • Territory • Intraspecific Behavior • Nest Chronology • Prey • Directional Exposure • Raptors • A. verreauxi • Rhodesia
3273	**Density** • Status • A. nipalensis • Caucasus
3281	**Density** • Raptors • A. pomarina • Belorussia
3304	**Density** • Nest Survey • Directional Exposure • Reproductive Success • Territory • A. chrysaetos • Scotland
23	**Descriptive Summary** • Field Id • Status • Raptors • A. rapax • A. clanga • A. nipalensis • A. heliaca • India
24	**Descriptive Summary** • Prey • Field Id • Raptors • A. chrysaetos • A. heliaca • India
25	**Descriptive Summary** • Field Id • Distribution • Habitat • Raptors • A. rapax • India
26	**Descriptive Summary** • Id Key • Etymology • Status • Distribution • Field Id • Raptors • Illus. • A. chrysaetos • A. heliaca • A. rapax • A. nipalensis • A. clanga • A. pomarina • India
46	**Descriptive Summary** • Field Id • A. chrysaetos • A. heliaca • A. pomarina • A. nipalensis • A. clanga • Iraq
56	**Descriptive Summary** • Etymology • Clutch Size • Plumage • Habitat • Museum Specimen • Measure • A. rapax • A. pomarina • India
65	**Descriptive Summary** • Museum Specimen • Wintering Population • Migration • Status • A. chrysaetos • Iowa
133	**Descriptive Summary** • Raptors • A. chrysaetos • A. clanga • Poland
144	**Descriptive Summary** • Raptors • A. audax
184	**Descriptive Summary** • Educational Material • A. chrysaetos • Great Britain
193	**Descriptive Summary** • Id Key • Etymology • Habitat • Egg • Raptors • Illus. • A. nipalensis • A. rapax • A. verreauxi • Somalia
210	**Descriptive Summary** • A. chrysaetos • North America
215	**Descriptive Summary** • Etymology • A. chrysaetos • Labrador • Newfoundland
218	**Descriptive Summary** • Rarity • Prey • Distribution • Etymology • Raptors • A. chrysaetos • A. heliaca • Japan
236	**Descriptive Summary** • Raptors • A. chrysaetos • Iowa
242	**Descriptive Summary** • Id Key • Etymology • Measure • Plumage • Raptors • A. chrysaetos • A. heliaca • A. nipalensis • A. rapax • A. clanga • A. pomarina • India
251	**Descriptive Summary** • Nest • Egg • Measure • Plumage • Taxonomy • Photo Account • Raptors • A. chrysaetos • A. rapax • Algeria
261	**Descriptive Summary** • Id Key • Distribution • Prey • Etymology • A. rapax • A. wahlbergi • Gambia • Sierra Leone • Nigeria
262	**Descriptive Summary** • Etymology • Status • Nest • Behavior • Plumage • A. heliaca • A. nipalensis • A. clanga • A. chrysaetos • Cyprus
263	**Descriptive Summary** • Field Id • Status • Raptors • A. heliaca • A. clanga • A. pomarina • Cyprus
296	**Descriptive Summary** • Courtship • Popular Monograph • A. chrysaetos • Montana
322	**Descriptive Summary** • Field Id • Raptors • A. chrysaetos • British Columbia
404	**Descriptive Summary** • Symbolism • Etymology • Raptors • A. chrysaetos • North America
408	**Descriptive Summary** • Plumage • Measure • Taxonomy • A. chrysaetos • A. heliaca • A. clanga • A. pomarina
476	**Descriptive Summary** • Prey • Nest • Raptors • A. chrysaetos • Maine
532	**Descriptive Summary** • Raptors • A. verreauxi
539	**Descriptive Summary** • Raptors • A. verreauxi
541	**Descriptive Summary** • Photo Account • A. chrysaetos • A. heliaca • A. clanga • A. rapax • A. verreauxi • A. pomarina • A. nipalensis • A. wahlbergi • Africa
545	**Descriptive Summary** • A. chrysaetos • A. heliaca • A. pomarina • A. clanga • A. nipalensis • A. wahlbergi • A. verreauxi • A. rapax • A. audax • A. gurneyi
591	**Descriptive Summary** • Photo Account • Popular Monograph • A. clanga • West Germany
592	**Descriptive Summary** • Popular Monograph • A. wahlbergi • East Germany
678	**Descriptive Summary** • Etymology • Status • Raptors • A. clanga • A. heliaca • A. chrysaetos • Japan
744	**Descriptive Summary** • Id Key • Plumage • Raptors • A. chrysaetos • Wisconsin • Illinois
748	**Descriptive Summary** • Etymology • Raptors • A. chrysaetos • North America
770	**Descriptive Summary** • Field Id • Monograph • Distribution • Status • Migration • Prey • Plumage • Measure • Etymology • Vocalization • A. pomarina • A. clanga • A. nipalensis • A. heliaca • A. chrysaetos • A. verreauxi • Africa • Europe
864	**Descriptive Summary** • Distribution • Taxonomy • Raptors • A. chrysaetos • A. heliaca • A. nipalensis • A. pomarina • A. clanga • Europe • Asia
883	**Descriptive Summary** • Etymology • Behavior • Prey • Display • A. heliaca • A. rapax • A. pomarina • A. clanga • India

Master No.	Keywords
900	**Descriptive Summary** • Prey • Hunting Behavior • Nest • Interspecific Behavior • Mortality • Vocalization • Raptors • Mt. Mckinley • A. chrysaetos • Alaska
901	**Descriptive Summary** • (Title Only) • A. pomarina • A. clanga
907	**Descriptive Summary** • Field Id • Plumage • Prey • Measure • Raptors • A. chrysaetos • A. heliaca • A. nipalensis • A. rapax • A. clanga • India
941	**Descriptive Summary** • Status • Distribution • Management • Literature Review • Prey • A. chrysaetos • Washington
987	**Descriptive Summary** • Field Id • Distribution • Nest • Status • Taxonomy • Egg • Illus. • A. chrysaetos • A. heliaca • A. nipalensis • A. rapax • A. clanga • A. pomarina • A. verreauxi • Africa
988	**Descriptive Summary** • Distribution • Nest • Raptors • A. nipalensis • A. clanga • A. heliaca • A. chrysaetos • China
993	**Descriptive Summary** • Field Id • Life History • Photo Account • Popular Monograph • A. chrysaetos • Scotland
998	**Descriptive Summary** • Photo Account • Raptors • A. chrysaetos • A. heliaca • A. nipalensis • A. verreauxi • A. gurneyi • A. wahlbergi • A. rapax
1002	**Descriptive Summary** • Raptors • Photo Account • A. chrysaetos • Utah
1008	**Descriptive Summary** • Etymology • Status • Nest Chronology • Prey • Raptors • A. chrysaetos • A. heliaca • A. pomarina • A. clanga • Switzerland
1039	**Descriptive Summary** • Raptors • Photo Account • Popular Monograph • A. chrysaetos • A. clanga
1042	**Descriptive Summary** • Measure • Hunting Behavior • Prey • Animal Damage • Raptors • A. chrysaetos • North America
1051	**Descriptive Summary** • Raptors • A. chrysaetos • Great Britain
1064	**Descriptive Summary** • Hunting Behavior • Status • Folklore • Raptors • A. chrysaetos
1105	**Descriptive Summary** • Etymology • Plumage • Nest • Distribution • A. chrysaetos • Alaska
1212	**Descriptive Summary** • Taxonomy • Etymology • Nest • Prey • A. chrysaetos • Switzerland
1225	**Descriptive Summary** • Raptors • Photo Account • A. chrysaetos • Switzerland
1231	**Descriptive Summary** • Status • Egg • Incubation • Raptors • A. pomarina • A. clanga • A. chrysaetos • West Germany
1244	**Descriptive Summary** • Nest • Photo Account • Popular Monograph • A. chrysaetos • Scotland
1263	**Descriptive Summary** • Nest • Photo Account • Popular Monograph • A. chrysaetos • Scotland
1276	**Descriptive Summary** • Status • Habitat • Raptors • A. chrysaetos • A. heliaca • A. clanga • Korea
1305	**Descriptive Summary** • Monograph • Anthropology • A. chrysaetos • A. clanga • A. rapax • A. audax • A. verreauxi • A. gurneyi • A. wahlbergi • A. pomarina • A. clanga • A. heliaca
1339	**Descriptive Summary** • Economics • Attitude • Raptors • A. verreauxi • A. rapax • South Africa
1343	**Descriptive Summary** • Behavior • (Title Only) • A. clanga • West Germany
1363	**Descriptive Summary** • Field Id • Migration • Status • Prey • Raptors • A. chrysaetos • Wisconsin
1423	**Descriptive Summary** • Raptors • A. rapax • A. nipalensis
1435	**Descriptive Summary** • Raptors • Photo Account • A. chrysaetos • New Jersey
1458	**Descriptive Summary** • Raptors • A. chrysaetos • North America
1506	**Descriptive Summary** • Protection • Nest • Prey • Raptors • Illus. • A. audax • Australia
1509	**Descriptive Summary** • Biology • (Title Only) • A. rapax
1518	**Descriptive Summary** • (Title Only) • A. pomarina • West Germany
1526	**Descriptive Summary** • Distribution • Status • Measure • A. verreauxi • A. nipalensis • A. rapax • A. pomarina • A. wahlbergi • Southwest Africa
1563	**Descriptive Summary** • Distribution • Nest Survey • Raptors • A. chrysaetos • Spain
1586	**Descriptive Summary** • Measure • Distribution • Nest • Photo Account • Illus. • Raptors • A. chrysaetos • A. heliaca • A. nipalensis • A. clanga • A. pomarina • Czechoslovakia
1614	**Descriptive Summary** • Field Id • Raptors • A. chrysaetos • Alabama
1624	**Descriptive Summary** • Distribution • Plumage • Raptors • A. audax • A. gurneyi • New Guinea
1641	**Descriptive Summary** • Id Key • Distribution • Status • A. verreauxi • A. nipalensis • A. rapax • A. pomarina • A. wahlbergi • Kenya • Uganda
1670	**Descriptive Summary** • Etymology • Measure • Trapping • Raptors • A. chrysaetos • Washington
1671	**Descriptive Summary** • Etymology • A. chrysaetos • A. heliaca • A. nipalensis • A. clanga
1676	**Descriptive Summary** • Management • Raptors • A. chrysaetos • Idaho
1693	**Descriptive Summary** • Museum Specimen • Raptors • A. chrysaetos • A. heliaca • A. nipalensis • A. clanga • Georgia Ssr
1761	**Descriptive Summary** • Raptors • Photo Account • A. chrysaetos • A. heliaca • A. clanga • Japan
1769	**Descriptive Summary** • Cain/Abel • Nest • Photo Account • Popular Monograph • A. chrysaetos • Scotland
1785	**Descriptive Summary** • Etymology • Status • Collect • Raptors • A. chrysaetos • A. heliaca • A. pomarina • A. clanga • Egypt
1836	**Descriptive Summary** • (Title Only) • A. clanga • West Germany
1847	**Descriptive Summary** • Plumage • Measure • Distribution • Raptors • A. chrysaetos • A. heliaca • A. nipalensis • A. clanga • China
1912	**Descriptive Summary** • Habitat • Nest Chronology • Nest • A. chrysaetos • France
1914	**Descriptive Summary** • Distribution • Habitat • Nest • Status • Clutch Size • Raptors • A. chrysaetos • New Mexico

Master No.	Keywords
2000	**Descriptive Summary** • A. chrysaetos • Great Britain
2002	**Descriptive Summary** • A. chrysaetos • West Germany
2006	**Descriptive Summary** • Field Id • Egg • Status • Distribution • A. nipalensis • A. rapax • A. pomarina • A. wahlbergi • Africa
2045	**Descriptive Summary** • Measure • Plumage • Raptors • A. chrysaetos • A. heliaca • A. nipalensis • A. pomarina • A. clanga • Italy
2092	**Descriptive Summary** • Field Id • Hunting Behavior • Animal Damage • Nest • Egg • Attitude • Piracy • Raptors • A. audax • Australia
2120	**Descriptive Summary** • Distribution • Plumage • Taxonomy • Hunting Behavior • Raptors • A. chrysaetos • A. nipalensis • A. rapax • A. clanga • Saudi Arabia
2136	**Descriptive Summary** • Distribution • Status • Population Decline • Raptors • A. chrysaetos • A. clanga • A. pomarina • Finland
2169	**Descriptive Summary** • Protection • Hunting Behavior • Illus. • Popular Monograph • A. chrysaetos • Wyoming
2186	**Descriptive Summary** • Prey • Banding • Raptors • A. wahlbergi • A. rapax • South Africa
2340	**Descriptive Summary** • Plumage • Distribution • Prey • Nest • Egg • Photo Account • Raptors • A. audax • Australia
2341	**Descriptive Summary** • Nest • Field Id • Raptors • A. verreauxi • A. rapax • A. wahlbergi • Kenya
2353	**Descriptive Summary** • Economics • Shooting • Eagles • A. chrysaetos • North America
2409	**Descriptive Summary** • Nesting Behavior • Photo Account • Popular Monograph • A. chrysaetos • Scotland
2413	**Descriptive Summary** • Attitude • Field Id • Folklore • Raptors • A. chrysaetos • United States
2415	**Descriptive Summary** • Nest • A. chrysaetos • Maine
2504	**Descriptive Summary** • Raptors • A. heliaca
2552	**Descriptive Summary** • Distribution • Taxonomy • Plumage • Habitat • Measure • Raptors • A. gurneyi • New Guinea
2564	**Descriptive Summary** • Raptors • A. chrysaetos
2581	**Descriptive Summary** • Etymology • Raptors • Photo Account • A. verreauxi • A. rapax • Lesotho
2637	**Descriptive Summary** • Field Id • Distribution • Etymology • Prey • Raptors • A. verreauxi • A. rapax • A. nipalensis • A. pomarina • A. wahlbergi • South Africa
2640	**Descriptive Summary** • Id Key • Raptors • A. chrysaetos • Minnesota
2706	**Descriptive Summary** • Plumage • Distribution • Museum Specimen • A. chrysaetos • Molluca Islands
2728	**Descriptive Summary** • Popular Monograph • (Title Only) • A. verreauxi • Africa
2732	**Descriptive Summary** • Distribution • Prey • Habitat • Raptors • A. chrysaetos • A. nipalensis • Tibet
2842	**Descriptive Summary** • Habitat • Raptors • A. wahlbergi • A. rapax • South Africa
2879	**Descriptive Summary** • Etymology • Status • Distribution • Raptors • A. chrysaetos • Canada
2888	**Descriptive Summary** • Migration • Raptors • A. chrysaetos • A. heliaca • A. clanga • A. pomarina • Italy
2922	**Descriptive Summary** • Etymology • Raptors • A. chrysaetos • North America
2942	**Descriptive Summary** • Field Id • Photo Account • A. pomarina • West Germany
3035	**Descriptive Summary** • Raptors • A. chrysaetos • California
3036	**Descriptive Summary** • Distribution • Status • Prey • Measure • Raptors • A. chrysaetos • Oklahoma
3082	**Descriptive Summary** • Raptors • A. chrysaetos • Canada
3118	**Descriptive Summary** • Distribution • Habitat • Prey • Density • Interspecific Competition • Eagles • A. chrysaetos • France
3165	**Descriptive Summary** • Nest • Raptors • A. pomarina • A. clanga • Poland
3166	**Descriptive Summary** • Status • Raptors • A. clanga • A. pomarina • A. heliaca • A. chrysaetos • Poland
3184	**Descriptive Summary** • Nest • Photo Account • Raptors • Popular Monograph • A. chrysaetos • United States
3203	**Descriptive Summary** • Conservation • Raptors • SRBPNA • A. chrysaetos • Idaho
3218	**Descriptive Summary** • Prey • Hunting Behavior • Habitat • Raptors • A. chrysaetos • A. heliaca • A. nipalensis • A. clanga • A. pomarina • West Germany
3249	**Descriptive Summary** • (Title Only) • A. heliaca
3257	**Descriptive Summary** • Taxonomy • Distribution • Habitat • Raptors • A. rapax • A. nipalensis • A. heliaca • A. clanga • A. pomarina • A. verreauxi
3335	**Descriptive Summary** • Lift • Prey • Raptors • A. chrysaetos • North America
3397	**Descriptive Summary** • Pellet • Behavior • Distribution • Plumage • Measure • Raptors • Illus. • A. chrysaetos • A. clanga • Great Britain
3437	**Descriptive Summary** • Measure • Illus. • Raptors • A. chrysaetos • A. heliaca • A. clanga • Japan
1881	**Digestion** • Pellet • Physiology • Raptors • A. nipalensis
1056	**Dimorphism** • Growth • Falconry • Aging • Parental Behavior • Measure • Plumage • Prey • Photo Account • A. audax • Australia
1505	**Dimorphism** • Taxonomy • Sexing • Museum Specimen • Measure • Raptors • A. chrysaetos • North America
1639	**Dimorphism** • Taxonomy • Measure • Plumage • Distribution • Prey • Behavior • Management • Habitat • Territory • Mortality • Life History • Economics • Literature Review • A. chrysaetos • Oregon • Washington
2880	**Dimorphism** • Food Requirements • Behavior • Theory • Raptors • A. chrysaetos
2933	**Dimorphism** • Anatomy • Paired Ovaries • Raptors • Gonads • A. chrysaetos • North America
614	**Directional Exposure** • Monograph • Nest Survey • Reproductive Success • Habitat • Behavior • A. chrysaetos • Utah
714	**Directional Exposure** • Food Requirements • Breeding Biology • Predator/Prey • Cain/Abel • Growth • Sex Ratio • Behavior • A. chrysaetos • Idaho

Master No.	Keywords
725	**Directional Exposure** • Nest • Prey • A. chrysaetos • California
2088	**Directional Exposure** • Breeding Biology • Prey • Banding • Migration • Mortality • A. chrysaetos • Montana
2238	**Directional Exposure** • Nest • Nest Survey • Mortality • Nestling • Parental Behavior • Hyperthermia • A. chrysaetos • United States
2269	**Directional Exposure** • Nest Survey • Reproductive Success • Density • Raptors • A. chrysaetos • Utah
2464	**Directional Exposure** • Management • Habitat • Disturbance • Mortality • Nest Survey • Density • Nest Chronology • Prey • Reproductive Success • Prog. Rep. • A. chrysaetos • Wyoming
2761	**Directional Exposure** • Breeding Biology • Prey • Vocalization • A. chrysaetos • West Germany
2854	**Directional Exposure** • Population Dynamics • Nest • Density • Prey • Breeding Biology • Distribution • Habitat • Reproductive Success • Mortality • Abandonment • Disturbance • Interspecific Behavior • Territory • A. chrysaetos • Utah
3259	**Directional Exposure** • Nest Survey • Habitat • Territory • Intraspecific Behavior • Density • Nest Chronology • Prey • Raptors • A. verreauxi • Rhodesia
3304	**Directional Exposure** • Nest Survey • Density • Reproductive Success • Territory • A. chrysaetos • Scotland
3360	**Directional Exposure** • Nest Survey • Prey • Habitat • Eagles • A. chrysaetos • Saskatchewan
736	**Disease** • Captivity • Vaccination • Raptors • A. rapax • Africa
750	**Disease** • Pox Virus • A. chrysaetos • California
955	**Disease** • Behavior • Pesticides • Parental Behavior • Vocalization • Disturbance • Measure • Breeding Biology • Parasite • Terminology • Growth • A. chrysaetos • Montana
957	**Disease** • Monograph • Behavior • Growth • Pesticides • Parental Behavior • Vocalization • Disturbance • Measure • Breeding Biology • Photography • Captivity • Parasite • Terminology • Photo Account • A. chrysaetos • Montana
1290	**Disease** • Literature Review • Raptors • A. chrysaetos • Great Britain
1748	**Disease** • Autopsy • Raptors • A. rapax • A. verreauxi • A. chrysaetos • A. audax • A. heliaca
2056	**Disease** • A. chrysaetos • Washington
2196	**Disease** • Pox Virus • A. chrysaetos • British Columbia
2237	**Disease** • Rehabilitation • Shooting • Conservation • Photo Account • Raptors • A. chrysaetos • California
2295	**Disease** • A. chrysaetos • United States
2576	**Disease** • Rehabilitation • Shooting • Trapping • Anaesthesia • Blood Sample • Raptors • A. chrysaetos • United States
2614	**Disease** • Blood Sample • Raptors • A. chrysaetos • California
2658	**Disease** • A. chrysaetos • California
2924	**Disease** • Raptors • A. chrysaetos
2937	**Disease** • Rehabilitation • Physiology • Riboflavin • A. chrysaetos • Idaho
3302	**Disease** • Autopsy • Nest • A. chrysaetos • Great Britain
3306	**Disease** • Pesticides • Autopsy • Mortality • A. chrysaetos • Great Britain
324	**Dispersal** • Reproductive Success • Food Requirements • Territory • Abandonment • Mortality • Cain/Abel • Density • Breeding Biology • Nest Survey • A. chrysaetos • Idaho
1150	**Dispersal** • Density • Territory • Nest • Survey • Intraspecific Behavior • Marking • Reproductive Success • Display • A. verreauxi • Rhodesia
1900	**Dispersal** • Mortality • Prey • Nest • Pellet • Habitat • Animal Damage • Bounty • Protection • Photo Account • A. audax • Australia
2612	**Dispersal** • Breeding Biology • Prey • Habitat • A. audax • Australia
2871	**Dispersal** • Marking • Home Range • Territory • Raptors • Photo Account • A. rapax • A. wahlbergi • South Africa
800	**Display** • Behavior • A. chrysaetos • Sweden
822	**Display** • A. audax • Australia
850	**Display** • Flight • Raptors • A. chrysaetos • California
883	**Display** • Descriptive Summary • Etymology • Behavior • Prey • A. heliaca • A. rapax • A. pomarina • A. clanga • India
1144	**Display** • Nest Survey • Status • Behavior • Fledging • Mortality • Banding • Marking • Cain/Abel • Reproductive Success • Breeding Biology • Interspecific Behavior • Hunting Behavior • Courtship • Raptors • A. verreauxi • Rhodesia
1145	**Display** • Nest • Parental Behavior • Breeding Biology • Interspecific Behavior • Intraspecific Behavior • Feeding Behavior • Hunting Behavior • Fledging • Prey • Raptors • A. verreauxi • Rhodesia
1150	**Display** • Density • Territory • Dispersal • Nest • Survey • Intraspecific Behavior • Marking • Reproductive Success • A. verreauxi • Rhodesia
1351	**Display** • Flight • A. chrysaetos • Switzerland
1379	**Display** • Flight • A. chrysaetos • Scotland
2085	**Display** • Telemetry • Territory • Agonistic Behavior • Behavior • Nest • A. chrysaetos • Wyoming • Montana
2152	**Display** • Habitat • Growth • Parental Behavior • Prey • Breeding Biology • Interspecific Behavior • Courtship • Photo Account • Raptors • A. heliaca • Spain
2243	**Display** • Status • Nest • Raptors • A. pomarina • A. clanga • A. chrysaetos • A. heliaca • Bulgaria
2434	**Display** • Flight • Status • A. chrysaetos • Scotland
2814	**Display** • Intraspecific Behavior • A. chrysaetos • Texas
3079	**Display** • Prey • Nest Survey • Clutch Size • Reproductive Success • Hunting Behavior • Density • Migration • A. wahlbergi • South Africa
3151	**Display** • A. heliaca • Portugal
3303	**Display** • Intraspecific Behavior • A. chrysaetos • Scotland
5	**Distribution** • Raptors • A. clanga • Egypt
17	**Distribution** • A. chrysaetos • Sweden

Master No.	Keywords
25	**Distribution** • Descriptive Summary • Field Id • Habitat • Raptors • A. rapax • India
26	**Distribution** • Id Key • Descriptive Summary • Etymology • Status • Field Id • Raptors • Illus. • A. chrysaetos • A. heliaca • A. rapax • A. nipalensis • A. clanga • A. pomarina • India
42	**Distribution** • (Title Only) • A. nipalensis • Kenya
47	**Distribution** • Population Decline • Historical Account • Status • Management • Raptors • A. chrysaetos • Tennessee
48	**Distribution** • Taxonomy • Raptors • A. audax • A. gurneyi • Australia
51	**Distribution** • A. chrysaetos • Denmark
54	**Distribution** • Museum Specimen • Plumage • Measure • Prey • A. heliaca • A. clanga • A. rapax • India
67	**Distribution** • Taxonomy • Egg • Plumage • Prey • Raptors • A. verreauxi • A. rapax • Southwest Africa
68	**Distribution** • Taxonomy • Museum Specimen • Plumage • A. nipalensis • A. clanga • A. pomarina • France
80	**Distribution** • Nest • Raptors • A. pomarina • A. clanga • A. chrysaetos • A. heliaca • West Germany
195	**Distribution** • Nest Survey • Wintering Population • Habitat • Attitude • Raptors • A. chrysaetos • Arizona
214	**Distribution** • A. chrysaetos • Austria
218	**Distribution** • Descriptive Summary • Rarity • Prey • Etymology • Raptors • A. chrysaetos • A. heliaca • Japan
219	**Distribution** • Status • A. chrysaetos • France
220	**Distribution** • Status • Raptors • A. heliaca • A. chrysaetos • A. clanga • A. pomarina • Moldavia
223	**Distribution** • Nest • Raptors • A. verreauxi • South Africa
229	**Distribution** • Status • Nest • Prey • A. audax • Australia
240	**Distribution** • Nest • Captivity • A. chrysaetos • Quebec
261	**Distribution** • Id Key • Descriptive Summary • Prey • Etymology • A. rapax • A. wahlbergi • Gambia • Sierra Leone • Nigeria
285	**Distribution** • A. chrysaetos • North America • Europe • Asia
305	**Distribution** • Anatomy • Breeding Biology • Cain/Abel • A. pomarina • A. clanga • South Africa
308	**Distribution** • Status • Raptors • A. audax • Australia
334	**Distribution** • Status • Raptors • A. clanga • Russia
335	**Distribution** • Nest • Captivity • Vocalization • Prey • Raptors • A. chrysaetos • Oregon
350	**Distribution** • A. heliaca • Yugoslavia
351	**Distribution** • Hunting Behavior • Measure • Life History • Courtship • Behavior • Vocalization • Plumage • Raptors • Photo Account • A. chrysaetos • North America
355	**Distribution** • Raptors • A. clanga • Hungary
371	**Distribution** • Raptors • A. audax • Australia
380	**Distribution** • Prey • Raptors • A. chrysaetos • India
386	**Distribution** • Bounty • Disturbance • Habitat Loss • Protection • Shooting • Status • Pesticides • Migration • Population Decline • Raptors • A. chrysaetos • A. heliaca • A. pomarina • A. clanga • Europe
398	**Distribution** • Raptors • A. chrysaetos • Finland • Norway
402	**Distribution** • Field Id • A. chrysaetos • Mexico
415	**Distribution** • Raptors • A. chrysaetos • A. heliaca • A. rapax • A. clanga • A. pomarina • India
438	**Distribution** • Nest • Raptors • A. audax • Australia
439	**Distribution** • Habitat • A. chrysaetos • Washington
445	**Distribution** • Status • A. verreauxi • South Africa
449	**Distribution** • Raptors • A. chrysaetos • France
466	**Distribution** • Attitude • Egg Collection • Prey • Raptors • A. audax • Australia
467	**Distribution** • A. pomarina • Denmark
475	**Distribution** • Status • Prey • Historical Account • Hunting Behavior • Raptors • A. chrysaetos • Maine
489	**Distribution** • Nest • Prey • Measure • Raptors • A. chrysaetos • Sardinia
495	**Distribution** • Wintering Population • Prey • Hunting Behavior • Age Structure • A. nipalensis • Africa
497	**Distribution** • Raptors • A. verreauxi • Angola
500	**Distribution** • Prey • Raptors • A. audax • Australia
502	**Distribution** • Status • Raptors • A. audax • Australia
504	**Distribution** • A. chrysaetos • West Virginia
508	**Distribution** • (Title Only) • A. heliaca • India
544	**Distribution** • Taxonomy • Status • Field Id • Flight • Habitat • Terminology • Mortality • Behavior • Life History • Longevity • Territory • Photo Account • Raptors • A. chrysaetos • A. clanga • Great Britain • Ireland • Scotland
586	**Distribution** • Raptors • A. chrysaetos • Wisconsin
607	**Distribution** • Raptors • A. chrysaetos • North Carolina
612	**Distribution** • Raptors • A. chrysaetos • North America
637	**Distribution** • Raptors • A. chrysaetos • Italy
640	**Distribution** • Raptors • A. wahlbergi • Gambia
644	**Distribution** • Raptors • A. chrysaetos • Czechoslovakia
656	**Distribution** • Measure • Raptors • A. chrysaetos • United States
664	**Distribution** • Status • A. chrysaetos • A. heliaca • A. nipalensis • A. clanga • China
680	**Distribution** • Wintering Population • Field Id • Raptors • A. chrysaetos • A. heliaca • A. nipalensis • A. pomarina • Iraq
694	**Distribution** • Raptors • A. clanga • South Africa
696	**Distribution** • Habitat • Status • Raptors • A. audax • Australia
702	**Distribution** • Life History • Food Habits • Popular Monograph • A. chrysaetos • Missouri
738	**Distribution** • Nest • Raptors • A. audax • Australia
742	**Distribution** • Raptors • A. audax • Australia
754	**Distribution** • Raptors • A. audax • Australia

Master No.	Keywords
770	**Distribution** • Field Id • Descriptive Summary • Monograph • Status • Migration • Prey • Plumage • Measure • Etymology • Vocalization • A. pomarina • A. clanga • A. nipalensis • A. heliaca • A. chrysaetos • A. verreauxi • Africa • Europe
792	**Distribution** • Protection • Eagles • A. chrysaetos • Sweden
795	**Distribution** • A. heliaca • Hungary
797	**Distribution** • Nest • Egg Collection • Etymology • Raptors • A. chrysaetos • A. heliaca • A. clanga • Turkey
798	**Distribution** • Raptors • A. chrysaetos • A. heliaca • Turkey
801	**Distribution** • Raptors • A. clanga • Siberia
823	**Distribution** • Wintering Population • Survey • A. chrysaetos • New Mexico
824	**Distribution** • Field Id • Etymology • A. chrysaetos • Mexico
830	**Distribution** • Survey • Raptors • A. audax • Australia
864	**Distribution** • Descriptive Summary • Taxonomy • Raptors • A. chrysaetos • A. heliaca • A. nipalensis • A. pomarina • A. clanga • Europe • Asia
871	**Distribution** • Etymology • Life History • Habitat • Molt • Prey • Measure • Field Id • Migration • Status • Raptors • A. chrysaetos • A. heliaca • A. rapax • A. clanga • A. pomarina • United Soviet Socialists Republic
874	**Distribution** • Nest • (Title Only) • A. chrysaetos • Europe
892	**Distribution** • Raptors • A. rapax • Algeria
904	**Distribution** • Raptors • A. chrysaetos • A. heliaca • A. nipalensis • A. clanga • Kazakhstan
914	**Distribution** • Raptors • A. chrysaetos • Morocco
920	**Distribution** • A. chrysaetos • Colorado
931	**Distribution** • Habitat • Raptors • A. rapax • Senegal
941	**Distribution** • Descriptive Summary • Status • Management • Literature Review • Prey • A. chrysaetos • Washington
943	**Distribution** • Prey • Raptors • A. audax • Australia
975	**Distribution** • Raptors • A. heliaca • A. rapax • A. nipalensis • Aden
977	**Distribution** • Status • Raptors • A. chrysaetos • Switzerland
987	**Distribution** • Descriptive Summary • Field Id • Nest • Status • Taxonomy • Egg • Illus. • A. chrysaetos • A. heliaca • A. nipalensis • A. rapax • A. clanga • A. pomarina • A. verreauxi • Africa
988	**Distribution** • Descriptive Summary • Nest • Raptors • A. nipalensis • A. clanga • A. heliaca • A. chrysaetos • China
997	**Distribution** • Status • A. chrysaetos • Scotland
1000	**Distribution** • Raptors • A. audax • Australia
1026	**Distribution** • Raptors • A. pomarina • Hungary
1065	**Distribution** • Raptors • A. audax • Australia
1070	**Distribution** • Nest • Raptors • A. chrysaetos • Alaska
1080	**Distribution** • Collect • Nest • Etymology • Literature Review • A. chrysaetos • A. rapax • Tunisia
1085	**Distribution** • A. chrysaetos • Austria
1094	**Distribution** • Raptors • A. verreauxi • Tanzania
1105	**Distribution** • Descriptive Summary • Etymology • Plumage • Nest • A. chrysaetos • Alaska
1128	**Distribution** • Status • Raptors • A. chrysaetos • A. heliaca • A. nipalensis • United Soviet Socialists Republic
1129	**Distribution** • Status • Raptors • A. nipalensis • A. clanga • A. pomarina • A. heliaca • A. chrysaetos • United Soviet Socialists Republic
1137	**Distribution** • Nest • Disturbance • A. heliaca • Hungary
1177	**Distribution** • Status • Nest Survey • Reproductive Success • Protection • Photo Account • A. pomarina • East Germany
1228	**Distribution** • Monograph • Vocalization • Prey • Plumage • Id Key • A. pomarina • A. chrysaetos • A. nipalensis • Europe
1301	**Distribution** • Habitat • Raptors • A. chrysaetos • California
1352	**Distribution** • (Title Only) • A. clanga • Czechoslovakia
1384	**Distribution** • Field Id • Raptors • A. chrysaetos • A. heliaca • A. nipalensis • A. clanga • A. pomarina • Europe
1392	**Distribution** • Collect • A. gurneyi • Molluca Islands
1445	**Distribution** • Field Id • Raptors • A. pomarina • A. clanga • A. heliaca • A. chrysaetos • A. verreauxi • Europe
1451	**Distribution** • Taxonomy • Raptors • A. chrysaetos • North America
1526	**Distribution** • Descriptive Summary • Status • Measure • A. verreauxi • A. nipalensis • A. rapax • A. pomarina • A. wahlbergi • Southwest Africa
1563	**Distribution** • Descriptive Summary • Nest Survey • Raptors • A. chrysaetos • Spain
1586	**Distribution** • Descriptive Summary • Measure • Nest • Photo Account • Illus. • Raptors • A. chrysaetos • A. heliaca • A. nipalensis • A. clanga • A. pomarina • Czechoslovakia
1589	**Distribution** • Population Decline • Status • Conservation • Raptors • Popular Monograph • A. clanga • A. heliaca • Europe
1591	**Distribution** • A. chrysaetos • Alaska
1624	**Distribution** • Descriptive Summary • Plumage • Raptors • A. audax • A. gurneyi • New Guinea
1639	**Distribution** • Taxonomy • Dimorphism • Measure • Plumage • Prey • Behavior • Management • Habitat • Territory • Mortality • Life History • Economics • Literature Review • A. chrysaetos • Oregon • Washington
1641	**Distribution** • Descriptive Summary • Id Key • Status • A. verreauxi • A. nipalensis • A. rapax • A. pomarina • A. wahlbergi • Kenya • Uganda
1642	**Distribution** • A. pomarina • Botswana
1697	**Distribution** • Plumage • Status • Taxonomy • Etymology • Field Id • Literature Review • Eagles • A. pomarina • A. clanga • France • Belgium • Switzerland • Italy
1722	**Distribution** • Raptors • A. pomarina • Poland
1735	**Distribution** • Status • Raptors • A. verreauxi • A. wahlbergi • A. pomarina • A. rapax • South Africa
1737	**Distribution** • Nest • Status • Raptors • A. nipalensis • A. rapax • A. wahlbergi • A. verreauxi • South Africa

Master No.	Keywords
1738	**Distribution** • A. clanga • Israel
1744	**Distribution** • Raptors • A. pomarina • A. chrysaetos • Hungary
1750	**Distribution** • Nest • Raptors • A. nipalensis • United Soviet Socialists Republic
1752	**Distribution** • Roadside Survey • Density • Raptors • A. chrysaetos • California
1754	**Distribution** • Nest • Wintering Population • Habitat • Status • Raptors • A. chrysaetos • Colorado
1756	**Distribution** • Roadside Survey • Raptors • A. chrysaetos • California
1759	**Distribution** • Prey • Raptors • A. rapax • Africa
1773	**Distribution** • Winter Survey • Population Dynamics • Prey • Age Structure • Wintering Population • Eagles • A. chrysaetos • Washington
1774	**Distribution** • Wintering Population • Prey • Raptors • Popular Monograph • A. chrysaetos • Washington
1791	**Distribution** • Nest • Raptors • A. chrysaetos • Ukraine
1792	**Distribution** • Status • Raptors • A. chrysaetos • A. clanga • Ukraine
1797	**Distribution** • Raptors • A. chrysaetos • A. nipalensis • Ukraine
1800	**Distribution** • Raptors • A. chrysaetos • A. nipalensis • A. clanga • China
1801	**Distribution** • Taxonomy • Raptors • A. nipalensis • A. rapax • United Soviet Socialists Republic
1809	**Distribution** • Raptors • A. chrysaetos • Yugoslavia
1847	**Distribution** • Descriptive Summary • Plumage • Measure • Raptors • A. chrysaetos • A. heliaca • A. nipalensis • A. clanga • China
1852	**Distribution** • Density • Raptors • A. chrysaetos • Siberia
1864	**Distribution** • Nest • A. nipalensis • Kazakhstan
1870	**Distribution** • Raptors • A. audax • Australia
1871	**Distribution** • A. chrysaetos • Hungary
1872	**Distribution** • Habitat • Photo Account • Popular Monograph • A. chrysaetos • Europe
1885	**Distribution** • Population Estimate • Raptors • A. chrysaetos • Colorado
1914	**Distribution** • Habitat • Nest • Status • Clutch Size • Descriptive Summary • Raptors • A. chrysaetos • New Mexico
1916	**Distribution** • Etymology • Raptors • A. chrysaetos • A. heliaca • A. clanga • Spain
1929	**Distribution** • Nest • Raptors • A. chrysaetos • Siberia
1935	**Distribution** • Territory • A. heliaca • Kazakhstan
1938	**Distribution** • Roadside Survey • Raptors • A. chrysaetos • Nebraska
1939	**Distribution** • Roadside Survey • Raptors • A. chrysaetos • Nebraska
1968	**Distribution** • Raptors • A. clanga • A. chrysaetos • Russia
1975	**Distribution** • Raptors • A. chrysaetos • A. clanga • Poland
1994	**Distribution** • Nest • Prey • Hunting Behavior • Raptors • A. verreauxi • South Africa
2006	**Distribution** • Descriptive Summary • Field Id • Egg • Status • A. nipalensis • A. rapax • A. pomarina • A. wahlbergi • Africa
2026	**Distribution** • A. chrysaetos • Hungary
2053	**Distribution** • Raptors • A. audax • Australia
2054	**Distribution** • Raptors • A. audax • Australia
2079	**Distribution** • Prey • Field Id • Attitude • Raptors • A. chrysaetos • United States
2097	**Distribution** • Habitat • Status • Nest • Raptors • A. chrysaetos • North Dakota
2105	**Distribution** • Raptors • A. chrysaetos • Belgium
2111	**Distribution** • Raptors • A. heliaca • Israel
2120	**Distribution** • Descriptive Summary • Plumage • Taxonomy • Hunting Behavior • Raptors • A. chrysaetos • A. nipalensis • A. rapax • A. clanga • Saudi Arabia
2131	**Distribution** • Status • Plumage • Raptors • A. chrysaetos • A. heliaca • A. clanga • A. pomarina • Russia
2136	**Distribution** • Descriptive Summary • Status • Population Decline • Raptors • A. chrysaetos • A. clanga • A. pomarina • Finland
2148	**Distribution** • Field Id • Management • Taxonomy • Habitat • Nest Survey • Prey • Breeding Biology • Photo Account • Eagles • A. pomarina • Europe
2171	**Distribution** • A. verreauxi • Angola
2181	**Distribution** • A. wahlbergi • Senegal
2184	**Distribution** • Aerial Survey • Reproductive Success • Management • Survey • Clutch Size • Habitat • Nest Survey • Prey • Raptors • A. chrysaetos • Arizona
2190	**Distribution** • Nest • Raptors • A. chrysaetos • Italy
2218	**Distribution** • Raptors • A. chrysaetos • Great Britain
2220	**Distribution** • Prey • A. verreauxi • Egypt
2221	**Distribution** • Monograph • Migration • Raptors • A. heliaca • A. pomarina • A. clanga • A. nipalensis • Africa
2314	**Distribution** • (Title Only) • A. chrysaetos • Austria
2340	**Distribution** • Descriptive Summary • Plumage • Prey • Nest • Egg • Photo Account • Raptors • A. audax • Australia
2356	**Distribution** • (Title Only) • A. chrysaetos • West Germany
2393	**Distribution** • Prey • Beneficial Effect • Raptors • A. heliaca • A. nipalensis • Kazakhstan
2408	**Distribution** • Status • Nest Survey • Photo Account • A. chrysaetos • Czechoslovakia
2422	**Distribution** • Status • Pesticides • Raptors • A. chrysaetos • Great Britain
2430	**Distribution** • Raptors • A. audax • Australia
2456	**Distribution** • Field Id • Etymology • A. chrysaetos • A. heliaca • A. rapax • A. clanga • A. pomarina • Europe
2461	**Distribution** • Status • Raptors • A. chrysaetos • Arizona
2491	**Distribution** • Density • Habitat • Status • Survey • Raptors • A. chrysaetos • New Mexico
2525	**Distribution** • Raptors • A. clanga • Russia
2535	**Distribution** • Field Id • Status • Behavior • Raptors • A. verreauxi • A. rapax • A. nipalensis • A. pomarina • A. wahlbergi • South Africa
2552	**Distribution** • Descriptive Summary • Taxonomy • Plumage • Habitat • Measure • Raptors • A. gurneyi • New Guinea
2557	**Distribution** • Raptors • A. chrysaetos • A. pomarina • Estonia
2567	**Distribution** • (Title Only) • A. chrysaetos • West Germany

Master No.	Keywords
2619	**Distribution** • Raptors • A. chrysaetos • A. heliaca • A. nipalensis • A. rapax • A. clanga • A. pomarina • India • Pakistan
2626	**Distribution** • Raptors • A. audax • Australia
2630	**Distribution** • Field Id • Raptors • A. chrysaetos • North America
2637	**Distribution** • Descriptive Summary • Field Id • Etymology • Prey • Raptors • A. verreauxi • A. rapax • A. nipalensis • A. pomarina • A. wahlbergi • South Africa
2671	**Distribution** • Raptors • A. audax • Australia
2706	**Distribution** • Descriptive Summary • Plumage • Museum Specimen • A. chrysaetos • Molluca Islands
2732	**Distribution** • Descriptive Summary • Prey • Habitat • Raptors • A. chrysaetos • A. nipalensis • Tibet
2754	**Distribution** • Raptors • A. heliaca • Hungary
2780	**Distribution** • Raptors • A. audax • Australia
2781	**Distribution** • Raptors • A. chrysaetos • Russia
2782	**Distribution** • Raptors • A. chrysaetos • A. heliaca • A. pomarina • A. clanga • A. nipalensis • A. rapax • Russia
2784	**Distribution** • Raptors • A. chrysaetos • A. heliaca • A. nipalensis • A. clanga • Caucasus
2785	**Distribution** • Museum Specimen • Raptors • A. clanga • Iran
2811	**Distribution** • Habitat • Status • A. chrysaetos • Great Britain • Ireland • Scotland
2815	**Distribution** • Raptors • A. heliaca • A. clanga • Egypt
2838	**Distribution** • Raptors • A. clanga • A. heliaca • India
2846	**Distribution** • Protection • Prey • Nest Survey • Photo Account • A. pomarina • Czechoslovakia
2854	**Distribution** • Population Dynamics • Nest • Density • Prey • Breeding Biology • Habitat • Directional Exposure • Reproductive Success • Mortality • Abandonment • Disturbance • Interspecific Behavior • Territory • A. chrysaetos • Utah
2875	**Distribution** • Raptors • A. chrysaetos • Morocco
2878	**Distribution** • A. chrysaetos • Quebec • Ontario
2879	**Distribution** • Etymology • Descriptive Summary • Status • Raptors • A. chrysaetos • Canada
2893	**Distribution** • Raptors • A. chrysaetos • Hungary
2927	**Distribution** • (Title Only) • A. chrysaetos • West Germany
2948	**Distribution** • Status • A. chrysaetos • A. heliaca • A. pomarina • Switzerland
2950	**Distribution** • A. chrysaetos • Switzerland
2955	**Distribution** • Status • (Title Only) • A. chrysaetos • Switzerland
2997	**Distribution** • Wintering Population • A. pomarina • Europe
3036	**Distribution** • Descriptive Summary • Status • Prey • Measure • Raptors • A. chrysaetos • Oklahoma
3080	**Distribution** • Cain/Abel • Educational Material • Field Id • Habitat • Hunting Behavior • Prey • Artificial Nest • Nest • Incubation • Wintering Population • Conservation • Photo Account • Eagles • A. chrysaetos • A. wahlbergi • A. verreauxi • A. nipalensis • A. pomarina • South Africa
3103	**Distribution** • Status • Field Id • Conservation • Photo Account • Raptors • A. chrysaetos • France
3106	**Distribution** • Status • Raptors • A. audax • Australia
3110	**Distribution** • Nest Survey • Status • Density • Habitat Loss • Density • Reproductive Success • Raptors • Photo Account • A. chrysaetos • California
3111	**Distribution** • Status • Raptors • A. chrysaetos • A. rapax • A. heliaca • Morocco
3118	**Distribution** • Habitat • Prey • Descriptive Summary • Density • Interspecific Competition • Eagles • A. chrysaetos • France
3128	**Distribution** • Roadside Survey • Status • Habitat • A. chrysaetos • A. pomarina • A. clanga • A. wahlbergi • A. rapax • A. verreauxi • Ivory Coast • Upper Volta • Mali
3136	**Distribution** • Migration • A. rapax • A. wahlbergi • France • Africa
3143	**Distribution** • (Title Only) • A. chrysaetos • Ireland
3181	**Distribution** • Raptors • A. chrysaetos • A. heliaca • A. nipalensis • A. clanga • Israel
3205	**Distribution** • Management • Animal Damage • Status • Wintering Population • A. chrysaetos • United States
3220	**Distribution** • Status • Habitat Loss • Raptors • A. chrysaetos • A. heliaca • A. clanga • A. pomarina • Greece
3233	**Distribution** • Raptors • A. chrysaetos • Mexico
3245	**Distribution** • Prey • Habitat • Raptors • A. nipalensis • A. heliaca • A. chrysaetos • United Soviet Socialists Republic
3257	**Distribution** • Taxonomy • Descriptive Summary • Habitat • Raptors • A. rapax • A. nipalensis • A. heliaca • A. clanga • A. pomarina • A. verreauxi
3283	**Distribution** • Breeding Biology • Prey • Conservation • Mortality • Photo Account • A. chrysaetos • Czechoslovakia
3292	**Distribution** • Measure • Prey • Raptors • A. gurneyi • New Guinea
3338	**Distribution** • Raptors • A. audax • Australia
3374	**Distribution** • Status • Protection • Conservation • Raptors • A. chrysaetos • Norway
3397	**Distribution** • Descriptive Summary • Pellet • Behavior • Plumage • Measure • Raptors • Illus. • A. chrysaetos • A. clanga • Great Britain
3400	**Distribution** • Raptors • A. chrysaetos • Poland
3405	**Distribution** • Taxonomy • Plumage • Subspecies • A. chrysaetos
3412	**Distribution** • Symbolism • Nest • Raptors • A. chrysaetos • Utah
3422	**Distribution** • Raptors • A. chrysaetos • Malawi
3444	**Distribution** • Breeding Biology • Nest Survey • Habitat • Protection • Prey • Reproductive Success • Photo Account • A. chrysaetos • Estonia
1	**Disturbance** • Artificial Incubation • A. chrysaetos • California
31	**Disturbance** • Environmental Impact • Nuclear Test • Raptors • A. chrysaetos • Wyoming
198	**Disturbance** • Prey • Monograph • Nest Survey • Habitat • Reproductive Success • Literature Review • Pellet • Animal Damage • A. chrysaetos • Utah

Master No.	Keywords
300	**Disturbance** • Status • Raptors • A. heliaca • A. pomarina • A. chrysaetos • Austria
386	**Disturbance** • Bounty • Habitat Loss • Protection • Shooting • Status • Pesticides • Migration • Distribution • Population Decline • Raptors • A. chrysaetos • A. heliaca • A. pomarina • A. clanga • Europe
462	**Disturbance** • Habitat • Attitude • Conservation • Flight Speed • Eagles • Photo Account • Popular Monograph • A. chrysaetos • North America
653	**Disturbance** • Status • A. heliaca • Spain
887	**Disturbance** • Behavior • A. verreauxi • Rhodesia
954	**Disturbance** • Abnormal Behavior • Interspecific Behavior • A. chrysaetos • Colorado
955	**Disturbance** • Behavior • Pesticides • Parental Behavior • Vocalization • Measure • Breeding Biology • Parasite • Disease • Terminology • Growth • A. chrysaetos • Montana
957	**Disturbance** • Monograph • Behavior • Growth • Pesticides • Parental Behavior • Vocalization • Measure • Breeding Biology • Photography • Captivity • Parasite • Disease • Terminology • Photo Account • A. chrysaetos • Montana
960	**Disturbance** • Electrocution • Artificial Nest • Raptors • A. chrysaetos • North America
995	**Disturbance** • Nest Survey • Reproductive Success • A. chrysaetos • Scotland
996	**Disturbance** • Survey • Pesticides • Reproductive Success • Status • A. chrysaetos • Scotland
1071	**Disturbance** • Raptors • A. chrysaetos • Alaska
1072	**Disturbance** • Habitat • Raptors • A. chrysaetos • Alaska
1073	**Disturbance** • Habitat • Raptors • A. chrysaetos • Alaska
1098	**Disturbance** • Survey • Banding • Nest • A. chrysaetos • Canada
1101	**Disturbance** • Status • Habitat Loss • Pesticides • Raptors • A. chrysaetos • Canada
1102	**Disturbance** • Abandonment • Egg • Mortality • Nestling • Photography • Banding • Raptors • Tech. • A. chrysaetos • North America
1137	**Disturbance** • Distribution • Nest • A. heliaca • Hungary
1141	**Disturbance** • Nest Survey • Habitat • Density • Reproductive Success • Clutch Size • Raptors • A. verreauxi • Rhodesia
1164	**Disturbance** • Status • Pesticides • Conservation • Population Estimate • Raptors • A. heliaca • A. chrysaetos • Spain
1366	**Disturbance** • Status • Mortality • Protection • Management • Eagles • A. chrysaetos • North America
1653	**Disturbance** • Punishment • Conservation • Nest Robbing • Popular Monograph • A. chrysaetos • Scotland
1903	**Disturbance** • Nest • Prey • Habitat • Conservation • Mortality • Hunting Behavior • A. chrysaetos • Israel
1945	**Disturbance** • Mining • Telemetry • Reproductive Success • Nest Survey • Raptors • A. chrysaetos • Montana • Wyoming
1949	**Disturbance** • Reproductive Success • Pesticides • Nest Failure • Egg • A. chrysaetos • Scotland
2246	**Disturbance** • Nest • Habitat • Raptors • A. pomarina • Poland
2261	**Disturbance** • Nest Survey • Prey • Population Dynamics • Management • Status • Animal Damage • Artificial Nest • Predator/Prey • A. chrysaetos • Utah
2311	**Disturbance** • Attitude • Status • Population Estimate • Mortality • Protection • Prey • Animal Damage • Raptors • A. chrysaetos • Great Britain
2316	**Disturbance** • Status • Population Decline • Banding • Nest Survey • Attitude • A. chrysaetos • Europe
2464	**Disturbance** • Management • Habitat • Directional Exposure • Mortality • Nest Survey • Density • Nest Chronology • Prey • Reproductive Success • Prog. Rep. • A. chrysaetos • Wyoming
2465	**Disturbance** • Habitat • Mining • Nest Survey • Marking • Prey • Time Lapse • Raptors • Proj. Out. • A. chrysaetos • Wyoming • Montana
2519	**Disturbance** • Status • A. chrysaetos • Great Britain
2569	**Disturbance** • Nest • Behavior • Prey • A. chrysaetos • Colorado
2712	**Disturbance** • Nest Survey • Reproductive Success • A. chrysaetos • Scotland
2854	**Disturbance** • Population Dynamics • Nest • Density • Prey • Breeding Biology • Distribution • Habitat • Directional Exposure • Reproductive Success • Mortality • Abandonment • Interspecific Behavior • Territory • A. chrysaetos • Utah
2930	**Disturbance** • Nest • Gas Explor. • Raptors • A. chrysaetos • Colorado
3027	**Disturbance** • Population Decline • Habitat Loss • A. nipalensis • Caucasus
3031	**Disturbance** • Management • Environmental Impact • Habitat Protection • Raptors • A. chrysaetos • North America
3069	**Disturbance** • Nest • Raptors • A. heliaca • Yugoslavia
3202	**Disturbance** • Protection • Aircraft • Mitigation • Environmental Impact • A. chrysaetos • California
3207	**Disturbance** • Refuge • Habitat • Nest Survey • Density • Prey • Predator/Prey • Telemetry • Territory • Reproductive Success • Raptors • SRBPNA • Photo Account • A. chrysaetos • Idaho
3349	**Disturbance** • Aircraft • Aerial Survey • Raptors • A. chrysaetos • Alaska
153	**Economics** • Shooting • Conservation • Protection • Aircraft • Reward • Attitude • A. chrysaetos • United States
164	**Economics** • Trade • Protection • Raptors • A. rapax • Great Britain • India
201	**Economics** • Animal Damage • A. chrysaetos • North America
231	**Economics** • Animal Damage • (Title Only) • A. chrysaetos • Texas • New Mexico
434	**Economics** • Animal Damage • Attitude • Management • Survey • A. chrysaetos • North America
572	**Economics** • Shooting • Protection • Eagles • A. chrysaetos • United States

Master No.	Keywords
606	**Economics** • Animal Damage • Attitude • Prey • Conservation • Raptors • A. chrysaetos • North America
651	**Economics** • Nest • Capture • Folklore • A. chrysaetos • Tennessee
1090	**Economics** • Attitude • Conservation • Animal Damage • Mortality • Popular Monograph • A. chrysaetos • Wyoming
1339	**Economics** • Attitude • Descriptive Summary • Raptors • A. verreauxi • A. rapax • South Africa
1507	**Economics** • Conservation • Protection • Mortality • Reward • Pesticides • Poisoning • Eagles • A. chrysaetos • Wyoming • Colorado
1639	**Economics** • Taxonomy • Dimorphism • Measure • Plumage • Distribution • Prey • Behavior • Management • Habitat • Territory • Mortality • Life History • Literature Review • A. chrysaetos • Oregon • Washington
1942	**Economics** • Food Habits • Nest • Time Lapse • Behavior • Animal Damage • Reproductive Success • Status • A. chrysaetos • Texas
2133	**Economics** • Attitude • A. chrysaetos • A. clanga • Belgium
2200	**Economics** • Prey • Animal Damage • Nest Survey • Trapping • A. chrysaetos • New Mexico • Texas
2280	**Economics** • Trapping • Protection • Raptors • A. chrysaetos • Nevada
2289	**Economics** • Animal Damage • A. chrysaetos • California
2322	**Economics** • Animal Damage • Questionnaire • A. chrysaetos • Utah
2349	**Economics** • Feeding Behavior • Animal Damage • Capture • Predator/Prey • Protection • Management • A. chrysaetos • Montana
2350	**Economics** • Animal Damage • A. chrysaetos • Montana
2353	**Economics** • Descriptive Summary • Shooting • Eagles • A. chrysaetos • North America
2399	**Economics** • Bounty • Conservation • Mortality • Eagles • Popular Monograph • A. chrysaetos • North America
2803	**Economics** • Attitude • Shooting • Aircraft • Animal Damage • Popular Monograph • A. chrysaetos • Texas
2804	**Economics** • Bounty • Shooting • Attitude • Protection • A. audax • Australia
3153	**Economics** • Animal Damage • A. chrysaetos • Wyoming
3375	**Economics** • Protection • Symbolism • Shooting • Attitude • Conservation • A. chrysaetos • United States
3378	**Economics** • Animal Damage • Protection • Popular Monograph • A. chrysaetos • North America
184	**Educational Material** • Descriptive Summary • A. chrysaetos • Great Britain
942	**Educational Material** • Conservation • Eagles • A. chrysaetos
1625	**Educational Material** • Status • Shooting • Poisoning • Prey • Refuge • Protection • Raptors • A. chrysaetos • Spain
2018	**Educational Material** • Field Id • Status • Raptors • A. chrysaetos • California
2019	**Educational Material** • Management • Falconry • Rehabilitation • Survey • Raptors • A. chrysaetos • California
2057	**Educational Material** • Wintering Population • Raptors • A. chrysaetos • Minnesota
2109	**Educational Material** • Conservation • Refuge • Raptors • A. chrysaetos • Idaho
2654	**Educational Material** • Symbolism • Life History • Eagles • Children's Book • Popular Monograph • A. chrysaetos
3080	**Educational Material** • Cain/Abel • Distribution • Field Id • Habitat • Hunting Behavior • Prey • Artificial Nest • Nest • Incubation • Wintering Population • Conservation • Photo Account • Eagles • A. chrysaetos • A. wahlbergi • A. verreauxi • A. nipalensis • A. pomarina • South Africa
3359	**Educational Material** • Historical Account • Eagles • Popular Monograph • A. chrysaetos • United States
12	**Egg** • Status • Nest • A. chrysaetos • Lebanon • Syria
18	**Egg** • Nest • Raptors • A. chrysaetos • Washington
67	**Egg** • Distribution • Taxonomy • Plumage • Prey • Raptors • A. verreauxi • A. rapax • Southwest Africa
79	**Egg** • Nest • Collect • A. pomarina • West Germany
193	**Egg** • Id Key • Descriptive Summary • Etymology • Habitat • Raptors • Illus. • A. nipalensis • A. rapax • A. verreauxi • Somalia
241	**Egg** • A. rapax • A. nipalensis
251	**Egg** • Descriptive Summary • Nest • Measure • Plumage • Taxonomy • Photo Account • Raptors • A. chrysaetos • A. rapax • Algeria
399	**Egg** • Nest • Eagles • A. heliaca • Spain
465	**Egg** • Behavior • Nest • Abandonment • Raptors • A. audax • Australia
468	**Egg** • Nest Chronology • Nest • Prey • A. nipalensis • Russia
574	**Egg** • Status • Nest • Shooting • Raptors • A. heliaca • Cyprus
616	**Egg** • Nest • Habitat • Prey • Vocalization • Growth • Trapping • Poisoning • Behavior • Lift • Photo Account • A. chrysaetos • Montana
617	**Egg** • Behavior • Mortality • Raptors • A. chrysaetos • Montana
635	**Egg** • Nest • Measure • Prey • A. audax • Australia
642	**Egg** • Field Id • Nest • Raptors • A. audax • Australia
652	**Egg** • A. chrysaetos • Great Britain
659	**Egg** • A. chrysaetos • Great Britain
710	**Egg** • Nest • A. verreauxi • Zambia
722	**Egg** • Nest • Behavior • Raptors • A. chrysaetos • A. pomarina • Rumania
752	**Egg** • Photo Account • A. chrysaetos • West Germany
803	**Egg** • Nest • Raptors • A. chrysaetos • Yugoslavia
856	**Egg** • Nest • Prey • Shooting • A. chrysaetos • Austria
893	**Egg** • Nest • Habitat • Photo Account • A. chrysaetos • California
894	**Egg** • Intraspecific Behavior • Nest • Habitat • Measure • Prey • Photo Account • A. chrysaetos • California

Master No.	Keywords
895	Egg • Nest • Territory • Measure • Intraspecific Behavior • Courtship • Interspecific Behavior • Hunting Behavior • A. chrysaetos • California
959	Egg • Telemetry • Tech. • Equipment • A. chrysaetos • North America
965	Egg • Status • Nest • Taxonomy • Raptors • A. chrysaetos • A. heliaca • A. nipalensis • A. clanga • Yugoslavia • Bulgaria
986	Egg • Nest • A. pomarina • Hungary
987	Egg • Descriptive Summary • Field Id • Distribution • Nest • Status • Taxonomy • Illus. • A. chrysaetos • A. heliaca • A. nipalensis • A. rapax • A. clanga • A. pomarina • A. verreauxi • Africa
1001	Egg • Status • Collect • Raptors • A. audax • Australia
1003	Egg • Nest • Habitat • Photo Account • A. chrysaetos • Scotland
1031	Egg • Residues • Prey • Raptors • Mercury • A. chrysaetos • Alberta • Saskatchewan
1034	Egg • Nest • Prey • Measure • Raptors • A. verreauxi • South Africa
1047	Egg • Longevity • Captivity • A. chrysaetos • Great Britain
1102	Egg • Abandonment • Disturbance • Mortality • Nestling • Photography • Banding • Raptors • Tech. • A. chrysaetos • North America
1149	Egg • Parental Behavior • A. verreauxi
1152	Egg • Nest • Parental Behavior • Incubation • A. verreauxi • Rhodesia
1231	Egg • Descriptive Summary • Status • Incubation • Raptors • A. pomarina • A. clanga • A. chrysaetos • West Germany
1304	Egg • Albinism • Abnormality • A. chrysaetos • North America
1320	Egg • Captivity • Raptors • A. audax • Australia
1321	Egg • Captivity • Raptors • A. audax • Australia
1433	Egg • Incubation • Measure • Theory • Raptors • A. heliaca • A. nipalensis • A. pomarina
1462	Egg • A. chrysaetos • West Germany
1496	Egg • Measure • Status • Pesticides • Raptors • A. chrysaetos • California
1503	Egg • Captivity • Longevity • Plumage • A. rapax • Tunisia
1607	Egg • Pigmentation • A. chrysaetos • British Columbia
1662	Egg • Nest • A. chrysaetos • New Mexico
1667	Egg • Nest • A. rapax • India
1668	Egg • Status • Etymology • Prey • Nest • Raptors • A. heliaca • A. nipalensis • A. rapax • A. clanga • A. pomarina • India
1703	Egg • Egg Collection • A. verreauxi • A. heliaca • A. nipalensis • A. clanga • A. chrysaetos • A. rapax • Europe
1706	Egg • Clutch Size • A. chrysaetos • Scotland
1716	Egg • Nest • Plumage • Clutch Size • A. pomarina • West Germany
1730	Egg • Pesticides • Heavy Metals • Raptors • DDE • Mercury • A. chrysaetos • North America
1806	Egg • Nest • Clutch Size • Measure • Raptors • A. wahlbergi • Rhodesia
1814	Egg • Nest • Eagles • A. pomarina • West Germany
1835	Egg • A. rapax • A. verreauxi • South Africa
1892	Egg • Nest Survey • Clutch Size • Measure • A. wahlbergi • Rhodesia
1949	Egg • Reproductive Success • Pesticides • Nest Failure • Disturbance • A. chrysaetos • Scotland
2006	Egg • Descriptive Summary • Field Id • Status • Distribution • A. nipalensis • A. rapax • A. pomarina • A. wahlbergi • Africa
2092	Egg • Field Id • Descriptive Summary • Hunting Behavior • Animal Damage • Nest • Attitude • Piracy • Raptors • A. audax • Australia
2229	Egg • Nest • Raptors • A. chrysaetos • A. clanga • Great Britain • Scotland
2274	Egg • A. chrysaetos • Great Britain
2275	Egg • A. chrysaetos • Europe
2290	Egg • Nest • Habitat • Raptors • A. chrysaetos • Wyoming
2339	Egg • Nest • Raptors • A. audax • Australia
2340	Egg • Descriptive Summary • Plumage • Distribution • Prey • Nest • Photo Account • Raptors • A. audax • Australia
2360	Egg • Nest • Survey • Photo Account • A. chrysaetos • Europe
2392	Egg • Nest • A. chrysaetos • California
2406	Egg • Nest • Raptors • A. chrysaetos • A. wahlbergi • Malawi
2410	Egg • Pellet • Nest • Photo Account • A. chrysaetos • Scotland
2473	Egg • Nest • Photo Account • A. chrysaetos • California
2485	Egg • Nest • Habitat • A. wahlbergi • Africa
2507	Egg • Nest • Clutch Size • Photo Account • Popular Monograph • A. nipalensis • United Soviet Socialists Republic
2518	Egg • Raptors • A. chrysaetos
2560	Egg • Nest Failure • Raptors • A. chrysaetos • Great Britain
2562	Egg • Pesticides • Raptors • A. chrysaetos • Great Britain
2565	Egg • Egg Measure • Pesticides • Raptors • A. chrysaetos • Scotland
2578	Egg • Nest • Raptors • A. chrysaetos • North America
2588	Egg • A. chrysaetos • Scotland
2599	Egg • Abandonment • Illus. • A. chrysaetos • Switzerland
2664	Egg • Raptors • A. verreauxi • Somalia
2665	Egg • Raptors • A. chrysaetos • A. nipalensis • A. heliaca • A. rapax • A. pomarina • A. clanga • A. audax
2709	Egg • Nest • Collect • Etymology • Raptors • A. chrysaetos • A. clanga • A. rapax • Algeria
2762	Egg • A. chrysaetos • West Germany
2764	Egg • Raptors • A. rapax • Southwest Africa
2765	Egg • Nest • Collect • A. rapax • Southwest Africa
2795	Egg • Nest • Status • Raptors • A. rapax • Nigeria
2836	Egg • Nest • A. heliaca • Yugoslavia
2837	Egg • Nest • Habitat • Raptors • A. clanga • Greece
2847	Egg • Nest • Growth • (Title Only) • A. pomarina • Czechoslovakia
2970	Egg • Nest Survey • Clutch Size • Measure • A. wahlbergi • Rhodesia
3000	Egg • Museum Specimen • Prey • Raptors • A. chrysaetos • A. nipalensis • China
3055	Egg • Etymology • Nest • Raptors • A. rapax • A. wahlbergi • Rhodesia

Master No.	Keywords
3058	**Egg** • Habitat • Prey • Prey Behavior • Nest • Animal Damage • Shooting • Behavior • A. verreauxi • South Africa
3066	**Egg** • A. pomarina
3067	**Egg** • Taxonomy • Anatomy • Egg Measure • Raptors • A. chrysaetos • A. heliaca • A. pomarina • A. clanga • A. nipalensis
3068	**Egg** • Taxonomy • Anatomy • Egg Measure • Raptors • A. chrysaetos • A. heliaca • A. pomarina • A. clanga • A. nipalensis
3235	**Egg** • Growth • Nest • Photography • Prey • Behavior • Hunting Behavior • Vocalization • Nest Adornment • Raptors • Popular Monograph • A. rapax • Kenya
3239	**Egg** • Temperature • Telemetry • Measure • Incubation • Tech. • A. chrysaetos • United States
3272	**Egg** • Nest • Habitat • Raptors • A. verreauxi • A. rapax • A. nipalensis • Rhodesia
3300	**Egg** • A. chrysaetos • Scotland
3369	**Egg** • Nest Survey • Vocalization • Habitat • Behavior • Hunting Behavior • Photo Account • A. chrysaetos • Arizona
3388	**Egg** • Photo Account • Raptors • A. verreauxi • A. rapax • A. wahlbergi • South Africa
3404	**Egg** • A. chrysaetos • A. heliaca • A. rapax • A. nipalensis • A. pomarina • A. wahlbergi • A. audax
3406	**Egg** • Habitat • Nest • Nestling • Raptors • A. chrysaetos • Korea
6	**Egg Collection** • Nest • Measure • Raptors • A. chrysaetos • California
102	**Egg Collection** • Conservation • A. chrysaetos • California
254	**Egg Collection** • Raptors • A. chrysaetos • West Germany
270	**Egg Collection** • Nest • Measure • A. chrysaetos • California
271	**Egg Collection** • Behavior • Nest • Vocalization • Behavior • Photo Account • Popular Monograph • A. chrysaetos • California
275	**Egg Collection** • Photo Account • A. chrysaetos • North America
277	**Egg Collection** • Nest • Photo Account • A. chrysaetos • California
278	**Egg Collection** • Photo Account • A. chrysaetos • North America
316	**Egg Collection** • Photography • A. chrysaetos • California
412	**Egg Collection** • Nest • Behavior • Cain/Abel • A. verreauxi • South Africa
442	**Egg Collection** • Nest • Raptors • A. verreauxi • A. chrysaetos • A. pomarina • A. heliaca • A. nipalensis • A. clanga • A. rapax
464	**Egg Collection** • Nest • Habitat • Behavior • Vocalization • Animal Damage • A. audax • Australia
466	**Egg Collection** • Distribution • Attitude • Prey • Raptors • A. audax • Australia
600	**Egg Collection** • A. chrysaetos • California
647	**Egg Collection** • A. chrysaetos • California
648	**Egg Collection** • A. chrysaetos • California
649	**Egg Collection** • Measure • A. chrysaetos • California
658	**Egg Collection** • Measure • A. chrysaetos • Great Britain
675	**Egg Collection** • Habitat • A. chrysaetos • California
724	**Egg Collection** • Nest • Raptors • A. chrysaetos • A. pomarina • Rumania
797	**Egg Collection** • Distribution • Nest • Etymology • Raptors • A. chrysaetos • A. heliaca • A. clanga • Turkey
815	**Egg Collection** • Status • Collect • Raptors • A. clanga • A. rapax • A. heliaca • India
919	**Egg Collection** • Nest • A. chrysaetos • A. pomarina • United Soviet Socialists Republic
1216	**Egg Collection** • A. chrysaetos • Scotland
1584	**Egg Collection** • A. chrysaetos • Wyoming
1585	**Egg Collection** • A. chrysaetos • United States
1643	**Egg Collection** • Poisoning • Measure • Raptors • A. audax • Australia
1701	**Egg Collection** • Raptors • A. heliaca • A. rapax • A. clanga • A. pomarina
1703	**Egg Collection** • Egg • A. verreauxi • A. heliaca • A. nipalensis • A. clanga • A. chrysaetos • A. rapax • Europe
1844	**Egg Collection** • Nest • Collect • Raptors • A. chrysaetos • China
1878	**Egg Collection** • Raptors • A. verreauxi • South Africa
1886	**Egg Collection** • Prey • A. verreauxi • Kenya
1891	**Egg Collection** • Nest Chronology • Behavior • Nest • Raptors • A. wahlbergi • Rhodesia
1898	**Egg Collection** • Status • Nest • A. chrysaetos • North Carolina
2235	**Egg Collection** • Nest • A. chrysaetos • Colorado
2337	**Egg Collection** • Measure • A. chrysaetos • California
2338	**Egg Collection** • Raptors • A. verreauxi • A. heliaca • A. nipalensis • A. rapax • A. audax • A. pomarina • A. clanga
2550	**Egg Collection** • Nest • A. audax • Australia
2633	**Egg Collection** • A. rapax • South Africa
2634	**Egg Collection** • Nest • A. wahlbergi • South Africa
2636	**Egg Collection** • Collect • Raptors • A. wahlbergi • A. rapax • South Africa
2645	**Egg Collection** • Conservation • Shooting • Poisoning • Nest Robbing • Protection • Trade • Trapping • Raptors • A. chrysaetos • Great Britain
2805	**Egg Collection** • Measure • A. chrysaetos • California
2851	**Egg Collection** • Nest • Habitat • Measure • Photo Account • A. chrysaetos • California
2869	**Egg Collection** • Nest • Eagles • A. chrysaetos • United States
2943	**Egg Collection** • Nest • Measure • A. chrysaetos • California
2947	**Egg Collection** • Nest • Behavior • A. chrysaetos • California
3091	**Egg Collection** • Nest • A. chrysaetos • California
3094	**Egg Collection** • Nest • Behavior • A. chrysaetos • California
3107	**Egg Collection** • Habitat • Nest • Photo Account • A. chrysaetos • Colorado
3355	**Egg Collection** • A. audax • Australia
3371	**Egg Collection** • Nest • Raptors • A. chrysaetos • California
61	**Egg Measure** • Raptors • A. chrysaetos • North America
325	**Egg Measure** • Monograph • Breeding Biology • Density • Mortality • Cain/Abel • Prey • Abnormal Behavior • Marking • A. chrysaetos • Idaho

Permuted List of Keywords (Cont.)

Master No.	Keywords
1293	**Egg Measure** • Captive Breeding • Behavior • Artificial Insemination • Artificial Incubation • Photography Account • A. chrysaetos • United States
1369	**Egg Measure** • Raptors • A. chrysaetos • North America
2563	**Egg Measure** • Abnormal Behavior • Nest Failure • Pesticides • Raptors • A. chrysaetos • Great Britain
2565	**Egg Measure** • Egg • Pesticides • Raptors • A. chrysaetos • Scotland
3067	**Egg Measure** • Taxonomy • Egg • Anatomy • Raptors • A. chrysaetos • A. heliaca • A. pomarina • A. clanga • A. nipalensis
3068	**Egg Measure** • Taxonomy • Egg • Anatomy • Raptors • A. chrysaetos • A. heliaca • A. pomarina • A. clanga • A. nipalensis
3195	**Egg Measure** • Taxonomy • Photo Account • Raptors • A. chrysaetos • A. heliaca • A. rapax
3341	**Egg Measure** • Status • Raptors • A. heliaca • A. nipalensis • A. rapax • India
66	**Electrocution** • Mortality • Management • Nest • A. chrysaetos • Idaho
169	**Electrocution** • Management • A. chrysaetos • United States
187	**Electrocution** • Management • Raptors • A. chrysaetos • Idaho
349	**Electrocution** • Behavior • Raptors • A. chrysaetos • North America
352	**Electrocution** • Mortality • Management • A. chrysaetos • North America
392	**Electrocution** • Prey • Wintering Population • A. chrysaetos • Denmark
423	**Electrocution** • Aerial Survey • Wintering Population • Nest Survey • Banding • Roadside Survey • A. chrysaetos • North America
426	**Electrocution** • Eagles • A. chrysaetos • North America
888	**Electrocution** • A. chrysaetos • North America
947	**Electrocution** • Roadside Survey • Roost • Nest Survey • Prey • Hunting Behavior • Capture • Marking • Interspecific Behavior • Shooting • A. chrysaetos • Utah
960	**Electrocution** • Artificial Nest • Disturbance • Raptors • A. chrysaetos • North America
1014	**Electrocution** • A. chrysaetos • Italy
1580	**Electrocution** • Management • Mortality • Monograph • A. chrysaetos • North America
1660	**Electrocution** • Captivity • A. chrysaetos • California
1776	**Electrocution** • Nest Survey • Refuge • Winter Survey • Density • Shooting • Mortality • Prey • Predator/Prey • Raptors • SRBPNA • A. chrysaetos • Idaho
1777	**Electrocution** • Predator/Prey • Aerial Survey • A. chrysaetos • North America
1879	**Electrocution** • Trapping • Management • Eagles • A. chrysaetos • United States
1963	**Electrocution** • A. chrysaetos • California
2040	**Electrocution** • Management • Mortality • Raptors • A. chrysaetos • Idaho
2255	**Electrocution** • Density • Predator/Prey • Nest • Prey • A. nipalensis • Russia
2263	**Electrocution** • Management • Habitat Loss • A. chrysaetos • North America
2292	**Electrocution** • Historical Account • Management • Raptors • A. chrysaetos • North America
2297	**Electrocution** • Artificial Nest • Raptors • A. chrysaetos • United States
2298	**Electrocution** • Mortality • Shooting • Artificial Nest. • Eagles • A. chrysaetos • Idaho
2299	**Electrocution** • Artificial Nest • Conservation • Management • Shooting • Eagles • Popular Monograph • A. chrysaetos • Idaho
2369	**Electrocution** • Prey • Folklore • Lift • Animal Damage • Conservation • Eagles • Popular Monograph • A. chrysaetos • North America
2377	**Electrocution** • Conservation • Management • Status • Literature Review • Behavior • Eagles • A. chrysaetos • North America
2432	**Electrocution** • Mortality • Raptors • A. chrysaetos • North America
2818	**Electrocution** • Mortality • A. nipalensis • Russia
2832	**Electrocution** • Mortality • A. chrysaetos • United States
2855	**Electrocution** • Raptors • A. chrysaetos • Utah
2928	**Electrocution** • Mortality • Habitat Loss • Nest • Survey • Raptors • A. chrysaetos • Colorado
2987	**Electrocution** • A. chrysaetos • California
3191	**Electrocution** • Conservation • Shooting • Poisoning • Attitude • Eagles • A. chrysaetos • United States
3192	**Electrocution** • Protection • Poisoning • Shooting • Attitude • Popular Monograph • A. chrysaetos • Wyoming
3201	**Electrocution** • Management • Eagles • A. chrysaetos • United States
3293	**Electrocution** • Trapping • Raptors • A. chrysaetos • Vermont
1741	**Energy Development** • Environmental Impact • Management • Raptors • A. chrysaetos • North America
31	**Environmental Impact** • Disturbance • Nuclear Test • Raptors • A. chrysaetos • Wyoming
1160	**Environmental Impact** • Conservation • (Title Only) • A. heliaca • Spain
1741	**Environmental Impact** • Energy Development • Management • Raptors • A. chrysaetos • North America
2264	**Environmental Impact** • Nest • Management • Photo Account • Raptors • A. chrysaetos • Utah
2265	**Environmental Impact** • Nest • Management • Photo Account • Raptors • A. chrysaetos • Utah
2266	**Environmental Impact** • Nest • Management • Photo Account • Raptors • A. chrysaetos • Utah
2267	**Environmental Impact** • Nest • Management • Photo Account • Raptors • A. chrysaetos • Utah
3031	**Environmental Impact** • Management • Disturbance • Habitat Protection • Raptors • A. chrysaetos • North America
3202	**Environmental Impact** • Disturbance • Protection • Aircraft • Mitigation • A. chrysaetos • California
16	**Etymology** • Taxonomy • A. pomarina • West Germany
26	**Etymology** • Id Key • Descriptive Summary • Status • Distribution • Field Id • Raptors • Illus. • A. chrysaetos • A. heliaca • A. rapax • A. nipalensis • A. clanga • A. pomarina • India
56	**Etymology** • Descriptive Summary • Clutch Size • Plumage • Habitat • Museum Specimen • Measure • A. rapax • A. pomarina • India

Master No.	Keywords
193	**Etymology** • Id Key • Descriptive Summary • Habitat • Egg • Raptors • Illus. • A. nipalensis • A. rapax • A. verreauxi • Somalia
215	**Etymology** • Descriptive Summary • A. chrysaetos • Labrador • Newfoundland
216	**Etymology** • Status • Raptors • A. chrysaetos • A. heliaca • A. clanga • Korea
218	**Etymology** • Descriptive Summary • Rarity • Prey • Distribution • Raptors • A. chrysaetos • A. heliaca • Japan
242	**Etymology** • Descriptive Summary • Id Key • Measure • Plumage • Raptors • A. chrysaetos • A. heliaca • A. nipalensis • A. rapax • A. clanga • A. pomarina • India
261	**Etymology** • Id Key • Descriptive Summary • Distribution • Prey • A. rapax • A. wahlbergi • Gambia • Sierra Leone • Nigeria
262	**Etymology** • Descriptive Summary • Status • Nest • Behavior • Plumage • A. heliaca • A. nipalensis • A. clanga • A. chrysaetos • Cyprus
264	**Etymology** • Migration • Measure • Raptors • A. chrysaetos • A. pomarina • Malawi
315	**Etymology** • Raptors • A. chrysaetos • Pennsylvania
403	**Etymology** • Trade • Raptors • A. chrysaetos • Japan
404	**Etymology** • Descriptive Summary • Symbolism • Raptors • A. chrysaetos • North America
405	**Etymology** • Status • A. wahlbergi • A. rapax • Africa
430	**Etymology** • Status • Raptors • A. chrysaetos • A. pomarina • A. heliaca • Bulgaria
443	**Etymology** • Eagles • A. verreauxi • South Africa
447	**Etymology** • Raptors • A. chrysaetos • A. heliaca • A. nipalensis • A. pomarina • A. clanga • France
509	**Etymology** • Taxonomy • Plumage • A. nipalensis • A. clanga • A. rapax • A. pomarina • India
510	**Etymology** • Taxonomy • Plumage • A. heliaca • A. nipalensis • A. clanga • A. pomarina • India
511	**Etymology** • Taxonomy • Plumage • A. rapax • India
513	**Etymology** • Taxonomy • Plumage • A. clanga • A. pomarina • India
514	**Etymology** • Taxonomy • Plumage • Raptors • A. clanga • A. pomarina • A. rapax
534	**Etymology** • Habitat • Density • Nest Survey • Breeding Biology • Interspecific Competition • Prey • Reproductive Success • Plumage • Vocalization • Nest Adornment • Raptors • A. verreauxi • A. wahlbergi • Kenya
535	**Etymology** • Nest Survey • Breeding Biology • Prey • Reproductive Success • Interspecific Competition • Raptors • Photo Account • A. wahlbergi • A. verreauxi • Kenya
666	**Etymology** • A. chrysaetos • A. heliaca • A. nipalensis • A. pomarina • A. clanga • Hungary
677	**Etymology** • A. chrysaetos
678	**Etymology** • Descriptive Summary • Status • Raptors • A. clanga • A. heliaca • A. chrysaetos • Japan
718	**Etymology** • Raptors • A. audax • Australia
729	**Etymology** • Symbolism • A. chrysaetos • Minnesota
746	**Etymology** • A. chrysaetos
748	**Etymology** • Descriptive Summary • Raptors • A. chrysaetos • North America
770	**Etymology** • Field Id • Descriptive Summary • Monograph • Distribution • Status • Migration • Prey • Plumage • Measure • Vocalization • A. pomarina • A. clanga • A. nipalensis • A. heliaca • A. chrysaetos • A. verreauxi • Africa • Europe
779	**Etymology** • A. chrysaetos • A. heliaca • A. pomarina • Yugoslavia
797	**Etymology** • Distribution • Nest • Egg Collection • Raptors • A. chrysaetos • A. heliaca • A. clanga • Turkey
824	**Etymology** • Distribution • Field Id • A. chrysaetos • Mexico
871	**Etymology** • Distribution • Life History • Habitat • Molt • Prey • Measure • Field Id • Migration • Status • Raptors • A. chrysaetos • A. heliaca • A. rapax • A. clanga • A. pomarina • United Soviet Socialists Republic
879	**Etymology** • Rarity • Raptors • A. heliaca • A. chrysaetos • A. pomarina • Malawi
880	**Etymology** • Status • Raptors • A. chrysaetos • A. pomarina • A. heliaca • Malawi
882	**Etymology** • Taxonomy • A. chrysaetos • A. clanga • A. rapax • A. audax • A. verreauxi • A. gurneyi • A. wahlbergi • A. pomarina • A. clanga • A. heliaca
883	**Etymology** • Descriptive Summary • Behavior • Prey • Display • A. heliaca • A. rapax • A. pomarina • A. clanga • India
1008	**Etymology** • Descriptive Summary • Status • Nest Chronology • Prey • Raptors • A. chrysaetos • A. heliaca • A. pomarina • A. clanga • Switzerland
1080	**Etymology** • Collect • Nest • Literature Review • Distribution • A. chrysaetos • A. rapax • Tunisia
1105	**Etymology** • Descriptive Summary • Plumage • Nest • Distribution • A. chrysaetos • Alaska
1173	**Etymology** • Status • Raptors • A. chrysaetos • A. heliaca • A. clanga • Russia
1212	**Etymology** • Descriptive Summary • Taxonomy • Nest • Prey • A. chrysaetos • Switzerland
1312	**Etymology** • Raptors • A. chrysaetos • North America
1395	**Etymology** • Status • Prey • Nest • Raptors • A. chrysaetos • A. heliaca • A. rapax • Morocco
1534	**Etymology** • A. chrysaetos • North America
1612	**Etymology** • Clutch Size • Nest • A. chrysaetos • Israel
1626	**Etymology** • Migration • Measure • Trapping • A. chrysaetos • Alaska
1634	**Etymology** • Raptors • A. chrysaetos • A. heliaca • A. clanga • Yugoslavia
1649	**Etymology** • Status • A. chrysaetos • Japan
1666	**Etymology** • Nest • Status • Raptors • A. rapax • India
1668	**Etymology** • Status • Prey • Nest • Egg • Raptors • A. heliaca • A. nipalensis • A. rapax • A. clanga • A. pomarina • India
1670	**Etymology** • Descriptive Summary • Measure • Trapping • Raptors • A. chrysaetos • Washington
1671	**Etymology** • Descriptive Summary • A. chrysaetos • A. heliaca • A. nipalensis • A. clanga

Master No.	Keywords
1697	**Etymology** • Plumage • Status • Distribution • Taxonomy • Field Id • Literature Review • Eagles • A. pomarina • A. clanga • France • Belgium • Switzerland • Italy
1700	**Etymology** • Nest • Nest Adornment • Raptors • A. chrysaetos • Corsica
1712	**Etymology** • Rarity • Raptors • A. chrysaetos • A. clanga • Netherlands
1785	**Etymology** • Descriptive Summary • Status • Collect • Raptors • A. chrysaetos • A. heliaca • A. pomarina • A. clanga • Egypt
1808	**Etymology** • Taxonomy • Raptors • A. chrysaetos • A. heliaca • A. clanga • Yugoslavia
1821	**Etymology** • Status • Nest • Raptors • A. chrysaetos • A. pomarina • A. clanga • Poland
1916	**Etymology** • Distribution • Raptors • A. chrysaetos • A. heliaca • A. clanga • Spain
1924	**Etymology** • Status • A. chrysaetos • A. pomarina • A. clanga • A. heliaca • Yugoslavia
2135	**Etymology** • A. chrysaetos • Finland
2142	**Etymology** • Raptors • A. verreauxi • A. wahlbergi • A. rapax • South Africa
2180	**Etymology** • A. chrysaetos • Tunisia
2207	**Etymology** • Raptors • A. chrysaetos • A. heliaca • A. nipalensis • A. clanga • Italy
2208	**Etymology** • Status • Raptors • A. chrysaetos • Italy
2245	**Etymology** • Nest • Prey • Vocalization • Raptors • A. verreauxi • A. rapax • A. wahlbergi • Rhodesia
2249	**Etymology** • Raptors • A. verreauxi • South Africa
2342	**Etymology** • Status • Habitat • Interspecific Behavior • Collect • Nest • Raptors • A. verreauxi • A. nipalensis • A. rapax • Africa
2456	**Etymology** • Field Id • Distribution • A. chrysaetos • A. heliaca • A. rapax • A. clanga • A. pomarina • Europe
2581	**Etymology** • Descriptive Summary • Raptors • Photo Account • A. verreauxi • A. rapax • Lesotho
2637	**Etymology** • Descriptive Summary • Field Id • Distribution • Prey • Raptors • A. verreauxi • A. rapax • A. nipalensis • A. pomarina • A. wahlbergi • South Africa
2675	**Etymology** • Collect • Prey • Wintering Population • Status • Raptors • A. chrysaetos • A. clanga • Yugoslavia
2709	**Etymology** • Nest • Collect • Egg • Raptors • A. chrysaetos • A. clanga • A. rapax • Algeria
2776	**Etymology** • Collect • Falconry • Raptors • A. chrysaetos • China
2879	**Etymology** • Descriptive Summary • Status • Distribution • Raptors • A. chrysaetos • Canada
2922	**Etymology** • Descriptive Summary • Raptors • A. chrysaetos • North America
3055	**Etymology** • Nest • Egg • Raptors • A. rapax • A. wahlbergi • Rhodesia
3147	**Etymology** • Status • Raptors • A. chrysaetos • A. heliaca • A. nipalensis • A. rapax • A. clanga • India
3193	**Etymology** • Status • Raptors • A. chrysaetos • Labrador • Quebec
3342	**Etymology** • Status • Raptors • A. chrysaetos • A. heliaca • A. nipalensis • A. rapax • India
3442	**Etymology** • Status • Raptors • A. chrysaetos • A. clanga • Italy
3447	**Etymology** • Collect • Raptors • A. rapax • A. chrysaetos • Tunisia
143	**Falconry** • Conservation • Raptors • A. rapax • A. chrysaetos • New York
146	**Falconry** • Artificial Nest • Raptors • A. chrysaetos • California
186	**Falconry** • Abnormal Behavior • A. chrysaetos • Siberia
196	**Falconry** • Nest Survey • Habitat • Wintering Population • Attitude • Raptors • A. chrysaetos • Arizona
306	**Falconry** • Behavior • Raptors • Photo Account • A. chrysaetos • West Germany
323	**Falconry** • Hunting Behavior • Field Id • Plumage • Photography Account • Raptors • A. chrysaetos
559	**Falconry** • Story • Measure • Photo Account • A. chrysaetos • Kirgizia • Great Britain
631	**Falconry** • Conservation • Story • Trapping • Rehabilitation • Trade • Wintering Population • Captivity • Raptors • Popular Monograph • A. chrysaetos • New York
767	**Falconry** • Telemetry • Protection • Nest • Behavior • Pesticides • Photo Account • Popular Monograph • A. chrysaetos • United States
863	**Falconry** • Prey • Raptors • A. chrysaetos • Kirgizia
867	**Falconry** • Photo Account • A. chrysaetos • Turkmenia
868	**Falconry** • Illus. • A. chrysaetos • United Soviet Socialists Republic
869	**Falconry** • Raptors • A. chrysaetos • United Soviet Socialists Republic
870	**Falconry** • Raptors • A. chrysaetos • United Soviet Socialists Republic
990	**Falconry** • Raptors • Photo Account • A. chrysaetos
1040	**Falconry** • Monograph • Breeding Biology • Prey • Vocalization • Territory • Taxonomy • Migration • Mortality • Molt • A. chrysaetos • A. verreauxi • A. audax
1046	**Falconry** • Protection • Photo Account • Raptors • A. chrysaetos • A. heliaca
1055	**Falconry** • Intraspecific Behavior • A. audax • Australia
1056	**Falconry** • Growth • Aging • Parental Behavior • Measure • Plumage • Dimorphism • Prey • Photo Account • A. audax • Australia
1273	**Falconry** • Mortality • Prey • Hunting Behavior • Flight • Nest • Mortality • Photo Account • Popular Monograph • A. chrysaetos • Great Britain
1361	**Falconry** • Story • Captive Breeding • Artificial Insemination • Photography Account • Popular Monograph • A. chrysaetos • United States
1386	**Falconry** • Prey • Photography • Photo Account • Popular Monograph • A. chrysaetos • Scotland
1770	**Falconry** • Behavior • Story • Eagles • Photography Account • A. chrysaetos • Great Britain
1894	**Falconry** • A. chrysaetos • China
1967	**Falconry** • Status • Raptors • A. chrysaetos • China
2019	**Falconry** • Management • Educational Material • Rehabilitation • Survey • Raptors • A. chrysaetos • California
2025	**Falconry** • Behavior • Flight • Eagles • Popular Monograph • A. chrysaetos • North America

Master No.	Keywords
2124	**Falconry** • Monograph • Breeding Biology • Migration • Prey • Hunting Behavior • Attack • Folklore • Animal Damage • Popular Monograph • A. chrysaetos
2293	**Falconry** • Conservation • Prey • Hunting Behavior • Raptors • Popular Monograph • A. chrysaetos • United States
2743	**Falconry** • Raptors • A. chrysaetos • West Germany
2776	**Falconry** • Collect • Etymology • Raptors • A. chrysaetos • China
2905	**Falconry** • Rehabilitation • Raptors • A. chrysaetos • California
3018	**Falconry** • Story • Behavior • Illus. • A. chrysaetos • Great Britain
3182	**Falconry** • Story • Hunting Behavior • Rehabilitation • Measure • Popular Monograph • A. chrysaetos • Texas
3236	**Falconry** • Rehabilitation • Shooting • Wintering Population • Photo Account • A. chrysaetos • Missouri
3441	**Falconry** • Nest • Breeding Biology • Prey • A. chrysaetos • Kirgizia
1285	**Fat** • Collect • Parasite • Molt • Museum Specimen • A. chrysaetos • North Carolina
2234	**Fat** • Prey • Shooting • A. chrysaetos • Montana
3340	**Fat** • Collect • Nest • Raptors • A. nipalensis • A. rapax • India
124	**Feeding Behavior** • Raptors • A. rapax • Rhodesia
157	**Feeding Behavior** • A. audax • Australia
158	**Feeding Behavior** • Photo Account • A. chrysaetos • Massachusetts
273	**Feeding Behavior** • Wintering Population • Raptors • A. chrysaetos • A. nipalensis • Aden
1145	**Feeding Behavior** • Nest • Parental Behavior • Breeding Biology • Interspecific Behavior • Intraspecific Behavior • Hunting Behavior • Fledging • Prey • Display • Raptors • A. verreauxi • Rhodesia
1855	**Feeding Behavior** • Raptors • A. chrysaetos
1888	**Feeding Behavior** • Flight • A. chrysaetos • West Germany
1927	**Feeding Behavior** • A. chrysaetos • Oklahoma
2101	**Feeding Behavior** • Hunting Behavior • A. chrysaetos • California
2349	**Feeding Behavior** • Animal Damage • Capture • Predator/Prey • Protection • Management • Economics • A. chrysaetos • Montana
2672	**Feeding Behavior** • Animal Damage • Prey • Hunting Behavior • A. audax • Australia
3022	**Feeding Behavior** • Cain/Abel • Growth • Behavior • Vocalization • Pellet • Raptors • A. chrysaetos • California
3152	**Feeding Behavior** • Prey • A. heliaca • Wyoming
3367	**Feeding Behavior** • Animal Damage • Attitude • A. chrysaetos • Texas
21	**Field Id** • Raptors • A. chrysaetos • A. heliaca • A. nipalensis • A. clanga • A. pomarina • A. rapax • India
23	**Field Id** • Descriptive Summary • Status • Raptors • A. rapax • A. clanga • A. nipalensis • A. heliaca • India
24	**Field Id** • Descriptive Summary • Prey • Raptors • A. chrysaetos • A. heliaca • India
25	**Field Id** • Descriptive Summary • Distribution • Habitat • Raptors • A. rapax • India
26	**Field Id** • Id Key • Descriptive Summary • Etymology • Status • Distribution • Raptors • Illus. • A. chrysaetos • A. heliaca • A. rapax • A. nipalensis • A. clanga • A. pomarina • India
46	**Field Id** • Descriptive Summary • A. chrysaetos • A. heliaca • A. pomarina • A. nipalensis • A. clanga • Iraq
49	**Field Id** • A. chrysaetos • North America
145	**Field Id** • Eagles • A. chrysaetos • A. pomarina • A. clanga • A. heliaca • A. nipalensis • Europe
171	**Field Id** • Wintering Population • Raptors • A. clanga • A. pomarina • A. heliaca • Kenya
173	**Field Id** • Measure • Protection • Reward • Winter Survey • Eagles • A. chrysaetos • Arkansas
183	**Field Id** • A. chrysaetos • North America
263	**Field Id** • Descriptive Summary • Status • Raptors • A. heliaca • A. clanga • A. pomarina • Cyprus
322	**Field Id** • Descriptive Summary • Raptors • A. chrysaetos • British Columbia
323	**Field Id** • Falconry • Hunting Behavior • Plumage • Photography Account • Raptors • A. chrysaetos
329	**Field Id** • A. chrysaetos • Colorado
332	**Field Id** • Attitude • Prey • Raptors • Popular Monograph • A. audax • Australia
384	**Field Id** • Taxonomy • Capture • Aging • Photo Account • A. nipalensis • A. rapax • Southwest Africa
402	**Field Id** • Distribution • A. chrysaetos • Mexico
474	**Field Id** • Migration • Refuge • Conservation • Raptors • A. chrysaetos • Pennsylvania
498	**Field Id** • Taxonomy • Molt • Measure • Museum Specimen • Behavior • Wintering Population • Interspecific Competition • Conservation • Eagles • Photo Account • A. nipalensis • A. pomarina • Africa
544	**Field Id** • Taxonomy • Status • Flight • Habitat • Terminology • Distribution • Mortality • Behavior • Life History • Longevity • Territory • Photo Account • Raptors • A. chrysaetos • A. clanga • Great Britain • Ireland • Scotland
556	**Field Id** • Habitat • Raptors • A. chrysaetos • California
566	**Field Id** • Raptors • A. chrysaetos • A. heliaca • A. nipalensis • A. pomarina • A. clanga • Europe
567	**Field Id** • Raptors • A. chrysaetos • A. heliaca • A. nipalensis • A. clanga • A. pomarina • Europe
611	**Field Id** • Habitat • Nest • Nest Survey • Behavior • Photo Account • Raptors • A. chrysaetos • North America
642	**Field Id** • Nest • Egg • Raptors • A. audax • Australia
655	**Field Id** • Anatomy • Raptors • A. chrysaetos • North America
680	**Field Id** • Distribution • Wintering Population • Raptors • A. chrysaetos • A. heliaca • A. nipalensis • A. pomarina • Iraq
682	**Field Id** • Migration • Raptors • A. heliaca • A. clanga • Cyprus
683	**Field Id** • Rarity • A. heliaca • Denmark

Master No.	Keywords
685	**Field Id** • A. chrysaetos • A. heliaca • A. nipalensis • A. clanga • A. pomarina • Europe
761	**Field Id** • Status • Conservation • Raptors • A. chrysaetos • Colorado
770	**Field Id** • Descriptive Summary • Monograph • Distribution • Status • Migration • Prey • Plumage • Measure • Etymology • Vocalization • A. pomarina • A. clanga • A. nipalensis • A. heliaca • A. chrysaetos • A. verreauxi • Africa • Europe
816	**Field Id** • Plumage • Flight • Eagles • A. chrysaetos • A. nipalensis • A. wahlbergi • A. pomarina • South Africa
824	**Field Id** • Distribution • Etymology • A. chrysaetos • Mexico
871	**Field Id** • Distribution • Etymology • Life History • Habitat • Molt • Prey • Measure • Migration • Status • Raptors • A. chrysaetos • A. heliaca • A. rapax • A. clanga • A. pomarina • United Soviet Socialists Republic
907	**Field Id** • Descriptive Summary • Plumage • Prey • Measure • Raptors • A. chrysaetos • A. heliaca • A. nipalensis • A. rapax • A. clanga • India
939	**Field Id** • Nest • Habitat • Prey • Hunting Behavior • Raptors • A. clanga • A. pomarina • A. chrysaetos • Poland
987	**Field Id** • Descriptive Summary • Distribution • Nest • Status • Taxonomy • Egg • Illus. • A. chrysaetos • A. heliaca • A. nipalensis • A. rapax • A. clanga • A. pomarina • A. verreauxi • Africa
993	**Field Id** • Descriptive Summary • Life History • Photo Account • Popular Monograph • A. chrysaetos • Scotland
1059	**Field Id** • Vocalization • Raptors • A. chrysaetos • A. heliaca • A. rapax • A. nipalensis • A. clanga • A. pomarina • Nepal
1060	**Field Id** • Status • Conservation • Raptors • A. chrysaetos • Colorado
1241	**Field Id** • Status • Raptors • A. chrysaetos • Kansas
1363	**Field Id** • Migration • Descriptive Summary • Status • Prey • Raptors • A. chrysaetos • Wisconsin
1384	**Field Id** • Distribution • Raptors • A. chrysaetos • A. heliaca • A. nipalensis • A. clanga • A. pomarina • Europe
1402	**Field Id** • Eagles • A. verreauxi • A. rapax • A. wahlbergi • A. pomarina • A. nipalensis
1437	**Field Id** • Migration • Habitat • Photo Account • Raptors • Popular Monograph • A. chrysaetos • North America
1438	**Field Id** • Migration • Raptors • Photo Account • A. chrysaetos • North America
1445	**Field Id** • Distribution • Raptors • A. pomarina • A. clanga • A. heliaca • A. chrysaetos • A. verreauxi • Europe
1488	**Field Id** • A. chrysaetos • Switzerland
1520	**Field Id** • Rarity • Eagles • A. chrysaetos • A. clanga • Switzerland
1614	**Field Id** • Descriptive Summary • Raptors • A. chrysaetos • Alabama
1652	**Field Id** • Taxonomy • A. rapax • A. nipalensis
1687	**Field Id** • Monograph • Molt • Breeding Biology • Literature Review • A. chrysaetos • Colorado
1697	**Field Id** • Plumage • Status • Distribution • Taxonomy • Etymology • Literature Review • Eagles • A. pomarina • A. clanga • France • Belgium • Switzerland • Italy
1869	**Field Id** • Migration • Raptors • A. clanga • Denmark
2006	**Field Id** • Descriptive Summary • Egg • Status • Distribution • A. nipalensis • A. rapax • A. pomarina • A. wahlbergi • Africa
2013	**Field Id** • Symbolism • Conservation • Popular Monograph • A. chrysaetos • Colorado
2014	**Field Id** • Survey • Raptors • A. chrysaetos • California
2018	**Field Id** • Status • Educational Material • Raptors • A. chrysaetos • California
2034	**Field Id** • Rarity • Eagles • A. heliaca • Sweden
2062	**Field Id** • Density • Nest • Photo Account • Popular Monograph • A. pomarina • East Germany
2079	**Field Id** • Prey • Distribution • Attitude • Raptors • A. chrysaetos • United States
2092	**Field Id** • Descriptive Summary • Hunting Behavior • Animal Damage • Nest • Egg • Attitude • Piracy • Raptors • A. audax • Australia
2148	**Field Id** • Distribution • Management • Taxonomy • Habitat • Nest Survey • Prey • Breeding Biology • Photo Account • Eagles • A. pomarina • Europe
2341	**Field Id** • Descriptive Summary • Nest • Raptors • A. verreauxi • A. rapax • A. wahlbergi • Kenya
2413	**Field Id** • Attitude • Folklore • Descriptive Summary • Raptors • A. chrysaetos • United States
2437	**Field Id** • Wintering Population • A. pomarina • A. nipalensis • Kenya
2454	**Field Id** • Raptors • A. chrysaetos • North America
2455	**Field Id** • A. chrysaetos • Texas
2456	**Field Id** • Distribution • Etymology • A. chrysaetos • A. heliaca • A. rapax • A. clanga • A. pomarina • Europe
2479	**Field Id** • Migration • A. clanga • Morocco
2505	**Field Id** • Protection • Photo Account • A. chrysaetos • A. clanga • A. heliaca • A. nipalensis • A. pomarina • Europe
2535	**Field Id** • Status • Distribution • Behavior • Raptors • A. verreauxi • A. rapax • A. nipalensis • A. pomarina • A. wahlbergi • South Africa
2630	**Field Id** • Distribution • Raptors • A. chrysaetos • North America
2637	**Field Id** • Descriptive Summary • Distribution • Etymology • Prey • Raptors • A. verreauxi • A. rapax • A. nipalensis • A. pomarina • A. wahlbergi • South Africa
2796	**Field Id** • Raptors • A. wahlbergi • A. rapax • Africa
2860	**Field Id** • Wintering Population • Migration • A. nipalensis • A. pomarina • A. clanga • A. heliaca • Africa
2909	**Field Id** • Nest • Trapping • Eagles • Popular Monograph • A. chrysaetos • Wyoming
2942	**Field Id** • Descriptive Summary • Photo Account • A. pomarina • West Germany

Master No.	Keywords
2969	**Field Id** • Nest Survey • Habitat • Prey • Breeding Biology • Reproductive Success • Growth • Vocalization • Migration • Behavior • Eagles • Photo Account • A. wahlbergi • A. rapax • Rhodesia
3040	**Field Id** • Plumage • Measure • Eagles • Photo Account • A. chrysaetos • A. pomarina • Europe
3080	**Field Id** • Cain/Abel • Distribution • Educational Material • Habitat • Hunting Behavior • Prey • Artificial Nest • Nest • Incubation • Wintering Population • Conservation • Photo Account • Eagles • A. chrysaetos • A. wahlbergi • A. verreauxi • A. nipalensis • A. pomarina • South Africa
3083	**Field Id** • Raptors • A. chrysaetos • Canada
3103	**Field Id** • Distribution • Status • Conservation • Photo Account • Raptors • A. chrysaetos • France
3160	**Field Id** • Wintering Population • Survey • Territory • Age Structure • Supplemental Feeding • Illus. • A. chrysaetos • Sweden
3212	**Field Id** • Rarity • (Title Only) • A. heliaca • A. nipalensis • Sweden
1144	**Fledging** • Nest Survey • Status • Behavior • Mortality • Banding • Marking • Cain/Abel • Reproductive Success • Breeding Biology • Interspecific Behavior • Hunting Behavior • Display • Courtship • Raptors • A. verreauxi • Rhodesia
1145	**Fledging** • Nest • Parental Behavior • Breeding Biology • Interspecific Behavior • Intraspecific Behavior • Feeding Behavior • Hunting Behavior • Prey • Display • Raptors • A. verreauxi • Rhodesia
1148	**Fledging** • Banding • Marking • A. verreauxi • Rhodesia
1202	**Fledging** • A. chrysaetos • Great Britain
1267	**Fledging** • Flight • Behavior • A. chrysaetos • Great Britain
1497	**Fledging** • Flight • Behavior • A. chrysaetos • Idaho • Oregon
2669	**Fledging** • Nest • Parental Behavior • Food Habits • Breeding Biology • A. chrysaetos • Tanzania
619	**Fledgling** • Behavior • Nest • Photo Account • Popular Monograph • A. chrysaetos • Montana
2848	**Fledgling** • Biology • (Title Only) • A. pomarina • West Germany
60	**Flight** • Interspecific Behavior • Abnormal Behavior • A. chrysaetos • New Mexico
75	**Flight** • Intraspecific Behavior • (Title Only) • A. chrysaetos • North America
86	**Flight** • Historical Account • Status • Attitude • Prey • Photo Account • A. chrysaetos • Great Britain
94	**Flight** • Folklore • Banding • Capture • A. chrysaetos • A. pomarina • Europe
107	**Flight** • Nest Adornment • Interspecific Behavior • Photography • A. chrysaetos • Europe
123	**Flight** • Trapping • Captivity • Photo Account • A. chrysaetos • North America
221	**Flight** • Attack • Lift • Altitude • Prey Behavior • A. chrysaetos • A. pomarina
395	**Flight** • Interspecific Behavior • Raptors • Popular Monograph • A. chrysaetos • Montana
544	**Flight** • Taxonomy • Status • Field Id • Habitat • Terminology • Distribution • Mortality • Behavior • Life History • Longevity • Territory • Photo Account • Raptors • A. chrysaetos • A. clanga • Great Britain • Ireland • Scotland
615	**Flight** • Altitude • A. audax • Australia
618	**Flight** • Prey • Nest • Attitude • Behavior • Breeding Biology • Poisoning • Molt • Hunting Behavior • Vocalization • Folklore • Lift • Photo Account • A. chrysaetos • Montana
708	**Flight** • A. chrysaetos • California
731	**Flight** • Behavior • A. chrysaetos • Great Britain
816	**Flight** • Field Id • Plumage • Eagles • A. chrysaetos • A. nipalensis • A. wahlbergi • A. pomarina • South Africa
850	**Flight** • Display • Raptors • A. chrysaetos • California
972	**Flight** • Anatomy • Raptors • A. chrysaetos • North America
976	**Flight** • A. chrysaetos • Japan
1033	**Flight** • Habitat • Prey • Nest • Plumage • Raptors • A. wahlbergi • A. rapax • South Africa
1185	**Flight** • Intraspecific Behavior • Plumage • Illus. • A. clanga • Yugoslavia
1254	**Flight** • Courtship • Intraspecific Behavior • Interspecific Behavior • A. chrysaetos • Great Britain
1259	**Flight** • Poetry • Captivity • Attitude • Protection • Folklore • Raptors • A. chrysaetos • Great Britain
1267	**Flight** • Fledging • Behavior • A. chrysaetos • Great Britain
1268	**Flight** • Intraspecific Behavior • Courtship • Courtship • A. chrysaetos • Scotland
1271	**Flight** • Incubation • Behavior • A. chrysaetos • Scotland
1273	**Flight** • Mortality • Falconry • Prey • Hunting Behavior • Nest • Mortality • Photo Account • Popular Monograph • A. chrysaetos • Great Britain
1284	**Flight** • Migration • A. chrysaetos • Denmark
1313	**Flight** • Behavior • A. chrysaetos • Switzerland
1351	**Flight** • Display • A. chrysaetos • Switzerland
1379	**Flight** • Display • A. chrysaetos • Scotland
1461	**Flight** • Raptors • A. heliaca • Spain
1497	**Flight** • Fledging • Behavior • A. chrysaetos • Idaho • Oregon
1863	**Flight** • Hunting Behavior • Prey • Flight Speed • Photo Account • Raptors • A. chrysaetos • Scotland
1888	**Flight** • Feeding Behavior • A. chrysaetos • West Germany
1977	**Flight** • A. chrysaetos • West Germany
1985	**Flight** • Habitat • Nest • Behavior • Prey • Photo Account • A. chrysaetos • Scotland
2025	**Flight** • Behavior • Falconry • Eagles • Popular Monograph • A. chrysaetos • North America
2163	**Flight** • Interspecific Behavior • A. chrysaetos • California
2173	**Flight** • Folklore • A. chrysaetos • California
2194	**Flight** • Aircraft • Altitude • Raptors • A. heliaca • Turkey
2214	**Flight** • Nest • Habitat • Vocalization • A. chrysaetos • Idaho
2343	**Flight** • Interspecific Behavior • A. chrysaetos • Great Britain

Master No.	Keywords
2434	**Flight** • Display • Status • A. chrysaetos • Scotland
2445	**Flight** • Agonistic Behavior • Aircraft • Prey • Raptors • A. rapax • Kenya • Tanzania
2603	**Flight** • Anatomy • A. chrysaetos
2653	**Flight** • Interspecific Behavior • Mobbing • Raptors • A. chrysaetos • Oregon
2659	**Flight** • Intraspecific Behavior • Nest • A. chrysaetos • Scotland
3105	**Flight** • Interspecific Behavior • Raptors • Popular Monograph • A. chrysaetos • Pennsylvania
3180	**Flight** • Raptors • A. rapax • Algeria
3291	**Flight** • Folklore • Measure • Nest • Weight • A. chrysaetos • California
3333	**Flight** • A. verreauxi • South Africa
3409	**Flight** • Intraspecific Behavior • A. chrysaetos • Scotland
490	**Flight Id** • Aging • Wintering Population • Eagles • A. nipalensis • A. pomarina • A. rapax • A. wahlbergi • A. verreauxi • Africa
686	**Flight Id** • A. chrysaetos • A. nipalensis • A. pomarina • A. heliaca • A. clanga • Europe
908	**Flight Id** • Collect • Photo Account • A. chrysaetos • A. nipalensis • A. heliaca • A. rapax • India
2889	**Flight Id** • A. clanga • A. pomarina • A. heliaca • A. rapax • A. chrysaetos • Europe
462	**Flight Speed** • Habitat • Attitude • Conservation • Disturbance • Eagles • Photo Account • Popular Monograph • A. chrysaetos • North America
528	**Flight Speed** • Raptors • A. chrysaetos • Pennsylvania
805	**Flight Speed** • Raptors • A. chrysaetos • Scotland
1230	**Flight Speed** • Interspecific Behavior • A. chrysaetos • Canada
1863	**Flight Speed** • Flight • Hunting Behavior • Prey • Photo Account • Raptors • A. chrysaetos • Scotland
1966	**Flight Speed** • A. chrysaetos • California
2021	**Flight Speed** • Conservation • Popular Monograph • A. chrysaetos • Ireland
3385	**Flight Speed** • Prey • Status • Raptors • A. wahlbergi • Rhodesia
73	**Folklore** • Historical Account • Symbolism • Poetry • Eagles • A. chrysaetos
91	**Folklore** • A. chrysaetos • Illinois
94	**Folklore** • Flight • Banding • Capture • A. chrysaetos • A. pomarina • Europe
114	**Folklore** • A. chrysaetos • Minnesota
115	**Folklore** • A. chrysaetos • United States
117	**Folklore** • A. chrysaetos • United States
125	**Folklore** • Protection • Shooting • A. chrysaetos • Switzerland
200	**Folklore** • A. chrysaetos • Texas
319	**Folklore** • Shooting • Animal Damage • Attitude • Aircraft • Popular Monograph • A. chrysaetos • Texas
618	**Folklore** • Prey • Nest • Attitude • Behavior • Breeding Biology • Flight • Poisoning • Molt • Hunting Behavior • Vocalization • Lift • Photo Account • A. chrysaetos • Montana
651	**Folklore** • Nest • Capture • Economics • A. chrysaetos • Tennessee
813	**Folklore** • Attitude • A. chrysaetos • Switzerland
946	**Folklore** • Eagles • Popular Monograph • A. chrysaetos • North America
1064	**Folklore** • Descriptive Summary • Hunting Behavior • Status • Raptors • A. chrysaetos
1131	**Folklore** • Hunting Behavior • Trapping • Animal Damage • Protection • Eagles • A. chrysaetos • Tennessee
1259	**Folklore** • Poetry • Captivity • Attitude • Protection • Flight • Raptors • A. chrysaetos • Great Britain
1281	**Folklore** • Prey • Behavior • A. chrysaetos • Texas
1335	**Folklore** • A. chrysaetos • Great Britain
1485	**Folklore** • A. chrysaetos • Switzerland
1538	**Folklore** • Symbolism • Photo Account • A. chrysaetos • Canada
2008	**Folklore** • Behavior • Mortality • A. chrysaetos • California
2103	**Folklore** • Conservation • Habitat • Shooting • Animal Damage • A. chrysaetos • United States
2124	**Folklore** • Monograph • Breeding Biology • Migration • Prey • Hunting Behavior • Attack • Falconry • Animal Damage • Popular Monograph • A. chrysaetos
2173	**Folklore** • Flight • A. chrysaetos • California
2179	**Folklore** • Museum Display • Attitude • Animal Damage • Popular Monograph • A. audax • Australia
2369	**Folklore** • Prey • Lift • Animal Damage • Conservation • Electrocution • Eagles • Popular Monograph • A. chrysaetos • North America
2413	**Folklore** • Attitude • Field Id • Descriptive Summary • Raptors • A. chrysaetos • United States
2921	**Folklore** • Conservation • Eagles • A. chrysaetos • United States
2986	**Folklore** • A. chrysaetos • California
3141	**Folklore** • A. chrysaetos • Mexico
3252	**Folklore** • Lift • A. chrysaetos • Switzerland
3291	**Folklore** • Flight • Measure • Nest • Weight • A. chrysaetos • California
702	**Food Habits** • Life History • Distribution • Popular Monograph • A. chrysaetos • Missouri
745	**Food Habits** • Pellet • Raptors • A. chrysaetos • Utah
1176	**Food Habits** • Breeding Biology • Nest • Courtship • A. pomarina • West Germany
1865	**Food Habits** • Measure • A. chrysaetos • Minnesota
1942	**Food Habits** • Nest • Time Lapse • Behavior • Economics • Animal Damage • Reproductive Success • Status • A. chrysaetos • Texas
2400	**Food Habits** • Animal Damage • Predator/Prey • Aerial Survey • Habitat • Status • Density • Population Dynamics • Nest • Interspecific Competition • Literature Review • A. chrysaetos • Texas • New Mexico
2669	**Food Habits** • Nest • Fledging • Parental Behavior • Breeding Biology • A. chrysaetos • Tanzania
2946	**Food Habits** • Nest • A. chrysaetos • Colorado
3301	**Food Habits** • A. nipalensis • United Soviet Socialists Republic
233	**Food Requirements** • Attitude • Reproductive Success • Density • Wintering Population • Habitat • Management • Aerial Survey • Nest Survey • Raptors • A. chrysaetos • Montana
324	**Food Requirements** • Reproductive Success • Territory • Abandonment • Mortality • Cain/Abel • Density • Dispersal • Breeding Biology • Nest Survey • A. chrysaetos • Idaho

Master No.	Keywords
538	**Food Requirements** • Nest Survey • Status • Population Estimate • Reproductive Success • Prey • A. chrysaetos • Scotland
540	**Food Requirements** • Nest Failure • Reproductive Success • Home Range • Breeding Biology • Parental Behavior • Eagles • Theory • A. wahlbergi • A. verreauxi • Africa
552	**Food Requirements** • Nest Survey • Prey • Habitat • Behavior • Status • Reproductive Success • Animal Damage • Intraspecific Behavior • Predator/Prey • A. chrysaetos • Scotland
713	**Food Requirements** • Growth • Raptors • SRBPNA • A. chrysaetos • Idaho
714	**Food Requirements** • Breeding Biology • Predator/Prey • Cain/Abel • Directional Exposure • Growth • Sex Ratio • Behavior • A. chrysaetos • Idaho
1027	**Food Requirements** • Raptors • A. chrysaetos • Montana
1618	**Food Requirements** • Mortality • Cain/Abel • Raptors • A. chrysaetos • A. verreauxi • A. pomarina
1772	**Food Requirements** • Prey • Nest • Predator/Prey • A. chrysaetos • Washington
1856	**Food Requirements** • Prey • Interspecific Competition • Raptors • A. chrysaetos • A. pomarina
1943	**Food Requirements** • Aerial Survey • Time Lapse • Animal Damage • Nest Survey • Prey • Habitat • Winter Survey • Capture • Age Structure • Shooting • A. chrysaetos • Texas
2086	**Food Requirements** • Monograph • Mortality • Banding • Migration • Breeding Biology • Prey • A. chrysaetos • Montana
2087	**Food Requirements** • A. chrysaetos • Montana
2125	**Food Requirements** • Poisoning • Pesticides • Status • Captivity • Density • Habitat • Protection • Predator/Prey • Raptors • Photography Account • PCB • A. clanga • A. heliaca • A. nipalensis • A. verreauxi • A. pomarina • Israel
2326	**Food Requirements** • Capture • Banding • Marking • Animal Damage • Sexing • Management • Translocation • Prey • Measure • Mortality • Captive Care • Aging • A. chrysaetos • Montana
2373	**Food Requirements** • Prey • Literature Review • Animal Damage • A. chrysaetos • Colorado
2870	**Food Requirements** • Banding • Survey • Marking • Nest • Capture • Prey • Raptors • A. chrysaetos • A. wahlbergi • South Africa
2880	**Food Requirements** • Dimorphism • Behavior • Theory • Raptors • A. chrysaetos
3021	**Food Requirements** • Captivity • Growth • Measure • Molt • Pellet • Temperature • Photo Account • Raptors • A. chrysaetos • California
3316	**Food Requirements** • Animal Damage • Prey • Management • A. chrysaetos • Scotland
174	**Forestry** • Management • Protection • A. chrysaetos • Washington
2159	**Forestry** • Habitat Loss • Conservation • Habitat Protection • Status • Raptors • A. heliaca • Spain
2801	**Forestry** • Nest • Prey • Mortality • Habitat • A. chrysaetos • Washington
302	**Fostering** • Prey • Nest • Habitat • Raptors • A. chrysaetos • Bulgaria
1364	**Fostering** • Hacking • Captive Care • Raptors • A. chrysaetos • A. pomarina
2151	**Fostering** • Clutch Size • Nestling • A. heliaca • Spain
2157	**Fostering** • Conservation • Reproductive Success • Cain/Abel • Photo Account • A. heliaca • Spain
587	**Growth** • Nest • Prey • Behavior • Vocalization • Nest Adornment • Photo Account • Illus. • A. chrysaetos • Switzerland
588	**Growth** • Behavior • Raptors • A. chrysaetos • North America
616	**Growth** • Nest • Habitat • Egg • Prey • Vocalization • Trapping • Poisoning • Behavior • Lift • Photo Account • A. chrysaetos • Montana
713	**Growth** • Food Requirements • Raptors • SRBPNA • A. chrysaetos • Idaho
714	**Growth** • Food Requirements • Breeding Biology • Predator/Prey • Cain/Abel • Directional Exposure • Sex Ratio • Behavior • A. chrysaetos • Idaho
944	**Growth** • Nest • Nest Adornment • Prey • Behavior • A. audax • Australia
955	**Growth** • Behavior • Pesticides • Parental Behavior • Vocalization • Disturbance • Measure • Breeding Biology • Parasite • Disease • Terminology • A. chrysaetos • Montana
957	**Growth** • Monograph • Behavior • Pesticides • Parental Behavior • Vocalization • Disturbance • Measure • Breeding Biology • Photography • Captivity • Parasite • Disease • Terminology • Photo Account • A. chrysaetos • Montana
963	**Growth** • Nest • Plumage • Prey • Vocalization • Behavior • Photo Account • A. audax • Australia
1037	**Growth** • Nest • Photography • Prey • A. chrysaetos • California
1056	**Growth** • Falconry • Aging • Parental Behavior • Measure • Plumage • Dimorphism • Prey • Photo Account • A. audax • Australia
1091	**Growth** • Nest • Prey • Photo Account • A. chrysaetos • Switzerland
1139	**Growth** • Cain/Abel • Photo Account • A. verreauxi • Rhodesia
1153	**Growth** • Cain/Abel • Parental Behavior • Clutch Size • Behavior • Eagles • A. verreauxi • A. pomarina • Rhodesia
1522	**Growth** • Nest • Nest Chronology • Photo Account • A. chrysaetos • California
1775	**Growth** • Monograph • Status • Nest Survey • Reproductive Success • Cain/Abel • Banding • Marking • Mortality • Pesticides • Prey • Residues • A. chrysaetos • Idaho
1998	**Growth** • Reproductive Success • Prey • Behavior • Photography Account • Popular Monograph • A. chrysaetos • Scotland
2152	**Growth** • Habitat • Parental Behavior • Prey • Breeding Biology • Interspecific Behavior • Courtship • Display • Photo Account • Raptors • A. heliaca • Spain
2230	**Growth** • Nest • A. audax • Australia
2598	**Growth** • Nest • A. chrysaetos • Switzerland
2670	**Growth** • Cain/Abel • Prey • Parental Behavior • Breeding Biology • A. verreauxi • Africa

Master No.	Keywords
2830	**Growth** • Breeding Biology • Habitat • Cain/Abel • Nest Adornment • Vocalization • Plumage • Prey • Hunting Behavior • Illus. • A. pomarina • West Germany
2847	**Growth** • Nest • Egg • (Title Only) • A. pomarina • Czechoslovakia
2969	**Growth** • Nest Survey • Habitat • Prey • Breeding Biology • Reproductive Success • Field Id • Vocalization • Migration • Behavior • Eagles • Photo Account • A. wahlbergi • A. rapax • Rhodesia
3022	**Growth** • Cain/Abel • Behavior • Vocalization • Pellet • Feeding Behavior • Raptors • A. chrysaetos • California
3019	**Growth** • Clutch Size • Photo Account • Raptors • A. chrysaetos • California
3020	**Growth** • Nest • Behavior • Clutch Size • Banding • Photo Account • A. chrysaetos • California
3021	**Growth** • Captivity • Food Requirements • Measure • Molt • Pellet • Temperature • Photo Account • Raptors • A. chrysaetos • California
3214	**Growth** • Collect • Nest • Habitat • Behavior • Captivity • A. chrysaetos • India
3235	**Growth** • Nest • Photography • Prey • Behavior • Hunting Behavior • Vocalization • Nest Adornment • Egg • Raptors • Popular Monograph • A. rapax • Kenya
25	**Habitat** • Descriptive Summary • Field Id • Distribution • Raptors • A. rapax • India
30	**Habitat** • Roadside Survey • Status • A. chrysaetos • Texas
56	**Habitat** • Etymology • Descriptive Summary • Clutch Size • Plumage • Museum Specimen • Measure • A. rapax • A. pomarina • India
58	**Habitat** • Nest • Clutch Size • Prey • Collect • A. pomarina • India
62	**Habitat** • Measure • Management • Eagles • A. chrysaetos • Oregon
64	**Habitat** • Nest • Eagles • A. chrysaetos • Washington
188	**Habitat** • Status • Raptors • A. heliaca • A. pomarina • A. clanga • A. chrysaetos • Yugoslavia
193	**Habitat** • Id Key • Descriptive Summary • Etymology • Egg • Raptors • Illus. • A. nipalensis • A. rapax • A. verreauxi • Somalia
195	**Habitat** • Nest Survey • Wintering Population • Distribution • Attitude • Raptors • A. chrysaetos • Arizona
196	**Habitat** • Nest Survey • Wintering Population • Attitude • Falconry • Raptors • A. chrysaetos • Arizona
198	**Habitat** • Prey • Monograph • Nest Survey • Reproductive Success • Literature Review • Disturbance • Pellet • Animal Damage • A. chrysaetos • Utah
233	**Habitat** • Attitude • Reproductive Success • Density • Wintering Population • Food Requirements • Management • Aerial Survey • Nest Survey • Raptors • A. chrysaetos • Montana
244	**Habitat** • Nest Survey • Prey • Artificial Nest • Raptors • A. chrysaetos • California
298	**Habitat** • Rarity • A. pomarina • A. heliaca • Austria
302	**Habitat** • Fostering • Prey • Nest • Raptors • A. chrysaetos • Bulgaria
372	**Habitat** • Nest • Wintering Population • Hunting Behavior • Territory • A. chrysaetos • France
375	**Habitat** • Hunting Behavior • Eagles • A. chrysaetos • France
439	**Habitat** • Distribution • A. chrysaetos • Washington
462	**Habitat** • Attitude • Conservation • Disturbance • Flight Speed • Eagles • Photo Account • Popular Monograph • A. chrysaetos • North America
464	**Habitat** • Nest • Egg Collection • Behavior • Vocalization • Animal Damage • A. audax • Australia
473	**Habitat** • Nest • A. chrysaetos • Colorado
477	**Habitat** • Status • Collect • Raptors • A. chrysaetos • A. clanga • Italy
501	**Habitat** • Prey • Pellet • Attack • Hunting Behavior • Animal Damage • Cache • A. audax • Australia
534	**Habitat** • Density • Nest Survey • Breeding Biology • Interspecific Competition • Prey • Reproductive Success • Etymology • Plumage • Vocalization • Nest Adornment • Raptors • A. verreauxi • A. wahlbergi • Kenya
537	**Habitat** • Nest • Prey • Interspecific Behavior • Raptors • A. verreauxi • A. wahlbergi • Kenya
542	**Habitat** • Nest • Abnormal Behavior • Cain/Abel • A. verreauxi • Rhodesia
544	**Habitat** • Taxonomy • Status • Field Id • Flight • Terminology • Distribution • Mortality • Behavior • Life History • Longevity • Territory • Photo Account • Raptors • A. chrysaetos • A. clanga • Great Britain • Ireland • Scotland
552	**Habitat** • Nest Survey • Prey • Behavior • Status • Reproductive Success • Food Requirements • Animal Damage • Intraspecific Behavior • Predator/Prey • A. chrysaetos • Scotland
556	**Habitat** • Field Id • Raptors • A. chrysaetos • California
610	**Habitat** • Artificial Nest • Management • Photo Account • Raptors • A. chrysaetos • North America
611	**Habitat** • Nest • Field Id • Nest Survey • Behavior • Photo Account • Raptors • A. chrysaetos • North America
614	**Habitat** • Monograph • Nest Survey • Reproductive Success • Directional Exposure • Behavior • A. chrysaetos • Utah
616	**Habitat** • Nest • Egg • Prey • Vocalization • Growth • Trapping • Poisoning • Behavior • Lift • Photo Account • A. chrysaetos • Montana
622	**Habitat** • Nest • A. chrysaetos • Alaska
628	**Habitat** • Hunting Behavior • Prey • Nest • Photography Account • A. chrysaetos • California
675	**Habitat** • Egg Collection • A. chrysaetos • California
696	**Habitat** • Distribution • Status • Raptors • A. audax • Australia
789	**Habitat** • Monograph • Prey • Nest Survey • Photo Account • A. chrysaetos • Colorado
871	**Habitat** • Distribution • Etymology • Life History • Molt • Prey • Measure • Field Id • Migration • Status • Raptors • A. chrysaetos • A. heliaca • A. rapax • A. clanga • A. pomarina • United Soviet Socialists Republic

Master No.	Keywords
893	**Habitat** • Nest • Egg • Photo Account • A. chrysaetos • California
894	**Habitat** • Intraspecific Behavior • Nest • Egg • Measure • Prey • Photo Account • A. chrysaetos • California
931	**Habitat** • Distribution • Raptors • A. rapax • Senegal
939	**Habitat** • Nest • Prey • Field Id • Hunting Behavior • Raptors • A. clanga • A. pomarina • A. chrysaetos • Poland
940	**Habitat** • Nest • Prey • Habitat Loss • Raptors • A. chrysaetos • India
979	**Habitat** • Rarity • Raptors • A. chrysaetos • France
1003	**Habitat** • Egg • Nest • Photo Account • A. chrysaetos • Scotland
1006	**Habitat** • Nest • Raptors • A. chrysaetos • A. heliaca • A. clanga • Bulgaria
1020	**Habitat** • Status • Population Decline • Mortality • Raptors • A. chrysaetos • Europe
1025	**Habitat** • Hunting Behavior • Nest • Prey • Density • Photo Account • Popular Monograph • A. heliaca • Spain
1033	**Habitat** • Prey • Nest • Flight • Plumage • Raptors • A. wahlbergi • A. rapax • South Africa
1041	**Habitat** • Status • Prey • Nest • Behavior • Photo Account • A. chrysaetos • Bulgaria
1072	**Habitat** • Disturbance • Raptors • A. chrysaetos • Alaska
1073	**Habitat** • Disturbance • Raptors • A. chrysaetos • Alaska
1099	**Habitat** • Nest Survey • Photo Account • A. chrysaetos • Saskatchewan
1109	**Habitat** • A. chrysaetos • A. heliaca • Hungary
1124	**Habitat** • Density • Raptors • A. clanga • Russia
1141	**Habitat** • Nest Survey • Density • Disturbance • Reproductive Success • Clutch Size • Raptors • A. verreauxi • Rhodesia
1154	**Habitat** • Nest • Density • Prey • Status • Habitat Protection • Reproductive Success • A. verreauxi • Rhodesia
1156	**Habitat** • Breeding Biology • Prey • Density • Nest Survey • A. verreauxi • Rhodesia
1162	**Habitat** • Status • Nest Survey • Protection • Conservation • Shooting • Raptors • A. chrysaetos • A. heliaca • Spain
1187	**Habitat** • Protection • Nest • Vocalization • Raptors • A. chrysaetos • A. heliaca • Yugoslavia
1189	**Habitat** • Nest • Prey • Raptors • A. chrysaetos • France
1276	**Habitat** • Descriptive Summary • Status • Raptors • A. chrysaetos • A. heliaca • A. clanga • Korea
1294	**Habitat** • Nest Survey • Population Dynamics • Raptors • Tech. • A. chrysaetos • North Dakota
1301	**Habitat** • Distribution • Raptors • A. chrysaetos • California
1336	**Habitat** • Wintering Population • Raptors • A. clanga • Thailand
1355	**Habitat** • Popular Monograph • Refuge • A. chrysaetos • Idaho • Switzerland
1405	**Habitat** • Status • Raptors • A. audax • Australia
1409	**Habitat** • Nest • Raptors • A. chrysaetos • Alaska
1410	**Habitat** • Nest • Raptors • A. chrysaetos • Alaska
1431	**Habitat** • Poisoning • Telemetry • Toxicology • Raptors • Cmpd. 1080 • A. chrysaetos • California
1437	**Habitat** • Migration • Field Id • Photo Account • Raptors • Popular Monograph • A. chrysaetos • North America
1513	**Habitat** • Nest Failure • Prey • Behavior • A. chrysaetos • Alaska
1548	**Habitat** • Prey • Nest • A. chrysaetos • Alaska
1597	**Habitat** • Prey • Collect • Behavior • Raptors • A. nipalensis • A. clanga • A. pomarina • India
1639	**Habitat** • Taxonomy • Dimorphism • Measure • Plumage • Distribution • Prey • Behavior • Management • Territory • Mortality • Life History • Economics • Literature Review • A. chrysaetos • Oregon • Washington
1709	**Habitat** • Nest • Conservation • Raptors • A. chrysaetos • A. pomarina • A. clanga • Hungary
1729	**Habitat** • Prey • Mortality • Conservation • A. chrysaetos • Japan
1754	**Habitat** • Distribution • Nest • Wintering Population • Status • Raptors • A. chrysaetos • Colorado
1781	**Habitat** • Nest Survey • Reproductive Success • Marking • Nest Chronology • Behavior • Prey • Banding • Territory • Population Dynamics • Raptors • SRBPNA • A. chrysaetos • Idaho
1849	**Habitat** • Raptors • A. rapax • A. wahlbergi • South Africa
1872	**Habitat** • Distribution • Photo Account • Popular Monograph • A. chrysaetos • Europe
1900	**Habitat** • Mortality • Prey • Nest • Dispersal • Pellet • Animal Damage • Bounty • Protection • Photo Account • A. audax • Australia
1903	**Habitat** • Nest • Prey • Disturbance • Conservation • Mortality • Hunting Behavior • A. chrysaetos • Israel
1912	**Habitat** • Descriptive Summary • Nest Chronology • Nest • A. chrysaetos • France
1914	**Habitat** • Distribution • Nest • Status • Clutch Size • Descriptive Summary • Raptors • A. chrysaetos • New Mexico
1943	**Habitat** • Aerial Survey • Time Lapse • Animal Damage • Nest Survey • Prey • Winter Survey • Capture • Age Structure • Food Requirements • Shooting • A. chrysaetos • Texas
1964	**Habitat** • Collect • A. verreauxi • Uganda
1972	**Habitat** • Nest • Prey • A. chrysaetos • Ontario
1985	**Habitat** • Nest • Behavior • Prey • Flight • Photo Account • A. chrysaetos • Scotland
2003	**Habitat** • Status • Raptors • A. audax • Australia
2029	**Habitat** • Nest • Status • Prey • Eagles • A. heliaca • A. chrysaetos • A. clanga • A. pomarina • Yugoslavia
2097	**Habitat** • Distribution • Status • Nest • Raptors • A. chrysaetos • North Dakota
2103	**Habitat** • Conservation • Folklore • Shooting • Animal Damage • A. chrysaetos • United States
2125	**Habitat** • Poisoning • Pesticides • Status • Captivity • Density • Food Requirements • Protection • Predator/Prey • Raptors • Photography Account • PCB • A. clanga • A. heliaca • A. nipalensis • A. verreauxi • A. pomarina • Israel
2145	**Habitat** • Breeding Biology • Cain/Abel • Density • Behavior • Prey • Nest Survey • Photo Account • A. pomarina • Czechoslovakia

Master No.	Keywords
2148	**Habitat** • Distribution • Field Id • Management • Taxonomy • Nest Survey • Prey • Breeding Biology • Photo Account • Eagles • A. pomarina • Europe
2152	**Habitat** • Growth • Parental Behavior • Prey • Breeding Biology • Interspecific Behavior • Courtship • Display • Photo Account • Raptors • A. heliaca • Spain
2183	**Habitat** • Nest Survey • Population Decline • Status • Raptors • A. heliaca • Morocco
2184	**Habitat** • Aerial Survey • Reproductive Success • Management • Survey • Clutch Size • Distribution • Nest Survey • Prey • Raptors • A. chrysaetos • Arizona
2214	**Habitat** • Nest • Flight • Vocalization • A. chrysaetos • Idaho
2222	**Habitat** • Nest • Raptors • A. rapax • Senegal
2239	**Habitat** • Survey • Territory • Nest Failure • Nest • Migration • Management • Reproductive Success • Prey • Roadside Survey • Status • Winter Survey • Uinta Nf • Wasatch Mt. • Raptors • A. chrysaetos • Utah
2246	**Habitat** • Nest • Disturbance • Raptors • A. pomarina • Poland
2252	**Habitat** • Status • Raptors • A. clanga • India
2290	**Habitat** • Nest • Egg • Raptors • A. chrysaetos • Wyoming
2312	**Habitat** • Refuge • Raptors • A. chrysaetos • A. heliaca • Europe
2342	**Habitat** • Etymology • Status • Interspecific Behavior • Collect • Nest • Raptors • A. verreauxi • A. nipalensis • A. rapax • Africa
2368	**Habitat** • Nest Survey • Density • Clutch Size • Raptors • A. chrysaetos • Colorado
2370	**Habitat** • Status • Nest • Nest Survey • Monograph • Mortality • Breeding Biology • Nest Chronology • Courtship • Density • Prey • Management • A. chrysaetos • Colorado
2371	**Habitat** • Popular Monograph • Story • Breeding Biology • Behavior • Nest Survey • Hunting Behavior • Prey • Raptors • Illus. • A. chrysaetos • Colorado
2372	**Habitat** • Nest Survey • Density • Status • Raptors • A. chrysaetos • Colorado
2376	**Habitat** • Management • Artificial Nest • Nest Chronology • A. chrysaetos • North America
2400	**Habitat** • Animal Damage • Predator/Prey • Aerial Survey • Status • Density • Population Dynamics • Food Habits • Nest • Interspecific Competition • Literature Review • A. chrysaetos • Texas • New Mexico
2401	**Habitat** • Nest Survey • A. chrysaetos • Nevada
2464	**Habitat** • Management • Directional Exposure • Disturbance • Mortality • Nest Survey • Density • Nest Chronology • Prey • Reproductive Success • Prog. Rep. • A. chrysaetos • Wyoming
2465	**Habitat** • Mining • Nest Survey • Disturbance • Marking • Prey • Time Lapse • Raptors • Proj. Out. • A. chrysaetos • Wyoming • Montana
2485	**Habitat** • Egg • Nest • A. wahlbergi • Africa
2491	**Habitat** • Distribution • Density • Status • Survey • Raptors • A. chrysaetos • New Mexico
2526	**Habitat** • Prey • Nest • A. clanga • Russia
2544	**Habitat** • Collect • Behavior • Prey • A. clanga • West Germany
2552	**Habitat** • Descriptive Summary • Distribution • Taxonomy • Plumage • Measure • Raptors • A. gurneyi • New Guinea
2558	**Habitat** • Nest Survey • Reproductive Success • Protection • Clutch Size • Eagles • A. chrysaetos • Estonia
2577	**Habitat** • Status • Raptors • A. chrysaetos • A. heliaca • A. nipalensis • A. clanga • Afghanistan
2583	**Habitat** • Collect • Prey • Status • Raptors • A. heliaca • A. nipalensis • A. clanga • A. rapax • A. pomarina • India
2602	**Habitat** • Nest • Photography • Prey • Photo Account • A. chrysaetos • Switzerland
2612	**Habitat** • Breeding Biology • Prey • Dispersal • A. audax • Australia
2677	**Habitat** • Nest • A. pomarina • A. clanga • Yugoslavia
2701	**Habitat** • Nest Survey • Raptors • A. chrysaetos • Alaska
2713	**Habitat** • Management • A. chrysaetos • North Carolina
2732	**Habitat** • Descriptive Summary • Distribution • Prey • Raptors • A. chrysaetos • A. nipalensis • Tibet
2753	**Habitat** • Monograph • Nest Survey • Behavior • Prey • Pellet • Abandonment • A. chrysaetos • Wyoming
2788	**Habitat** • Nest Survey • Aerial Survey • Density • Photo Account • A. chrysaetos • Nevada
2801	**Habitat** • Nest • Prey • Forestry • Mortality • A. chrysaetos • Washington
2811	**Habitat** • Distribution • Status • A. chrysaetos • Great Britain • Ireland • Scotland
2830	**Habitat** • Breeding Biology • Cain/Abel • Nest Adornment • Growth • Vocalization • Plumage • Prey • Hunting Behavior • Illus. • A. pomarina • West Germany
2834	**Habitat** • Nest • Raptors • A. clanga • A. heliaca • Turkey
2837	**Habitat** • Nest • Egg • Raptors • A. clanga • Greece
2842	**Habitat** • Descriptive Summary • Raptors • A. wahlbergi • A. rapax • South Africa
2851	**Habitat** • Egg Collection • Nest • Measure • Photo Account • A. chrysaetos • California
2852	**Habitat** • Roadside Survey • Prey • Monograph • Density • Interspecific Competition • Taxonomy • Vocalization • Hunting Behavior • Nest Chronology • Photo Account • Raptors • A. wahlbergi • A. rapax • A. nipalensis • A. heliaca • Kenya
2854	**Habitat** • Population Dynamics • Nest • Density • Prey • Breeding Biology • Distribution • Directional Exposure • Reproductive Success • Mortality • Abandonment • Disturbance • Interspecific Behavior • Territory • A. chrysaetos • Utah
2856	**Habitat** • Breeding Biology • Nest Survey • Hunting Behavior • Territory • Interspecific Behavior • Prey • Raptors • A. chrysaetos • Utah
2873	**Habitat** • Literature Review • Life History • Management • Population Estimate • A. chrysaetos • United States

Master No.	Keywords
2883	**Habitat** • Prey • Raptors • A. heliaca • Kazakhstan
2932	**Habitat** • Roadside Survey • Wintering Population • Raptors • A. chrysaetos • Colorado
2969	**Habitat** • Nest Survey • Prey • Breeding Biology • Reproductive Success • Growth • Field Id • Vocalization • Migration • Behavior • Eagles • Photo Account • A. wahlbergi • A. rapax • Rhodesia
3058	**Habitat** • Prey • Prey Behavior • Nest • Animal Damage • Shooting • Egg • Behavior • A. verreauxi • South Africa
3080	**Habitat** • Cain/Abel • Distribution • Educational Material • Field Id • Hunting Behavior • Prey • Artificial Nest • Nest • Incubation • Wintering Population • Conservation • Photo Account • Eagles • A. chrysaetos • A. wahlbergi • A. verreauxi • A. nipalensis • A. pomarina • South Africa
3107	**Habitat** • Nest • Egg Collection • Photo Account • A. chrysaetos • Colorado
3118	**Habitat** • Distribution • Prey • Descriptive Summary • Density • Interspecific Competition • Eagles • A. chrysaetos • France
3122	**Habitat** • Status • Density • Raptors • A. rapax • A. wahlbergi • A. nipalensis • A. pomarina • Chad
3127	**Habitat** • Hunting Behavior • Interspecific Competition • Raptors • A. wahlbergi • Ivory Coast
3128	**Habitat** • Roadside Survey • Status • Distribution • A. chrysaetos • A. pomarina • A. clanga • A. wahlbergi • A. rapax • A. verreauxi • Ivory Coast • Upper Volta • Mali
3130	**Habitat** • Roadside Survey • Predator/Prey • A. rapax • A. nipalensis • A. pomarina • A. wahlbergi • Uganda
3131	**Habitat** • Niche • Prey • Raptors • A. wahlbergi • A. rapax • Ivory Coast
3133	**Habitat** • Wintering Population • Prey • Density • Raptors • A. wahlbergi • A. rapax • A. nipalensis • A. pomarina • Cambodia
3134	**Habitat** • Roadside Survey • Raptors • A. rapax • A. clanga • A. nipalensis • A. pomarina • A. wahlbergi • A. verreauxi • Uganda
3164	**Habitat** • Nest • Photo Account • A. chrysaetos • Poland
3179	**Habitat** • Nest • Raptors • A. chrysaetos • Algeria
3190	**Habitat** • Density • Raptors • A. pomarina • A. clanga • Czechoslovakia
3204	**Habitat** • Refuge • Conservation • Raptors • SRBPNA • A. chrysaetos • Idaho
3207	**Habitat** • Refuge • Nest Survey • Density • Prey • Predator/Prey • Disturbance • Telemetry • Territory • Reproductive Success • Raptors • SRBPNA • Photo Account • A. chrysaetos • Idaho
3209	**Habitat** • Status • Raptors • A. heliaca • Hungary
3214	**Habitat** • Collect • Nest • Behavior • Growth • Captivity • A. chrysaetos • India
3218	**Habitat** • Descriptive Summary • Prey • Hunting Behavior • Raptors • A. chrysaetos • A. heliaca • A. nipalensis • A. clanga • A. pomarina • West Germany
3225	**Habitat** • Nest • Prey • Raptors • A. heliaca • Spain
3240	**Habitat** • Prey • Raptors • A. heliaca • A. nipalensis • Kazakhstan
3241	**Habitat** • Survey • Status • Raptors • A. heliaca • A. nipalensis • United Soviet Socialists Republic
3243	**Habitat** • Hunting Behavior • Prey • Raptors • A. chrysaetos • A. heliaca • A. nipalensis • United Soviet Socialists Republic
3245	**Habitat** • Prey • Distribution • Raptors • A. nipalensis • A. heliaca • A. chrysaetos • United Soviet Socialists Republic
3257	**Habitat** • Taxonomy • Descriptive Summary • Distribution • Raptors • A. rapax • A. nipalensis • A. heliaca • A. clanga • A. pomarina • A. verreauxi
3259	**Habitat** • Nest Survey • Territory • Intraspecific Behavior • Density • Nest Chronology • Prey • Directional Exposure • Raptors • A. verreauxi • Rhodesia
3272	**Habitat** • Nest • Egg • Raptors • A. verreauxi • A. rapax • A. nipalensis • Rhodesia
3305	**Habitat** • Status • Prey • Raptors • A. chrysaetos • Scotland
3308	**Habitat** • A. chrysaetos • California
3330	**Habitat** • Status • Management • A. chrysaetos • Washington
3352	**Habitat** • Status • Raptors • A. verreauxi • A. rapax • A. wahlbergi • Rhodesia
3360	**Habitat** • Nest Survey • Prey • Directional Exposure • Eagles • A. chrysaetos • Saskatchewan
3369	**Habitat** • Nest Survey • Egg • Vocalization • Behavior • Hunting Behavior • Photo Account • A. chrysaetos • Arizona
3406	**Habitat** • Nest • Egg • Nestling • Raptors • A. chrysaetos • Korea
3444	**Habitat** • Breeding Biology • Nest Survey • Protection • Prey • Distribution • Reproductive Success • Photo Account • A. chrysaetos • Estonia
3457	**Habitat** • Prey • Raptors • A. chrysaetos • Washington
386	**Habitat Loss** • Bounty • Disturbance • Protection • Shooting • Status • Pesticides • Migration • Distribution • Population Decline • Raptors • A. chrysaetos • A. heliaca • A. pomarina • A. clanga • Europe
543	**Habitat Loss** • Status • Wintering Population • Raptors • AF • Africa
548	**Habitat Loss** • Status • Poisoning • Conservation • A. pomarina • A. clanga • A. heliaca • East Africa • Ethiopia
940	**Habitat Loss** • Nest • Prey • Habitat • Raptors • A. chrysaetos • India
1101	**Habitat Loss** • Status • Pesticides • Disturbance • Raptors • A. chrysaetos • Canada
1468	**Habitat Loss** • Status • Protection • Raptors • A. chrysaetos • Switzerland
2153	**Habitat Loss** • Status • Raptors • Photo Account • A. heliaca • Spain
2159	**Habitat Loss** • Conservation • Forestry • Habitat Protection • Status • Raptors • A. heliaca • Spain
2263	**Habitat Loss** • Management • Electrocution • A. chrysaetos • North America
2787	**Habitat Loss** • Status • Conservation • Protection • Mortality • Raptors • A. chrysaetos • A. clanga • Sweden

Permuted List of Keywords (Cont.)

Master No.	Keywords
2839	**Habitat Loss** • Nest Survey • Status • Raptors • Photo Account • A. chrysaetos • Nevada
2928	**Habitat Loss** • Mortality • Electrocution • Nest • Survey • Raptors • A. chrysaetos • Colorado
3027	**Habitat Loss** • Population Decline • Disturbance • A. nipalensis • Caucasus
3110	**Habitat Loss** • Nest Survey • Distribution • Status • Density • Density • Reproductive Success • Raptors • Photo Account • A. chrysaetos • California
3220	**Habitat Loss** • Status • Distribution • Raptors • A. chrysaetos • A. heliaca • A. clanga • A. pomarina • Greece
1154	**Habitat Protection** • Nest • Density • Prey • Status • Habitat • Reproductive Success • A. verreauxi • Rhodesia
2159	**Habitat Protection** • Habitat Loss • Conservation • Forestry • Status • Raptors • A. heliaca • Spain
2374	**Habitat Protection** • Management • Conservation • Refuge • Theory • Tech. • Raptors • A. chrysaetos • United States
3031	**Habitat Protection** • Management • Environmental Impact • Disturbance • Raptors • A. chrysaetos • North America
827	**Hacking** • Rehabilitation • Raptors • A. chrysaetos
1364	**Hacking** • Captive Care • Fostering • Raptors • A. chrysaetos • A. pomarina
2296	**Hacking** • Experiment • A. chrysaetos • North America • Great Britain
1730	**Heavy Metals** • Pesticides • Egg • Raptors • DDE • Mercury • A. chrysaetos • North America
47	**Historical Account** • Population Decline • Status • Management • Distribution • Raptors • A. chrysaetos • Tennessee
73	**Historical Account** • Folklore • Symbolism • Poetry • Eagles • A. chrysaetos
86	**Historical Account** • Flight • Status • Attitude • Prey • Photo Account • A. chrysaetos • Great Britain
239	**Historical Account** • A. chrysaetos • Virginia
268	**Historical Account** • Wintering Population • Status • A. chrysaetos • Kentucky
475	**Historical Account** • Status • Distribution • Prey • Hunting Behavior • Raptors • A. chrysaetos • Maine
527	**Historical Account** • Migration • Shooting • Conservation • Population Decline • Hawk Mt. • Raptors • Popular Monograph • A. chrysaetos • Pennsylvania
838	**Historical Account** • Status • Population Decline • Nest • Prey • A. chrysaetos • Ireland
966	**Historical Account** • Symbolism • Eagles • A. chrysaetos
1288	**Historical Account** • Nest • A. chrysaetos • Oklahoma
1489	**Historical Account** • Shooting • A. chrysaetos • Switzerland
1494	**Historical Account** • Nest • Raptors • A. chrysaetos • A. clanga • West Germany
1728	**Historical Account** • Nest • A. chrysaetos • West Germany
2080	**Historical Account** • Nest • A. chrysaetos • Washington
2292	**Historical Account** • Electrocution • Management • Raptors • A. chrysaetos • North America
2420	**Historical Account** • Symbolism • Capture • Anthropology • Raptors • Photo Account • A. chrysaetos • South Dakota
2425	**Historical Account** • Capture • Prey • Lift • A. chrysaetos • Scotland
2474	**Historical Account** • Survey • Trapping • A. chrysaetos • California
2683	**Historical Account** • Migration • Status • Raptors • A. chrysaetos • Sweden
3006	**Historical Account** • Nest • A. chrysaetos • Great Britain
3227	**Historical Account** • Raptors • A. chrysaetos • A. nipalensis • A. clanga • A. pomarina • Lithuania
3313	**Historical Account** • A. chrysaetos • West Germany
3359	**Historical Account** • Educational Material • Eagles • Popular Monograph • A. chrysaetos • United States
109	**History** • Symbolism • A. chrysaetos • United States
765	**History** • Nest • Management • Conservation • Raptors • Popular Monograph • A. chrysaetos • United States
540	**Home Range** • Nest Failure • Food Requirements • Reproductive Success • Breeding Biology • Parental Behavior • Eagles • Theory • A. wahlbergi • A. verreauxi • Africa
2871	**Home Range** • Marking • Territory • Dispersal • Raptors • Photo Account • A. rapax • A. wahlbergi • South Africa
22	**Hunting Behavior** • Animal Damage • Prey • Talon Wounds • A. chrysaetos • Texas
63	**Hunting Behavior** • A. chrysaetos • Washington
89	**Hunting Behavior** • Behavior • Nest • Popular Monograph • A. chrysaetos • California
90	**Hunting Behavior** • A. chrysaetos • Switzerland
126	**Hunting Behavior** • A. chrysaetos • Saskatchewan
207	**Hunting Behavior** • Prey • A. chrysaetos • California
323	**Hunting Behavior** • Falconry • Field Id • Plumage • Photography Account • Raptors • A. chrysaetos
351	**Hunting Behavior** • Measure • Life History • Courtship • Behavior • Distribution • Vocalization • Plumage • Raptors • Photo Account • A. chrysaetos • North America
366	**Hunting Behavior** • Prey • Raptors • A. chrysaetos • A. heliaca • Spain
372	**Hunting Behavior** • Habitat • Nest • Wintering Population • Territory • A. chrysaetos • France
375	**Hunting Behavior** • Habitat • Eagles • A. chrysaetos • France
376	**Hunting Behavior** • A. chrysaetos • Czechoslovakia
381	**Hunting Behavior** • Status • Raptors • A. chrysaetos • India
460	**Hunting Behavior** • A. chrysaetos • Kansas
475	**Hunting Behavior** • Status • Distribution • Prey • Historical Account • Raptors • A. chrysaetos • Maine
492	**Hunting Behavior** • Raptors • A. pomarina • Zambia
495	**Hunting Behavior** • Wintering Population • Prey • Distribution • Age Structure • A. nipalensis • Africa
501	**Hunting Behavior** • Prey • Pellet • Attack • Animal Damage • Habitat • Cache • A. audax • Australia

Master No.	Keywords
546	**Hunting Behavior** • Anatomy • Status • A. chrysaetos • A. heliaca • A. audax • A. gurneyi • A. rapax • A. nipalensis • A. clanga • A. verreauxi • A. wahlbergi • A. pomarina
565	**Hunting Behavior** • Prey • Prey Behavior • A. chrysaetos • Alberta
578	**Hunting Behavior** • Shooting • Attitude • Animal Damage • A. chrysaetos • Texas
605	**Hunting Behavior** • Animal Damage • Poisoning • A. audax • Australia
618	**Hunting Behavior** • Prey • Nest • Attitude • Behavior • Breeding Biology • Flight • Poisoning • Molt • Vocalization • Folklore • Lift • Photo Account • A. chrysaetos • Montana
621	**Hunting Behavior** • A. chrysaetos • Washington
628	**Hunting Behavior** • Prey • Nest • Habitat • Photography Account • A. chrysaetos • California
662	**Hunting Behavior** • Raptors • A. rapax • Ethiopia
698	**Hunting Behavior** • Prey Behavior • Vocalization • A. chrysaetos • California
716	**Hunting Behavior** • Prey • A. chrysaetos • Oregon
732	**Hunting Behavior** • Prey Behavior • A. chrysaetos • Scotland
782	**Hunting Behavior** • A. chrysaetos • Alabama
783	**Hunting Behavior** • Protection • Conservation • Prey • Animal Damage • Prey Behavior • Raptors • A. chrysaetos • Sweden
829	**Hunting Behavior** • Photo Account • Popular Monograph • A. chrysaetos • Scotland
842	**Hunting Behavior** • Courtship • Animal Damage • Prey Beh. • A. audax • Australia
853	**Hunting Behavior** • Prey • A. chrysaetos • Spain
895	**Hunting Behavior** • Nest • Territory • Egg • Measure • Intraspecific Behavior • Courtship • Interspecific Behavior • A. chrysaetos • California
898	**Hunting Behavior** • Prey • A. chrysaetos • Alaska
900	**Hunting Behavior** • Descriptive Summary • Prey • Nest • Interspecific Behavior • Mortality • Vocalization • Raptors • Mt. Mckinley • A. chrysaetos • Alaska
939	**Hunting Behavior** • Nest • Habitat • Prey • Field Id • Raptors • A. clanga • A. pomarina • A. chrysaetos • Poland
947	**Hunting Behavior** • Roadside Survey • Roost • Nest Survey • Prey • Capture • Marking • Interspecific Behavior • Shooting • Electrocution • A. chrysaetos • Utah
1009	**Hunting Behavior** • A. chrysaetos • Switzerland
1025	**Hunting Behavior** • Nest • Prey • Density • Habitat • Photo Account • Popular Monograph • A. heliaca • Spain
1042	**Hunting Behavior** • Descriptive Summary • Measure • Prey • Animal Damage • Raptors • A. chrysaetos • North America
1064	**Hunting Behavior** • Descriptive Summary • Status • Folklore • Raptors • A. chrysaetos
1066	**Hunting Behavior** • Prey • A. chrysaetos • California • Nevada
1103	**Hunting Behavior** • Status • Prey • Poisoning • Raptors • A. chrysaetos • California
1131	**Hunting Behavior** • Folklore • Trapping • Animal Damage • Protection • Eagles • A. chrysaetos • Tennessee
1144	**Hunting Behavior** • Nest Survey • Status • Behavior • Fledging • Mortality • Banding • Marking • Cain/Abel • Reproductive Success • Breeding Biology • Interspecific Behavior • Display • Courtship • Raptors • A. verreauxi • Rhodesia
1145	**Hunting Behavior** • Nest • Parental Behavior • Breeding Biology • Interspecific Behavior • Intraspecific Behavior • Feeding Behavior • Fledging • Prey • Display • Raptors • A. verreauxi • Rhodesia
1168	**Hunting Behavior** • Prey • Animal Damage • A. audax • Australia
1219	**Hunting Behavior** • A. chrysaetos • West Germany
1243	**Hunting Behavior** • Prey • Prey Behavior • A. chrysaetos • Wyoming
1246	**Hunting Behavior** • Behavior • Lift • Trapping • Attack • Photography • Behavior • Photo Account • Popular Monograph • A. chrysaetos • Scotland
1250	**Hunting Behavior** • Popular Monograph • A. chrysaetos • Scotland
1273	**Hunting Behavior** • Mortality • Falconry • Prey • Flight • Nest • Mortality • Photo Account • Popular Monograph • A. chrysaetos • Great Britain
1298	**Hunting Behavior** • Prey • A. chrysaetos • Montana
1350	**Hunting Behavior** • Prey Behavior • A. chrysaetos • Texas
1388	**Hunting Behavior** • Prey Behavior • A. chrysaetos • Norway
1389	**Hunting Behavior** • Popular Monograph • (Title Only) • A. chrysaetos • Scotland
1404	**Hunting Behavior** • Prey • A. chrysaetos • Manitoba
1454	**Hunting Behavior** • Animal Damage • Raptors • A. chrysaetos • A. nipalensis
1467	**Hunting Behavior** • A. chrysaetos • Switzerland
1514	**Hunting Behavior** • A. audax • Australia
1517	**Hunting Behavior** • Interspecific Behavior • Predator/Prey • Fox • A. chrysaetos • Alaska
1547	**Hunting Behavior** • Status • A. wahlbergi • Gold Coast
1561	**Hunting Behavior** • A. rapax • South Africa
1564	**Hunting Behavior** • Nest • Vocalization • Behavior • Raptors • A. verreauxi • South Africa
1673	**Hunting Behavior** • Prey Behavior • A. chrysaetos • Saskatchewan
1717	**Hunting Behavior** • Monograph • Prey • Animal Damage • Eagles • A. chrysaetos • North America
1732	**Hunting Behavior** • Prey • A. chrysaetos • Nevada
1733	**Hunting Behavior** • A. wahlbergi • South Africa
1794	**Hunting Behavior** • Shooting • A. chrysaetos • West Germany
1810	**Hunting Behavior** • Prey • Population Decline • Poisoning • A. chrysaetos • Yugoslavia
1818	**Hunting Behavior** • Breeding Biology • Prey • Raptors • A. clanga • Siberia
1839	**Hunting Behavior** • A. chrysaetos • Northwest Territory
1848	**Hunting Behavior** • Prey • A. chrysaetos • Massachusetts
1863	**Hunting Behavior** • Flight • Prey • Flight Speed • Photo Account • Raptors • A. chrysaetos • Scotland

Master No.	Keywords
1867	**Hunting Behavior** • Prey • Animal Damage • A. audax • Australia
1884	**Hunting Behavior** • A. audax • Australia
1896	**Hunting Behavior** • Predator/Prey • A. chrysaetos • Colorado
1903	**Hunting Behavior** • Nest • Prey • Habitat • Disturbance • Conservation • Mortality • A. chrysaetos • Israel
1928	**Hunting Behavior** • Prey • Pellet • Nest • Interspecific Behavior • Reproductive Success • Eagles • A. chrysaetos • Oklahoma
1988	**Hunting Behavior** • Popular Monograph • (Title Only) • A. chrysaetos • Great Britain
1994	**Hunting Behavior** • Distribution • Nest • Prey • Raptors • A. verreauxi • South Africa
2032	**Hunting Behavior** • Raptors • A. chrysaetos • Colorado
2091	**Hunting Behavior** • Animal Damage • Poisoning • A. audax • Australia
2092	**Hunting Behavior** • Field Id • Descriptive Summary • Animal Damage • Nest • Egg • Attitude • Piracy • Raptors • A. audax • Australia
2093	**Hunting Behavior** • Animal Damage • A. audax • Australia
2098	**Hunting Behavior** • A. chrysaetos • Scotland
2100	**Hunting Behavior** • Prey Behavior • A. chrysaetos • North America
2101	**Hunting Behavior** • Feeding Behavior • A. chrysaetos • California
2110	**Hunting Behavior** • Status • Collect • Raptors • A. chrysaetos • A. nipalensis • Pakistan
2114	**Hunting Behavior** • Raptors • A. chrysaetos • India
2116	**Hunting Behavior** • Prey • Raptors • A. chrysaetos • A. heliaca • A. nipalensis
2120	**Hunting Behavior** • Descriptive Summary • Distribution • Plumage • Taxonomy • Raptors • A. chrysaetos • A. nipalensis • A. rapax • A. clanga • Saudi Arabia
2121	**Hunting Behavior** • Prey Behavior • Raptors • A. chrysaetos • A. heliaca • A. nipalensis • A. rapax • A. verreauxi • A. pomarina • A. clanga
2124	**Hunting Behavior** • Monograph • Breeding Biology • Migration • Prey • Attack • Folklore • Falconry • Animal Damage • Popular Monograph • A. chrysaetos
2134	**Hunting Behavior** • A. chrysaetos • France
2143	**Hunting Behavior** • Prey • A. audax • Australia
2169	**Hunting Behavior** • Protection • Descriptive Summary • Illus. • Popular Monograph • A. chrysaetos • Wyoming
2187	**Hunting Behavior** • Prey • A. chrysaetos • Colorado
2253	**Hunting Behavior** • A. chrysaetos • British Columbia
2293	**Hunting Behavior** • Conservation • Falconry • Prey • Raptors • Popular Monograph • A. chrysaetos • United States
2294	**Hunting Behavior** • Lift • Behavior • Conservation • A. chrysaetos • North America
2344	**Hunting Behavior** • A. chrysaetos • Scotland
2357	**Hunting Behavior** • Nest • Photo Account • A. chrysaetos • West Germany
2371	**Hunting Behavior** • Popular Monograph • Story • Breeding Biology • Behavior • Habitat • Nest Survey • Prey • Raptors • Illus. • A. chrysaetos • Colorado
2397	**Hunting Behavior** • Status • Popular Monograph • A. chrysaetos • Wyoming
2433	**Hunting Behavior** • A. chrysaetos • British Columbia
2444	**Hunting Behavior** • A. verreauxi • Scotland
2478	**Hunting Behavior** • A. chrysaetos • Scotland
2484	**Hunting Behavior** • Prey • A. chrysaetos • West Germany
2489	**Hunting Behavior** • Prey • Interspecific Behavior • A. chrysaetos • Utah
2536	**Hunting Behavior** • Raptors • A. rapax • Botswana
2616	**Hunting Behavior** • Prey Behavior • A. chrysaetos • California
2672	**Hunting Behavior** • Animal Damage • Prey • Feeding Behavior • A. audax • Australia
2688	**Hunting Behavior** • Prey • Animal Damage • Eagles • A. chrysaetos • Michigan
2714	**Hunting Behavior** • Popular Monograph • Shooting • Prey • Interspecific Behavior • Eagles • A. chrysaetos • North America
2806	**Hunting Behavior** • Shooting • Predator/Prey • A. chrysaetos • Montana
2819	**Hunting Behavior** • Prey • Raptors • A. nipalensis • Kazakhstan
2830	**Hunting Behavior** • Breeding Biology • Habitat • Cain/Abel • Nest Adornment • Growth • Vocalization • Plumage • Prey • Illus. • A. pomarina • West Germany
2852	**Hunting Behavior** • Roadside Survey • Prey • Monograph • Density • Habitat • Interspecific Competition • Taxonomy • Vocalization • Nest Chronology • Photo Account • Raptors • A. wahlbergi • A. rapax • A. nipalensis • A. heliaca • Kenya
2856	**Hunting Behavior** • Breeding Biology • Nest Survey • Habitat • Territory • Interspecific Behavior • Prey • Raptors • A. chrysaetos • Utah
2867	**Hunting Behavior** • Prey Behavior • A. clanga • A. nipalensis • Ethiopia
2868	**Hunting Behavior** • Raptors • A. audax • Australia
2882	**Hunting Behavior** • Prey • A. heliaca • United Soviet Socialists Republic
2885	**Hunting Behavior** • Prey • A. chrysaetos • Oregon
2904	**Hunting Behavior** • A. chrysaetos • Colorado
2931	**Hunting Behavior** • Wintering Population • Raptors • A. chrysaetos • Colorado
3013	**Hunting Behavior** • Prey • Raptors • A. clanga • A. chrysaetos • A. heliaca • Turkmenia
3079	**Hunting Behavior** • Prey • Nest Survey • Clutch Size • Reproductive Success • Display • Density • Migration • A. wahlbergi • South Africa
3080	**Hunting Behavior** • Cain/Abel • Distribution • Educational Material • Field Id • Habitat • Prey • Artificial Nest • Nest • Incubation • Wintering Population • Conservation • Photo Account • Eagles • A. chrysaetos • A. wahlbergi • A. verreauxi • A. nipalensis • A. pomarina • South Africa
3085	**Hunting Behavior** • A. verreauxi • South Africa
3101	**Hunting Behavior** • A. chrysaetos • Northwest Territory
3119	**Hunting Behavior** • Interspecific Competition • Raptors • A. wahlbergi • A. rapax • Africa
3123	**Hunting Behavior** • Status • Nest • Raptors • A. wahlbergi • A. rapax • Ivory Coast
3127	**Hunting Behavior** • Habitat • Interspecific Competition • Raptors • A. wahlbergi • Ivory Coast

Master No.	Keywords
3139	**Hunting Behavior** • Prey Behavior • A. chrysaetos • Texas
3144	**Hunting Behavior** • A. chrysaetos • Montana
3182	**Hunting Behavior** • Story • Rehabilitation • Falconry • Measure • Popular Monograph • A. chrysaetos • Texas
3218	**Hunting Behavior** • Descriptive Summary • Prey • Habitat • Raptors • A. chrysaetos • A. heliaca • A. nipalensis • A. clanga • A. pomarina • West Germany
3235	**Hunting Behavior** • Growth • Nest • Photography • Prey • Behavior • Vocalization • Nest Adornment • Egg • Raptors • Popular Monograph • A. rapax • Kenya
3243	**Hunting Behavior** • Habitat • Prey • Raptors • A. chrysaetos • A. heliaca • A. nipalensis • United Soviet Socialists Republic
3284	**Hunting Behavior** • Prey • A. chrysaetos • United States
3321	**Hunting Behavior** • Aerial Survey • Capture • Animal Damage • A. chrysaetos • Texas • New Mexico
3345	**Hunting Behavior** • Prey • Eagles • A. chrysaetos • A. nipalensis • A. rapax • India
3363	**Hunting Behavior** • Capture • Attack • A. audax • Australia
3369	**Hunting Behavior** • Nest Survey • Egg • Vocalization • Habitat • Behavior • Photo Account • A. chrysaetos • Arizona
3370	**Hunting Behavior** • Trapping • Eagles • A. chrysaetos • Arizona
3382	**Hunting Behavior** • Shooting • Raptors • A. chrysaetos • Switzerland
3399	**Hunting Behavior** • Prey • A. audax • Australia
3408	**Hunting Behavior** • Measure • Animal Damage • A. chrysaetos • California
3413	**Hunting Behavior** • Bounty • Prey • Lift • A. chrysaetos • Montana
3427	**Hunting Behavior** • Nest • A. chrysaetos • Washington
2238	**Hyperthermia** • Nest • Nest Survey • Mortality • Nestling • Parental Behavior • Directional Exposure • A. chrysaetos • United States
26	**Id Key** • Descriptive Summary • Etymology • Status • Distribution • Field Id • Raptors • Illus. • A. chrysaetos • A. heliaca • A. rapax • A. nipalensis • A. clanga • A. pomarina • India
193	**Id Key** • Descriptive Summary • Etymology • Habitat • Egg • Raptors • Illus. • A. nipalensis • A. rapax • A. verreauxi • Somalia
242	**Id Key** • Descriptive Summary • Etymology • Measure • Plumage • Raptors • A. chrysaetos • A. heliaca • A. nipalensis • A. rapax • A. clanga • A. pomarina • India
261	**Id Key** • Descriptive Summary • Distribution • Prey • Etymology • A. rapax • A. wahlbergi • Gambia • Sierra Leone • Nigeria
440	**Id Key** • Status • Raptors • Illus. • A. chrysaetos • North America
744	**Id Key** • Plumage • Descriptive Summary • Raptors • A. chrysaetos • Wisconsin • Illinois
1228	**Id Key** • Monograph • Vocalization • Distribution • Prey • Plumage • A. pomarina • A. chrysaetos • A. nipalensis • Europe
1632	**Id Key** • A. chrysaetos • A. heliaca • A. nipalensis • A. pomarina • A. clanga • United Soviet Socialists Republic
1641	**Id Key** • Descriptive Summary • Distribution • Status • A. verreauxi • A. nipalensis • A. rapax • A. pomarina • A. wahlbergi • Kenya • Uganda
2059	**Id Key** • Raptors • A. heliaca • A. chrysaetos • A. nipalensis • A. clanga • A. pomarina
2060	**Id Key** • Raptors • A. chrysaetos • A. nipalensis • A. heliaca • A. pomarina • A. clanga • West Germany
2202	**Id Key** • Measure • Prey • Raptors • Photo Account • A. chrysaetos • A. nipalensis • A. clanga • Italy
2640	**Id Key** • Descriptive Summary • Raptors • A. chrysaetos • Minnesota
924	**Incubation** • Raptors • A. chrysaetos • A. heliaca • A. clanga
1152	**Incubation** • Nest • Egg • Parental Behavior • A. verreauxi • Rhodesia
1231	**Incubation** • Descriptive Summary • Status • Egg • Raptors • A. pomarina • A. clanga • A. chrysaetos • West Germany
1240	**Incubation** • Nest • Raptors • A. chrysaetos • Great Britain
1271	**Incubation** • Behavior • Flight • A. chrysaetos • Scotland
1433	**Incubation** • Measure • Egg • Theory • Raptors • A. heliaca • A. nipalensis • A. pomarina
1705	**Incubation** • A. chrysaetos • West Germany
2309	**Incubation** • Raptors • A. chrysaetos
2638	**Incubation** • Photo Account • Popular Monograph • A. chrysaetos • Scotland
2966	**Incubation** • Prey • Behavior • Cain/Abel • Reproductive Success • A. rapax • South Africa
2975	**Incubation** • Prey Behavior • Cain/Abel • Reproductive Success • A. rapax • South Africa
2994	**Incubation** • Nest Chronology • Courtship • Nest Adornment • Nest Construction • A. chrysaetos • Texas
3080	**Incubation** • Cain/Abel • Distribution • Educational Material • Field Id • Habitat • Hunting Behavior • Prey • Artificial Nest • Nest • Wintering Population • Conservation • Photo Account • Eagles • A. chrysaetos • A. wahlbergi • A. verreauxi • A. nipalensis • A. pomarina • South Africa
3239	**Incubation** • Egg • Temperature • Telemetry • Measure • Tech. • A. chrysaetos • United States
3323	**Incubation** • Territory • Cain/Abel • A. pomarina • West Germany
2993	**Infertility** • Nest • A. chrysaetos • Texas
60	**Interspecific Behavior** • Abnormal Behavior • Flight • A. chrysaetos • New Mexico
107	**Interspecific Behavior** • Flight • Nest Adornment • Photography • A. chrysaetos • Europe
395	**Interspecific Behavior** • Flight • Raptors • Popular Monograph • A. chrysaetos • Montana
433	**Interspecific Behavior** • A. chrysaetos • California
458	**Interspecific Behavior** • Raptors • A. chrysaetos • Scotland
524	**Interspecific Behavior** • Raptors • A. chrysaetos • Pennsylvania
537	**Interspecific Behavior** • Nest • Prey • Habitat • Raptors • A. verreauxi • A. wahlbergi • Kenya

Master No.	Keywords
570	**Interspecific Behavior** • A. chrysaetos • California
608	**Interspecific Behavior** • Prey • Eagles • A. audax • Australia
645	**Interspecific Behavior** • Rarity • A. clanga • Czechoslovakia
646	**Interspecific Behavior** • Mortality • Mobbing • A. chrysaetos • New York
687	**Interspecific Behavior** • A. chrysaetos • Scotland
699	**Interspecific Behavior** • Mobbing • Raptors • A. audax • Australia
711	**Interspecific Behavior** • Prey • Lift • Raptors • A. chrysaetos • Scotland
818	**Interspecific Behavior** • A. chrysaetos • Ireland
847	**Interspecific Behavior** • Migration • Raptors • A. chrysaetos • Alberta
895	**Interspecific Behavior** • Nest • Territory • Egg • Measure • Intraspecific Behavior • Courtship • Hunting Behavior • A. chrysaetos • California
897	**Interspecific Behavior** • Piracy • A. chrysaetos • Alaska
900	**Interspecific Behavior** • Descriptive Summary • Prey • Hunting Behavior • Nest • Mortality • Vocalization • Raptors • Mt. Mckinley • A. chrysaetos • Alaska
947	**Interspecific Behavior** • Roadside Survey • Roost • Nest Survey • Prey • Hunting Behavior • Capture • Marking • Shooting • Electrocution • A. chrysaetos • Utah
954	**Interspecific Behavior** • Disturbance • Abnormal Behavior • A. chrysaetos • Colorado
971	**Interspecific Behavior** • A. chrysaetos • Colorado
980	**Interspecific Behavior** • Migration • Raptors • A. chrysaetos • A. clanga • Libya
1050	**Interspecific Behavior** • Raptors • A. chrysaetos • California
1052	**Interspecific Behavior** • Prey • Raptors • A. chrysaetos • Washington
1144	**Interspecific Behavior** • Nest Survey • Status • Behavior • Fledging • Mortality • Banding • Marking • Cain/Abel • Reproductive Success • Breeding Biology • Hunting Behavior • Display • Courtship • Raptors • A. verreauxi • Rhodesia
1145	**Interspecific Behavior** • Nest • Parental Behavior • Breeding Biology • Intraspecific Behavior • Feeding Behavior • Hunting Behavior • Fledging • Prey • Display • Raptors • A. verreauxi • Rhodesia
1172	**Interspecific Behavior** • Raptors • A. audax • Australia
1193	**Interspecific Behavior** • A. chrysaetos • Switzerland
1230	**Interspecific Behavior** • Flight Speed • A. chrysaetos • Canada
1254	**Interspecific Behavior** • Courtship • Flight • Intraspecific Behavior • A. chrysaetos • Great Britain
1517	**Interspecific Behavior** • Hunting Behavior • Predator/Prey • Fox • A. chrysaetos • Alaska
1627	**Interspecific Behavior** • A. chrysaetos • Alaska
1696	**Interspecific Behavior** • A. chrysaetos • Saskatchewan
1928	**Interspecific Behavior** • Prey • Pellet • Nest • Hunting Behavior • Reproductive Success • Eagles • A. chrysaetos • Oklahoma
1952	**Interspecific Behavior** • Prey • Raptors • A. chrysaetos • A. heliaca • Rumania
2035	**Interspecific Behavior** • Mobbing • Prey Behavior • Monograph • Raptors • A. chrysaetos • A. clanga • Sweden
2152	**Interspecific Behavior** • Habitat • Growth • Parental Behavior • Prey • Breeding Biology • Courtship • Display • Photo Account • Raptors • A. heliaca • Spain
2163	**Interspecific Behavior** • Flight • A. chrysaetos • California
2257	**Interspecific Behavior** • Prey Behavior • Prey • Pellet • A. chrysaetos • Alaska
2342	**Interspecific Behavior** • Etymology • Status • Habitat • Collect • Nest • Raptors • A. verreauxi • A. nipalensis • A. rapax • Africa
2343	**Interspecific Behavior** • Flight • A. chrysaetos • Great Britain
2389	**Interspecific Behavior** • Eagles • A. chrysaetos • Italy
2411	**Interspecific Behavior** • Nest • Behavior • Photography • Prey • Nest Adornment • Talon Grappling • Cain/Abel • Photo Account • Popular Monograph • A. chrysaetos • Scotland
2489	**Interspecific Behavior** • Hunting Behavior • Prey • A. chrysaetos • Utah
2512	**Interspecific Behavior** • Wintering Population • Eagles • A. chrysaetos • East Germany
2554	**Interspecific Behavior** • Conservation • Nest • Behavior • Prey • Trapping • Animal Damage • Rehabilitation • Photo Account • Popular Monograph • A. chrysaetos • Wyoming
2593	**Interspecific Behavior** • Behavior • Talon Grappling • Raptors • A. chrysaetos • Washington
2653	**Interspecific Behavior** • Flight • Mobbing • Raptors • A. chrysaetos • Oregon
2714	**Interspecific Behavior** • Popular Monograph • Shooting • Prey • Hunting Behavior • Eagles • A. chrysaetos • North America
2854	**Interspecific Behavior** • Population Dynamics • Nest • Density • Prey • Breeding Biology • Distribution • Habitat • Directional Exposure • Reproductive Success • Mortality • Abandonment • Disturbance • Territory • A. chrysaetos • Utah
2856	**Interspecific Behavior** • Breeding Biology • Nest Survey • Habitat • Hunting Behavior • Territory • Prey • Raptors • A. chrysaetos • Utah
2973	**Interspecific Behavior** • Life History • Prey • Plumage • Cain/Abel • Banding • Reproductive Success • Piracy • Photo Account • Raptors • A. chrysaetos • Rhodesia
2974	**Interspecific Behavior** • Talon Grappling • Raptors • A. verreauxi • Rhodesia
3105	**Interspecific Behavior** • Flight • Raptors • Popular Monograph • A. chrysaetos • Pennsylvania
3135	**Interspecific Behavior** • Predator/Prey • Migration • Raptors • A. chrysaetos • A. heliaca • A. nipalensis • A. pomarina • A. clanga • Nepal
3196	**Interspecific Behavior** • Raptors • A. chrysaetos • United States
3206	**Interspecific Behavior** • A. chrysaetos • New Mexico
3253	**Interspecific Behavior** • Eagles • A. chrysaetos • France
3255	**Interspecific Behavior** • Raptors • A. chrysaetos • China
3395	**Interspecific Behavior** • Status • Raptors • A. chrysaetos • A. heliaca • Spain

Master No.	Keywords
498	**Interspecific Competition** • Taxonomy • Field Id • Molt • Measure • Museum Specimen • Behavior • Wintering Population • Conservation • Eagles • Photo Account • A. nipalensis • A. pomarina • Africa
534	**Interspecific Competition** • Habitat • Density • Nest Survey • Breeding Biology • Prey • Reproductive Success • Etymology • Plumage • Vocalization • Nest Adornment • Raptors • A. verreauxi • A. wahlbergi • Kenya
535	**Interspecific Competition** • Nest Survey • Breeding Biology • Etymology • Prey • Reproductive Success • Raptors • Photo Account • A. wahlbergi • A. verreauxi • Kenya
671	**Interspecific Competition** • Territory • Eagles • A. chrysaetos • France • Morocco
673	**Interspecific Competition** • Prey • Raptors • A. chrysaetos • France
1062	**Interspecific Competition** • Prey • A. chrysaetos • Alberta
1151	**Interspecific Competition** • Reproductive Success • Density • Population Dynamics • Prey • Nest • Climatology • Territory • Mortality • Clutch Size • Pesticides • Eagles • A. verreauxi • Rhodesia
1238	**Interspecific Competition** • Prey • Raptors • A. pomarina • Belorussia
1853	**Interspecific Competition** • Prey • Raptors • A. chrysaetos • Siberia
1856	**Interspecific Competition** • Food Requirements • Prey • Raptors • A. chrysaetos • A. pomarina
2036	**Interspecific Competition** • Nest • A. chrysaetos • Scotland • Great Britain
2081	**Interspecific Competition** • Eagles • A. chrysaetos • Montana
2400	**Interspecific Competition** • Animal Damage • Predator/Prey • Aerial Survey • Habitat • Status • Density • Population Dynamics • Food Habits • Nest • Literature Review • A. chrysaetos • Texas • New Mexico
2470	**Interspecific Competition** • Nest • Raptors • A. chrysaetos • Colorado
2561	**Interspecific Competition** • Nest • Raptors • A. chrysaetos • Scotland
2657	**Interspecific Competition** • Eagles • (Title Only) • A. chrysaetos • Sweden
2852	**Interspecific Competition** • Roadside Survey • Prey • Monograph • Density • Habitat • Taxonomy • Vocalization • Hunting Behavior • Nest Chronology • Photo Account • Raptors • A. wahlbergi • A. rapax • A. nipalensis • A. heliaca • Kenya
3118	**Interspecific Competition** • Distribution • Habitat • Prey • Descriptive Summary • Density • Eagles • A. chrysaetos • France
3119	**Interspecific Competition** • Hunting Behavior • Raptors • A. wahlbergi • A. rapax • Africa
3126	**Interspecific Competition** • Migration • Wintering Population • Raptors • A. nipalensis • A. pomarina • A. clanga • A. rapax • A. heliaca • Africa
3127	**Interspecific Competition** • Habitat • Hunting Behavior • Raptors • A. wahlbergi • Ivory Coast
3446	**Interspecific Competition** • Nest Failure • Raptors • A. chrysaetos • Tunisia
75	**Intraspecific Behavior** • Flight • (Title Only) • A. chrysaetos • North America
127	**Intraspecific Behavior** • Talon Grappling • A. chrysaetos • Scotland • Switzerland
304	**Intraspecific Behavior** • Plumage • Aging • Behavior • A. heliaca • A. nipalensis • A. chrysaetos • A. pomarina • A. clanga • Europe
487	**Intraspecific Behavior** • Talon Grappling • A. verreauxi • South Africa
552	**Intraspecific Behavior** • Nest Survey • Prey • Habitat • Behavior • Status • Reproductive Success • Food Requirements • Animal Damage • Predator/Prey • A. chrysaetos • Scotland
894	**Intraspecific Behavior** • Nest • Habitat • Egg • Measure • Prey • Photo Account • A. chrysaetos • California
895	**Intraspecific Behavior** • Nest • Territory • Egg • Measure • Courtship • Interspecific Behavior • Hunting Behavior • A. chrysaetos • California
1055	**Intraspecific Behavior** • Falconry • A. audax • Australia
1145	**Intraspecific Behavior** • Nest • Parental Behavior • Breeding Biology • Interspecific Behavior • Feeding Behavior • Hunting Behavior • Fledging • Prey • Display • Raptors • A. verreauxi • Rhodesia
1150	**Intraspecific Behavior** • Density • Territory • Dispersal • Nest • Survey • Marking • Reproductive Success • Display • A. verreauxi • Rhodesia
1171	**Intraspecific Behavior** • Nest • Prey • A. pomarina • Armenia
1185	**Intraspecific Behavior** • Flight • Plumage • Illus. • A. clanga • Yugoslavia
1254	**Intraspecific Behavior** • Courtship • Flight • Interspecific Behavior • A. chrysaetos • Great Britain
1268	**Intraspecific Behavior** • Flight • Courtship • Courtship • A. chrysaetos • Scotland
1635	**Intraspecific Behavior** • Density • Nest • Conservation • Prey • A. chrysaetos • Belorussia
1955	**Intraspecific Behavior** • Nest • Prey • Raptors • A. chrysaetos • California
2391	**Intraspecific Behavior** • Talon Grappling • A. chrysaetos • Scotland
2659	**Intraspecific Behavior** • Flight • Nest • A. chrysaetos • Scotland
2814	**Intraspecific Behavior** • Display • A. chrysaetos • Texas
2989	**Intraspecific Behavior** • A. wahlbergi • Rhodesia
3259	**Intraspecific Behavior** • Nest Survey • Habitat • Territory • Density • Nest Chronology • Prey • Directional Exposure • Raptors • A. verreauxi • Rhodesia
3303	**Intraspecific Behavior** • Display • A. chrysaetos • Scotland
3409	**Intraspecific Behavior** • Flight • A. chrysaetos • Scotland
2460	**Invertebrates** • Parasite • Mortality • Raptors • A. chrysaetos • United States
831	**Karyotype** • Raptors • A. audax
1527	**Karyotype** • Raptors • A. chrysaetos
3076	**Karyotype** • Raptors • A. chrysaetos • A. heliaca

Master No.	Keywords
547	**Knowledge Status** • Breeding Biology • Conservation • Behavior • A. chrysaetos • A. heliaca • A. audax • A. rapax • A. nipalensis • A. pomarina • A. clanga • A. gurneyi • A. verreauxi • A. wahlbergi
549	**Knowledge Status** • A. verreauxi • A. rapax • A. wahlbergi • Africa
217	**Life History** • Symbolism • Raptors • A. chrysaetos
351	**Life History** • Hunting Behavior • Measure • Courtship • Behavior • Distribution • Vocalization • Plumage • Raptors • Photo Account • A. chrysaetos • North America
544	**Life History** • Taxonomy • Status • Field Id • Flight • Habitat • Terminology • Distribution • Mortality • Behavior • Longevity • Territory • Photo Account • Raptors • A. chrysaetos • A. clanga • Great Britain • Ireland • Scotland
702	**Life History** • Food Habits • Distribution • Popular Monograph • A. chrysaetos • Missouri
871	**Life History** • Distribution • Etymology • Habitat • Molt • Prey • Measure • Field Id • Migration • Status • Raptors • A. chrysaetos • A. heliaca • A. rapax • A. clanga • A. pomarina • United Soviet Socialists Republic
993	**Life History** • Field Id • Descriptive Summary • Photo Account • Popular Monograph • A. chrysaetos • Scotland
1245	**Life History** • Monograph • A. chrysaetos • Great Britain
1639	**Life History** • Taxonomy • Dimorphism • Measure • Plumage • Distribution • Prey • Behavior • Management • Habitat • Territory • Mortality • Economics • Literature Review • A. chrysaetos • Oregon • Washington
2654	**Life History** • Symbolism • Educational Material • Eagles • Children's Book • Popular Monograph • A. chrysaetos
2873	**Life History** • Literature Review • Habitat • Management • Population Estimate • A. chrysaetos • United States
2973	**Life History** • Prey • Plumage • Cain/Abel • Banding • Reproductive Success • Interspecific Behavior • Piracy • Photo Account • Raptors • A. chrysaetos • Rhodesia
221	**Lift** • Attack • Flight • Altitude • Prey Behavior • A. chrysaetos • A. pomarina
294	**Lift** • Animal Damage • Poisoning • Shooting • Captivity • A. audax • Australia
365	**Lift** • Prey • Raptors • A. chrysaetos • Norway
616	**Lift** • Nest • Habitat • Egg • Prey • Vocalization • Growth • Trapping • Poisoning • Behavior • Photo Account • A. chrysaetos • Montana
618	**Lift** • Prey • Nest • Attitude • Behavior • Breeding Biology • Flight • Poisoning • Molt • Hunting Behavior • Vocalization • Folklore • Photo Account • A. chrysaetos • Montana
711	**Lift** • Prey • Interspecific Behavior • Raptors • A. chrysaetos • Scotland
739	**Lift** • Prey • A. rapax • Africa
855	**Lift** • Prey • A. chrysaetos • India
1203	**Lift** • Prey • A. chrysaetos • Great Britain
1204	**Lift** • Prey • Animal Damage • A. chrysaetos • Great Britain
1208	**Lift** • Prey • A. chrysaetos • Scotland
1246	**Lift** • Behavior • Hunting Behavior • Trapping • Attack • Photography • Behavior • Photo Account • Popular Monograph • A. chrysaetos • Scotland
1260	**Lift** • Prey • A. chrysaetos • Great Britain
1413	**Lift** • Prey • A. chrysaetos • Switzerland
1592	**Lift** • A. chrysaetos
1758	**Lift** • Captive Breeding • A. chrysaetos • North America
2294	**Lift** • Hunting Behavior • Behavior • Conservation • A. chrysaetos • North America
2369	**Lift** • Prey • Folklore • Animal Damage • Conservation • Electrocution • Eagles • Popular Monograph • A. chrysaetos • North America
2425	**Lift** • Capture • Prey • Historical Account • A. chrysaetos • Scotland
2477	**Lift** • Prey • A. chrysaetos • Great Britain
2628	**Lift** • Animal Damage • A. chrysaetos • Ohio
3252	**Lift** • Folklore • A. chrysaetos • Switzerland
3290	**Lift** • Captivity • Nest • Photo Account • Popular Monograph • A. chrysaetos • California
3335	**Lift** • Descriptive Summary • Prey • Raptors • A. chrysaetos • North America
3398	**Lift** • A. chrysaetos • Scotland
3413	**Lift** • Bounty • Prey • Hunting Behavior • A. chrysaetos • Montana
34	**Literature Review** • Bibliography • Eagles • A. chrysaetos • Washington
198	**Literature Review** • Prey • Monograph • Nest Survey • Habitat • Reproductive Success • Disturbance • Pellet • Animal Damage • A. chrysaetos • Utah
941	**Literature Review** • Descriptive Summary • Status • Distribution • Management • Prey • A. chrysaetos • Washington
1080	**Literature Review** • Collect • Nest • Etymology • Distribution • A. chrysaetos • A. rapax • Tunisia
1081	**Literature Review** • Taxonomy • Collect • A. rapax • Africa
1290	**Literature Review** • Disease • Raptors • A. chrysaetos • Great Britain
1508	**Literature Review** • Prey • Pellet • Shooting • Predator/Prey • Nest Survey • Attitude • Reproductive Success • A. chrysaetos • North America
1549	**Literature Review** • Rarity • A. chrysaetos • Ontario
1687	**Literature Review** • Monograph • Molt • Breeding Biology • Field Id • A. chrysaetos • Colorado
1697	**Literature Review** • Plumage • Status • Distribution • Taxonomy • Etymology • Field Id • Eagles • A. pomarina • A. clanga • France • Belgium • Switzerland • Italy
2168	**Literature Review** • Prey • Owls • A. chrysaetos • Europe
2373	**Literature Review** • Prey • Animal Damage • Food Requirements • A. chrysaetos • Colorado
2375	**Literature Review** • Bibliography • Raptors • A. chrysaetos
2377	**Literature Review** • Conservation • Electrocution • Management • Status • Behavior • Eagles • A. chrysaetos • North America

Master No.	Keywords
2400	**Literature Review** • Animal Damage • Predator/Prey • Aerial Survey • Habitat • Status • Density • Population Dynamics • Food Habits • Nest • Interspecific Competition • A. chrysaetos • Texas • New Mexico
2873	**Literature Review** • Life History • Habitat • Management • Population Estimate • A. chrysaetos • United States
78	**Longevity** • A. heliaca • West Germany
137	**Longevity** • A. chrysaetos • Scotland
147	**Longevity** • Captivity • A. audax • A. chrysaetos • A. heliaca
531	**Longevity** • A. pomarina
544	**Longevity** • Taxonomy • Status • Field Id • Flight • Habitat • Terminology • Distribution • Mortality • Behavior • Life History • Territory • Photo Account • Raptors • A. chrysaetos • A. clanga • Great Britain • Ireland • Scotland
630	**Longevity** • Courtship • A. chrysaetos • California
1047	**Longevity** • Egg • Captivity • A. chrysaetos • Great Britain
1048	**Longevity** • Captivity • A. chrysaetos • Great Britain
1063	**Longevity** • Captivity • A. chrysaetos • A. audax • A. verreauxi • A. heliaca • A. rapax • A. pomarina • Europe • Australia
1175	**Longevity** • Raptors • Popular Monograph • A. chrysaetos • East Germany
1224	**Longevity** • Banding • A. chrysaetos • Switzerland
1317	**Longevity** • Captivity • A. heliaca • Spain
1332	**Longevity** • Captivity • Raptors • A. chrysaetos • A. heliaca
1333	**Longevity** • A. heliaca • Spain
1503	**Longevity** • Captivity • Egg • Plumage • A. rapax • Tunisia
1647	**Longevity** • Banding • A. chrysaetos • Switzerland
1648	**Longevity** • Banding • A. chrysaetos • Switzerland
2082	**Longevity** • Nest • A. verreauxi • Rhodesia
2697	**Longevity** • A. pomarina • East Germany
2698	**Longevity** • A. chrysaetos • A. pomarina
2752	**Longevity** • Banding • A. pomarina • West Germany
2992	**Longevity** • Raptors • A. chrysaetos • United States
3023	**Longevity** • Banding • Raptors • A. chrysaetos • California
3261	**Longevity** • Mortality • Theory • A. verreauxi • Rhodesia
3286	**Longevity** • Captivity • A. chrysaetos
47	**Management** • Population Decline • Historical Account • Status • Distribution • Raptors • A. chrysaetos • Tennessee
62	**Management** • Habitat • Measure • Eagles • A. chrysaetos • Oregon
66	**Management** • Electrocution • Mortality • Nest • A. chrysaetos • Idaho
169	**Management** • Electrocution • A. chrysaetos • United States
174	**Management** • Forestry • Protection • A. chrysaetos • Washington
187	**Management** • Electrocution • Raptors • A. chrysaetos • Idaho
233	**Management** • Attitude • Reproductive Success • Density • Wintering Population • Food Requirements • Habitat • Aerial Survey • Nest Survey • Raptors • A. chrysaetos • Montana
352	**Management** • Electrocution • Mortality • A. chrysaetos • North America
358	**Management** • Status • Population Estimate • Protection • Artificial Nest • Raptors • A. chrysaetos • A. clanga • Finland
425	**Management** • Aerial Survey • Animal Damage • Wintering Population • A. chrysaetos • Texas • New Mexico
434	**Management** • Animal Damage • Economics • Attitude • Survey • A. chrysaetos • North America
610	**Management** • Artificial Nest • Habitat • Photo Account • Raptors • A. chrysaetos • North America
765	**Management** • Nest • History • Conservation • Raptors • Popular Monograph • A. chrysaetos • United States
941	**Management** • Descriptive Summary • Status • Distribution • Literature Review • Prey • A. chrysaetos • Washington
999	**Management** • Conservation • Refuge • Wintering Population • Raptors • A. chrysaetos • Great Britain
1035	**Management** • Conservation • Raptors • A. chrysaetos • United States
1366	**Management** • Status • Mortality • Protection • Disturbance • Eagles • A. chrysaetos • North America
1580	**Management** • Mortality • Monograph • Electrocution • A. chrysaetos • North America
1593	**Management** • Wintering Population • A. chrysaetos • North Carolina
1639	**Management** • Taxonomy • Dimorphism • Measure • Plumage • Distribution • Prey • Behavior • Habitat • Territory • Mortality • Life History • Economics • Literature Review • A. chrysaetos • Oregon • Washington
1676	**Management** • Descriptive Summary • Raptors • A. chrysaetos • Idaho
1741	**Management** • Energy Development • Environmental Impact • Raptors • A. chrysaetos • North America
1879	**Management** • Electrocution • Trapping • Eagles • A. chrysaetos • United States
1889	**Management** • Artificial Nest • Aerial Survey • Raptors • A. chrysaetos • North America
1946	**Management** • Mortality • Protection • Photo Account • A. chrysaetos • Scotland
1950	**Management** • Animal Damage • Nest Survey • Abandonment • Prey • A. chrysaetos • Scotland
2019	**Management** • Educational Material • Falconry • Rehabilitation • Survey • Raptors • A. chrysaetos • California
2040	**Management** • Mortality • Electrocution • Raptors • A. chrysaetos • Idaho
2146	**Management** • Cain/Abel • A. pomarina • Czechoslovakia
2148	**Management** • Distribution • Field Id • Taxonomy • Habitat • Nest Survey • Prey • Breeding Biology • Photo Account • Eagles • A. pomarina • Europe
2154	**Management** • Cain/Abel • Mortality • Reproductive Success • Experiment • A. chrysaetos • A. heliaca
2155	**Management** • Cain/Abel • Experiment • Theory • A. pomarina • A. heliaca • A. verreauxi • A. chrysaetos

Master No.	Keywords
2156	**Management** • Cain/Abel • Clutch Size • Translocation • Eagles • Tech. • A. pomarina • A. verreauxi • A. rapax • A. chrysaetos • A. heliaca
2161	**Management** • Cain/Abel • A. pomarina • A. verreauxi • A. heliaca • East Germany
2184	**Management** • Aerial Survey • Reproductive Success • Survey • Clutch Size • Distribution • Habitat • Nest Survey • Prey • Raptors • A. chrysaetos • Arizona
2212	**Management** • Nest • Reproductive Success • Shooting • Prey • Density • Raptors • A. chrysaetos • Washington
2239	**Management** • Survey • Territory • Nest Failure • Nest • Migration • Habitat • Reproductive Success • Prey • Roadside Survey • Status • Winter Survey • Uinta Nf • Wasatch Mt. • Raptors • A. chrysaetos • Utah
2261	**Management** • Nest Survey • Disturbance • Prey • Population Dynamics • Status • Animal Damage • Artificial Nest • Predator/Prey • A. chrysaetos • Utah
2262	**Management** • Conservation • Animal Damage • Raptors • A. chrysaetos • United States
2263	**Management** • Electrocution • Habitat Loss • A. chrysaetos • North America
2264	**Management** • Environmental Impact • Nest • Photo Account • Raptors • A. chrysaetos • Utah
2265	**Management** • Environmental Impact • Nest • Photo Account • Raptors • A. chrysaetos • Utah
2266	**Management** • Environmental Impact • Nest • Photo Account • Raptors • A. chrysaetos • Utah
2267	**Management** • Environmental Impact • Nest • Photo Account • Raptors • A. chrysaetos • Utah
2268	**Management** • Nest Survey • Raptors • Photo Account • A. chrysaetos • Utah
2292	**Management** • Electrocution • Historical Account • Raptors • A. chrysaetos • North America
2299	**Management** • Electrocution • Artificial Nest • Conservation • Shooting • Eagles • Popular Monograph • A. chrysaetos • Idaho
2324	**Management** • Capture • Animal Damage • Banding • Prey • Sexing • Translocation • Marking • Measure • Captive Care • Tech. • Eagles • A. chrysaetos • Montana
2325	**Management** • Capture • Banding • Marking • Animal Damage • Sexing • Translocation • Prey • Survey • Measure • Mortality • Captive Care • Aging • A. chrysaetos • Montana
2326	**Management** • Capture • Banding • Marking • Animal Damage • Sexing • Translocation • Prey • Measure • Mortality • Capt. Care • Aging • Food Requirements • A. chrysaetos • Montana
2349	**Management** • Feeding Behavior • Animal Damage • Capture • Predator/Prey • Protection • Economics • A. chrysaetos • Montana
2370	**Management** • Status • Nest • Nest Survey • Monograph • Habitat • Mortality • Breeding Biology • Nest Chronology • Courtship • Density • Prey • A. chrysaetos • Colorado
2374	**Management** • Conservation • Habitat Protection • Refuge • Theory • Tech. • Raptors • A. chrysaetos • United States
2376	**Management** • Habitat • Artificial Nest • Nest Chronology • A. chrysaetos • North America
2377	**Management** • Conservation • Electrocution • Status • Literature Review • Behavior • Eagles • A. chrysaetos • North America
2464	**Management** • Habitat • Directional Exposure • Disturbance • Mortality • Nest Survey • Density • Nest Chronology • Prey • Reproductive Success • Prog. Rep. • A. chrysaetos • Wyoming
2590	**Management** • Status • A. chrysaetos • Idaho
2713	**Management** • Habitat • A. chrysaetos • North Carolina
2873	**Management** • Literature Review • Life History • Habitat • Population Estimate • A. chrysaetos • United States
2874	**Management** • Survey • Raptors • A. chrysaetos • Idaho
2914	**Management** • Conservation • A. chrysaetos • United States
3031	**Management** • Environmental Impact • Disturbance • Habitat Protection • Raptors • A. chrysaetos • North America
3201	**Management** • Electrocution • Eagles • A. chrysaetos • United States
3205	**Management** • Animal Damage • Status • Distribution • Wintering Population • A. chrysaetos • United States
3316	**Management** • Animal Damage • Prey • Food Requirements • A. chrysaetos • Scotland
3330	**Management** • Status • Habitat • A. chrysaetos • Washington
180	**Marking** • A. wahlbergi • South Africa
181	**Marking** • Rehabilitation • Eagles • A. chrysaetos • Washington
189	**Marking** • Migration • A. wahlbergi • South Africa
325	**Marking** • Monograph • Breeding Biology • Density • Mortality • Cain/Abel • Prey • Abnormal Behavior • Egg Measure • A. chrysaetos • Idaho
947	**Marking** • Roadside Survey • Roost • Nest Survey • Prey • Hunting Behavior • Capture • Interspecific Behavior • Shooting • Electrocution • A. chrysaetos • Utah
958	**Marking** • Behavior • Dye • Tech. • A. chrysaetos
974	**Marking** • Rehabilitation • Shooting • Behavior • Eagles • A. chrysaetos • United States
1144	**Marking** • Nest Survey • Status • Behavior • Fledging • Mortality • Banding • Cain/Abel • Reproductive Success • Breeding Biology • Interspecific Behavior • Hunting Behavior • Display • Courtship • Raptors • A. verreauxi • Rhodesia
1148	**Marking** • Banding • Fledging • A. verreauxi • Rhodesia
1150	**Marking** • Density • Territory • Dispersal • Nest • Survey • Intraspecific Behavior • Reproductive Success • Display • A. verreauxi • Rhodesia
1736	**Marking** • Captivity • Raptors • A. wahlbergi • South Africa
1775	**Marking** • Monograph • Status • Nest Survey • Growth • Reproductive Success • Cain/Abel • Banding • Mortality • Pesticides • Prey • Residues • A. chrysaetos • Idaho

Master No.	Keywords
894	**Measure** • Intraspecific Behavior • Nest • Habitat • Egg • Prey • Photo Account • A. chrysaetos • California
895	**Measure** • Nest • Territory • Egg • Intraspecific Behavior • Courtship • Interspecific Behavior • Hunting Behavior • A. chrysaetos • California
907	**Measure** • Descriptive Summary • Field Id • Plumage • Prey • Raptors • A. chrysaetos • A. heliaca • A. nipalensis • A. rapax • A. clanga • India
916	**Measure** • Taxonomy • Plumage • A. nipalensis • A. clanga • A. pomarina
951	**Measure** • Trapping • A. chrysaetos • Ontario
955	**Measure** • Behavior • Pesticides • Parental Behavior • Vocalization • Disturbance • Breeding Biology • Parasite • Disease • Terminology • Growth • A. chrysaetos • Montana
956	**Measure** • Capture • Aircraft • Tech. • A. chrysaetos • Montana
957	**Measure** • Monograph • Behavior • Growth • Pesticides • Parental Behavior • Vocalization • Disturbance • Breeding Biology • Photography • Captivity • Parasite • Disease • Terminology • Photo Account • A. chrysaetos • Montana
978	**Measure** • Museum Specimen • A. chrysaetos • France
989	**Measure** • Taxonomy • Plumage • Museum Specimen • A. clanga • A. pomarina • A. rapax • A. nipalensis • France
1013	**Measure** • Rarity • Captivity • Poisoning • Plumage • Photo Account • A. heliaca • Italy
1019	**Measure** • Captive Breeding • Artificial Incubation • Raptors • A. chrysaetos • A. rapax • A. nipalensis • West Germany
1028	**Measure** • Banding • Migration • Raptors • Photo Account • A. chrysaetos • Ontario
1034	**Measure** • Nest • Prey • Egg • Raptors • A. verreauxi • South Africa
1042	**Measure** • Descriptive Summary • Hunting Behavior • Prey • Animal Damage • Raptors • A. chrysaetos • North America
1045	**Measure** • Osteology • A. chrysaetos • North America
1054	**Measure** • Captive Breeding • Nest • Behavior • Plumage • Photo Account • Popular Monograph • A. audax • Australia
1056	**Measure** • Growth • Falconry • Aging • Parental Behavior • Plumage • Dimorphism • Prey • Photo Account • A. audax • Australia
1095	**Measure** • Museum Specimen • Shooting • Plumage • Taxonomy • A. clanga • Albania
1196	**Measure** • Museum Specimen • Raptors • A. chrysaetos • Italy
1197	**Measure** • Shooting • A. clanga • Italy
1198	**Measure** • Nest • Collect • A. clanga • A. chrysaetos • Italy
1210	**Measure** • Collect • A. clanga • Italy
1211	**Measure** • Rarity • Collect • Plumage • Photo Account • A. pomarina • Italy
1299	**Measure** • Trapping • Raptors • A. chrysaetos • Alaska
1314	**Measure** • Taxonomy • Plumage • A. clanga • Europe
1329	**Measure** • Plumage • Rarity • A. clanga • A. pomarina • Great Britain
1345	**Measure** • Rarity • Shooting • Plumage • Prey • A. chrysaetos • West Germany
1396	**Measure** • Taxonomy • Plumage • Museum Specimen • A. rapax • A. nipalensis • Great Britain
1399	**Measure** • Physiology • Raptors • A. chrysaetos
1418	**Measure** • Taxonomy • Plumage • Photo Account • A. clanga • A. pomarina • Belgium
1427	**Measure** • Prey • Behavior • Shooting • A. chrysaetos • Georgia
1433	**Measure** • Incubation • Egg • Theory • Raptors • A. heliaca • A. nipalensis • A. pomarina
1457	**Measure** • Shooting • A. chrysaetos • Ohio
1496	**Measure** • Egg • Status • Pesticides • Raptors • A. chrysaetos • California
1505	**Measure** • Dimorphism • Taxonomy • Sexing • Museum Specimen • Raptors • A. chrysaetos • North America
1526	**Measure** • Descriptive Summary • Distribution • Status • A. verreauxi • A. nipalensis • A. rapax • A. pomarina • A. wahlbergi • Southwest Africa
1575	**Measure** • Paleontology • Anatomy • A. chrysaetos • North America
1586	**Measure** • Descriptive Summary • Distribution • Nest • Photo Account • Illus. • Raptors • A. chrysaetos • A. heliaca • A. nipalensis • A. clanga • A. pomarina • Czechoslovakia
1598	**Measure** • Prey • Raptors • A. chrysaetos • A. heliaca • A. clanga • A. rapax • India
1600	**Measure** • Plumage • Raptors • A. heliaca • A. nipalensis • A. clanga • India
1616	**Measure** • Raptors • A. chrysaetos • Kansas
1626	**Measure** • Migration • Etymology • Trapping • A. chrysaetos • Alaska
1639	**Measure** • Taxonomy • Dimorphism • Plumage • Distribution • Prey • Behavior • Management • Habitat • Territory • Mortality • Life History • Economics • Literature Review • A. chrysaetos • Oregon • Washington
1643	**Measure** • Egg Collection • Poisoning • Raptors • A. audax • Australia
1670	**Measure** • Descriptive Summary • Etymology • Trapping • Raptors • A. chrysaetos • Washington
1677	**Measure** • A. chrysaetos • New York
1702	**Measure** • Shooting • A. chrysaetos • Great Britain
1760	**Measure** • Taxonomy • A. chrysaetos • Japan
1806	**Measure** • Nest • Clutch Size • Egg • Raptors • A. wahlbergi • Rhodesia
1841	**Measure** • Rarity • Collect • Plumage • A. heliaca • France
1847	**Measure** • Descriptive Summary • Plumage • Distribution • Raptors • A. chrysaetos • A. heliaca • A. nipalensis • A. clanga • China
1865	**Measure** • Food Habits • A. chrysaetos • Minnesota
1883	**Measure** • Albinism • Trapping • A. audax • Australia
1892	**Measure** • Nest Survey • Clutch Size • Egg • A. wahlbergi • Rhodesia
1997	**Measure** • Abnormal Behavior • Prey • A. chrysaetos • Scotland
2020	**Measure** • Rarity • Museum Specimen • Plumage • Photo Account • A. heliaca • Italy

Master No.	Keywords
2045	**Measure** • Descriptive Summary • Plumage • Raptors • A. chrysaetos • A. heliaca • A. nipalensis • A. pomarina • A. clanga • Italy
2202	**Measure** • Id Key • Prey • Raptors • Photo Account • A. chrysaetos • A. nipalensis • A. clanga • Italy
2206	**Measure** • Collect • Prey • A. clanga • Italy
2223	**Measure** • A. audax • Australia
2273	**Measure** • Shooting • Animal Damage • Wintering Population • Eagles • A. chrysaetos • Illinois
2281	**Measure** • Prey • Trapping • A. chrysaetos • Maine
2324	**Measure** • Capture • Animal Damage • Banding • Management • Prey • Sexing • Translocation • Marking • Captive Care • Tech. • Eagles • A. chrysaetos • Montana
2325	**Measure** • Capture • Banding • Marking • Animal Damage • Sexing • Management • Translocation • Prey • Survey • Mortality • Captive Care • Aging • A. chrysaetos • Montana
2326	**Measure** • Capture • Banding • Marking • Animal Damage • Sexing • Management • Translocation • Prey • Mortality • Captive Care • Aging • Food Requirements • A. chrysaetos • Montana
2337	**Measure** • Egg Collection • A. chrysaetos • California
2365	**Measure** • Rarity • A. clanga • Great Britain
2386	**Measure** • Shooting • A. chrysaetos • Italy
2450	**Measure** • Nest • Behavior • A. chrysaetos • Spain
2452	**Measure** • Wing Span • A. audax • Australia
2500	**Measure** • Wing Loading • Anatomy • Raptors • A. chrysaetos • North America
2506	**Measure** • Anatomy • Raptors • A. chrysaetos
2513	**Measure** • Rarity • Prey • A. clanga • Great Britain
2552	**Measure** • Descriptive Summary • Distribution • Taxonomy • Plumage • Habitat • Raptors • A. gurneyi • New Guinea
2639	**Measure** • Mortality • A. chrysaetos • Quebec
2644	**Measure** • Refuge • Conservation • Nest • SRBPNA • Photo Account • Popular Monograph • A. chrysaetos • Idaho
2649	**Measure** • Anatomy • Raptors • A. chrysaetos • A. rapax
2652	**Measure** • Rarity • Plumage • A. clanga • Great Britain
2721	**Measure** • Prey • A. chrysaetos • California
2794	**Measure** • Shooting • Plumage • A. chrysaetos • Pennsylvania
2805	**Measure** • Egg Collection • A. chrysaetos • California
2840	**Measure** • Altitude • Mortality • A. nipalensis • India
2851	**Measure** • Egg Collection • Nest • Habitat • Photo Account • A. chrysaetos • California
2907	**Measure** • Wintering Population • Trapping • Plumage • Molt • A. chrysaetos • Tennessee
2908	**Measure** • Wintering Population • Trapping • Plumage • A. chrysaetos • Tennessee
2926	**Measure** • Banding • Raptors • A. chrysaetos
2934	**Measure** • Anatomy • Physiology • Raptors • Ovary • Testes • A. chrysaetos
2943	**Measure** • Egg Collection • Nest • A. chrysaetos • California
2963	**Measure** • Rarity • Animal Damage • Captivity • Behavior • A. chrysaetos • Georgia
2967	**Measure** • A. chrysaetos • Alabama
2970	**Measure** • Nest Survey • Clutch Size • Egg • A. wahlbergi • Rhodesia
2985	**Measure** • Prey • Shooting • Photo Account • Eagles • A. chrysaetos • New York
3021	**Measure** • Captivity • Growth • Food Requirements • Molt • Pellet • Temperature • Photo Account • Raptors • A. chrysaetos • California
3026	**Measure** • Shooting • A. chrysaetos • Virginia
3034	**Measure** • Trapping • Shooting • A. chrysaetos • Pennsylvania
3036	**Measure** • Descriptive Summary • Distribution • Status • Prey • Raptors • A. chrysaetos • Oklahoma
3040	**Measure** • Field Id • Plumage • Eagles • Photo Account • A. chrysaetos • A. pomarina • Europe
3045	**Measure** • Taxonomy • Museum Specimen • Rarity • A. rapax • Rumania
3182	**Measure** • Story • Hunting Behavior • Rehabilitation • Falconry • Popular Monograph • A. chrysaetos • Texas
3239	**Measure** • Egg • Temperature • Telemetry • Incubation • Tech. • A. chrysaetos • United States
3248	**Measure** • Museum Specimen • Raptors • A. chrysaetos • A. clanga • A. pomarina • Hungary
3269	**Measure** • Status • Raptors • A. pomarina • Chad
3291	**Measure** • Flight • Folklore • Nest • Weight • A. chrysaetos • California
3292	**Measure** • Distribution • Prey • Raptors • A. gurneyi • New Guinea
3397	**Measure** • Descriptive Summary • Pellet • Behavior • Distribution • Plumage • Raptors • Illus. • A. chrysaetos • A. clanga • Great Britain
3408	**Measure** • Hunting Behavior • Animal Damage • A. chrysaetos • California
3417	**Measure** • Shooting • A. chrysaetos • New York
3426	**Measure** • Trapping • A. chrysaetos • Ireland
3437	**Measure** • Descriptive Summary • Illus. • Raptors • A. chrysaetos • A. heliaca • A. clanga • Japan
14	**Melanism** • Capture • Captivity • A. chrysaetos • Colorado
3	**Migration** • Status • Raptors • A. nipalensis • Turkey
27	**Migration** • Raptors • A. chrysaetos • A. heliaca • A. nipalensis • A. pomarina • A. clanga • Italy
65	**Migration** • Museum Specimen • Wintering Population • Descriptive Summary • Status • A. chrysaetos • Iowa
134	**Migration** • A. wahlbergi • Africa
142	**Migration** • A. pomarina • South Africa
189	**Migration** • Marking • A. wahlbergi • South Africa
194	**Migration** • Raptors • A. heliaca • A. nipalensis • A. pomarina • Israel
203	**Migration** • A. nipalensis • A. rapax • A. pomarina • A. wahlbergi • Kenya
204	**Migration** • Raptors • A. chrysaetos • France
234	**Migration** • A. nipalensis • A. pomarina • Egypt
248	**Migration** • Raptors • A. pomarina • A. chrysaetos • A. heliaca • Turkey
264	**Migration** • Etymology • Measure • Raptors • A. chrysaetos • A. pomarina • Malawi
311	**Migration** • Raptors • A. pomarina • Malawi
314	**Migration** • Raptors • A. chrysaetos • Maryland
328	**Migration** • Raptors • A. chrysaetos • Denmark

Master No.	Keywords
386	**Migration** • Bounty • Disturbance • Habitat Loss • Protection • Shooting • Status • Pesticides • Distribution • Population Decline • Raptors • A. chrysaetos • A. heliaca • A. pomarina • A. clanga • Europe
389	**Migration** • Raptors • A. chrysaetos • California
455	**Migration** • A. pomarina • Kenya
474	**Migration** • Field Id • Refuge • Conservation • Raptors • A. chrysaetos • Pennsylvania
479	**Migration** • Shooting • Protection • Refuge • Conservation • Raptors • Hawk Mt. • A. chrysaetos • Pennsylvania
522	**Migration** • Banding • Raptors • A. chrysaetos • Pennsylvania
523	**Migration** • Raptors • A. chrysaetos • Pennsylvania
525	**Migration** • Hawk Mt. • Popular Monograph • A. chrysaetos • Pennsylvania
526	**Migration** • Refuge • Hawk Mt. • Raptors • A. chrysaetos • Pennsylvania
527	**Migration** • Shooting • Conservation • Population Decline • Historical Account • Hawk Mt. • Raptors • Popular Monograph • A. chrysaetos • Pennsylvania
558	**Migration** • Prey • Raptors • A. clanga • A. pomarina • Poland • South Africa
620	**Migration** • Raptors • A. clanga • A. pomarina • A. chrysaetos • Turkey • Lebanon
679	**Migration** • Raptors • A. nipalensis • Sudan
681	**Migration** • Raptors • A. chrysaetos • Denmark • Sweden
682	**Migration** • Field Id • Raptors • A. heliaca • A. clanga • Cyprus
695	**Migration** • A. chrysaetos • New Jersey
712	**Migration** • Raptors • A. heliaca • A. nipalensis • A. pomarina • A. clanga • Turkey
770	**Migration** • Field Id • Descriptive Summary • Monograph • Distribution • Status • Prey • Plumage • Measure • Etymology • Vocalization • A. pomarina • A. clanga • A. nipalensis • A. heliaca • A. chrysaetos • A. verreauxi • Africa • Europe
826	**Migration** • Raptors • A. nipalensis • China
846	**Migration** • Raptors • Photo Account • A. nipalensis • A. pomarina • Tunisia
847	**Migration** • Interspecific Behavior • Raptors • A. chrysaetos • Alberta
871	**Migration** • Distribution • Etymology • Life History • Habitat • Molt • Prey • Measure • Field Id • Status • Raptors • A. chrysaetos • A. heliaca • A. rapax • A. clanga • A. pomarina • United Soviet Socialists Republic
875	**Migration** • Status • A. chrysaetos • Europe
876	**Migration** • Banding • A. chrysaetos • Europe
910	**Migration** • Raptors • A. pomarina • A. nipalensis • A. clanga • A. heliaca • Israel
911	**Migration** • Raptors • A. pomarina • A. nipalensis • A. heliaca • A. clanga • A. chrysaetos • Israel
945	**Migration** • Hawk Mt. • Raptors • A. chrysaetos • Pennsylvania
952	**Migration** • Status • Raptors • A. wahlbergi • Nigeria
980	**Migration** • Interspecific Behavior • Raptors • A. chrysaetos • A. clanga • Libya
985	**Migration** • Raptors • A. clanga • A. chrysaetos • Finland
1004	**Migration** • Status • Raptors • A. chrysaetos • New Jersey
1018	**Migration** • Raptors • A. nipalensis • A. pomarina • A. clanga • Iran
1028	**Migration** • Banding • Measure • Raptors • Photo Account • A. chrysaetos • Ontario
1029	**Migration** • Banding • Raptors • A. chrysaetos • Ontario
1030	**Migration** • Banding • Raptors • A. chrysaetos • Ontario
1040	**Migration** • Monograph • Breeding Biology • Prey • Vocalization • Territory • Taxonomy • Mortality • Molt • Falconry • A. chrysaetos • A. verreauxi • A. audax
1084	**Migration** • Raptors • A. chrysaetos • Pennsylvania
1242	**Migration** • Raptors • A. nipalensis • A. heliaca • A. clanga • Egypt
1284	**Migration** • Flight • A. chrysaetos • Denmark
1363	**Migration** • Field Id • Descriptive Summary • Status • Prey • Raptors • A. chrysaetos • Wisconsin
1378	**Migration** • Raptors • A. chrysaetos • Denmark
1403	**Migration** • Story • Raptors • Popular Monograph • A. chrysaetos • United States
1408	**Migration** • Raptors • A. chrysaetos • New York
1411	**Migration** • A. chrysaetos • New York
1434	**Migration** • Age Structure • Raptors • A. chrysaetos • Pennsylvania
1436	**Migration** • Raptors • A. chrysaetos • Pennsylvania
1437	**Migration** • Habitat • Field Id • Photo Account • Raptors • Popular Monograph • A. chrysaetos • North America
1438	**Migration** • Field Id • Raptors • Photo Account • A. chrysaetos • North America
1439	**Migration** • Age Structure • Raptors • Bake Oven • A. chrysaetos • Pennsylvania
1440	**Migration** • A. chrysaetos • Pennsylvania
1441	**Migration** • Age Structure • Bake Oven • Raptors • A. chrysaetos • Pennsylvania
1442	**Migration** • Age Structure • Bake Oven • Raptors • A. chrysaetos • Pennsylvania
1443	**Migration** • Age Structure • Raptors • Bake Oven • A. chrysaetos • Pennsylvania
1444	**Migration** • Age Structure • Raptors • Bake Oven • A. chrysaetos • Pennsylvania
1535	**Migration** • Raptors • A. chrysaetos • Minnesota
1537	**Migration** • Raptors • A. chrysaetos • A. heliaca • A. clanga • Libya
1551	**Migration** • Banding • Raptors • A. chrysaetos • Pennsylvania
1603	**Migration** • Raptors • A. chrysaetos • West Virginia
1604	**Migration** • Raptors • A. chrysaetos • West Virginia
1626	**Migration** • Etymology • Measure • Trapping • A. chrysaetos • Alaska
1658	**Migration** • Raptors • A. chrysaetos • New York
1731	**Migration** • Raptors • A. heliaca • A. pomarina • Rumania
1743	**Migration** • Raptors • A. chrysaetos • New York
1762	**Migration** • Shooting • Raptors • Photo Account • A. pomarina • Turkey • Greece
1830	**Migration** • Raptors • A. heliaca • A. clanga • A. pomarina • Turkey
1869	**Migration** • Field Id • Raptors • A. clanga • Denmark
1910	**Migration** • A. nipalensis • A. pomarina • A. rapax • Kenya

Master No.	Keywords
1911	**Migration** • Raptors • A. rapax • A. pomarina • A. nipalensis • A. wahlbergi • Kenya
1919	**Migration** • Banding • Raptors • A. pomarina • Europe
1965	**Migration** • Banding • A. pomarina • Bulgaria • West Germany
2030	**Migration** • A. nipalensis • A. heliaca • A. pomarina • A. clanga • Iraq
2049	**Migration** • Status • Shooting • Pesticides • Protection • Poisoning • A. chrysaetos • Siberia
2065	**Migration** • Raptors • A. pomarina • A. clanga • A. nipalensis • Turkey
2086	**Migration** • Monograph • Mortality • Banding • Breeding Biology • Prey • Food Requirements • A. chrysaetos • Montana
2088	**Migration** • Breeding Biology • Prey • Banding • Directional Exposure • Mortality • A. chrysaetos • Montana
2119	**Migration** • Raptors • A. heliaca • Israel
2124	**Migration** • Monograph • Breeding Biology • Prey • Hunting Behavior • Attack • Folklore • Falconry • Animal Damage • Popular Monograph • A. chrysaetos
2221	**Migration** • Monograph • Distribution • Raptors • A. heliaca • A. pomarina • A. clanga • A. nipalensis • Africa
2239	**Migration** • Survey • Territory • Nest Failure • Nest • Habitat • Management • Reproductive Success • Prey • Roadside Survey • Status • Winter Survey • Uinta Nf • Wasatch Mt. • Raptors • A. chrysaetos • Utah
2247	**Migration** • Raptors • A. chrysaetos • Wisconsin
2248	**Migration** • Raptors • A. chrysaetos • Wisconsin
2271	**Migration** • Raptors • A. chrysaetos • Europe
2278	**Migration** • Refuge • Conservation • Survey • Shooting • Hawk Mt. • A. chrysaetos • Pennsylvania
2306	**Migration** • Raptors • A. chrysaetos • A. wahlbergi • South Africa
2318	**Migration** • Raptors • A. heliaca • A. clanga • A. nipalensis • A. pomarina • Iran
2319	**Migration** • Raptors • A. pomarina • A. clanga • Turkey • Lebanon
2320	**Migration** • Raptors • A. heliaca • A. chrysaetos • Iran
2321	**Migration** • Raptors • A. heliaca • A. nipalensis • A. pomarina • A. clanga • Lebanon
2330	**Migration** • A. pomarina • A. clanga • Turkey
2331	**Migration** • Raptors • A. clanga • A. pomarina • Turkey
2367	**Migration** • Raptors • A. chrysaetos • Poland
2476	**Migration** • Raptors • A. heliaca • A. nipalensis • A. pomarina • Israel
2479	**Migration** • Field Id • A. clanga • Morocco
2480	**Migration** • Raptors • A. chrysaetos • A. heliaca • A. clanga • Spain • Morocco
2482	**Migration** • Raptors • A. chrysaetos • A. heliaca • Morocco
2503	**Migration** • Raptors • A. pomarina • A. heliaca • A. clanga • A. nipalensis • Turkey
2579	**Migration** • A. chrysaetos • Denmark
2580	**Migration** • A. chrysaetos • Minnesota
2681	**Migration** • Raptors • A. chrysaetos • Sweden
2682	**Migration** • Population Decline • Shooting • Raptors • A. chrysaetos • Sweden

Master No.	Keywords
2683	**Migration** • Status • Historical Account • Raptors • A. chrysaetos • Sweden
2699	**Migration** • Raptors • A. heliaca • A. nipalensis • A. pomarina • A. clanga • Israel
2737	**Migration** • Raptors • A. clanga • Czechoslovakia
2738	**Migration** • A. clanga • Hungary
2792	**Migration** • Wintering Population • Eagles • A. chrysaetos • Russia
2817	**Migration** • A. nipalensis • Israel
2833	**Migration** • Raptors • A. nipalensis • A. heliaca • A. pomarina • Egypt
2859	**Migration** • Raptors • A. chrysaetos • New York
2860	**Migration** • Field Id • Wintering Population • A. nipalensis • A. pomarina • A. clanga • A. heliaca • Africa
2863	**Migration** • Raptors • A. nipalensis • A. clanga • Ethiopia
2865	**Migration** • Raptors • A. pomarina • A. clanga • A. nipalensis • Ethiopia
2888	**Migration** • Descriptive Summary • Raptors • A. chrysaetos • A. heliaca • A. clanga • A. pomarina • Italy
2891	**Migration** • Banding • Raptors • A. chrysaetos • New Jersey
2892	**Migration** • Banding • Raptors • A. chrysaetos • New Jersey
2915	**Migration** • Banding • Shooting • A. chrysaetos • North America
2916	**Migration** • Shooting • A. chrysaetos • North America
2917	**Migration** • Population Decline • Hawk Mt. • A. chrysaetos • Pennsylvania
2944	**Migration** • Raptors • A. clanga • A. pomarina • A. chrysaetos • A. heliaca • A. nipalensis
2968	**Migration** • Status • Raptors • A. chrysaetos • Maryland
2969	**Migration** • Nest Survey • Habitat • Prey • Breeding Biology • Reproductive Success • Growth • Field Id • Vocalization • Behavior • Eagles • Photo Account • A. wahlbergi • A. rapax • Rhodesia
3017	**Migration** • Status • Shooting • Conservation • Raptors • A. pomarina • Malawi
3048	**Migration** • Raptors • A. heliaca • A. clanga • A. pomarina • A. nipalensis • Israel
3078	**Migration** • Status • Nest • Photo Account • A. heliaca • Hungary
3079	**Migration** • Prey • Nest Survey • Clutch Size • Reproductive Success • Display • Hunting Behavior • Density • A. wahlbergi • South Africa
3109	**Migration** • Raptors • A. pomarina • Malawi
3115	**Migration** • Monograph • Raptors • A. chrysaetos • Switzerland
3116	**Migration** • Monograph • Raptors • A. chrysaetos • Switzerland
3117	**Migration** • Monograph • Raptors • A. chrysaetos • Switzerland
3124	**Migration** • A. wahlbergi • Uganda • Rwanda
3126	**Migration** • Wintering Population • Interspecific Competition • Raptors • A. nipalensis • A. pomarina • A. clanga • A. rapax • A. heliaca • Africa
3129	**Migration** • Raptors • A. pomarina • Tunisia

Master No.	Keywords
3135	**Migration** • Predator/Prey • Interspecific Behavior • Raptors • A. chrysaetos • A. heliaca • A. nipalensis • A. pomarina • A. clanga • Nepal
3136	**Migration** • Distribution • A. rapax • A. wahlbergi • France • Africa
3138	**Migration** • Raptors • A. heliaca • Spain
3156	**Migration** • Raptors • A. chrysaetos • New Jersey
3162	**Migration** • Raptors • A. heliaca • A. nipalensis • A. clanga • A. pomarina • Lebanon
3173	**Migration** • Raptors • A. nipalensis • A. pomarina • Rhodesia
3174	**Migration** • Status • Raptors • A. pomarina • A. nipalensis • A. wahlbergi • Rhodesia
3176	**Migration** • Raptors • A. pomarina • A. nipalensis • Zambia
3210	**Migration** • Rarity • Raptors • A. clanga • Sweden
3211	**Migration** • A. wahlbergi • Tanzania
3221	**Migration** • Rarity • A. chrysaetos • Tunisia
3237	**Migration** • Status • Raptors • A. chrysaetos • A. pomarina • A. wahlbergi • A. rapax • Rwanda
1945	**Mining** • Telemetry • Disturbance • Reproductive Success • Nest Survey • Raptors • A. chrysaetos • Montana • Wyoming
2465	**Mining** • Habitat • Nest Survey • Disturbance • Marking • Prey • Time Lapse • Raptors • Proj. Out. • A. chrysaetos • Wyoming • Montana
3202	**Mitigation** • Disturbance • Protection • Aircraft • Environmental Impact • A. chrysaetos • California
364	**Mobbing** • A. chrysaetos • Poland
388	**Mobbing** • Behavior • Photography Account • A. chrysaetos • France
646	**Mobbing** • Interspecific Behavior • Mortality • A. chrysaetos • New York
699	**Mobbing** • Interspecific Behavior • Raptors • A. audax • Australia
751	**Mobbing** • A. chrysaetos • Portugal
2035	**Mobbing** • Interspecific Behavior • Prey Behavior • Monograph • Raptors • A. chrysaetos • A. clanga • Sweden
2653	**Mobbing** • Interspecific Behavior • Flight • Raptors • A. chrysaetos • Oregon
2862	**Mobbing** • Raptors • A. verreauxi • Rhodesia
3038	**Mobbing** • A. audax • Australia
3280	**Mobbing** • A. chrysaetos • Switzerland
3294	**Mobbing** • Raptors • A. heliaca • Spain
292	**Molt** • Captivity • Raptors • A. chrysaetos • Hungary
498	**Molt** • Taxonomy • Field Id • Measure • Museum Specimen • Behavior • Wintering Population • Interspecific Competition • Conservation • Eagles • Photo Account • A. nipalensis • A. pomarina • Africa
618	**Molt** • Prey • Nest • Attitude • Behavior • Breeding Biology • Flight • Poisoning • Hunting Behavior • Vocalization • Folklore • Lift • Photo Account • A. chrysaetos • Montana
871	**Molt** • Distribution • Etymology • Life History • Habitat • Prey • Measure • Field Id • Migration • Status • Raptors • A. chrysaetos • A. heliaca • A. rapax • A. clanga • A. pomarina • United Soviet Socialists Republic
1040	**Molt** • Monograph • Breeding Biology • Prey • Vocalization • Territory • Taxonomy • Migration • Mortality • Falconry • A. chrysaetos • A. verreauxi • A. audax
1285	**Molt** • Collect • Parasite • Fat • Museum Specimen • A. chrysaetos • North Carolina
1688	**Molt** • Plumage • Aging • A. chrysaetos • North America
2800	**Molt** • Telemetry • Captivity • Eagles • A. chrysaetos • Washington
2907	**Molt** • Wintering Population • Trapping • Plumage • Measure • A. chrysaetos • Tennessee
3001	**Molt** • Raptors • A. chrysaetos • A. heliaca • A. audax • A. wahlbergi • A. pomarina
3021	**Molt** • Captivity • Growth • Food Requirements • Measure • Pellet • Temperature • Photo Account • Raptors • A. chrysaetos • California
3028	**Molt** • Taxonomy • Plumage • A. chrysaetos • A. verreauxi • A. heliaca • A. rapax • A. clanga
1687	**Molt** • Monograph • Breeding Biology • Field Id • Literature Review • A. chrysaetos • Colorado
198	**Monograph** • Prey • Nest Survey • Habitat • Reproductive Success • Literature Review • Disturbance • Pellet • Animal Damage • A. chrysaetos • Utah
325	**Monograph** • Breeding Biology • Density • Mortality • Cain/Abel • Prey • Abnormal Behavior • Egg Measure • Marking • A. chrysaetos • Idaho
551	**Monograph** • Conservation • Taxonomy • Courtship • Territory • A. chrysaetos • A. heliaca • A. clanga • A. rapax • A. verreauxi • A. wahlbergi • A. audax • A. gurneyi • A. nipalensis • A. pomarina
614	**Monograph** • Nest Survey • Reproductive Success • Habitat • Directional Exposure • Behavior • A. chrysaetos • Utah
770	**Monograph** • Field Id • Descriptive Summary • Distribution • Status • Migration • Prey • Plumage • Measure • Etymology • Vocalization • A. pomarina • A. clanga • A. nipalensis • A. heliaca • A. chrysaetos • A. verreauxi • Africa • Europe
789	**Monograph** • Habitat • Prey • Nest Survey • Photo Account • A. chrysaetos • Colorado
957	**Monograph** • Behavior • Growth • Pesticides • Parental Behavior • Vocalization • Disturbance • Measure • Breeding Biology • Photography • Captivity • Parasite • Disease • Terminology • Photo Account • A. chrysaetos • Montana
1040	**Monograph** • Breeding Biology • Prey • Vocalization • Territory • Taxonomy • Migration • Mortality • Molt • Falconry • A. chrysaetos • A. verreauxi • A. audax
1228	**Monograph** • Vocalization • Distribution • Prey • Plumage • Id Key • A. pomarina • A. chrysaetos • A. nipalensis • Europe
1245	**Monograph** • Life History • A. chrysaetos • Great Britain
1305	**Monograph** • Anthropology • Descriptive Summary • A. chrysaetos • A. clanga • A. rapax • A. audax • A. verreauxi • A. gurneyi • A. wahlbergi • A. pomarina • A. clanga • A. heliaca

Master No.	Keywords
1580	**Monograph** • Management • Mortality • Electrocution • A. chrysaetos • North America
1687	**Monograph** • Molt • Breeding Biology • Field Id • Literature Review • A. chrysaetos • Colorado
1717	**Monograph** • Prey • Hunting Behavior • Animal Damage • Eagles • A. chrysaetos • North America
1775	**Monograph** • Status • Nest Survey • Growth • Reproductive Success • Cain/Abel • Banding • Marking • Mortality • Pesticides • Prey • Residues • A. chrysaetos • Idaho
2035	**Monograph** • Mobbing • Interspecific Behavior • Prey Beh. • Raptors • A. chrysaetos • A. clanga • Sweden
2086	**Monograph** • Mortality • Banding • Migration • Breeding Biology • Prey • Food Requirements • A. chrysaetos • Montana
2124	**Monograph** • Breeding Biology • Migration • Prey • Hunting Behavior • Attack • Folklore • Falconry • Animal Damage • Popular Monograph • A. chrysaetos
2221	**Monograph** • Migration • Distribution • Raptors • A. heliaca • A. pomarina • A. clanga • A. nipalensis • Africa
2308	**Monograph** • A. chrysaetos • A. rapax • A. clanga • A. pomarina • A. nipalensis • A. heliaca • A. wahlbergi • A. gurneyi • A. verreauxi • A. audax
2370	**Monograph** • Status • Nest • Nest Survey • Habitat • Mortality • Breeding Biology • Nest Chronology • Courtship • Density • Prey • Management • A. chrysaetos • Colorado
2594	**Monograph** • Pesticides • Density • Prey • Nest Survey • Mortality • Reproductive Success • A. chrysaetos • Montana
2753	**Monograph** • Habitat • Nest Survey • Behavior • Prey • Pellet • Abandonment • A. chrysaetos • Wyoming
2789	**Monograph** • Pesticides • Raptors • A. chrysaetos • Montana
2852	**Monograph** • Roadside Survey • Prey • Density • Habitat • Interspecific Competition • Taxonomy • Vocalization • Hunting Behavior • Nest Chronology • Photo Account • Raptors • A. wahlbergi • A. rapax • A. nipalensis • A. heliaca • Kenya
2954	**Monograph** • A. chrysaetos • Switzerland
3046	**Monograph** • A. chrysaetos • A. heliaca • A. clanga • A. rapax • A. nipalensis • A. pomarina • A. wahlbergi • A. verreauxi • A. audax • A. gurneyi
3115	**Monograph** • Migration • Raptors • A. chrysaetos • Switzerland
3116	**Monograph** • Migration • Raptors • A. chrysaetos • Switzerland
3117	**Monograph** • Migration • Raptors • A. chrysaetos • Switzerland
3327	**Monograph** • Breeding Biology • Prey • A. pomarina • A. clanga
66	**Mortality** • Electrocution • Management • Nest • A. chrysaetos • Idaho
104	**Mortality** • Aircraft • A. chrysaetos • Nevada
138	**Mortality** • Shooting • A. audax • Australia
151	**Mortality** • Poisoning • Wintering Population • Conservation • Attitude • Punishment • A. chrysaetos • Wyoming

Master No.	Keywords
159	**Mortality** • Trapping • Capture • Raptors • A. chrysaetos • Nevada
161	**Mortality** • Shooting • Protection • Reward • A. chrysaetos • Idaho
163	**Mortality** • Banding • Raptors • A. audax • Australia
172	**Mortality** • Rehabilitation • Protection • Raptors • A. chrysaetos • Utah
324	**Mortality** • Reproductive Success • Food Requirements • Territory • Abandonment • Cain/Abel • Density • Dispersal • Breeding Biology • Nest Survey • A. chrysaetos • Idaho
325	**Mortality** • Monograph • Breeding Biology • Density • Cain/Abel • Prey • Abnormal Behavior • Egg Measure • Marking • A. chrysaetos • Idaho
352	**Mortality** • Electrocution • Management • A. chrysaetos • North America
394	**Mortality** • Museum Specimen • Captivity • A. chrysaetos • Illinois
544	**Mortality** • Taxonomy • Status • Field Id • Flight • Habitat • Terminology • Distribution • Behavior • Life History • Longevity • Territory • Photo Account • Raptors • A. chrysaetos • A. clanga • Great Britain • Ireland • Scotland
613	**Mortality** • Poisoning • Protection • Eagles • A. chrysaetos • Wyoming
617	**Mortality** • Behavior • Egg • Raptors • A. chrysaetos • Montana
646	**Mortality** • Interspecific Behavior • Mobbing • A. chrysaetos • New York
690	**Mortality** • Poisoning • (Title Only) • A. chrysaetos • Wyoming
720	**Mortality** • Status • Conservation • Raptors • A. audax • Australia
723	**Mortality** • A. chrysaetos • Scotland
819	**Mortality** • Animal Damage • Shooting • Poisoning • Eagles • A. chrysaetos • Ohio
836	**Mortality** • Pesticides • Autopsy • Toxicology • Eagles • Popular Monograph • A. chrysaetos • North America
852	**Mortality** • Clutch Size • Nestling • Prey • Pesticides • Cain/Abel • A. chrysaetos • Spain
872	**Mortality** • Reproductive Success • A. chrysaetos • Scotland
889	**Mortality** • Status • Raptors • A. chrysaetos • Sardinia
900	**Mortality** • Descriptive Summary • Prey • Hunting Behavior • Nest • Interspecific Behavior • Vocalization • Raptors • Mt. Mckinley • A. chrysaetos • Alaska
961	**Mortality** • Raptors • A. chrysaetos • Utah
1020	**Mortality** • Status • Habitat • Population Decline • Raptors • A. chrysaetos • Europe
1040	**Mortality** • Monograph • Breeding Biology • Prey • Vocalization • Territory • Taxonomy • Migration • Molt • Falconry • A. chrysaetos • A. verreauxi • A. audax
1090	**Mortality** • Economics • Attitude • Conservation • Animal Damage • Popular Monograph • A. chrysaetos • Wyoming
1102	**Mortality** • Abandonment • Disturbance • Egg • Nestling • Photography • Banding • Raptors • Tech. • A. chrysaetos • North America

Master No.	Keywords
1138	**Mortality** • (Title Only) • A. verreauxi • Rhodesia
1144	**Mortality** • Nest Survey • Status • Behavior • Fledging • Banding • Marking • Cain/Abel • Reproductive Success • Breeding Biology • Interspecific Behavior • Hunting Behavior • Display • Courtship • Raptors • A. verreauxi • Rhodesia
1151	**Mortality** • Reproductive Success • Density • Population Dynamics • Prey • Nest • Climatology • Territory • Interspecific Competition • Clutch Size • Pesticides • Eagles • A. verreauxi • Rhodesia
1273	**Mortality** • Falconry • Prey • Hunting Behavior • Flight • Nest • Mortality • Photo Account • Popular Monograph • A. chrysaetos • Great Britain
1273	**Mortality** • Mortality • Falconry • Prey • Hunting Behavior • Flight • Nest • Photo Account • Popular Monograph • A. chrysaetos • Great Britain
1366	**Mortality** • Status • Protection • Management • Disturbance • Eagles • A. chrysaetos • North America
1424	**Mortality** • Status • Nest Chronology • A. chrysaetos • Utah
1447	**Mortality** • Eagles • A. chrysaetos • Sweden
1507	**Mortality** • Conservation • Protection • Economics • Reward • Pesticides • Poisoning • Eagles • A. chrysaetos • Wyoming • Colorado
1558	**Mortality** • Wintering Population • Prey • Raptors • A. chrysaetos • Alberta
1580	**Mortality** • Management • Monograph • Electrocution • A. chrysaetos • North America
1618	**Mortality** • Cain/Abel • Food Requirements • Raptors • A. chrysaetos • A. verreauxi • A. pomarina
1639	**Mortality** • Taxonomy • Dimorphism • Measure • Plumage • Distribution • Prey • Behavior • Management • Habitat • Territory • Life History • Economics • Literature Review • A. chrysaetos • Oregon • Washington
1681	**Mortality** • Shooting • Bounty • A. chrysaetos • Texas
1720	**Mortality** • Status • Nest • Prey • Raptors • A. chrysaetos • A. clanga • A. heliaca • Yugoslavia
1729	**Mortality** • Prey • Habitat • Conservation • A. chrysaetos • Japan
1775	**Mortality** • Monograph • Status • Nest Survey • Growth • Reproductive Success • Cain/Abel • Banding • Marking • Pesticides • Prey • Residues • A. chrysaetos • Idaho
1776	**Mortality** • Nest Survey • Refuge • Winter Survey • Density • Electrocution • Shooting • Prey • Predator/Prey • Raptors • SRBPNA • A. chrysaetos • Idaho
1789	**Mortality** • Capture • A. chrysaetos • Hungary
1866	**Mortality** • Prey • A. chrysaetos • Minnesota
1897	**Mortality** • Conservation • Attitude • Popular Monograph • A. chrysaetos • Wyoming
1900	**Mortality** • Prey • Nest • Dispersal • Pellet • Habitat • Animal Damage • Bounty • Protection • Photo Account • A. audax • Australia
1903	**Mortality** • Nest • Prey • Habitat • Disturbance • Conservation • Hunting Behavior • A. chrysaetos • Israel
1926	**Mortality** • A. chrysaetos • New York
1946	**Mortality** • Protection • Management • Photo Account • A. chrysaetos • Scotland
1948	**Mortality** • Pesticides • (Title Only) • A. chrysaetos • Scotland
1978	**Mortality** • Raptors • Popular Monograph • A. pomarina • East Germany
2008	**Mortality** • Folklore • Behavior • A. chrysaetos • California
2040	**Mortality** • Management • Electrocution • Raptors • A. chrysaetos • Idaho
2086	**Mortality** • Monograph • Banding • Migration • Breeding Biology • Prey • Food Requirements • A. chrysaetos • Montana
2088	**Mortality** • Breeding Biology • Prey • Banding • Migration • Directional Exposure • A. chrysaetos • Montana
2126	**Mortality** • Status • Nest Survey • Wintering Population • Rehabilitation • Raptors • A. chrysaetos • A. heliaca • A. pomarina • A. nipalensis • A. clanga • Israel
2128	**Mortality** • Pesticides • Prey • Rehabilitation • Wintering Population • Azodrin • Raptors • A. heliaca • A. clanga • A. pomarina • Israel
2150	**Mortality** • Cain/Abel • Theory • Photo Account • A. chrysaetos • A. pomarina • A. verreauxi • A. heliaca • Czechoslovakia • South Africa • Spain
2154	**Mortality** • Cain/Abel • Reproductive Success • Management • Experiment • A. chrysaetos • A. heliaca
2215	**Mortality** • Status • Protection • Rehabilitation • Raptors • A. audax • Australia
2238	**Mortality** • Nest • Nest Survey • Nestling • Parental Behavior • Directional Exposure • Hyperthermia • A. chrysaetos • United States
2241	**Mortality** • Refuge • Photography • Behavior • Nest Survey • Reproductive Success • Photo Account • Raptors • Popular Monograph • A. heliaca • Spain
2298	**Mortality** • Electrocution • Shooting • Artificial Nest. • Eagles • A. chrysaetos • Idaho
2307	**Mortality** • Status • Density • Conservation • Pesticides • Raptors • A. chrysaetos • Scotland
2311	**Mortality** • Attitude • Status • Population Estimate • Protection • Prey • Animal Damage • Disturbance • Raptors • A. chrysaetos • Great Britain
2325	**Mortality** • Capture • Banding • Marking • Animal Damage • Sexing • Management • Translocation • Prey • Survey • Measure • Captive Care • Aging • A. chrysaetos • Montana
2326	**Mortality** • Capture • Banding • Marking • Animal Damage • Sexing • Management • Translocation • Prey • Measure • Capt. Care • Aging • Food Requirements • A. chrysaetos • Montana
2327	**Mortality** • Prey • A. chrysaetos • Norway
2370	**Mortality** • Status • Nest • Nest Survey • Monograph • Habitat • Breeding Biology • Nest Chronology • Courtship • Density • Prey • Management • A. chrysaetos • Colorado
2399	**Mortality** • Economics • Bounty • Conservation • Eagles • Popular Monograph • A. chrysaetos • North America

Master No.	Keywords
2432	**Mortality** • Electrocution • Raptors • A. chrysaetos • North America
2460	**Mortality** • Parasite • Invertebrates • Raptors • A. chrysaetos • United States
2464	**Mortality** • Management • Habitat • Directional Exposure • Disturbance • Nest Survey • Density • Nest Chronology • Prey • Reproductive Success • Prog. Rep. • A. chrysaetos • Wyoming
2530	**Mortality** • Nest • Eagles • A. verreauxi • South Africa
2584	**Mortality** • Pesticides • A. chrysaetos • North America
2594	**Mortality** • Monograph • Pesticides • Density • Prey • Nest Survey • Reproductive Success • A. chrysaetos • Montana
2596	**Mortality** • Nest • Photo Account • A. chrysaetos • Switzerland
2639	**Mortality** • Measure • A. chrysaetos • Quebec
2642	**Mortality** • Conservation • Photo Account • Raptors • A. chrysaetos • United States
2690	**Mortality** • Pesticides • A. pomarina • West Germany
2723	**Mortality** • Nest Survey • Banding • Status • Reproductive Success • Raptors • A. chrysaetos • Finland
2787	**Mortality** • Status • Conservation • Protection • Habitat Loss • Raptors • A. chrysaetos • A. clanga • Sweden
2801	**Mortality** • Nest • Prey • Forestry • Habitat • A. chrysaetos • Washington
2818	**Mortality** • Electrocution • A. nipalensis • Russia
2832	**Mortality** • Electrocution • A. chrysaetos • United States
2840	**Mortality** • Altitude • Measure • A. nipalensis • India
2854	**Mortality** • Population Dynamics • Nest • Density • Prey • Breeding Biology • Distribution • Habitat • Directional Exposure • Reproductive Success • Abandonment • Disturbance • Interspecific Behavior • Territory • A. chrysaetos • Utah
2881	**Mortality** • Museum Display • A. chrysaetos • Virginia
2906	**Mortality** • Attitude • Animal Damage • Prey • Raptors • A. chrysaetos • Texas
2912	**Mortality** • Status • A. chrysaetos • North America
2928	**Mortality** • Electrocution • Habitat Loss • Nest • Survey • Raptors • A. chrysaetos • Colorado
2979	**Mortality** • Cain/Abel • Raptors • Theory • A. chrysaetos • A. clanga • A. pomarina • A. verreauxi • A. heliaca • A. rapax
2990	**Mortality** • Conservation • Attitude • Protection • Photo Account • Popular Monograph • A. chrysaetos • North America
3002	**Mortality** • Capture • Attitude • A. chrysaetos • Illinois
3039	**Mortality** • Nest Survey • Reproductive Success • A. heliaca • A. pomarina • Czechoslovakia
3102	**Mortality** • Population Decline • Abandonment • Raptors • A. chrysaetos • France
3155	**Mortality** • Protection • Conservation • A. chrysaetos • North America
3261	**Mortality** • Longevity • Theory • A. verreauxi • Rhodesia
3277	**Mortality** • Nest • Protection • Popular Monograph • A. heliaca • South Africa
3283	**Mortality** • Breeding Biology • Prey • Conservation • Distribution • Photo Account • A. chrysaetos • Czechoslovakia
3288	**Mortality** • A. chrysaetos • South Dakota
3289	**Mortality** • A. chrysaetos • South Dakota
3306	**Mortality** • Pesticides • Autopsy • Disease • A. chrysaetos • Great Britain
3372	**Mortality** • Protection • Population Decline • Conservation • Animal Damage • Bounty • A. chrysaetos • Norway
3373	**Mortality** • Status • Bounty • Conservation • Eagles • A. chrysaetos • Norway
3379	**Mortality** • Capture • Captivity • Symbolism • Anthropology • A. chrysaetos • North America
3425	**Mortality** • A. chrysaetos • United States
39	**Museum Display** • A. chrysaetos • New York
2179	**Museum Display** • Attitude • Folklore • Animal Damage • Popular Monograph • A. audax • Australia
2881	**Museum Display** • Mortality • A. chrysaetos • Virginia
3407	**Museum Display** • Raptors • A. chrysaetos • Poland
2	**Museum Specimen** • Measure • Raptors • A. chrysaetos • A. heliaca • A. rapax • A. nipalensis • A. clanga • India
54	**Museum Specimen** • Plumage • Distribution • Measure • Prey • A. heliaca • A. clanga • A. rapax • India
56	**Museum Specimen** • Etymology • Descriptive Summary • Clutch Size • Plumage • Habitat • Measure • A. rapax • A. pomarina • India
65	**Museum Specimen** • Wintering Population • Migration • Descriptive Summary • Status • A. chrysaetos • Iowa
68	**Museum Specimen** • Taxonomy • Plumage • Distribution • A. nipalensis • A. clanga • A. pomarina • France
69	**Museum Specimen** • Albinism • A. nipalensis • Russia
77	**Museum Specimen** • Measure • A. chrysaetos • California
224	**Museum Specimen** • Collect • Prey • Raptors • A. wahlbergi • A. verreauxi • South Africa
250	**Museum Specimen** • Clutch Size • Raptors • A. chrysaetos • A. heliaca • A. rapax • Tunisia • Morocco
260	**Museum Specimen** • Collect • Raptors • A. rapax • Somalia
291	**Museum Specimen** • A. chrysaetos • Hungary
333	**Museum Specimen** • Shooting • A. chrysaetos • Pennsylvania
378	**Museum Specimen** • A. chrysaetos • Louisiana
394	**Museum Specimen** • Mortality • Captivity • A. chrysaetos • Illinois
406	**Museum Specimen** • Anatomy • A. chrysaetos • A. pomarina • A. heliaca • A. rapax • A. audax • A. verreauxi • A. wahlbergi • A. clanga • Great Britain
418	**Museum Specimen** • A. rapax • A. wahlbergi • Portugal

Master No.	Keywords
498	**Museum Specimen** • Taxonomy • Field Id • Molt • Measure • Behavior • Wintering Population • Interspecific Competition • Conservation • Eagles • Photo Account • A. nipalensis • A. pomarina • Africa
576	**Museum Specimen** • Raptors • A. heliaca • Cyprus
636	**Museum Specimen** • Measure • A. pomarina • Tunisia
688	**Museum Specimen** • Status • A. clanga • West Germany
841	**Museum Specimen** • Raptors • A. rapax • Kenya
926	**Museum Specimen** • Photo Account • A. chrysaetos • Belgium
930	**Museum Specimen** • Nest • Raptors • A. chrysaetos • A. rapax • Algeria
978	**Museum Specimen** • Measure • A. chrysaetos • France
981	**Museum Specimen** • Rarity • A. pomarina • France
989	**Museum Specimen** • Taxonomy • Plumage • Measure • A. clanga • A. pomarina • A. rapax • A. nipalensis • France
1058	**Museum Specimen** • Taxonomy • Collect • Behavior • Raptors • A. pomarina • A. clanga • A. nipalensis • Nepal
1087	**Museum Specimen** • Collect • Plumage • Raptors • A. rapax • Ethiopia • Kenya
1088	**Museum Specimen** • Plumage • Raptors • A. rapax • A. wahlbergi • Africa
1095	**Museum Specimen** • Shooting • Plumage • Measure • Taxonomy • A. clanga • Albania
1182	**Museum Specimen** • Prey • Collect • Raptors • A. wahlbergi • Cambodia
1196	**Museum Specimen** • Measure • Raptors • A. chrysaetos • Italy
1201	**Museum Specimen** • Status • A. chrysaetos • A. pomarina • Italy
1285	**Museum Specimen** • Collect • Parasite • Fat • Molt • A. chrysaetos • North Carolina
1289	**Museum Specimen** • Taxonomy • Type Spec. • Raptors • A. heliaca • A. rapax • A. nipalensis • A. clanga • A. pomarina • A. audax
1326	**Museum Specimen** • Taxonomy • Plumage • A. nipalensis • Turkey
1328	**Museum Specimen** • Plumage • Taxonomy • Captivity • Eagles • A. chrysaetos • A. heliaca • A. rapax • A. clanga • A. nipalensis
1330	**Museum Specimen** • A. chrysaetos • A. rapax • A. clanga • A. pomarina • A. audax • A. gurneyi • A. heliaca • A. wahlbergi • A. verreauxi • A. nipalensis
1394	**Museum Specimen** • Taxonomy • Raptors • A. heliaca • A. rapax • A. clanga • A. pomarina
1396	**Museum Specimen** • Taxonomy • Plumage • Measure • A. rapax • A. nipalensis • Great Britain
1397	**Museum Specimen** • Taxonomy • Plumage • A. audax • Australia
1491	**Museum Specimen** • Raptors • A. chrysaetos • A. pomarina • West Germany
1505	**Museum Specimen** • Dimorphism • Taxonomy • Sexing • Measure • Raptors • A. chrysaetos • North America
1550	**Museum Specimen** • A. chrysaetos • Louisiana
1683	**Museum Specimen** • Anatomy • Taxonomy • Osteology • Raptors • A. chrysaetos • A. rapax
1684	**Museum Specimen** • Anatomy • Taxonomy • Osteology • Raptors • A. chrysaetos • A. rapax
1685	**Museum Specimen** • Anatomy • Taxonomy • Osteology • Raptors • A. chrysaetos • A. rapax
1686	**Museum Specimen** • Anatomy • Taxonomy • Osteology • Raptors • A. chrysaetos • A. rapax
1693	**Museum Specimen** • Descriptive Summary • Raptors • A. chrysaetos • A. heliaca • A. nipalensis • A. clanga • Georgia Ssr
1824	**Museum Specimen** • Raptors • A. clanga • Lebanon
1828	**Museum Specimen** • Status • Raptors • A. chrysaetos • A. heliaca • A. pomarina • A. clanga • Turkey
1887	**Museum Specimen** • A. chrysaetos • France
1907	**Museum Specimen** • Rarity • Raptors • A. clanga • France
1917	**Museum Specimen** • Raptors • A. heliaca • A. chrysaetos • Spain
1982	**Museum Specimen** • Status • Raptors • A. wahlbergi • A. rapax • A. nipalensis • A. verreauxi • Sudan
2020	**Museum Specimen** • Rarity • Measure • Plumage • Photo Account • A. heliaca • Italy
2042	**Museum Specimen** • Status • Raptors • A. heliaca • A. nipalensis • A. clanga • China
2063	**Museum Specimen** • Raptors • A. chrysaetos • A. heliaca • A. clanga • Yugoslavia
2072	**Museum Specimen** • Status • Raptors • A. pomarina • A. chrysaetos • France
2359	**Museum Specimen** • Raptors • A. chrysaetos • A. clanga • Italy
2428	**Museum Specimen** • Raptors • A. nipalensis • A. chrysaetos • A. pomarina • Hungary
2468	**Museum Specimen** • Status • Raptors • A. chrysaetos • A. clanga • A. pomarina • Italy
2592	**Museum Specimen** • Collect • A. pomarina • A. clanga • Italy
2608	**Museum Specimen** • A. chrysaetos
2618	**Museum Specimen** • Raptors • A. wahlbergi • Kenya
2648	**Museum Specimen** • Status • A. chrysaetos • A. clanga • France
2655	**Museum Specimen** • A. chrysaetos • Georgia
2706	**Museum Specimen** • Descriptive Summary • Plumage • Distribution • A. chrysaetos • Molluca Islands
2720	**Museum Specimen** • Status • Behavior • Raptors • A. chrysaetos • A. clanga • Switzerland
2735	**Museum Specimen** • Raptors • A. verreauxi • South Africa
2778	**Museum Specimen** • Collect • Raptors • A. chrysaetos • India
2785	**Museum Specimen** • Distribution • Raptors • A. clanga • Iran
2810	**Museum Specimen** • Raptors • A. rapax • South Africa
2853	**Museum Specimen** • Status • Raptors • A. chrysaetos • A. heliaca • Portugal
3000	**Museum Specimen** • Egg • Prey • Raptors • A. chrysaetos • A. nipalensis • China
3033	**Museum Specimen** • Agonistic Behavior • Sexing • A. chrysaetos • Switzerland
3045	**Museum Specimen** • Taxonomy • Rarity • Measure • A. rapax • Rumania
3075	**Museum Specimen** • Raptors • A. chrysaetos • A. heliaca • Portugal
3086	**Museum Specimen** • Plumage • Raptors • A. clanga • A. heliaca • Egypt

Master No.	Keywords
3088	**Museum Specimen** • Collect • Raptors • A. clanga • Egypt
3167	**Museum Specimen** • Raptors • A. chrysaetos • A. clanga • Italy
3226	**Museum Specimen** • Density • Reproductive Success • Sub-Adult • Plumage • Population Decline • Illus. • A. heliaca • Spain
3248	**Museum Specimen** • Measure • Raptors • A. chrysaetos • A. clanga • A. pomarina • Hungary
3263	**Museum Specimen** • Status • Shooting • Animal Damage • Raptors • A. verreauxi • A. nipalensis • A. rapax • South Africa
3366	**Museum Specimen** • Status • Plumage • Raptors • A. nipalensis • A. heliaca • Hungary
4	**Nest** • Status • Raptors • A. rapax • India
6	**Nest** • Egg Collection • Measure • Raptors • A. chrysaetos • California
9	**Nest** • A. pomarina • Hungary
10	**Nest** • A. pomarina • Hungary
11	**Nest** • A. pomarina • Hungary
12	**Nest** • Status • Egg • A. chrysaetos • Lebanon • Syria
15	**Nest** • Popular Monograph • A. chrysaetos • Wyoming
18	**Nest** • Egg • Raptors • A. chrysaetos • Washington
32	**Nest** • Status • A. chrysaetos • New York
33	**Nest** • A. chrysaetos • California
40	**Nest** • A. chrysaetos • Colorado
58	**Nest** • Clutch Size • Prey • Habitat • Collect • A. pomarina • India
64	**Nest** • Habitat • Eagles • A. chrysaetos • Washington
66	**Nest** • Electrocution • Mortality • Management • A. chrysaetos • Idaho
79	**Nest** • Egg • Collect • A. pomarina • West Germany
80	**Nest** • Distribution • Raptors • A. pomarina • A. clanga • A. chrysaetos • A. heliaca • West Germany
89	**Nest** • Behavior • Hunting Behavior • Popular Monograph • A. chrysaetos • California
93	**Nest** • A. chrysaetos • Switzerland
101	**Nest** • A. chrysaetos • Switzerland
108	**Nest** • Clutch Size • Plumage • Raptors • A. wahlbergi • A. rapax • A. verreauxi • Kenya • Uganda
111	**Nest** • Popular Monograph • (Title Only) • A. chrysaetos • Great Britain
176	**Nest** • Photo Account • Raptors • Popular Monograph • A. chrysaetos • United States
185	**Nest** • Protection • Photo Account • A. chrysaetos • Israel
192	**Nest** • Prey • Status • Raptors • A. chrysaetos • A. heliaca • Spain
197	**Nest** • Prey • Behavior • Captivity • A. chrysaetos • Algeria
199	**Nest** • A. chrysaetos • California
206	**Nest** • Photo Account • A. verreauxi • Rhodesia
223	**Nest** • Distribution • Raptors • A. verreauxi • South Africa
227	**Nest** • Rarity • Raptors • A. clanga • Finland
229	**Nest** • Distribution • Status • Prey • A. audax • Australia
237	**Nest** • Prey • A. chrysaetos • Alaska
238	**Nest** • Behavior • A. chrysaetos • New Mexico
240	**Nest** • Distribution • Captivity • A. chrysaetos • Quebec
249	**Nest** • Prey • Status • A. chrysaetos • Algeria • Morocco • Tunisia
251	**Nest** • Descriptive Summary • Egg • Measure • Plumage • Taxonomy • Photo Account • Raptors • A. chrysaetos • A. rapax • Algeria
262	**Nest** • Descriptive Summary • Etymology • Status • Behavior • Plumage • A. heliaca • A. nipalensis • A. clanga • A. chrysaetos • Cyprus
265	**Nest** • Prey • Attitude • Raptors • A. pomarina • United Soviet Socialists Republic
270	**Nest** • Egg Collection • Measure • A. chrysaetos • California
271	**Nest** • Behavior • Egg Collection • Vocalization • Behavior • Photo Account • Popular Monograph • A. chrysaetos • California
277	**Nest** • Egg Collection • Photo Account • A. chrysaetos • California
287	**Nest** • Prey • Behavior • Photo Account • Raptors • A. chrysaetos • Norway
289	**Nest** • Status • Raptors • A. chrysaetos • France
290	**Nest** • Shooting • Status • Conservation • A. chrysaetos • France
295	**Nest** • A. pomarina • West Germany
301	**Nest** • Status • Raptors • A. chrysaetos • Czechoslovakia
302	**Nest** • Fostering • Prey • Habitat • Raptors • A. chrysaetos • Bulgaria
317	**Nest** • Prey • A. heliaca • Hungary
321	**Nest** • Raptors • A. chrysaetos • Utah
327	**Nest** • Prey • Raptors • A. chrysaetos • Utah
335	**Nest** • Distribution • Captivity • Vocalization • Prey • Raptors • A. chrysaetos • Oregon
336	**Nest** • Prey • A. chrysaetos • Ireland
341	**Nest** • Collect • Raptors • A. verreauxi • A. wahlbergi • Malawi
345	**Nest** • Clutch Size • A. wahlbergi • Malawi
346	**Nest** • Owls • A. rapax • Zambia
362	**Nest** • Prey • Measure • A. audax • Australia
363	**Nest** • Prey • Clutch Size • A. audax • Australia
368	**Nest** • Eagles • A. chrysaetos • Alaska
369	**Nest** • Status • Raptors • A. chrysaetos • France
372	**Nest** • Habitat • Wintering Population • Hunting Behavior • Territory • A. chrysaetos • France
387	**Nest** • Photo Account • Popular Monograph • A. heliaca • Spain
390	**Nest** • Raptors • A. chrysaetos • Italy
396	**Nest** • Status • Raptors • A. chrysaetos • Alaska
397	**Nest** • Raptors • A. chrysaetos • United Soviet Socialists Republic
399	**Nest** • Egg • Eagles • A. heliaca • Spain
412	**Nest** • Egg Collection • Behavior • Cain/Abel • A. verreauxi • South Africa
438	**Nest** • Distribution • Raptors • A. audax • Australia
442	**Nest** • Egg Collection • Raptors • A. verreauxi • A. chrysaetos • A. pomarina • A. heliaca • A. nipalensis • A. clanga • A. rapax
464	**Nest** • Habitat • Egg Collection • Behavior • Vocalization • Animal Damage • A. audax • Australia
465	**Nest** • Behavior • Egg • Abandonment • Raptors • A. audax • Australia
468	**Nest** • Nest Chronology • Egg • Prey • A. nipalensis • Russia
473	**Nest** • Habitat • A. chrysaetos • Colorado
476	**Nest** • Descriptive Summary • Prey • Raptors • A. chrysaetos • Maine

Master No.	Keywords
480	Nest • Raptors • A. rapax • A. verreauxi • A. wahlbergi • Africa
481	Nest • Raptors • A. rapax • A. verreauxi • A. wahlbergi • Africa
485	Nest • A. chrysaetos • Israel
489	Nest • Distribution • Prey • Measure • Raptors • A. chrysaetos • Sardinia
533	Nest • Survey • Photo Account • Popular Monograph • A. chrysaetos • Scotland
537	Nest • Prey • Interspecific Behavior • Habitat • Raptors • A. verreauxi • A. wahlbergi • Kenya
542	Nest • Abnormal Behavior • Cain/Abel • Habitat • A. verreauxi • Rhodesia
550	Nest • Status • A. wahlbergi • Africa
568	Nest • Prey • Conservation • Photo Account • Popular Monograph • A. chrysaetos • South Dakota
574	Nest • Status • Shooting • Egg • Raptors • A. heliaca • Cyprus
579	Nest • Behavior • Photo Account • A. chrysaetos • Wyoming
581	Nest • A. chrysaetos • Switzerland
585	Nest • Sub-Adult • Collect • A. heliaca • Spain
587	Nest • Prey • Behavior • Growth • Vocalization • Nest Adornment • Photo Account • Illus. • A. chrysaetos • Switzerland
590	Nest • A. chrysaetos • Nebraska
599	Nest • Shooting • A. chrysaetos • Great Britain
611	Nest • Habitat • Field Id • Nest Survey • Behavior • Photo Account • Raptors • A. chrysaetos • North America
616	Nest • Habitat • Egg • Prey • Vocalization • Growth • Trapping • Poisoning • Behavior • Lift • Photo Account • A. chrysaetos • Montana
618	Nest • Prey • Attitude • Behavior • Breeding Biology • Flight • Poisoning • Molt • Hunting Behavior • Vocalization • Folklore • Lift • Photo Account • A. chrysaetos • Montana
619	Nest • Behavior • Fledgling • Photo Account • Popular Monograph • A. chrysaetos • Montana
622	Nest • Habitat • A. chrysaetos • Alaska
628	Nest • Hunting Behavior • Prey • Habitat • Photography Account • A. chrysaetos • California
629	Nest • Breeding Biology • (Title Only) • A. chrysaetos • West Germany
633	Nest • Poisoning • A. audax • Australia
635	Nest • Egg • Measure • Prey • A. audax • Australia
639	Nest • Shooting • A. chrysaetos • Italy
642	Nest • Field Id • Egg • Raptors • A. audax • Australia
651	Nest • Capture • Economics • Folklore • A. chrysaetos • Tennessee
654	Nest • Wintering Population • Plumage • Raptors • A. heliaca • A. chrysaetos • Spain
667	Nest • A. chrysaetos • Switzerland
672	Nest • Status • Prey • Raptors • A. chrysaetos • France
692	Nest • Collect • Raptors • A. wahlbergi • A. rapax • Rhodesia
693	Nest • A. clanga • Poland
701	Nest • A. audax • Australia
710	Nest • Egg • A. verreauxi • Zambia
722	Nest • Egg • Behavior • Raptors • A. chrysaetos • A. pomarina • Rumania
724	Nest • Egg Collection • Raptors • A. chrysaetos • A. pomarina • Rumania
725	Nest • Prey • Directional Exposure • A. chrysaetos • California
727	Nest • Status • Raptors • A. chrysaetos • Michigan
730	Nest • Behavior • Raptors • A. chrysaetos • California
738	Nest • Distribution • Raptors • A. audax • Australia
743	Nest • A. chrysaetos • Switzerland
747	Nest • A. chrysaetos • Arizona
753	Nest • Behavior • Shooting • Photography • Story • Banding • Photo Account • Raptors • A. verreauxi • South Africa
755	Nest • Prey • Raptors • A. audax • Australia
763	Nest • Roadside Survey • Wintering Population • Survey • Raptors • A. chrysaetos • Idaho
765	Nest • History • Management • Conservation • Raptors • Popular Monograph • A. chrysaetos • United States
766	Nest • Banding • Raptors • Popular Monograph • A. chrysaetos • Wyoming
767	Nest • Telemetry • Protection • Falconry • Behavior • Pesticides • Photo Account • Popular Monograph • A. chrysaetos • United States
773	Nest • Collect • Prey • Raptors • A. heliaca • A. pomarina • India
797	Nest • Distribution • Egg Collection • Etymology • Raptors • A. chrysaetos • A. heliaca • A. clanga • Turkey
803	Nest • Egg • Raptors • A. chrysaetos • Yugoslavia
812	Nest • Shooting • A. chrysaetos • Switzerland
821	Nest • Raptors • A. chrysaetos • Finland
828	Nest • Raptors • A. chrysaetos • Washington
833	Nest • (Title Only) • A. chrysaetos • Portugal
838	Nest • Historical Account • Status • Population Decline • Prey • A. chrysaetos • Ireland
851	Nest • Prey • Clutch Size • A. chrysaetos • Spain
856	Nest • Egg • Prey • Shooting • A. chrysaetos • Austria
857	Nest • A. chrysaetos • Austria
859	Nest • Reproductive Success • A. chrysaetos • Austria
860	Nest • A. chrysaetos • West Germany
874	Nest • Distribution • (Title Only) • A. chrysaetos • Europe
885	Nest • Status • Raptors • A. chrysaetos • Italy
890	Nest • A. heliaca • Spain
893	Nest • Habitat • Egg • Photo Account • A. chrysaetos • California
894	Nest • Intraspecific Behavior • Habitat • Egg • Measure • Prey • Photo Account • A. chrysaetos • California
895	Nest • Territory • Egg • Measure • Intraspecific Behavior • Courtship • Interspecific Behavior • Hunting Behavior • A. chrysaetos • California
900	Nest • Descriptive Summary • Prey • Hunting Behavior • Interspecific Behavior • Mortality • Vocalization • Raptors • Mt. Mckinley • A. chrysaetos • Alaska
919	Nest • Egg Collection • A. chrysaetos • A. pomarina • United Soviet Socialists Republic
930	Nest • Museum Specimen • Raptors • A. chrysaetos • A. rapax • Algeria
932	Nest • A. rapax • Senegal
933	Nest • A. chrysaetos • Italy

Master No.	Keywords
939	Nest • Habitat • Prey • Field Id • Hunting Behavior • Raptors • A. clanga • A. pomarina • A. chrysaetos • Poland
940	Nest • Prey • Habitat Loss • Habitat • Raptors • A. chrysaetos • India
944	Nest • Nest Adornment • Prey • Behavior • Growth • A. audax • Australia
963	Nest • Growth • Plumage • Prey • Vocalization • Behavior • Photo Account • A. audax • Australia
965	Nest • Status • Egg • Taxonomy • Raptors • A. chrysaetos • A. heliaca • A. nipalensis • A. clanga • Yugoslavia • Bulgaria
986	Nest • Egg • A. pomarina • Hungary
987	Nest • Descriptive Summary • Field Id • Distribution • Status • Taxonomy • Egg • Illus. • A. chrysaetos • A. heliaca • A. nipalensis • A. rapax • A. clanga • A. pomarina • A. verreauxi • Africa
988	Nest • Descriptive Summary • Distribution • Raptors • A. nipalensis • A. clanga • A. heliaca • A. chrysaetos • China
992	Nest • Behavior • A. chrysaetos • California
1003	Nest • Egg • Habitat • Photo Account • A. chrysaetos • Scotland
1006	Nest • Habitat • Raptors • A. chrysaetos • A. heliaca • A. clanga • Bulgaria
1025	Nest • Hunting Behavior • Prey • Density • Habitat • Photo Account • Popular Monograph • A. heliaca • Spain
1033	Nest • Habitat • Prey • Flight • Plumage • Raptors • A. wahlbergi • A. rapax • South Africa
1034	Nest • Prey • Egg • Measure • Raptors • A. verreauxi • South Africa
1036	Nest • Photography • Popular Monograph • A. chrysaetos • California
1037	Nest • Photography • Prey • Growth • A. chrysaetos • California
1038	Nest • Behavior • A. verreauxi • Kenya
1041	Nest • Status • Prey • Habitat • Behavior • Photo Account • A. chrysaetos • Bulgaria
1054	Nest • Captive Breeding • Behavior • Plumage • Measure • Photo Account • Popular Monograph • A. audax • Australia
1070	Nest • Distribution • Raptors • A. chrysaetos • Alaska
1080	Nest • Collect • Etymology • Literature Review • Distribution • A. chrysaetos • A. rapax • Tunisia
1091	Nest • Growth • Prey • Photo Account • A. chrysaetos • Switzerland
1092	Nest • Behavior • Pellet • A. chrysaetos • Switzerland
1098	Nest • Survey • Banding • Disturbance • A. chrysaetos • Canada
1104	Nest • A. chrysaetos • North Dakota
1105	Nest • Descriptive Summary • Etymology • Plumage • Distribution • A. chrysaetos • Alaska
1110	Nest • Raptors • A. chrysaetos • Hungary
1113	Nest • A. chrysaetos • Oman
1137	Nest • Distribution • Disturbance • A. heliaca • Hungary
1145	Nest • Parental Behavior • Breeding Biology • Interspecific Behavior • Intraspecific Behavior • Feeding Behavior • Hunting Behavior • Fledging • Prey • Display • Raptors • A. verreauxi • Rhodesia
1150	Nest • Density • Territory • Dispersal • Survey • Intraspecific Behavior • Marking • Reproductive Success • Display • A. verreauxi • Rhodesia
1151	Nest • Reproductive Success • Density • Population Dynamics • Prey • Climatology • Territory • Interspecific Competition • Mortality • Clutch Size • Pesticides • Eagles • A. verreauxi • Rhodesia
1152	Nest • Egg • Parental Behavior • Incubation • A. verreauxi • Rhodesia
1154	Nest • Density • Prey • Status • Habitat Protection • Habitat • Reproductive Success • A. verreauxi • Rhodesia
1155	Nest • Behavior • Physiology • Temperature • A. verreauxi • Rhodesia
1159	Nest • Raptors • A. chrysaetos • A. heliaca • Spain
1163	Nest • Behavior • Prey • Clutch Size • A. heliaca • Spain
1167	Nest • Photography • Prey • Behavior • Photo Account • A. audax • Australia
1171	Nest • Intraspecific Behavior • Prey • A. pomarina • Armenia
1174	Nest • Vocalization • Prey • Behavior • Photo Account • Popular Monograph • A. pomarina • West Germany
1176	Nest • Breeding Biology • Courtship • Food Habits • A. pomarina • West Germany
1179	Nest • Behavior • Raptors • A. audax • Australia
1187	Nest • Protection • Habitat • Vocalization • Raptors • A. chrysaetos • A. heliaca • Yugoslavia
1189	Nest • Prey • Habitat • Raptors • A. chrysaetos • France
1198	Nest • Measure • Collect • A. clanga • A. chrysaetos • Italy
1200	Nest • A. chrysaetos • Alaska
1212	Nest • Descriptive Summary • Taxonomy • Etymology • Prey • A. chrysaetos • Switzerland
1221	Nest • Breeding Biology • Biology • (Title Only) • A. chrysaetos • Texas
1232	Nest • Status • Clutch Size • Raptors • A. nipalensis • Ukraine
1234	Nest • Prey • Raptors • A. pomarina • Belorussia
1240	Nest • Incubation • Raptors • A. chrysaetos • Great Britain
1244	Nest • Descriptive Summary • Photo Account • Popular Monograph • A. chrysaetos • Scotland
1248	Nest • Behavior • A. chrysaetos • Scotland
1252	Nest • Photo Account • Popular Monograph • A. chrysaetos • Scotland
1255	Nest • Photo Account • A. chrysaetos • Scotland
1256	Nest • Photo Account • Popular Monograph • A. chrysaetos • Scotland
1258	Nest • Popular Monograph • (Title Only) • A. chrysaetos • Great Britain
1261	Nest • Prey • Behavior • Photo Account • Popular Monograph • A. chrysaetos • Scotland
1263	Nest • Descriptive Summary • Photo Account • Popular Monograph • A. chrysaetos • Scotland
1265	Nest • Photo Account • Popular Monograph • A. chrysaetos • Scotland

Master No.	Keywords
1273	Nest • Mortality • Falconry • Prey • Hunting Behavior • Flight • Mortality • Photo Account • Popular Monograph • A. chrysaetos • Great Britain
1288	Nest • Historical Account • A. chrysaetos • Oklahoma
1300	Nest • Status • A. chrysaetos • California
1303	Nest • Rarity • A. clanga • Finland
1307	Nest • A. chrysaetos • West Germany
1309	Nest • Clutch Size • Prey • A. nipalensis • Russia
1316	Nest • Popular Monograph • A. chrysaetos • Great Britain
1331	Nest • A. chrysaetos • Great Britain
1338	Nest • Prey Behavior • A. chrysaetos • Hungary
1342	Nest • Raptors • A. chrysaetos • Finland
1353	Nest • (Title Only) • A. chrysaetos • Sweden
1354	Nest • Raptors • A. wahlbergi • Rhodesia
1370	Nest • Behavior • A. chrysaetos • California
1373	Nest • A. chrysaetos • Switzerland
1380	Nest • Raptors • A. chrysaetos • Utah
1395	Nest • Status • Prey • Etymology • Raptors • A. chrysaetos • A. heliaca • A. rapax • Morocco
1401	Nest • Raptors • A. verreauxi • South Africa
1406	Nest • Prey • A. chrysaetos • Alaska
1409	Nest • Habitat • Raptors • A. chrysaetos • Alaska
1410	Nest • Habitat • Raptors • A. chrysaetos • Alaska
1415	Nest • A. chrysaetos • Switzerland
1417	Nest • A. chrysaetos • Switzerland
1419	Nest • A. chrysaetos • New Hampshire
1420	Nest • Prey • A. chrysaetos • Scotland
1426	Nest • Raptors • A. chrysaetos • North America
1464	Nest • Prey • A. chrysaetos • Spain
1480	Nest • A. chrysaetos • Switzerland
1483	Nest • A. chrysaetos • Switzerland
1492	Nest • Raptors • A. rapax • Ethiopia
1494	Nest • Historical Account • Raptors • A. chrysaetos • A. clanga • West Germany
1498	Nest • Raptors • A. chrysaetos • Idaho
1502	Nest • Photography • Photo Account • Popular Monograph • A. chrysaetos • Scotland
1506	Nest • Descriptive Summary • Protection • Prey • Raptors • Illus. • A. audax • Australia
1510	Nest • Status • A. chrysaetos • New York
1516	Nest • Prey • Parasite • A. audax • Australia
1521	Nest • Behavior • A. chrysaetos • California
1522	Nest • Growth • Nest Chronology • Photo Account • A. chrysaetos • California
1524	Nest • Photo Account • A. verreauxi • Southwest Africa
1528	Nest • A. clanga • West Germany
1529	Nest • A. clanga • West Germany
1532	Nest • A. clanga • West Germany
1533	Nest • Breeding Biology • (Title Only) • A. pomarina • Europe
1536	Nest • Raptors • A. rapax • Sudan
1548	Nest • Habitat • Prey • A. chrysaetos • Alaska
1555	Nest • A. pomarina • Switzerland
1562	Nest • Prey • Photography • Photo Account • A. chrysaetos • Scotland
1564	Nest • Vocalization • Behavior • Hunting Behavior • Raptors • A. verreauxi • South Africa
1565	Nest • Banding • Raptors • A. chrysaetos • Saskatchewan
1586	Nest • Descriptive Summary • Measure • Distribution • Photo Account • Illus. • Raptors • A. chrysaetos • A. heliaca • A. nipalensis • A. clanga • A. pomarina • Czechoslovakia
1602	Nest • Behavior • A. chrysaetos • California
1612	Nest • Clutch Size • Etymology • A. chrysaetos • Israel
1613	Nest • Status • A. chrysaetos • A. clanga • A. heliaca • United Soviet Socialists Republic
1629	Nest • Prey • Parental Behavior • A. chrysaetos • Switzerland
1630	Nest • Prey • A. chrysaetos • Switzerland
1631	Nest • Parental Behavior • Prey • A. nipalensis • Kazakhstan
1633	Nest • Raptors • A. pomarina • Caucasus
1635	Nest • Density • Conservation • Prey • Intraspecific Behavior • A. chrysaetos • Belorussia
1661	Nest • Capture • Symbolism • A. chrysaetos • New Mexico
1662	Nest • Egg • A. chrysaetos • New Mexico
1666	Nest • Etymology • Status • Raptors • A. rapax • India
1667	Nest • Egg • A. rapax • India
1668	Nest • Status • Etymology • Prey • Egg • Raptors • A. heliaca • A. nipalensis • A. rapax • A. clanga • A. pomarina • India
1669	Nest • Prey • Raptors • A. chrysaetos • Oregon
1680	Nest • Clutch Size • A. chrysaetos • Washington
1691	Nest • Prey • A. chrysaetos • Texas
1700	Nest • Nest Adornment • Etymology • Raptors • A. chrysaetos • Corsica
1708	Nest • Raptors • A. chrysaetos • Algeria
1709	Nest • Habitat • Conservation • Raptors • A. chrysaetos • A. pomarina • A. clanga • Hungary
1714	Nest • Eagles • Photo Account • A. chrysaetos • Estonia
1716	Nest • Plumage • Egg • Clutch Size • A. pomarina • West Germany
1720	Nest • Status • Prey • Mortality • Raptors • A. chrysaetos • A. clanga • A. heliaca • Yugoslavia
1721	Nest • Raptors • A. chrysaetos • A. pomarina • Poland
1728	Nest • Historical Account • A. chrysaetos • West Germany
1737	Nest • Distribution • Status • Raptors • A. nipalensis • A. rapax • A. wahlbergi • A. verreauxi • South Africa
1750	Nest • Distribution • Raptors • A. nipalensis • United Soviet Socialists Republic
1754	Nest • Distribution • Wintering Population • Habitat • Status • Raptors • A. chrysaetos • Colorado
1755	Nest • Status • A. pomarina • Hungary
1763	Nest • Prey • Behavior • Photo Account • Popular Monograph • A. chrysaetos • West Germany
1769	Nest • Cain/Abel • Descriptive Summary • Photo Account • Popular Monograph • A. chrysaetos • Scotland
1772	Nest • Prey • Food Requirements • Predator/Prey • A. chrysaetos • Washington
1791	Nest • Distribution • Raptors • A. chrysaetos • Ukraine
1795	Nest • Status • Raptors • A. chrysaetos • Alabama
1798	Nest • Prey • Raptors • A. chrysaetos • Kirgizia

Master No.	Keywords
1799	Nest • Nest Failure • Protection • Raptors • A. nipalensis • Caucasus
1805	Nest • Raptors • A. nipalensis • A. pomarina • West Germany
1806	Nest • Clutch Size • Egg • Measure • Raptors • A. wahlbergi • Rhodesia
1813	Nest • Status • Eagles • A. pomarina • A. clanga • West Germany
1814	Nest • Egg • Eagles • A. pomarina • West Germany
1815	Nest • Density • Raptors • A. chrysaetos • Kirgizia
1821	Nest • Status • Etymology • Raptors • A. chrysaetos • A. pomarina • A. clanga • Poland
1844	Nest • Egg Collection • Collect • Raptors • A. chrysaetos • China
1854	Nest • Captivity • Shooting • A. chrysaetos • Texas
1864	Nest • Distribution • A. nipalensis • Kazakhstan
1877	Nest • Status • Raptors • A. audax • Australia
1890	Nest • Parasite • A. chrysaetos • Texas
1891	Nest • Egg Collection • Nest Chronology • Behavior • Raptors • A. wahlbergi • Rhodesia
1898	Nest • Status • Egg Collection • A. chrysaetos • North Carolina
1900	Nest • Mortality • Prey • Dispersal • Pellet • Habitat • Animal Damage • Bounty • Protection • Photo Account • A. audax • Australia
1902	Nest • Status • Conservation • Prey • Behavior • A. chrysaetos • Israel
1903	Nest • Prey • Habitat • Disturbance • Conservation • Mortality • Hunting Behavior • A. chrysaetos • Israel
1906	Nest • Prey • Behavior • A. heliaca • Spain
1908	Nest • Breeding Biology • Behavior • Clutch Size • Prey • A. nipalensis • Bulgaria
1909	Nest • Breeding Biology • A. clanga • Kazakhstan
1912	Nest • Descriptive Summary • Habitat • Nest Chronology • A. chrysaetos • France
1914	Nest • Distribution • Habitat • Status • Clutch Size • Descriptive Summary • Raptors • A. chrysaetos • New Mexico
1915	Nest • Density • Raptors • A. clanga • A. pomarina • Russia
1925	Nest • A. heliaca • Hungary
1928	Nest • Prey • Pellet • Hunting Behavior • Interspecific Behavior • Reproductive Success • Eagles • A. chrysaetos • Oklahoma
1929	Nest • Distribution • Raptors • A. chrysaetos • Siberia
1934	Nest • Breeding Biology • Density • Territory • A. heliaca • Kazakhstan
1937	Nest • Status • Raptors • A. chrysaetos • Siberia
1942	Nest • Food Habits • Time Lapse • Behavior • Economics • Animal Damage • Reproductive Success • Status • A. chrysaetos • Texas
1955	Nest • Prey • Intraspecific Behavior • Raptors • A. chrysaetos • California
1962	Nest • A. chrysaetos • Kansas
1972	Nest • Prey • Habitat • A. chrysaetos • Ontario
1979	Nest • Raptors • A. chrysaetos • Denmark
1980	Nest • Raptors • A. chrysaetos • Spain
1981	Nest • Collect • Raptors • A. chrysaetos • Morocco
1984	Nest • A. chrysaetos • Switzerland
1985	Nest • Habitat • Behavior • Prey • Flight • Photo Account • A. chrysaetos • Scotland
1986	Nest • Clutch Size • Prey • Captivity • Nest Survey • Collect • A. chrysaetos • Canada
1990	Nest • Raptors • A. verreauxi • Southwest Africa
1994	Nest • Distribution • Prey • Hunting Behavior • Raptors • A. verreauxi • South Africa
2001	Nest • A. chrysaetos • Great Britain
2009	Nest • Status • A. heliaca • Yugoslavia
2010	Nest • Status • A. pomarina • Yugoslavia
2029	Nest • Status • Habitat • Prey • Eagles • A. heliaca • A. chrysaetos • A. clanga • A. pomarina • Yugoslavia
2036	Nest • Interspecific Competition • A. chrysaetos • Scotland • Great Britain
2039	Nest • Status • Collect • Prey • Raptors • A. wahlbergi • Rhodesia
2044	Nest • Prey • Raptors • A. verreauxi • South Africa
2047	Nest • Captivity • A. chrysaetos • Switzerland
2062	Nest • Density • Field Id • Photo Account • Popular Monograph • A. pomarina • East Germany
2077	Nest • Raptors • A. chrysaetos • New Hampshire
2080	Nest • Historical Account • A. chrysaetos • Washington
2082	Nest • Longevity • A. verreauxi • Rhodesia
2083	Nest • Status • A. chrysaetos • Wyoming
2085	Nest • Telemetry • Territory • Display • Agonistic Behavior • Behavior • A. chrysaetos • Wyoming • Montana
2090	Nest • Status • Raptors • A. audax • Australia
2092	Nest • Field Id • Descriptive Summary • Hunting Behavior • Animal Damage • Egg • Attitude • Piracy • Raptors • A. audax • Australia
2094	Nest • Raptors • A. heliaca • A. clanga • A. chrysaetos • Turkey
2097	Nest • Distribution • Habitat • Status • Raptors • A. chrysaetos • North Dakota
2107	Nest • Raptors • A. chrysaetos • Morocco
2108	Nest • Shooting • A. chrysaetos • New York
2117	Nest • Raptors • A. chrysaetos • Saudi Arabia
2130	Nest • Raptors • A. chrysaetos • Morocco
2132	Nest • Raptors • A. heliaca • Hungary
2138	Nest • Behavior • Prey • A. chrysaetos • Switzerland
2144	Nest • Parental Behavior • Prey • Vocalization • Nest Adornment • Photo Account • A. chrysaetos • West Germany
2149	Nest • Breeding Biology • Cain/Abel • Taxonomy • Prey • Photo Account • Popular Monograph • A. pomarina • West Germany
2158	Nest • Clutch Size • A. heliaca • Spain
2172	Nest • Behavior • Raptors • A. chrysaetos • Arizona
2189	Nest • Collect • Raptors • A. chrysaetos • A. heliaca • A. pomarina • A. clanga • Australia
2190	Nest • Distribution • Raptors • A. chrysaetos • Italy
2195	Nest • A. chrysaetos • New Mexico
2211	Nest • Raptors • A. chrysaetos • Saskatchewan
2212	Nest • Reproductive Success • Management • Shooting • Prey • Density • Raptors • A. chrysaetos • Washington
2214	Nest • Habitat • Flight • Vocalization • A. chrysaetos • Idaho
2217	Nest • Attitude • Prey • Photo Account • Popular Monograph • A. chrysaetos • Canada
2222	Nest • Habitat • Raptors • A. rapax • Senegal
2224	Nest • Banding • Prey • A. chrysaetos • South Dakota
2225	Nest • Status • Raptors • A. audax • Australia
2229	Nest • Egg • Raptors • A. chrysaetos • A. clanga • Great Britain • Scotland
2230	Nest • Growth • A. audax • Australia

Master No.	Keywords
2235	Nest • Egg Collection • A. chrysaetos • Colorado
2238	Nest • Nest Survey • Mortality • Nestling • Parental Behavior • Directional Exposure • Hyperthermia • A. chrysaetos • United States
2239	Nest • Survey • Territory • Nest Failure • Migration • Habitat • Management • Reproductive Success • Prey • Roadside Survey • Status • Winter Survey • Uinta Nf • Wasatch Mt. • Raptors • A. chrysaetos • Utah
2240	Nest • Prey • Photography • Behavior • Photo Account • A. heliaca • Spain
2243	Nest • Status • Display • Raptors • A. pomarina • A. clanga • A. chrysaetos • A. heliaca • Bulgaria
2245	Nest • Prey • Vocalization • Etymology • Raptors • A. verreauxi • A. rapax • A. wahlbergi • Rhodesia
2246	Nest • Habitat • Disturbance • Raptors • A. pomarina • Poland
2250	Nest • Captivity • A. chrysaetos • United States
2255	Nest • Electrocution • Density • Predator/Prey • Prey • A. nipalensis • Russia
2259	Nest • Conservation • Photo Account • Popular Monograph • A. chrysaetos • United States
2264	Nest • Environmental Impact • Management • Photo Account • Raptors • A. chrysaetos • Utah
2265	Nest • Environmental Impact • Management • Photo Account • Raptors • A. chrysaetos • Utah
2266	Nest • Environmental Impact • Management • Photo Account • Raptors • A. chrysaetos • Utah
2267	Nest • Environmental Impact • Management • Photo Account • Raptors • A. chrysaetos • Utah
2272	Nest • Prey • A. chrysaetos • Texas
2288	Nest • A. pomarina • West Germany
2290	Nest • Habitat • Egg • Raptors • A. chrysaetos • Wyoming
2333	Nest • Status • Prey • Raptors • A. heliaca • Spain
2339	Nest • Egg • Raptors • A. audax • Australia
2340	Nest • Descriptive Summary • Plumage • Distribution • Prey • Egg • Photo Account • Raptors • A. audax • Australia
2341	Nest • Descriptive Summary • Field Id • Raptors • A. verreauxi • A. rapax • A. wahlbergi • Kenya
2342	Nest • Etymology • Status • Habitat • Interspecific Behavior • Collect • Raptors • A. verreauxi • A. nipalensis • A. rapax • Africa
2345	Nest • Prey • Shooting • Density • Reproductive Success • Eagles • A. chrysaetos • Italy
2357	Nest • Hunting Behavior • Photo Account • A. chrysaetos • West Germany
2360	Nest • Egg • Survey • Photo Account • A. chrysaetos • Europe
2370	Nest • Status • Nest Survey • Monograph • Habitat • Mortality • Breeding Biology • Nest Chronology • Courtship • Density • Prey • Management • A. chrysaetos • Colorado
2378	Nest • Status • A. heliaca • Portugal
2383	Nest • Clutch Size • A. chrysaetos • Great Britain
2392	Nest • Egg • A. chrysaetos • California
2395	Nest • Status • Symbolism • Prey • Raptors • A. chrysaetos • South Dakota
2400	Nest • Animal Damage • Predator/Prey • Aerial Survey • Habitat • Status • Density • Population Dynamics • Food Habits • Interspecific Competition • Literature Review • A. chrysaetos • Texas • New Mexico
2404	Nest • Status • Nest Chronology • Clutch Size • Raptors • A. wahlbergi • A. rapax • Malawi
2405	Nest • Status • Raptors • A. rapax • A. verreauxi • Uganda
2406	Nest • Egg • Raptors • A. chrysaetos • A. wahlbergi • Malawi
2410	Nest • Pellet • Egg • Photo Account • A. chrysaetos • Scotland
2411	Nest • Interspecific Behavior • Behavior • Photography • Prey • Nest Adornment • Talon Grappling • Cain/Abel • Photo Account • Popular Monograph • A. chrysaetos • Scotland
2414	Nest • Wintering Population • Eagles • A. chrysaetos • Maine
2415	Nest • Descriptive Summary • A. chrysaetos • Maine
2439	Nest • Prey • Raptors • A. audax • Australia
2446	Nest • Raptors • Popular Monograph • A. chrysaetos • New York
2450	Nest • Measure • Behavior • A. chrysaetos • Spain
2451	Nest • Prey • Raptors • A. chrysaetos • Siberia
2469	Nest • Status • Taxonomy • Raptors • A. chrysaetos • A. pomarina • A. clanga • Yugoslavia
2470	Nest • Interspecific Competition • Raptors • A. chrysaetos • Colorado
2472	Nest • Photo Account • A. chrysaetos • California
2473	Nest • Egg • Photo Account • A. chrysaetos • California
2475	Nest • Photography Account • A. chrysaetos • North America
2485	Nest • Egg • Habitat • A. wahlbergi • Africa
2488	Nest • Abandonment • Prey • Raptors • A. chrysaetos • Utah
2493	Nest • A. clanga • United Soviet Socialists Republic
2494	Nest • Status • Raptors • A. chrysaetos • Italy
2496	Nest • A. verreauxi • Kenya
2507	Nest • Egg • Clutch Size • Photo Account • Popular Monograph • A. nipalensis • United Soviet Socialists Republic
2511	Nest • A. pomarina • West Germany
2515	Nest • Reproductive Success • Photo Account • A. chrysaetos • Switzerland
2517	Nest • Prey • A. pomarina • West Germany
2526	Nest • Prey • Habitat • A. clanga • Russia
2528	Nest • Raptors • A. chrysaetos • Oregon
2530	Nest • Mortality • Eagles • A. verreauxi • South Africa
2531	Nest • Prey • Behavior • A. verreauxi • South Africa
2533	Nest • Eagles • A. chrysaetos • Italy
2534	Nest • A. chrysaetos • New Hampshire
2537	Nest • Prey • Status • A. clanga • Russia
2539	Nest • Status • Prey • Raptors • A. chrysaetos • A. clanga • Russia
2543	Nest • A. chrysaetos • Scotland
2547	Nest • A. audax • Australia
2550	Nest • Egg Collection • A. audax • Australia
2553	Nest • Story • Shooting • Rehabilitation • Conservation • Popular Monograph • A. chrysaetos • Wyoming

Master No.	Keywords
2554	Nest • Conservation • Behavior • Interspecific Behavior • Prey • Trapping • Animal Damage • Rehabilitation • Photo Account • Popular Monograph • A. chrysaetos • Wyoming
2561	Nest • Interspecific Competition • Raptors • A. chrysaetos • Scotland
2566	Nest • Shooting • A. chrysaetos • Pennsylvania
2569	Nest • Behavior • Disturbance • Prey • A. chrysaetos • Colorado
2578	Nest • Egg • Raptors • A. chrysaetos • North America
2586	Nest • Attack • A. chrysaetos • Yugoslavia
2587	Nest • Shooting • A. chrysaetos • Switzerland
2589	Nest • A. heliaca • A. pomarina • Turkey
2595	Nest • Raptors • A. chrysaetos • Tennessee
2596	Nest • Mortality • Photo Account • A. chrysaetos • Switzerland
2597	Nest • Prey • Illus. • A. chrysaetos • Switzerland
2598	Nest • Growth • A. chrysaetos • Switzerland
2600	Nest • Pellet • A. chrysaetos • Switzerland
2601	Nest • Prey • Banding • Photo Account • A. chrysaetos • Switzerland
2602	Nest • Photography • Habitat • Prey • Photo Account • A. chrysaetos • Switzerland
2611	Nest • Prey • Banding • Conservation • Photo Account • Popular Monograph • A. audax • Australia
2625	Nest • Behavior • Raptors • Illus. • A. chrysaetos • France
2634	Nest • Egg Collection • A. wahlbergi • South Africa
2644	Nest • Refuge • Conservation • Measure • SRBPNA • Photo Account • Popular Monograph • A. chrysaetos • Idaho
2659	Nest • Flight • Intraspecific Behavior • A. chrysaetos • Scotland
2660	Nest • A. chrysaetos • United Soviet Socialists Republic
2669	Nest • Fledging • Parental Behavior • Food Habits • Breeding Biology • A. chrysaetos • Tanzania
2677	Nest • Habitat • A. pomarina • A. clanga • Yugoslavia
2680	Nest • Talon Grappling • Raptors • A. pomarina • Yugoslavia
2685	Nest • A. pomarina • West Germany
2691	Nest • Protection • A. chrysaetos • West Germany
2702	Nest • Raptors • A. chrysaetos • Alaska
2709	Nest • Collect • Egg • Etymology • Raptors • A. chrysaetos • A. clanga • A. rapax • Algeria
2710	Nest • Photo Account • Raptors • Popular Monograph • A. chrysaetos • Idaho
2717	Nest • Raptors • A. heliaca • A. chrysaetos • Spain
2745	Nest • A. chrysaetos • Switzerland
2751	Nest • Collect • Raptors • A. pomarina • West Germany
2757	Nest • Breeding Biology • (Title Only) • A. pomarina • West Germany
2760	Nest • Prey • Vocalization • Behavior • A. chrysaetos • West Germany
2765	Nest • Collect • Egg • A. rapax • Southwest Africa
2766	Nest • A. pomarina • West Germany
2779	Nest • Status • Prey • A. chrysaetos • Siberia
2795	Nest • Egg • Status • Raptors • A. rapax • Nigeria
2801	Nest • Prey • Forestry • Mortality • Habitat • A. chrysaetos • Washington
2813	Nest • Status • Raptors • A. chrysaetos • Alaska
2816	Nest • Collect • Prey • Status • Raptors • A. wahlbergi • South Africa
2834	Nest • Habitat • Raptors • A. clanga • A. heliaca • Turkey
2836	Nest • Egg • A. heliaca • Yugoslavia
2837	Nest • Habitat • Egg • Raptors • A. clanga • Greece
2844	Nest • Raptors • A. chrysaetos • Wyoming
2847	Nest • Egg • Growth • (Title Only) • A. pomarina • Czechoslovakia
2849	Nest • Prey • A. pomarina • Czechoslovakia
2851	Nest • Egg Collection • Habitat • Measure • Photo Account • A. chrysaetos • California
2854	Nest • Population Dynamics • Density • Prey • Breeding Biology • Distribution • Habitat • Directional Exposure • Reproductive Success • Mortality • Abandonment • Disturbance • Interspecific Behavior • Territory • A. chrysaetos • Utah
2869	Nest • Egg Collection • Eagles • A. chrysaetos • United States
2870	Nest • Banding • Survey • Marking • Capture • Prey • Food Requirements • Raptors • A. chrysaetos • A. wahlbergi • South Africa
2876	Nest • A. chrysaetos • California
2886	Nest • Prey • Behavior • Raptors • A. chrysaetos • Turkmenia
2897	Nest • Raptors • A. chrysaetos • United Soviet Socialists Republic
2898	Nest • Prey Behavior • Prey • A. chrysaetos • Colorado
2909	Nest • Field Id • Trapping • Eagles • Popular Monograph • A. chrysaetos • Wyoming
2910	Nest • A. chrysaetos • Quebec
2919	Nest • Photo Account • A. chrysaetos • New York
2928	Nest • Mortality • Electrocution • Habitat Loss • Survey • Raptors • A. chrysaetos • Colorado
2930	Nest • Disturbance • Gas Explor. • Raptors • A. chrysaetos • Colorado
2936	Nest • Raptors • A. chrysaetos • Yugoslavia
2938	Nest • Prey • Photo Account • A. chrysaetos • Kansas
2943	Nest • Egg Collection • Measure • A. chrysaetos • California
2946	Nest • Food Habits • A. chrysaetos • Colorado
2947	Nest • Egg Collection • Behavior • A. chrysaetos • California
2949	Nest • Status • Vocalization • Prey • Behavior • Photo Account • A. chrysaetos • Switzerland
2951	Nest • Prey • Photo Account • A. chrysaetos • Switzerland
2952	Nest • A. chrysaetos • Switzerland
2957	Nest • Raptors • A. chrysaetos • Kazakhstan
2961	Nest • Winter Survey • Roadside Survey • Raptors • A. chrysaetos • Kansas
2971	Nest • Photo Account • A. chrysaetos • Rhodesia
2980	Nest • Wintering Population • Raptors • A. chrysaetos • Wisconsin
2993	Nest • Infertility • A. chrysaetos • Texas
2998	Nest • Status • Raptors • A. chrysaetos • A. clanga • Crete
2999	Nest • A. chrysaetos • Italy
3006	Nest • Historical Account • A. chrysaetos • Great Britain
3011	Nest • Shooting • A. heliaca • Spain
3015	Nest • Prey • A. chrysaetos • Finland

Master No.	Keywords
3020	Nest • Growth • Behavior • Clutch Size • Banding • Photo Account • A. chrysaetos • California
3055	Nest • Etymology • Egg • Raptors • A. rapax • A. wahlbergi • Rhodesia
3057	Nest • Status • (Title Only) • A. chrysaetos • Sweden
3058	Nest • Habitat • Prey • Prey Behavior • Animal Damage • Shooting • Egg • Behavior • A. verreauxi • South Africa
3061	Nest • Collect • Vocalization • Eagles • A. heliaca • Hungary
3062	Nest • Collect • Photo Account • A. pomarina • Hungary
3063	Nest • Prey • A. pomarina • West Germany
3064	Nest • Status • Prey • A. pomarina • Australia
3065	Nest • Vocalization • Clutch Size • A. pomarina • Austria
3069	Nest • Disturbance • Raptors • A. heliaca • Yugoslavia
3070	Nest • Raptors • A. heliaca • Yugoslavia
3078	Nest • Status • Migration • Photo Account • A. heliaca • Hungary
3080	Nest • Cain/Abel • Distribution • Educational Material • Field Id • Habitat • Hunting Behavior • Prey • Artificial Nest • Incubation • Wintering Population • Conservation • Photo Account • Eagles • A. chrysaetos • A. wahlbergi • A. verreauxi • A. nipalensis • A. pomarina • South Africa
3084	Nest • Collect • Behavior • Raptors • A. wahlbergi • South Africa
3091	Nest • Egg Collection • A. chrysaetos • California
3092	Nest • A. chrysaetos • California
3093	Nest • A. chrysaetos • California
3094	Nest • Egg Collection • Behavior • A. chrysaetos • California
3095	Nest • Animal Damage • Prey • Behavior • A. verreauxi • South Africa
3107	Nest • Habitat • Egg Collection • Photo Account • A. chrysaetos • Colorado
3114	Nest • Banding • A. pomarina • West Germany
3123	Nest • Status • Hunting Behavior • Raptors • A. wahlbergi • A. rapax • Ivory Coast
3164	Nest • Habitat • Photo Account • A. chrysaetos • Poland
3165	Nest • Descriptive Summary • Raptors • A. pomarina • A. clanga • Poland
3169	Nest • Photo Account • A. chrysaetos • Austria
3170	Nest • Prey • Photo Account • Popular Monograph • A. nipalensis • West Germany
3179	Nest • Habitat • Raptors • A. chrysaetos • Algeria
3183	Nest • Prey • Photo Account • Popular Monograph • A. chrysaetos • Montana
3184	Nest • Descriptive Summary • Photo Account • Raptors • Popular Monograph • A. chrysaetos • United States
3214	Nest • Collect • Habitat • Behavior • Growth • Captivity • A. chrysaetos • India
3215	Nest • Illus. • A. chrysaetos • Yugoslavia
3216	Nest • Prey • Raptors • A. pomarina • West Germany
3217	Nest • Prey • A. pomarina • West Germany
3219	Nest • Raptors • A. verreauxi • South Africa
3225	Nest • Habitat • Prey • Raptors • A. heliaca • Spain
3232	Nest • Status • Raptors • A. heliaca • A. nipalensis • A. chrysaetos • A. pomarina • Rumania

Master No.	Keywords
3235	Nest • Growth • Photography • Prey • Behavior • Hunting Behavior • Vocalization • Nest Adornment • Egg • Raptors • Popular Monograph • A. rapax • Kenya
3251	Nest • Status • Raptors • A. chrysaetos • A. heliaca • A. rapax • Morocco
3254	Nest • Rarity • A. chrysaetos • Scotland
3260	Nest • Prey • Raptors • A. rapax • Rhodesia
3271	Nest • A. chrysaetos • Latvia
3272	Nest • Habitat • Egg • Raptors • A. verreauxi • A. rapax • A. nipalensis • Rhodesia
3274	Nest • Raptors • A. chrysaetos • Siberia
3277	Nest • Mortality • Protection • Popular Monograph • A. heliaca • South Africa
3278	Nest • Prey • A. wahlbergi • Ethiopia
3290	Nest • Lift • Captivity • Photo Account • Popular Monograph • A. chrysaetos • California
3291	Nest • Flight • Folklore • Measure • Weight • A. chrysaetos • California
3302	Nest • Disease • Autopsy • A. chrysaetos • Great Britain
3328	Nest • A. chrysaetos • Austria
3329	Nest • A. chrysaetos • Austria
3340	Nest • Fat • Collect • Raptors • A. nipalensis • A. rapax • India
3343	Nest • Status • Raptors • A. heliaca • A. nipalensis • A. clanga • A. rapax • India
3344	Nest • A. heliaca • A. clanga • A. pomarina • Albania
3347	Nest • Status • Raptors • A. chrysaetos • Nebraska
3350	Nest • Survey • Raptors • A. chrysaetos • Alaska
3362	Nest • Status • A. audax • Australia
3368	Nest • Sub-Adult • Vocalization • A. chrysaetos • Norway
3371	Nest • Egg Collection • Raptors • A. chrysaetos • California
3381	Nest • Status • Raptors • A. audax • Australia
3384	Nest • Status • Raptors • A. wahlbergi • Rhodesia
3406	Nest • Habitat • Egg • Nestling • Raptors • A. chrysaetos • Korea
3411	Nest • A. wahlbergi • South Africa
3412	Nest • Symbolism • Distribution • Raptors • A. chrysaetos • Utah
3414	Nest • Banding • Photo Account • Popular Monograph • A. chrysaetos • California
3427	Nest • Hunting Behavior • A. chrysaetos • Washington
3438	Nest • Predator/Prey • A. chrysaetos • Oregon
3439	Nest • Prey • Clutch Size • Nest Chronology • A. audax • Australia
3441	Nest • Falconry • Breeding Biology • Prey • A. chrysaetos • Kirgizia
3451	Nest • A. chrysaetos • Tunisia
3455	Nest • Raptors • A. chrysaetos • Russia
3456	Nest • Status • Raptors • A. chrysaetos • Italy
107	Nest Adornment • Flight • Interspecific Behavior • Photography • A. chrysaetos • Europe
235	Nest Adornment • Photography • Photo Account • Popular Monograph • A. chrysaetos • Colorado
534	Nest Adornment • Habitat • Density • Nest Survey • Breeding Biology • Interspecific Competition • Prey • Reproductive Success • Etymology • Plumage • Vocalization • Raptors • A. verreauxi • A. wahlbergi • Kenya

Master No.	Keywords
587	**Nest Adornment** • Nest • Prey • Behavior • Growth • Vocalization • Photo Account • Illus. • A. chrysaetos • Switzerland
944	**Nest Adornment** • Nest • Prey • Behavior • Growth • A. audax • Australia
1247	**Nest Adornment** • Behavior • A. chrysaetos • Scotland
1700	**Nest Adornment** • Nest • Etymology • Raptors • A. chrysaetos • Corsica
2144	**Nest Adornment** • Nest • Parental Behavior • Prey • Vocalization • Photo Account • A. chrysaetos • West Germany
2277	**Nest Adornment** • Predator/Prey • Raptors • A. chrysaetos • Norway
2411	**Nest Adornment** • Interspecific Behavior • Nest • Behavior • Photography • Prey • Talon Grappling • Cain/Abel • Photo Account • Popular Monograph • A. chrysaetos • Scotland
2830	**Nest Adornment** • Breeding Biology • Habitat • Cain/Abel • Growth • Vocalization • Plumage • Prey • Hunting Behavior • Illus. • A. pomarina • West Germany
2994	**Nest Adornment** • Nest Chronology • Incubation • Courtship • Nest Construction • A. chrysaetos • Texas
3235	**Nest Adornment** • Growth • Nest • Photography • Prey • Behavior • Hunting Behavior • Vocalization • Egg • Raptors • Popular Monograph • A. rapax • Kenya
191	**Nest Chronology** • A. chrysaetos • Spain
326	**Nest Chronology** • Raptors • A. rapax • Tanzania
348	**Nest Chronology** • Raptors • A. verreauxi • A. rapax • A. wahlbergi • Rhodesia • Malawi
367	**Nest Chronology** • A. chrysaetos • Spain
453	**Nest Chronology** • A. chrysaetos • Washington
468	**Nest Chronology** • Nest • Egg • Prey • A. nipalensis • Russia
494	**Nest Chronology** • Plumage • A. wahlbergi • Zambia
1008	**Nest Chronology** • Descriptive Summary • Etymology • Status • Prey • Raptors • A. chrysaetos • A. heliaca • A. pomarina • A. clanga • Switzerland
1161	**Nest Chronology** • A. chrysaetos • Spain
1424	**Nest Chronology** • Status • Mortality • A. chrysaetos • Utah
1515	**Nest Chronology** • Status • Raptors • A. audax • Australia
1522	**Nest Chronology** • Nest • Growth • Photo Account • A. chrysaetos • California
1778	**Nest Chronology** • Nest Survey • Prey • Reproductive Success • Abandonment • Marking • Roadside Survey • Territory • Population Dynamics • Raptors • SRBPNA • A. chrysaetos • Idaho
1780	**Nest Chronology** • Nest Survey • Prey • Reproductive Success • Abandonment • Marking • Roadside Survey • Territory • Population Dynamics • Raptors • SRBPNA • A. chrysaetos • Idaho
1781	**Nest Chronology** • Nest Survey • Reproductive Success • Marking • Habitat • Behavior • Prey • Banding • Territory • Population Dynamics • Raptors • SRBPNA • A. chrysaetos • Idaho
1891	**Nest Chronology** • Egg Collection • Behavior • Nest • Raptors • A. wahlbergi • Rhodesia
1912	**Nest Chronology** • Descriptive Summary • Habitat • Nest • A. chrysaetos • France
1960	**Nest Chronology** • Status • Raptors • A. audax • Australia
1992	**Nest Chronology** • Raptors • A. rapax • South Africa
2226	**Nest Chronology** • A. chrysaetos • Spain
2283	**Nest Chronology** • Nest Survey • Prey • Population Decline • A. chrysaetos • France
2370	**Nest Chronology** • Status • Nest • Nest Survey • Monograph • Habitat • Mortality • Breeding Biology • Courtship • Density • Prey • Management • A. chrysaetos • Colorado
2376	**Nest Chronology** • Habitat • Management • Artificial Nest • A. chrysaetos • North America
2404	**Nest Chronology** • Status • Nest • Clutch Size • Raptors • A. wahlbergi • A. rapax • Malawi
2464	**Nest Chronology** • Management • Habitat • Directional Exposure • Disturbance • Mortality • Nest Survey • Density • Prey • Reproductive Success • Prog. Rep. • A. chrysaetos • Wyoming
2715	**Nest Chronology** • Status • Raptors • A. chrysaetos • Montana
2763	**Nest Chronology** • Behavior • Prey • A. chrysaetos • West Germany
2852	**Nest Chronology** • Roadside Survey • Prey • Monograph • Density • Habitat • Interspecific Competition • Taxonomy • Vocalization • Hunting Behavior • Photo Account • Raptors • A. wahlbergi • A. rapax • A. nipalensis • A. heliaca • Kenya
2994	**Nest Chronology** • Incubation • Courtship • Nest Adornment • Nest Construction • A. chrysaetos • Texas
3238	**Nest Chronology** • Raptors • A. wahlbergi • Rwanda
3259	**Nest Chronology** • Nest Survey • Habitat • Territory • Intraspecific Behavior • Density • Prey • Directional Exposure • Raptors • A. verreauxi • Rhodesia
3322	**Nest Chronology** • Nest Survey • Cain/Abel • Breeding Biology • A. pomarina • West Germany
3439	**Nest Chronology** • Nest • Prey • Clutch Size • A. audax • Australia
2994	**Nest Construction** • Nest Chronology • Incubation • Courtship • Nest Adornment • A. chrysaetos • Texas
8	**Nest Failure** • Density • Prey • Reproductive Success • A. nipalensis • Russia
540	**Nest Failure** • Food Requirements • Reproductive Success • Home Range • Breeding Biology • Parental Behavior • Eagles • Theory • A. wahlbergi • A. verreauxi • Africa
1165	**Nest Failure** • Raptors • A. chrysaetos • Spain
1477	**Nest Failure** • A. chrysaetos • Switzerland
1513	**Nest Failure** • Prey • Habitat • Behavior • A. chrysaetos • Alaska
1799	**Nest Failure** • Nest • Protection • Raptors • A. nipalensis • Caucasus
1944	**Nest Failure** • Marking • Abnormal Behavior • Courtship • Telemetry • Raptors • A. chrysaetos • Wyoming • Idaho
1949	**Nest Failure** • Reproductive Success • Pesticides • Egg • Disturbance • A. chrysaetos • Scotland

Master No.	Keywords
2239	**Nest Failure** • Survey • Territory • Nest • Migration • Habitat • Management • Reproductive Success • Prey • Roadside Survey • Status • Winter Survey • Uinta Nf • Wasatch Mt. • Raptors • A. chrysaetos • Utah
2560	**Nest Failure** • Egg • Raptors • A. chrysaetos • Great Britain
2563	**Nest Failure** • Abnormal Behavior • Egg Measure • Pesticides • Raptors • A. chrysaetos • Great Britain
2896	**Nest Failure** • Status • Raptors • A. nipalensis • United Soviet Socialists Republic
3047	**Nest Failure** • Prey • Raptors • A. chrysaetos • Scotland
3446	**Nest Failure** • Interspecific Competition • Raptors • A. chrysaetos • Tunisia
1653	**Nest Robbing** • Punishment • Conservation • Disturbance • Popular Monograph • A. chrysaetos • Scotland
2645	**Nest Robbing** • Conservation • Shooting • Poisoning • Egg Collection • Protection • Trade • Trapping • Raptors • A. chrysaetos • Great Britain
53	**Nest Survey** • Reproductive Success • A. chrysaetos • Colorado
135	**Nest Survey** • A. chrysaetos • Scotland
156	**Nest Survey** • A. audax • Australia
195	**Nest Survey** • Wintering Population • Habitat • Distribution • Attitude • Raptors • A. chrysaetos • Arizona
196	**Nest Survey** • Habitat • Wintering Population • Attitude • Falconry • Raptors • A. chrysaetos • Arizona
198	**Nest Survey** • Prey • Monograph • Habitat • Reproductive Success • Literature Review • Disturbance • Pellet • Animal Damage • A. chrysaetos • Utah
233	**Nest Survey** • Attitude • Reproductive Success • Density • Wintering Population • Food Requirements • Habitat • Management • Aerial Survey • Raptors • A. chrysaetos • Montana
244	**Nest Survey** • Prey • Habitat • Artificial Nest • Raptors • A. chrysaetos • California
253	**Nest Survey** • Raptors • A. chrysaetos • New Mexico
324	**Nest Survey** • Reproductive Success • Food Requirements • Territory • Abandonment • Mortality • Cain/Abel • Density • Dispersal • Breeding Biology • A. chrysaetos • Idaho
337	**Nest Survey** • Raptors • A. chrysaetos • Nebraska
338	**Nest Survey** • Raptors • A. chrysaetos • Nebraska
339	**Nest Survey** • Raptors • A. chrysaetos • Nebraska
340	**Nest Survey** • Raptors • A. chrysaetos • Nebraska
422	**Nest Survey** • Aerial Survey • Aircraft • A. chrysaetos • New Mexico • Wyoming • Colorado
423	**Nest Survey** • Aerial Survey • Wintering Population • Banding • Roadside Survey • Electrocution • A. chrysaetos • North America
424	**Nest Survey** • Aerial Survey • Age Structure • A. chrysaetos • United States
427	**Nest Survey** • Aerial Survey • Age Structure • Prey • Banding • Wintering Population • A. chrysaetos • North America
499	**Nest Survey** • A. audax • Australia
534	**Nest Survey** • Habitat • Density • Breeding Biology • Interspecific Competition • Prey • Reproductive Success • Etymology • Plumage • Vocalization • Nest Adornment • Raptors • A. verreauxi • A. wahlbergi • Kenya
535	**Nest Survey** • Breeding Biology • Etymology • Prey • Reproductive Success • Interspecific Competition • Raptors • Photo Account • A. wahlbergi • A. verreauxi • Kenya
538	**Nest Survey** • Status • Population Estimate • Reproductive Success • Prey • Food Requirements • A. chrysaetos • Scotland
552	**Nest Survey** • Prey • Habitat • Behavior • Status • Reproductive Success • Food Requirements • Animal Damage • Intraspecific Behavior • Predator/Prey • A. chrysaetos • Scotland
611	**Nest Survey** • Habitat • Nest • Field Id • Behavior • Photo Account • Raptors • A. chrysaetos • North America
614	**Nest Survey** • Monograph • Reproductive Success • Habitat • Directional Exposure • Behavior • A. chrysaetos • Utah
741	**Nest Survey** • Prey • Pesticides • Reproductive Success • A. chrysaetos • Scotland
756	**Nest Survey** • Reproductive Success • Prey • Raptors • A. chrysaetos • Colorado
757	**Nest Survey** • Reproductive Success • Aerial Survey • Density • Raptors • A. chrysaetos • Colorado
758	**Nest Survey** • Density • Reproductive Success • Winter Survey • Wintering Population • Status • Aerial Survey • A. chrysaetos • Colorado
759	**Nest Survey** • Eagles • A. chrysaetos • Colorado
789	**Nest Survey** • Monograph • Habitat • Prey • Photo Account • A. chrysaetos • Colorado
858	**Nest Survey** • Conservation • Status • A. chrysaetos • Austria
886	**Nest Survey** • Status • Raptors • A. chrysaetos • Italy
947	**Nest Survey** • Roadside Survey • Roost • Prey • Hunting Behavior • Capture • Marking • Interspecific Behavior • Shooting • Electrocution • A. chrysaetos • Utah
995	**Nest Survey** • Reproductive Success • Disturbance • A. chrysaetos • Scotland
1007	**Nest Survey** • Territory • Reproductive Success • A. chrysaetos • Italy
1099	**Nest Survey** • Habitat • Photo Account • A. chrysaetos • Saskatchewan
1116	**Nest Survey** • Status • Raptors • A. clanga • Russia
1141	**Nest Survey** • Habitat • Density • Disturbance • Reproductive Success • Clutch Size • Raptors • A. verreauxi • Rhodesia
1144	**Nest Survey** • Status • Behavior • Fledging • Mortality • Banding • Marking • Cain/Abel • Reproductive Success • Breeding Biology • Interspecific Behavior • Hunting Behavior • Display • Courtship • Raptors • A. verreauxi • Rhodesia
1147	**Nest Survey** • Prey • Photo Account • Popular Monograph • A. verreauxi • Rhodesia
1156	**Nest Survey** • Breeding Biology • Habitat • Prey • Density • A. verreauxi • Rhodesia
1157	**Nest Survey** • Reproductive Success • A. verreauxi • Rhodesia
1158	**Nest Survey** • Reproductive Success • Cain/Abel • Raptors • A. verreauxi • Rhodesia

Master No.	Keywords
1162	**Nest Survey** • Status • Protection • Conservation • Shooting • Habitat • Raptors • A. chrysaetos • A. heliaca • Spain
1177	**Nest Survey** • Status • Distribution • Reproductive Success • Protection • Photo Account • A. pomarina • East Germany
1184	**Nest Survey** • A. chrysaetos • Switzerland
1227	**Nest Survey** • Status • A. chrysaetos • Switzerland
1294	**Nest Survey** • Habitat • Population Dynamics • Raptors • Tech. • A. chrysaetos • North Dakota
1508	**Nest Survey** • Literature Review • Prey • Pellet • Shooting • Predator/Prey • Attitude • Reproductive Success • A. chrysaetos • North America
1563	**Nest Survey** • Distribution • Descriptive Summary • Raptors • A. chrysaetos • Spain
1581	**Nest Survey** • Raptors • A. chrysaetos • Idaho
1582	**Nest Survey** • Raptors • A. chrysaetos • Idaho
1775	**Nest Survey** • Monograph • Status • Growth • Reproductive Success • Cain/Abel • Banding • Marking • Mortality • Pesticides • Prey • Residues • A. chrysaetos • Idaho
1776	**Nest Survey** • Refuge • Winter Survey • Density • Electrocution • Shooting • Mortality • Prey • Predator/Prey • Raptors • SRBPNA • A. chrysaetos • Idaho
1778	**Nest Survey** • Nest Chronology • Prey • Reproductive Success • Abandonment • Marking • Roadside Survey • Territory • Population Dynamics • Raptors • SRBPNA • A. chrysaetos • Idaho
1780	**Nest Survey** • Nest Chronology • Prey • Reproductive Success • Abandonment • Marking • Roadside Survey • Territory • Population Dynamics • Raptors • SRBPNA • A. chrysaetos • Idaho
1781	**Nest Survey** • Reproductive Success • Marking • Habitat • Nest Chronology • Behavior • Prey • Banding • Territory • Population Dynamics • Raptors • SRBPNA • A. chrysaetos • Idaho
1868	**Nest Survey** • Density • Raptors • A. chrysaetos • Oregon
1892	**Nest Survey** • Clutch Size • Egg • Measure • A. wahlbergi • Rhodesia
1940	**Nest Survey** • Rehabilitation • Raptors • A. chrysaetos • Nebraska
1943	**Nest Survey** • Aerial Survey • Time Lapse • Animal Damage • Prey • Habitat • Winter Survey • Capture • Age Structure • Food Requirements • Shooting • A. chrysaetos • Texas
1945	**Nest Survey** • Mining • Telemetry • Disturbance • Reproductive Success • Raptors • A. chrysaetos • Montana • Wyoming
1950	**Nest Survey** • Animal Damage • Abandonment • Prey • Management • A. chrysaetos • Scotland
1986	**Nest Survey** • Nest • Clutch Size • Prey • Captivity • Collect • A. chrysaetos • Canada
1996	**Nest Survey** • Rarity • Raptors • A. chrysaetos • A. clanga • Scotland
2126	**Nest Survey** • Status • Wintering Population • Mortality • Rehabilitation • Raptors • A. chrysaetos • A. heliaca • A. pomarina • A. nipalensis • A. clanga • Israel
2145	**Nest Survey** • Breeding Biology • Habitat • Cain/Abel • Density • Behavior • Prey • Photo Account • A. pomarina • Czechoslovakia
2148	**Nest Survey** • Distribution • Field Id • Management • Taxonomy • Habitat • Prey • Breeding Biology • Photo Account • Eagles • A. pomarina • Europe
2183	**Nest Survey** • Population Decline • Habitat • Status • Raptors • A. heliaca • Morocco
2184	**Nest Survey** • Aerial Survey • Reproductive Success • Management • Survey • Clutch Size • Distribution • Habitat • Prey • Raptors • A. chrysaetos • Arizona
2200	**Nest Survey** • Prey • Animal Damage • Economics • Trapping • A. chrysaetos • New Mexico • Texas
2238	**Nest Survey** • Nest • Mortality • Nestling • Parental Behavior • Directional Exposure • Hyperthermia • A. chrysaetos • United States
2241	**Nest Survey** • Refuge • Photography • Behavior • Reproductive Success • Mortality • Photo Account • Raptors • Popular Monograph • A. heliaca • Spain
2242	**Nest Survey** • Prey • Raptors • A. heliaca • Spain
2261	**Nest Survey** • Disturbance • Prey • Population Dynamics • Management • Status • Animal Damage • Artificial Nest • Predator/Prey • A. chrysaetos • Utah
2268	**Nest Survey** • Management • Raptors • Photo Account • A. chrysaetos • Utah
2269	**Nest Survey** • Directional Exposure • Reproductive Success • Density • Raptors • A. chrysaetos • Utah
2283	**Nest Survey** • Prey • Population Decline • Nest Chronology • A. chrysaetos • France
2316	**Nest Survey** • Status • Population Decline • Banding • Disturbance • Attitude • A. chrysaetos • Europe
2347	**Nest Survey** • Reproductive Success • Raptors • A. chrysaetos • South Dakota
2368	**Nest Survey** • Density • Habitat • Clutch Size • Raptors • A. chrysaetos • Colorado
2370	**Nest Survey** • Status • Nest • Monograph • Habitat • Mortality • Breeding Biology • Nest Chronology • Courtship • Density • Prey • Management • A. chrysaetos • Colorado
2371	**Nest Survey** • Popular Monograph • Story • Breeding Biology • Behavior • Habitat • Hunting Behavior • Prey • Raptors • Illus. • A. chrysaetos • Colorado
2372	**Nest Survey** • Density • Status • Habitat • Raptors • A. chrysaetos • Colorado
2401	**Nest Survey** • Habitat • A. chrysaetos • Nevada
2408	**Nest Survey** • Status • Distribution • Photo Account • A. chrysaetos • Czechoslovakia
2426	**Nest Survey** • Conservation • Raptors • A. heliaca • A. pomarina • Hungary
2464	**Nest Survey** • Management • Habitat • Directional Exposure • Disturbance • Mortality • Density • Nest Chronology • Prey • Reproductive Success • Prog. Rep. • A. chrysaetos • Wyoming

Master No.	Keywords
2465	**Nest Survey** • Habitat • Mining • Disturbance • Marking • Prey • Time Lapse • Raptors • Proj. Out. • A. chrysaetos • Wyoming • Montana
2492	**Nest Survey** • Density • Raptors • A. chrysaetos • Idaho
2522	**Nest Survey** • Reproductive Success • Raptors • A. chrysaetos • Idaho
2540	**Nest Survey** • Reproductive Success • Raptors • A. chrysaetos • South Dakota
2558	**Nest Survey** • Reproductive Success • Habitat • Protection • Clutch Size • Eagles • A. chrysaetos • Estonia
2594	**Nest Survey** • Monograph • Pesticides • Density • Prey • Mortality • Reproductive Success • A. chrysaetos • Montana
2701	**Nest Survey** • Habitat • Raptors • A. chrysaetos • Alaska
2712	**Nest Survey** • Reproductive Success • Disturbance • A. chrysaetos • Scotland
2723	**Nest Survey** • Banding • Mortality • Status • Reproductive Success • Raptors • A. chrysaetos • Finland
2753	**Nest Survey** • Monograph • Habitat • Behavior • Prey • Pellet • Abandonment • A. chrysaetos • Wyoming
2788	**Nest Survey** • Habitat • Aerial Survey • Density • Photo Account • A. chrysaetos • Nevada
2839	**Nest Survey** • Status • Habitat Loss • Raptors • Photo Account • A. chrysaetos • Nevada
2846	**Nest Survey** • Protection • Distribution • Prey • Photo Account • A. pomarina • Czechoslovakia
2850	**Nest Survey** • Status • Prey • Photo Account • A. heliaca • Czechoslovakia
2856	**Nest Survey** • Breeding Biology • Habitat • Hunting Behavior • Territory • Interspecific Behavior • Prey • Raptors • A. chrysaetos • Utah
2918	**Nest Survey** • Population Decline • Photo Account • A. chrysaetos • North America
2953	**Nest Survey** • Status • A. chrysaetos • Australia
2969	**Nest Survey** • Habitat • Prey • Breeding Biology • Reproductive Success • Growth • Field Id • Vocalization • Migration • Behavior • Eagles • Photo Account • A. wahlbergi • A. rapax • Rhodesia
2970	**Nest Survey** • Clutch Size • Egg • Measure • A. wahlbergi • Rhodesia
3039	**Nest Survey** • Reproductive Success • Mortality • A. heliaca • A. pomarina • Czechoslovakia
3050	**Nest Survey** • Raptors • A. chrysaetos • Alaska
3079	**Nest Survey** • Prey • Clutch Size • Reproductive Success • Display • Hunting Behavior • Density • Migration • A. wahlbergi • South Africa
3110	**Nest Survey** • Distribution • Status • Density • Habitat Loss • Density • Reproductive Success • Raptors • Photo Account • A. chrysaetos • California
3146	**Nest Survey** • Status • Population Dynamics • Prey • A. chrysaetos • Idaho
3187	**Nest Survey** • Prey • Banding • Raptors • A. wahlbergi • Rhodesia
3207	**Nest Survey** • Refuge • Habitat • Density • Prey • Predator/Prey • Disturbance • Telemetry • Territory • Reproductive Success • Raptors • SRBPNA • Photo Account • A. chrysaetos • Idaho
3259	**Nest Survey** • Habitat • Territory • Intraspecific Behavior • Density • Nest Chronology • Prey • Directional Exposure • Raptors • A. verreauxi • Rhodesia
3262	**Nest Survey** • Raptors • A. wahlbergi • Rhodesia
3304	**Nest Survey** • Density • Directional Exposure • Reproductive Success • Territory • A. chrysaetos • Scotland
3322	**Nest Survey** • Nest Chronology • Cain/Abel • Breeding Biology • A. pomarina • West Germany
3324	**Nest Survey** • Raptors • A. pomarina • West Germany
3351	**Nest Survey** • Aerial Survey • Behavior • Raptors • A. chrysaetos • Alaska
3360	**Nest Survey** • Prey • Habitat • Directional Exposure • Eagles • A. chrysaetos • Saskatchewan
3369	**Nest Survey** • Egg • Vocalization • Habitat • Behavior • Hunting Behavior • Photo Account • A. chrysaetos • Arizona
3444	**Nest Survey** • Breeding Biology • Habitat • Protection • Prey • Distribution • Reproductive Success • Photo Account • A. chrysaetos • Estonia
401	**Nesting Behavior** • Prey • Photo Account • Popular Monograph • A. chrysaetos • Scotland
1205	**Nesting Behavior** • Prey • Status • Photo Account • A. chrysaetos • Scotland
1274	**Nesting Behavior** • Trapping • Photo Account • Popular Monograph • A. chrysaetos • Scotland
2409	**Nesting Behavior** • Descriptive Summary • Photo Account • Popular Monograph • A. chrysaetos • Scotland
2412	**Nesting Behavior** • Photo Account • Popular Monograph • A. chrysaetos • Scotland
2606	**Nesting Behavior** • Photo Account • Popular Monograph • A. chrysaetos • Scotland
3319	**Nesting Behavior** • Photo Account • Popular Monograph • A. chrysaetos • Scotland
852	**Nestling** • Clutch Size • Mortality • Prey • Pesticides • Cain/Abel • A. chrysaetos • Spain
1102	**Nestling** • Abandonment • Disturbance • Egg • Mortality • Photography • Banding • Raptors • Tech. • A. chrysaetos • North America
1249	**Nestling** • A. chrysaetos • Great Britain
1257	**Nestling** • A. chrysaetos • Great Britain
2151	**Nestling** • Clutch Size • Fostering • A. heliaca • Spain
2238	**Nestling** • Nest • Nest Survey • Mortality • Parental Behavior • Directional Exposure • Hyperthermia • A. chrysaetos • United States
2524	**Nestling** • Plumage • A. clanga • Russia
2646	**Nestling** • (Title Only) • A. pomarina • West Germany
3376	**Nestling** • Parasite • Photo Account • Raptors • A. chrysaetos • Colorado • Wyoming
3406	**Nestling** • Habitat • Nest • Egg • Raptors • A. chrysaetos • Korea
3445	**Nestling** • (Title Only) • A. pomarina • West Germany
414	**Niche** • Predator/Prey • Raptors • A. chrysaetos • Europe

Master No.	Keywords
3131	**Niche** • Habitat • Prey • Raptors • A. wahlbergi • A. rapax • Ivory Coast
230	**Osteology** • Raptors • A. chrysaetos
1045	**Osteology** • Measure • A. chrysaetos • North America
1570	**Osteology** • A. chrysaetos • California
1683	**Osteology** • Anatomy • Taxonomy • Museum Specimen • Raptors • A. chrysaetos • A. rapax
1684	**Osteology** • Anatomy • Taxonomy • Museum Specimen • Raptors • A. chrysaetos • A. rapax
1685	**Osteology** • Anatomy • Taxonomy • Museum Specimen • Raptors • A. chrysaetos • A. rapax
1686	**Osteology** • Anatomy • Taxonomy • Museum Specimen • Raptors • A. chrysaetos • A. rapax
2213	**Osteology** • Paleontology • A. chrysaetos • California
2812	**Paired Ovaries** • Anatomy • Raptors • A. clanga • China
2877	**Paired Ovaries** • Anatomy • Raptors • A. chrysaetos • Canada
2933	**Paired Ovaries** • Anatomy • Dimorphism • Raptors • Gonads • A. chrysaetos • North America
110	**Paleontology** • Eagles • A. chrysaetos • California
202	**Paleontology** • Anatomy • A. chrysaetos • Cuba
948	**Paleontology** • A. chrysaetos • Nebraska
1086	**Paleontology** • Raptors • A. chrysaetos • Alaska
1287	**Paleontology** • Klamath Basin • Raptors • A. chrysaetos • California • Oregon
1571	**Paleontology** • Raptors • A. chrysaetos • California
1572	**Paleontology** • A. chrysaetos • California
1573	**Paleontology** • Eagles • A. chrysaetos • Nevada
1574	**Paleontology** • A. chrysaetos • California
1575	**Paleontology** • Anatomy • Measure • A. chrysaetos • North America
1576	**Paleontology** • Raptors • A. chrysaetos • California
1577	**Paleontology** • Raptors • A. chrysaetos • New Mexico
1578	**Paleontology** • A. chrysaetos • California
1579	**Paleontology** • Raptors • A. chrysaetos • New Mexico
1654	**Paleontology** • Raptors • A. chrysaetos • Hungary
1861	**Paleontology** • A. chrysaetos • Hungary
1862	**Paleontology** • A. chrysaetos • Europe
2174	**Paleontology** • Raptors • A. chrysaetos • California
2175	**Paleontology** • A. chrysaetos • California
2176	**Paleontology** • Raptors • A. chrysaetos • California
2177	**Paleontology** • Raptors • A. chrysaetos • California
2213	**Paleontology** • Osteology • A. chrysaetos • California
2417	**Paleontology** • Raptors • A. chrysaetos • France
2419	**Paleontology** • A. chrysaetos • Illinois
2820	**Paleontology** • A. chrysaetos • North America
2821	**Paleontology** • Oregon
2822	**Paleontology** • Taxonomy • Eagles • A. chrysaetos
2823	**Paleontology** • Taxonomy • Eagles • A. chrysaetos • Oregon
2824	**Paleontology** • Eagles • A. chrysaetos • Oregon
3334	**Paleontology** • Raptors • A. chrysaetos • Nebraska
3336	**Paleontology** • A. chrysaetos • Wyoming
3337	**Paleontology** • A. chrysaetos
256	**Parasite** • Collect • Raptors • A. rapax • Tanzania
385	**Parasite** • (Title Only) • A. chrysaetos
451	**Parasite** • A. chrysaetos • Switzerland
950	**Parasite** • A. chrysaetos
955	**Parasite** • Behavior • Pesticides • Parental Behavior • Vocalization • Disturbance • Measure • Breeding Biology • Disease • Terminology • Growth • A. chrysaetos • Montana
957	**Parasite** • Monograph • Behavior • Growth • Pesticides • Parental Behavior • Vocalization • Disturbance • Measure • Breeding Biology • Photography • Captivity • Disease • Terminology • Photo Account • A. chrysaetos • Montana
1285	**Parasite** • Collect • Fat • Molt • Museum Specimen • A. chrysaetos • North Carolina
1504	**Parasite** • Raptor • A. chrysaetos • California
1516	**Parasite** • Nest • Prey • A. audax • Australia
1890	**Parasite** • Nest • A. chrysaetos • Texas
2453	**Parasite** • Raptors • A. chrysaetos • North America
2460	**Parasite** • Mortality • Invertebrates • Raptors • A. chrysaetos • United States
2925	**Parasite** • Blood Sample • Raptors • A. chrysaetos • Colorado
3376	**Parasite** • Nestling • Photo Account • Raptors • A. chrysaetos • Colorado • Wyoming
1790	**Parasite** • A. heliaca • A. chrysaetos • A. pomarina • Europe
540	**Parental Behavior** • Nest Failure • Food Requirements • Reproductive Success • Home Range • Breeding Biology • Eagles • Theory • A. wahlbergi • A. verreauxi • Africa
955	**Parental Behavior** • Behavior • Pesticides • Vocalization • Disturbance • Measure • Breeding Biology • Parasite • Disease • Terminology • Growth • A. chrysaetos • Montana
957	**Parental Behavior** • Monograph • Behavior • Growth • Pesticides • Vocalization • Disturbance • Measure • Breeding Biology • Photography • Captivity • Parasite • Disease • Terminology • Photo Account • A. chrysaetos • Montana
962	**Parental Behavior** • Prey • A. audax • Australia
1056	**Parental Behavior** • Growth • Falconry • Aging • Measure • Plumage • Dimorphism • Prey • Photo Account • A. audax • Australia
1145	**Parental Behavior** • Nest • Breeding Biology • Interspecific Behavior • Intraspecific Behavior • Feeding Behavior • Hunting Behavior • Fledging • Prey • Display • Raptors • A. verreauxi • Rhodesia
1149	**Parental Behavior** • Egg • A. verreauxi
1152	**Parental Behavior** • Nest • Egg • Incubation • A. verreauxi • Rhodesia
1153	**Parental Behavior** • Cain/Abel • Clutch Size • Behavior • Growth • Eagles • A. verreauxi • A. pomarina • Rhodesia
1629	**Parental Behavior** • Nest • Prey • A. chrysaetos • Switzerland
1631	**Parental Behavior** • Nest • Prey • A. nipalensis • Kazakhstan
2144	**Parental Behavior** • Nest • Prey • Vocalization • Nest Adornment • Photo Account • A. chrysaetos • West Germany
2152	**Parental Behavior** • Habitat • Growth • Prey • Breeding Biology • Interspecific Behavior • Courtship • Display • Photo Account • Raptors • A. heliaca • Spain
2238	**Parental Behavior** • Nest • Nest Survey • Mortality • Nestling • Directional Exposure • Hyperthermia • A. chrysaetos • United States
2669	**Parental Behavior** • Nest • Fledging • Food Habits • Breeding Biology • A. chrysaetos • Tanzania
2670	**Parental Behavior** • Cain/Abel • Prey • Growth • Breeding Biology • A. verreauxi • Africa

Master No.	Keywords
198	**Pellet** • Prey • Monograph • Nest Survey • Habitat • Reproductive Success • Literature Review • Disturbance • Animal Damage • A. chrysaetos • Utah
501	**Pellet** • Prey • Attack • Hunting Behavior • Animal Damage • Habitat • Cache • A. audax • Australia
745	**Pellet** • Food Habits • Raptors • A. chrysaetos • Utah
1092	**Pellet** • Nest • Behavior • A. chrysaetos • Switzerland
1508	**Pellet** • Literature Review • Prey • Shooting • Predator/Prey • Nest Survey • Attitude • Reproductive Success • A. chrysaetos • North America
1881	**Pellet** • Physiology • Digestion • Raptors • A. nipalensis
1900	**Pellet** • Mortality • Prey • Nest • Dispersal • Habitat • Animal Damage • Bounty • Protection • Photo Account • A. audax • Australia
1928	**Pellet** • Prey • Nest • Hunting Behavior • Interspecific Behavior • Reproductive Success • Eagles • A. chrysaetos • Oklahoma
2257	**Pellet** • Prey Behavior • Prey • Interspecific Behavior • A. chrysaetos • Alaska
2410	**Pellet** • Egg • Nest • Photo Account • A. chrysaetos • Scotland
2600	**Pellet** • Nest • A. chrysaetos • Switzerland
2753	**Pellet** • Monograph • Habitat • Nest Survey • Behavior • Prey • Abandonment • A. chrysaetos • Wyoming
3022	**Pellet** • Cain/Abel • Growth • Behavior • Vocalization • Feeding Behavior • Raptors • A. chrysaetos • California
3021	**Pellet** • Captivity • Growth • Food Requirements • Measure • Molt • Temperature • Photo Account • Raptors • A. chrysaetos • California
3397	**Pellet** • Descriptive Summary • Behavior • Distribution • Plumage • Measure • Raptors • Illus. • A. chrysaetos • A. clanga • Great Britain
386	**Pesticides** • Bounty • Disturbance • Habitat Loss • Protection • Shooting • Status • Migration • Distribution • Population Decline • Raptors • A. chrysaetos • A. heliaca • A. pomarina • A. clanga • Europe
391	**Pesticides** • PCB • A. chrysaetos • California
741	**Pesticides** • Nest Survey • Prey • Reproductive Success • A. chrysaetos • Scotland
767	**Pesticides** • Telemetry • Protection • Nest • Falconry • Behavior • Photo Account • Popular Monograph • A. chrysaetos • United States
771	**Pesticides** • Raptors • A. chrysaetos • Great Britain
836	**Pesticides** • Mortality • Autopsy • Toxicology • Eagles • Popular Monograph • A. chrysaetos • North America
852	**Pesticides** • Clutch Size • Mortality • Nestling • Prey • Cain/Abel • A. chrysaetos • Spain
955	**Pesticides** • Behavior • Parental Behavior • Vocalization • Disturbance • Measure • Breeding Biology • Parasite • Disease • Terminology • Growth • A. chrysaetos • Montana
957	**Pesticides** • Monograph • Behavior • Growth • Parental Behavior • Vocalization • Disturbance • Measure • Breeding Biology • Photography • Captivity • Parasite • Disease • Terminology • Photo Account • A. chrysaetos • Montana
996	**Pesticides** • Survey • Reproductive Success • Status • Disturbance • A. chrysaetos • Scotland
1078	**Pesticides** • Raptors • A. rapax • Kenya
1101	**Pesticides** • Status • Habitat Loss • Disturbance • Raptors • A. chrysaetos • Canada
1151	**Pesticides** • Reproductive Success • Density • Population Dynamics • Prey • Nest • Climatology • Territory • Interspecific Competition • Mortality • Clutch Size • Eagles • A. verreauxi • Rhodesia
1164	**Pesticides** • Status • Conservation • Disturbance • Population Estimate • Raptors • A. heliaca • A. chrysaetos • Spain
1206	**Pesticides** • DDE • Raptors • PCB • A. chrysaetos • Canada
1496	**Pesticides** • Measure • Egg • Status • Raptors • A. chrysaetos • California
1507	**Pesticides** • Conservation • Protection • Economics • Mortality • Reward • Poisoning • Eagles • A. chrysaetos • Wyoming • Colorado
1710	**Pesticides** • Raptors • A. chrysaetos • Switzerland
1730	**Pesticides** • Egg • Heavy Metals • Raptors • DDE • Mercury • A. chrysaetos • North America
1775	**Pesticides** • Monograph • Status • Nest Survey • Growth • Reproductive Success • Cain/Abel • Banding • Marking • Mortality • Prey • Residues • A. chrysaetos • Idaho
1948	**Pesticides** • Mortality • (Title Only) • A. chrysaetos • Scotland
1949	**Pesticides** • Reproductive Success • Nest Failure • Egg • Disturbance • A. chrysaetos • Scotland
1951	**Pesticides** • Reproductive Success • A. chrysaetos • Scotland
2049	**Pesticides** • Status • Shooting • Migration • Protection • Poisoning • A. chrysaetos • Siberia
2125	**Pesticides** • Poisoning • Status • Captivity • Density • Food Requirements • Habitat • Protection • Predator/Prey • Raptors • Photography Account • PCB • A. clanga • A. heliaca • A. nipalensis • A. verreauxi • A. pomarina • Israel
2128	**Pesticides** • Mortality • Prey • Rehabilitation • Wintering Population • Azodrin • Raptors • A. heliaca • A. clanga • A. pomarina • Israel
2307	**Pesticides** • Status • Density • Conservation • Mortality • Raptors • A. chrysaetos • Scotland
2422	**Pesticides** • Status • Distribution • Raptors • A. chrysaetos • Great Britain
2562	**Pesticides** • Egg • Raptors • A. chrysaetos • Great Britain
2563	**Pesticides** • Abnormal Behavior • Egg Measure • Nest Failure • Raptors • A. chrysaetos • Great Britain
2565	**Pesticides** • Egg • Egg Measure • Raptors • A. chrysaetos • Scotland
2584	**Pesticides** • Mortality • A. chrysaetos • North America
2594	**Pesticides** • Monograph • Density • Prey • Nest Survey • Mortality • Reproductive Success • A. chrysaetos • Montana

Master No.	Keywords
2627	**Pesticides** • Status • Protection • Raptors • A. chrysaetos • Italy
2690	**Pesticides** • Mortality • A. pomarina • West Germany
2789	**Pesticides** • Monograph • Raptors • A. chrysaetos • Montana
2977	**Pesticides** • Raptors • A. chrysaetos • United States
2978	**Pesticides** • Raptors • A. chrysaetos • United States
3186	**Pesticides** • Toxicology • Comp. 1080 • Strychnine • Raptors • A. chrysaetos • United States
3306	**Pesticides** • Autopsy • Mortality • Disease • A. chrysaetos • Great Britain
103	**Photography** • (Title Only) • A. chrysaetos • Great Britain
107	**Photography** • Flight • Nest Adornment • Interspecific Behavior • A. chrysaetos • Europe
235	**Photography** • Nest Adornment • Photo Account • Popular Monograph • A. chrysaetos • Colorado
316	**Photography** • Egg Collection • A. chrysaetos • California
383	**Photography** • Banding • Measure • Raptors • A. nipalensis • A. wahlbergi • A. rapax • South Africa
753	**Photography** • Behavior • Nest • Shooting • Story • Banding • Photo Account • Raptors • A. verreauxi • South Africa
957	**Photography** • Monograph • Behavior • Growth • Pesticides • Parental Behavior • Vocalization • Disturbance • Measure • Breeding Biology • Captivity • Parasite • Disease • Terminology • Photo Account • A. chrysaetos • Montana
1036	**Photography** • Nest • Popular Monograph • A. chrysaetos • California
1037	**Photography** • Nest • Prey • Growth • A. chrysaetos • California
1102	**Photography** • Abandonment • Disturbance • Egg • Mortality • Nestling • Banding • Raptors • Tech. • A. chrysaetos • North America
1167	**Photography** • Nest • Prey • Behavior • Photo Account • A. audax • Australia
1246	**Photography** • Behavior • Hunting Behavior • Lift • Trapping • Attack • Behavior • Photo Account • Popular Monograph • A. chrysaetos • Scotland
1262	**Photography** • Photo Account • Popular Monograph • (Title Only) • A. chrysaetos • Scotland
1386	**Photography** • Falconry • Prey • Photo Account • Popular Monograph • A. chrysaetos • Scotland
1502	**Photography** • Nest • Photo Account • Popular Monograph • A. chrysaetos • Scotland
1562	**Photography** • Nest • Prey • Photo Account • A. chrysaetos • Scotland
1953	**Photography** • Raptors • Photo Account • Popular Monograph • A. chrysaetos • A. heliaca • Rumania
2240	**Photography** • Nest • Prey • Behavior • Photo Account • A. heliaca • Spain
2241	**Photography** • Refuge • Behavior • Nest Survey • Reproductive Success • Mortality • Photo Account • Raptors • Popular Monograph • A. heliaca • Spain
2411	**Photography** • Interspecific Behavior • Nest • Behavior • Prey • Nest Adornment • Talon Grappling • Cain/Abel • Photo Account • Popular Monograph • A. chrysaetos • Scotland
2602	**Photography** • Nest • Habitat • Prey • Photo Account • A. chrysaetos • Switzerland
2958	**Photography** • Photography Account • (Title Only) • A. chrysaetos • Great Britain
2972	**Photography** • Plumage • Story • Breeding Biology • Prey • Photo Account • Popular Monograph • A. verreauxi • A. wahlbergi • A. rapax • Rhodesia • South Africa
3235	**Photography** • Growth • Nest • Prey • Behavior • Hunting Behavior • Vocalization • Nest Adornment • Egg • Raptors • Popular Monograph • A. rapax • Kenya
604	**Physiology** • Anatomy • Captivity • Raptors • A. chrysaetos • A. verreauxi • A. rapax • A. wahlbergi
787	**Physiology** • Anatomy • Measure • Raptors • A. chrysaetos • Yugoslavia
1077	**Physiology** • Anaesthesia • Raptors • A. chrysaetos
1083	**Physiology** • Iron • A. pomarina • A. clanga
1155	**Physiology** • Behavior • Nest • Temperature • A. verreauxi • Rhodesia
1399	**Physiology** • Measure • Raptors • A. chrysaetos
1594	**Physiology** • Raptors • A. rapax
1881	**Physiology** • Pellet • Digestion • Raptors • A. nipalensis
1957	**Physiology** • Anatomy • Raptors • A. chrysaetos • North America
2348	**Physiology** • Blood Sample • A. chrysaetos • North America
2684	**Physiology** • Telemetry • Temperature • A. chrysaetos • Idaho
2934	**Physiology** • Anatomy • Measure • Raptors • Ovary • Testes • A. chrysaetos
2937	**Physiology** • Rehabilitation • Disease • Riboflavin • A. chrysaetos • Idaho
897	**Piracy** • Interspecific Behavior • A. chrysaetos • Alaska
1319	**Piracy** • Status • Raptors • A. rapax • South Africa
2092	**Piracy** • Field Id • Descriptive Summary • Hunting Behavior • Animal Damage • Nest • Egg • Attitude • Raptors • A. audax • Australia
2458	**Piracy** • Prey • Eagles • A. chrysaetos • Sweden
2962	**Piracy** • Prey • Density • Raptors • A. chrysaetos • Washington
2973	**Piracy** • Life History • Prey • Plumage • Cain/Abel • Banding • Reproductive Success • Interspecific Behavior • Photo Account • Raptors • A. chrysaetos • Rhodesia
3357	**Piracy** • Prey • Status • Collect • Raptors • A. heliaca • A. nipalensis • A. clanga • A. rapax • India
54	**Plumage** • Museum Specimen • Distribution • Measure • Prey • A. heliaca • A. clanga • A. rapax • India
55	**Plumage** • Taxonomy • A. pomarina • A. nipalensis • A. heliaca • India
56	**Plumage** • Etymology • Descriptive Summary • Clutch Size • Habitat • Museum Specimen • Measure • A. rapax • A. pomarina • India
59	**Plumage** • Taxonomy • Measure • A. nipalensis • A. heliaca • India

Master No.	Keywords
67	**Plumage** • Distribution • Taxonomy • Egg • Prey • Raptors • A. verreauxi • A. rapax • Southwest Africa
68	**Plumage** • Taxonomy • Museum Specimen • Distribution • A. nipalensis • A. clanga • A. pomarina • France
74	**Plumage** • Taxonomy • A. nipalensis • A. pomarina • A. clanga • West Germany
108	**Plumage** • Clutch Size • Nest • Raptors • A. wahlbergi • A. rapax • A. verreauxi • Kenya • Uganda
242	**Plumage** • Descriptive Summary • Id Key • Etymology • Measure • Raptors • A. chrysaetos • A. heliaca • A. nipalensis • A. rapax • A. clanga • A. pomarina • India
251	**Plumage** • Descriptive Summary • Nest • Egg • Measure • Taxonomy • Photo Account • Raptors • A. chrysaetos • A. rapax • Algeria
262	**Plumage** • Descriptive Summary • Etymology • Status • Nest • Behavior • A. heliaca • A. nipalensis • A. clanga • A. chrysaetos • Cyprus
304	**Plumage** • Aging • Intraspecific Behavior • Behavior • A. heliaca • A. nipalensis • A. chrysaetos • A. pomarina • A. clanga • Europe
323	**Plumage** • Falconry • Hunting Behavior • Field Id • Photography Account • Raptors • A. chrysaetos
351	**Plumage** • Hunting Behavior • Measure • Life History • Courtship • Behavior • Distribution • Vocalization • Raptors • Photo Account • A. chrysaetos • North America
374	**Plumage** • A. chrysaetos • France
408	**Plumage** • Descriptive Summary • Measure • Taxonomy • A. chrysaetos • A. heliaca • A. clanga • A. pomarina
413	**Plumage** • Rarity • A. heliaca • A. rapax • Morocco
494	**Plumage** • Nest Chronology • A. wahlbergi • Zambia
505	**Plumage** • Taxonomy • A. heliaca • A. clanga • India
506	**Plumage** • Taxonomy • Raptors • A. clanga • India
507	**Plumage** • Taxonomy • A. heliaca • A. nipalensis • India
509	**Plumage** • Taxonomy • Etymology • A. nipalensis • A. clanga • A. rapax • A. pomarina • India
510	**Plumage** • Taxonomy • Etymology • A. heliaca • A. nipalensis • A. clanga • A. pomarina • India
511	**Plumage** • Taxonomy • Etymology • A. rapax • India
512	**Plumage** • Taxonomy • A. rapax • A. nipalensis • A. pomarina • India
513	**Plumage** • Taxonomy • Etymology • A. clanga • A. pomarina • India
514	**Plumage** • Taxonomy • Etymology • Raptors • A. clanga • A. pomarina • A. rapax
518	**Plumage** • Taxonomy • A. clanga • A. rapax
534	**Plumage** • Habitat • Density • Nest Survey • Breeding Biology • Interspecific Competition • Prey • Reproductive Success • Etymology • Vocalization • Nest Adornment • Raptors • A. verreauxi • A. wahlbergi • Kenya
626	**Plumage** • Capture • Measure • A. pomarina • Tunisia
654	**Plumage** • Wintering Population • Nest • Raptors • A. heliaca • A. chrysaetos • Spain
744	**Plumage** • Id Key • Descriptive Summary • Raptors • A. chrysaetos • Wisconsin • Illinois
770	**Plumage** • Field Id • Descriptive Summary • Monograph • Distribution • Status • Migration • Prey • Measure • Etymology • Vocalization • A. pomarina • A. clanga • A. nipalensis • A. heliaca • A. chrysaetos • A. verreauxi • Africa • Europe
780	**Plumage** • Collect • Prey • Raptors • A. heliaca • A. clanga • Hungary
799	**Plumage** • Captivity • Status • Raptors • A. chrysaetos • A. heliaca • A. clanga • Rumania
816	**Plumage** • Field Id • Flight • Eagles • A. chrysaetos • A. nipalensis • A. wahlbergi • A. pomarina • South Africa
907	**Plumage** • Descriptive Summary • Field Id • Prey • Measure • Raptors • A. chrysaetos • A. heliaca • A. nipalensis • A. rapax • A. clanga • India
915	**Plumage** • Taxonomy • A. nipalensis • A. heliaca • A. clanga • A. pomarina • A. rapax • India
916	**Plumage** • Taxonomy • Measure • A. nipalensis • A. clanga • A. pomarina
963	**Plumage** • Growth • Nest • Prey • Vocalization • Behavior • Photo Account • A. audax • Australia
989	**Plumage** • Taxonomy • Measure • Museum Specimen • A. clanga • A. pomarina • A. rapax • A. nipalensis • France
1013	**Plumage** • Rarity • Captivity • Poisoning • Measure • Photo Account • A. heliaca • Italy
1032	**Plumage** • Raptors • A. wahlbergi • South Africa
1033	**Plumage** • Habitat • Prey • Nest • Flight • Raptors • A. wahlbergi • A. rapax • South Africa
1054	**Plumage** • Captive Breeding • Nest • Behavior • Measure • Photo Account • Popular Monograph • A. audax • Australia
1056	**Plumage** • Growth • Falconry • Aging • Parental Behavior • Measure • Dimorphism • Prey • Photo Account • A. audax • Australia
1087	**Plumage** • Collect • Museum Specimen • Raptors • A. rapax • Ethiopia • Kenya
1088	**Plumage** • Museum Specimen • Raptors • A. rapax • A. wahlbergi • Africa
1095	**Plumage** • Museum Specimen • Shooting • Measure • Taxonomy • A. clanga • Albania
1105	**Plumage** • Descriptive Summary • Etymology • Nest • Distribution • A. chrysaetos • Alaska
1185	**Plumage** • Flight • Intraspecific Behavior • Illus. • A. clanga • Yugoslavia
1186	**Plumage** • Vocalization • Illus. • A. pomarina • Yugoslavia
1209	**Plumage** • Taxonomy • Captivity • Behavior • Eagles • A. chrysaetos • Nova Scotia
1211	**Plumage** • Rarity • Collect • Measure • Photo Account • A. pomarina • Italy
1228	**Plumage** • Monograph • Vocalization • Distribution • Prey • Id Key • A. pomarina • A. chrysaetos • A. nipalensis • Europe
1314	**Plumage** • Taxonomy • Measure • A. clanga • Europe
1322	**Plumage** • Taxonomy • A. heliaca • France
1326	**Plumage** • Taxonomy • Museum Specimen • A. nipalensis • Turkey
1327	**Plumage** • Taxonomy • A. heliaca • A. nipalensis
1328	**Plumage** • Museum Specimen • Taxonomy • Captivity • Eagles • A. chrysaetos • A. heliaca • A. rapax • A. clanga • A. nipalensis

Master No.	Keywords
1329	**Plumage** • Measure • Rarity • A. clanga • A. pomarina • Great Britain
1345	**Plumage** • Rarity • Shooting • Prey • Measure • A. chrysaetos • West Germany
1347	**Plumage** • Anatomy • Behavior • Taxonomy • Raptors • Theory • A. chrysaetos
1393	**Plumage** • Collect • Raptors • A. rapax • A. wahlbergi • Nigeria
1396	**Plumage** • Taxonomy • Measure • Museum Specimen • A. rapax • A. nipalensis • Great Britain
1397	**Plumage** • Taxonomy • Museum Specimen • A. audax • Australia
1418	**Plumage** • Taxonomy • Measure • Photo Account • A. clanga • A. pomarina • Belgium
1503	**Plumage** • Captivity • Longevity • Egg • A. rapax • Tunisia
1600	**Plumage** • Measure • Raptors • A. heliaca • A. nipalensis • A. clanga • India
1617	**Plumage** • A. clanga • Great Britain
1624	**Plumage** • Descriptive Summary • Distribution • Raptors • A. audax • A. gurneyi • New Guinea
1639	**Plumage** • Taxonomy • Dimorphism • Measure • Distribution • Prey • Behavior • Management • Habitat • Territory • Mortality • Life History • Economics • Literature Review • A. chrysaetos • Oregon • Washington
1688	**Plumage** • Molt • Aging • A. chrysaetos • North America
1697	**Plumage** • Status • Distribution • Taxonomy • Etymology • Field Id • Literature Review • Eagles • A. pomarina • A. clanga • France • Belgium • Switzerland • Italy
1716	**Plumage** • Nest • Egg • Clutch Size • A. pomarina • West Germany
1734	**Plumage** • A. wahlbergi • South Africa
1765	**Plumage** • Taxonomy • A. clanga • Europe
1766	**Plumage** • Anatomy • A. clanga • A. pomarina • A. nipalensis • A. heliaca
1767	**Plumage** • A. heliaca • A. nipalensis
1841	**Plumage** • Rarity • Collect • Measure • A. heliaca • France
1847	**Plumage** • Descriptive Summary • Measure • Distribution • Raptors • A. chrysaetos • A. heliaca • A. nipalensis • A. clanga • China
1874	**Plumage** • Taxonomy • Status • Photo Account • A. chrysaetos • A. clanga • A. nipalensis • France
1918	**Plumage** • Collect • A. heliaca • Spain
2020	**Plumage** • Rarity • Museum Specimen • Measure • Photo Account • A. heliaca • Italy
2045	**Plumage** • Descriptive Summary • Measure • Raptors • A. chrysaetos • A. heliaca • A. nipalensis • A. pomarina • A. clanga • Italy
2048	**Plumage** • Vocalization • A. chrysaetos • Tunisia
2120	**Plumage** • Descriptive Summary • Distribution • Taxonomy • Hunting Behavior • Raptors • A. chrysaetos • A. nipalensis • A. rapax • A. clanga • Saudi Arabia
2131	**Plumage** • Distribution • Status • Raptors • A. chrysaetos • A. heliaca • A. clanga • A. pomarina • Russia
2254	**Plumage** • A. chrysaetos • Connecticut
2340	**Plumage** • Descriptive Summary • Distribution • Prey • Nest • Egg • Photo Account • Raptors • A. audax • Australia
2352	**Plumage** • Collect • Prey • A. nipalensis • A. heliaca • India
2355	**Plumage** • Collect • Taxonomy • Raptors • A. rapax • Kenya
2421	**Plumage** • Taxonomy • A. clanga
2524	**Plumage** • Nestling • A. clanga • Russia
2552	**Plumage** • Descriptive Summary • Distribution • Taxonomy • Habitat • Measure • Raptors • A. gurneyi • New Guinea
2622	**Plumage** • Collect • Rarity • A. clanga • West Germany
2652	**Plumage** • Rarity • Measure • A. clanga • Great Britain
2666	**Plumage** • Taxonomy • A. clanga
2706	**Plumage** • Descriptive Summary • Distribution • Museum Specimen • A. chrysaetos • Molluca Islands
2718	**Plumage** • Taxonomy • Raptors • A. chrysaetos • A. heliaca • A. clanga • Spain
2773	**Plumage** • Collect • Raptors • A. wahlbergi • A. heliaca • A. rapax • A. nipalensis • Sudan
2794	**Plumage** • Shooting • Measure • A. chrysaetos • Pennsylvania
2830	**Plumage** • Breeding Biology • Habitat • Cain/Abel • Nest Adornment • Growth • Vocalization • Prey • Hunting Behavior • Illus. • A. pomarina • West Germany
2894	**Plumage** • Collect • Raptors • A. wahlbergi • Rhodesia
2907	**Plumage** • Wintering Population • Trapping • Measure • Molt • A. chrysaetos • Tennessee
2908	**Plumage** • Wintering Population • Trapping • Measure • A. chrysaetos • Tennessee
2911	**Plumage** • A. chrysaetos • Tennessee
2913	**Plumage** • Shooting • Survey • Prey • Animal Damage • A. chrysaetos • Texas
2972	**Plumage** • Photography • Story • Breeding Biology • Prey • Photo Account • Popular Monograph • A. verreauxi • A. wahlbergi • A. rapax • Rhodesia • South Africa
2973	**Plumage** • Life History • Prey • Cain/Abel • Banding • Reproductive Success • Interspecific Behavior • Piracy • Photo Account • Raptors • A. chrysaetos • Rhodesia
3028	**Plumage** • Taxonomy • Molt • A. chrysaetos • A. verreauxi • A. heliaca • A. rapax • A. clanga
3040	**Plumage** • Field Id • Measure • Eagles • Photo Account • A. chrysaetos • A. pomarina • Europe
3044	**Plumage** • Taxonomy • A. chrysaetos
3086	**Plumage** • Museum Specimen • Raptors • A. clanga • A. heliaca • Egypt
3089	**Plumage** • Taxonomy • Captivity • Behavior • A. rapax
3140	**Plumage** • Captivity • A. chrysaetos • Iowa
3226	**Plumage** • Density • Reproductive Success • Sub-Adult • Museum Specimen • Population Decline • Illus. • A. heliaca • Spain
3366	**Plumage** • Status • Museum Specimen • Raptors • A. nipalensis • A. heliaca • Hungary
3397	**Plumage** • Descriptive Summary • Pellet • Behavior • Distribution • Measure • Raptors • Illus. • A. chrysaetos • A. clanga • Great Britain
3405	**Plumage** • Distribution • Taxonomy • Subspecies • A. chrysaetos

Master No.	Keywords
3436	**Plumage** • Supplemental Feeding • Behavior • Rarity • Photo Account • A. heliaca • Spain
3448	**Plumage** • Collect • Raptors • A. chrysaetos • A. rapax • Ethiopia
73	**Poetry** • Folklore • Historical Account • Symbolism • Eagles • A. chrysaetos
276	**Poetry** • A. chrysaetos • North America
1259	**Poetry** • Captivity • Attitude • Protection • Flight • Folklore • Raptors • A. chrysaetos • Great Britain
2629	**Poetry** • Shooting • A. chrysaetos • Ontario
151	**Poisoning** • Mortality • Wintering Population • Conservation • Attitude • Punishment • A. chrysaetos • Wyoming
155	**Poisoning** • Banding • Raptors • A. audax • Australia
167	**Poisoning** • A. chrysaetos • Quebec
294	**Poisoning** • Animal Damage • Shooting • Captivity • Lift • A. audax • Australia
313	**Poisoning** • Toxicology • Autopsy • Captivity • Residues • Thal. Sulfate • A. chrysaetos
382	**Poisoning** • A. chrysaetos • West Germany
469	**Poisoning** • Status • Raptors • A. audax • Australia
548	**Poisoning** • Status • Habitat Loss • Conservation • A. pomarina • A. clanga • A. heliaca • East Africa • Ethiopia
602	**Poisoning** • Status • Raptors • A. chrysaetos • Great Britain
605	**Poisoning** • Animal Damage • Hunting Behavior • A. audax • Australia
613	**Poisoning** • Protection • Mortality • Eagles • A. chrysaetos • Wyoming
616	**Poisoning** • Nest • Habitat • Egg • Prey • Vocalization • Growth • Trapping • Behavior • Lift • Photo Account • A. chrysaetos • Montana
618	**Poisoning** • Prey • Nest • Attitude • Behavior • Breeding Biology • Flight • Molt • Hunting Behavior • Vocalization • Folklore • Lift • Photo Account • A. chrysaetos • Montana
633	**Poisoning** • Nest • A. audax • Australia
690	**Poisoning** • Mortality • (Title Only) • A. chrysaetos • Wyoming
786	**Poisoning** • Shooting • Raptors • A. chrysaetos • Yugoslavia
819	**Poisoning** • Mortality • Animal Damage • Shooting • Eagles • A. chrysaetos • Ohio
1013	**Poisoning** • Rarity • Captivity • Measure • Plumage • Photo Account • A. heliaca • Italy
1044	**Poisoning** • Animal Damage • A. chrysaetos • Alabama
1089	**Poisoning** • Animal Damage • A. audax • Australia
1103	**Poisoning** • Status • Hunting Behavior • Prey • Raptors • A. chrysaetos • California
1431	**Poisoning** • Telemetry • Toxicology • Habitat • Raptors • Cmpd. 1080 • A. chrysaetos • California
1507	**Poisoning** • Conservation • Protection • Economics • Mortality • Reward • Pesticides • Eagles • A. chrysaetos • Wyoming • Colorado
1608	**Poisoning** • Conservation • A. chrysaetos • Scotland
1625	**Poisoning** • Status • Shooting • Prey • Refuge • Educational Material • Protection • Raptors • A. chrysaetos • Spain
1643	**Poisoning** • Egg Collection • Measure • Raptors • A. audax • Australia
1807	**Poisoning** • Status • Raptors • A. chrysaetos • Yugoslavia
1810	**Poisoning** • Prey • Hunting Behavior • Population Decline • A. chrysaetos • Yugoslavia
1904	**Poisoning** • Status • Conservation • A. chrysaetos • Israel
1923	**Poisoning** • Thallium • A. chrysaetos • California
1973	**Poisoning** • Winter Survey • Roost • Refuge • Photo Account • Popular Monograph • A. chrysaetos • Wyoming
1995	**Poisoning** • A. chrysaetos • Ontario
2049	**Poisoning** • Status • Shooting • Pesticides • Migration • Protection • A. chrysaetos • Siberia
2091	**Poisoning** • Animal Damage • Hunting Behavior • A. audax • Australia
2125	**Poisoning** • Pesticides • Status • Captivity • Density • Food Requirements • Habitat • Protection • Predator/Prey • Raptors • Photography Account • PCB • A. clanga • A. heliaca • A. nipalensis • A. verreauxi • A. pomarina • Israel
2129	**Poisoning** • Rehabilitation • Raptors • A. clanga • A. pomarina • A. heliaca • Israel
2645	**Poisoning** • Conservation • Shooting • Egg Collection • Nest Robbing • Protection • Trade • Trapping • Raptors • A. chrysaetos • Great Britain
2656	**Poisoning** • Winter Survey • Conservation • Shooting • Protection • Reward • Photo Account • Eagles • Popular Monograph • A. chrysaetos • Wyoming
2939	**Poisoning** • Punishment • Eagles • A. chrysaetos • Wyoming
3191	**Poisoning** • Conservation • Shooting • Electrocution • Attitude • Eagles • A. chrysaetos • United States
3192	**Poisoning** • Protection • Shooting • Electrocution • Attitude • Popular Monograph • A. chrysaetos • Wyoming
3299	**Poisoning** • Toxicology • Gophacide • Raptors • A. chrysaetos • United States
3310	**Poisoning** • Shooting • Bounty • Conservation • Punishment • Story • Eagles • Popular Monograph • A. chrysaetos • Wyoming
3317	**Poisoning** • Raptors • A. chrysaetos • Scotland
3318	**Poisoning** • A. chrysaetos • Scotland
119	**Popular Monograph** • Photo Account • A. chrysaetos • Great Britain
120	**Popular Monograph** • Eagles • Photo Account • A. audax
1114	**Popular Monograph** • Raptors • A. chrysaetos • A. clanga • United Soviet Socialists Republic
1355	**Popular Monograph** • Refuge • Habitat • A. chrysaetos • Idaho • Switzerland
2371	**Popular Monograph** • Story • Breeding Biology • Behavior • Habitat • Nest Survey • Hunting Behavior • Prey • Raptors • Illus. • A. chrysaetos • Colorado
2714	**Popular Monograph** • Shooting • Prey • Interspecific Behavior • Hunting Behavior • Eagles • A. chrysaetos • North America
3208	**Popular Monograph** • Eagles • A. chrysaetos • United States
47	**Population Decline** • Historical Account • Status • Management • Distribution • Raptors • A. chrysaetos • Tennessee

Master No.	Keywords
386	**Population Decline** • Bounty • Disturbance • Habitat Loss • Protection • Shooting • Status • Pesticides • Migration • Distribution • Raptors • A. chrysaetos • A. heliaca • A. pomarina • A. clanga • Europe
527	**Population Decline** • Migration • Shooting • Conservation • Historical Account • Hawk Mt. • Raptors • Popular Monograph • A. chrysaetos • Pennsylvania
632	**Population Decline** • Shooting • A. chrysaetos • Ireland
705	**Population Decline** • Art • Conservation • Refuge • A. heliaca • Spain
838	**Population Decline** • Historical Account • Status • Nest • Prey • A. chrysaetos • Ireland
1020	**Population Decline** • Status • Habitat • Mortality • Raptors • A. chrysaetos • Europe
1136	**Population Decline** • Status • A. heliaca • A. chrysaetos • A. nipalensis • A. clanga • A. pomarina • Moldavia
1348	**Population Decline** • Raptors • A. chrysaetos • Norway
1589	**Population Decline** • Status • Distribution • Conservation • Raptors • Popular Monograph • A. clanga • A. heliaca • Europe
1699	**Population Decline** • A. chrysaetos • Ireland
1810	**Population Decline** • Prey • Hunting Behavior • Poisoning • A. chrysaetos • Yugoslavia
1820	**Population Decline** • Conservation • Protection • Raptors • A. chrysaetos • A. clanga • A. pomarina • West Germany
1983	**Population Decline** • Status • A. nipalensis • Ukraine
2012	**Population Decline** • Status • Attitude • Raptors • A. chrysaetos • A. clanga • Russia
2136	**Population Decline** • Descriptive Summary • Distribution • Status • Raptors • A. chrysaetos • A. clanga • A. pomarina • Finland
2183	**Population Decline** • Nest Survey • Habitat • Status • Raptors • A. heliaca • Morocco
2283	**Population Decline** • Nest Survey • Prey • Nest Chronology • A. chrysaetos • France
2316	**Population Decline** • Status • Banding • Nest Survey • Disturbance • Attitude • A. chrysaetos • Europe
2682	**Population Decline** • Migration • Shooting • Raptors • A. chrysaetos • Sweden
2694	**Population Decline** • Prey • Raptors • A. nipalensis • Kazakhstan
2917	**Population Decline** • Migration • Hawk Mt. • A. chrysaetos • Pennsylvania
2918	**Population Decline** • Nest Survey • Photo Account • A. chrysaetos • North America
3027	**Population Decline** • Disturbance • Habitat Loss • A. nipalensis • Caucasus
3102	**Population Decline** • Mortality • Abandonment • Raptors • A. chrysaetos • France
3226	**Population Decline** • Density • Reproductive Success • Sub-Adult • Plumage • Museum Specimen • Illus. • A. heliaca • Spain
3372	**Population Decline** • Protection • Mortality • Conservation • Animal Damage • Bounty • A. chrysaetos • Norway
536	**Population Dynamics** • Prey • Photo Account • Popular Monograph • A. chrysaetos • Scotland
1146	**Population Dynamics** • Theory • A. verreauxi • Rhodesia
1151	**Population Dynamics** • Reproductive Success • Density • Prey • Nest • Climatology • Territory • Interspecific Competition • Mortality • Clutch Size • Pesticides • Eagles • A. verreauxi • Rhodesia
1294	**Population Dynamics** • Nest Survey • Habitat • Raptors • Tech. • A. chrysaetos • North Dakota
1773	**Population Dynamics** • Winter Survey • Distribution • Prey • Age Structure • Wintering Population • Eagles • A. chrysaetos • Washington
1778	**Population Dynamics** • Nest Survey • Nest Chronology • Prey • Reproductive Success • Abandonment • Marking • Roadside Survey • Territory • Raptors • SRBPNA • A. chrysaetos • Idaho
1779	**Population Dynamics** • Reproductive Success • Prey • Aerial Survey • Banding • Marking • Raptors • SRBPNA • A. chrysaetos • Idaho
1780	**Population Dynamics** • Nest Survey • Nest Chronology • Prey • Reproductive Success • Abandonment • Marking • Roadside Survey • Territory • Raptors • SRBPNA • A. chrysaetos • Idaho
1781	**Population Dynamics** • Nest Survey • Reproductive Success • Marking • Habitat • Nest Chronology • Behavior • Prey • Banding • Territory • Raptors • SRBPNA • A. chrysaetos • Idaho
1782	**Population Dynamics** • Marking • Reproductive Success • Prey • Aerial Survey • Banding • SRBPNA • Raptors • A. chrysaetos • Idaho
2261	**Population Dynamics** • Nest Survey • Disturbance • Prey • Management • Status • Animal Damage • Artificial Nest • Predator/Prey • A. chrysaetos • Utah
2400	**Population Dynamics** • Animal Damage • Predator/Prey • Aerial Survey • Habitat • Status • Density • Food Habits • Nest • Interspecific Competition • Literature Review • A. chrysaetos • Texas • New Mexico
2854	**Population Dynamics** • Nest • Density • Prey • Breeding Biology • Distribution • Habitat • Directional Exposure • Reproductive Success • Mortality • Abandonment • Disturbance • Interspecific Behavior • Territory • A. chrysaetos • Utah
3146	**Population Dynamics** • Status • Nest Survey • Prey • A. chrysaetos • Idaho
358	**Population Estimate** • Status • Management • Protection • Artificial Nest • Raptors • A. chrysaetos • A. clanga • Finland
538	**Population Estimate** • Nest Survey • Status • Reproductive Success • Prey • Food Requirements • A. chrysaetos • Scotland
1164	**Population Estimate** • Status • Pesticides • Conservation • Disturbance • Raptors • A. heliaca • A. chrysaetos • Spain
1885	**Population Estimate** • Distribution • Raptors • A. chrysaetos • Colorado
2260	**Population Estimate** • Status • Eagles • A. chrysaetos • North America
2311	**Population Estimate** • Attitude • Status • Mortality • Protection • Prey • Animal Damage • Disturbance • Raptors • A. chrysaetos • Great Britain

Master No.	Keywords
2873	**Population Estimate** • Literature Review • Life History • Habitat • Management • A. chrysaetos • United States
3419	**Population Estimate** • Winter Survey • Aerial Survey • Eagles • A. chrysaetos • Wyoming
269	**Predator/Prey** • (Title Only) • A. chrysaetos • New Mexico
414	**Predator/Prey** • Niche • Raptors • A. chrysaetos • Europe
435	**Predator/Prey** • Wintering Population • Raptors • A. chrysaetos • Spain
552	**Predator/Prey** • Nest Survey • Prey • Habitat • Behavior • Status • Reproductive Success • Food Requirements • Animal Damage • Intraspecific Behavior • A. chrysaetos • Scotland
663	**Predator/Prey** • Raptors • A. nipalensis • United Soviet Socialists Republic
714	**Predator/Prey** • Food Requirements • Breeding Biology • Cain/Abel • Directional Exposure • Growth • Sex Ratio • Behavior • A. chrysaetos • Idaho
749	**Predator/Prey** • (Title Only) • A. chrysaetos • Montana
1069	**Predator/Prey** • Raptors • Rodents • A. nipalensis • United Soviet Socialists Republic
1117	**Predator/Prey** • Density • Prey • A. clanga • Russia
1121	**Predator/Prey** • Raptors • A. clanga • United Soviet Socialists Republic
1122	**Predator/Prey** • Raptors • A. chrysaetos • United Soviet Socialists Republic
1130	**Predator/Prey** • Raptors • A. clanga • Russia
1199	**Predator/Prey** • Raptors • A. heliaca • Kazakhstan
1233	**Predator/Prey** • Raptors • A. pomarina • Belorussia
1237	**Predator/Prey** • Prey • Raptors • A. pomarina • Belorussia
1508	**Predator/Prey** • Literature Review • Prey • Pellet • Shooting • Nest Survey • Attitude • Reproductive Success • A. chrysaetos • North America
1517	**Predator/Prey** • Interspecific Behavior • Hunting Behavior • Fox • A. chrysaetos • Alaska
1656	**Predator/Prey** • A. chrysaetos • North America
1659	**Predator/Prey** • Animal Damage • A. chrysaetos • Scotland
1739	**Predator/Prey** • A. chrysaetos • New Mexico
1751	**Predator/Prey** • Raptors • A. nipalensis • Kazakhstan
1772	**Predator/Prey** • Prey • Food Requirements • Nest • A. chrysaetos • Washington
1776	**Predator/Prey** • Nest Survey • Refuge • Winter Survey • Density • Electrocution • Shooting • Mortality • Prey • Raptors • SRBPNA • A. chrysaetos • Idaho
1777	**Predator/Prey** • Electrocution • Aerial Survey • A. chrysaetos • North America
1788	**Predator/Prey** • Raptors • A. chrysaetos • North America
1896	**Predator/Prey** • Hunting Behavior • A. chrysaetos • Colorado
1921	**Predator/Prey** • Breeding Biology • Density • A. nipalensis • Russia
2125	**Predator/Prey** • Poisoning • Pesticides • Status • Captivity • Density • Food Requirements • Habitat • Protection • Raptors • Photography Account • PCB • A. clanga • A. heliaca • A. nipalensis • A. verreauxi • A. pomarina • Israel
2255	**Predator/Prey** • Electrocution • Density • Nest • Prey • A. nipalensis • Russia
2261	**Predator/Prey** • Nest Survey • Disturbance • Prey • Population Dynamics • Management • Status • Animal Damage • Artificial Nest • A. chrysaetos • Utah
2277	**Predator/Prey** • Nest Adornment • Raptors • A. chrysaetos • Norway
2349	**Predator/Prey** • Feeding Behavior • Animal Damage • Capture • Protection • Management • Economics • A. chrysaetos • Montana
2400	**Predator/Prey** • Animal Damage • Aerial Survey • Habitat • Status • Density • Population Dynamics • Food Habits • Nest • Interspecific Competition • Literature Review • A. chrysaetos • Texas • New Mexico
2407	**Predator/Prey** • Prey • Photo Account • A. pomarina • Czechoslovakia
2448	**Predator/Prey** • A. heliaca • Kazakhstan
2449	**Predator/Prey** • Raptors • A. heliaca • Kazakhstan
2793	**Predator/Prey** • A. nipalensis • Caucasus
2806	**Predator/Prey** • Hunting Behavior • Shooting • A. chrysaetos • Montana
3130	**Predator/Prey** • Roadside Survey • Habitat • A. rapax • A. nipalensis • A. pomarina • A. wahlbergi • Uganda
3135	**Predator/Prey** • Interspecific Behavior • Migration • Raptors • A. chrysaetos • A. heliaca • A. nipalensis • A. pomarina • A. clanga • Nepal
3198	**Predator/Prey** • Density • A. nipalensis • Kazakhstan
3207	**Predator/Prey** • Refuge • Habitat • Nest Survey • Density • Prey • Disturbance • Telemetry • Territory • Reproductive Success • Raptors • SRBPNA • Photo Account • A. chrysaetos • Idaho
3438	**Predator/Prey** • Nest • A. chrysaetos • Oregon
3443	**Predator/Prey** • Raptors • A. heliaca • A. nipalensis • Turkmenia
8	**Prey** • Density • Nest Failure • Reproductive Success • A. nipalensis • Russia
20	**Prey** • A. chrysaetos • Kansas
22	**Prey** • Hunting Behavior • Animal Damage • Talon Wounds • A. chrysaetos • Texas
24	**Prey** • Descriptive Summary • Field Id • Raptors • A. chrysaetos • A. heliaca • India
28	**Prey** • Collect • A. chrysaetos • France
29	**Prey** • Collect • A. chrysaetos • France
41	**Prey** • Animal Damage • A. chrysaetos • Nevada
54	**Prey** • Museum Specimen • Plumage • Distribution • Measure • A. heliaca • A. clanga • A. rapax • India
57	**Prey** • Collect • Behavior • A. heliaca • India
58	**Prey** • Nest • Clutch Size • Habitat • Collect • A. pomarina • India
67	**Prey** • Distribution • Taxonomy • Egg • Plumage • Raptors • A. verreauxi • A. rapax • Southwest Africa
71	**Prey** • Shooting • A. chrysaetos • Scotland
72	**Prey** • Rarity • A. chrysaetos • Great Britain

Master No.	Keywords
82	**Prey** • Animal Damage • A. chrysaetos • Great Britain
86	**Prey** • Historical Account • Flight • Status • Attitude • Photo Account • A. chrysaetos • Great Britain
129	**Prey** • Shooting • A. chrysaetos • Virginia
190	**Prey** • Raptors • A. heliaca • Spain
192	**Prey** • Nest • Status • Raptors • A. chrysaetos • A. heliaca • Spain
197	**Prey** • Nest • Behavior • Captivity • A. chrysaetos • Algeria
198	**Prey** • Monograph • Nest Survey • Habitat • Reproductive Success • Literature Review • Disturbance • Pellet • Animal Damage • A. chrysaetos • Utah
207	**Prey** • Hunting Behavior • A. chrysaetos • California
209	**Prey** • Raptors • A. audax • Australia
218	**Prey** • Descriptive Summary • Rarity • Distribution • Etymology • Raptors • A. chrysaetos • A. heliaca • Japan
222	**Prey** • Collect • Raptors • A. rapax • South Africa
224	**Prey** • Collect • Museum Specimen • Raptors • A. wahlbergi • A. verreauxi • South Africa
225	**Prey** • Collect • Raptors • A. rapax • South Africa
229	**Prey** • Distribution • Status • Nest • A. audax • Australia
237	**Prey** • Nest • A. chrysaetos • Alaska
244	**Prey** • Nest Survey • Habitat • Artificial Nest • Raptors • A. chrysaetos • California
246	**Prey** • Attitude • Protection • A. chrysaetos • Ohio
249	**Prey** • Status • Nest • A. chrysaetos • Algeria • Morocco • Tunisia
261	**Prey** • Id Key • Descriptive Summary • Distribution • Etymology • A. rapax • A. wahlbergi • Gambia • Sierra Leone • Nigeria
265	**Prey** • Nest • Attitude • Raptors • A. pomarina • United Soviet Socialists Republic
287	**Prey** • Behavior • Nest • Photo Account • Raptors • A. chrysaetos • Norway
288	**Prey** • A. chrysaetos • Norway
302	**Prey** • Fostering • Nest • Habitat • Raptors • A. chrysaetos • Bulgaria
310	**Prey** • A. chrysaetos • Utah
317	**Prey** • Nest • A. heliaca • Hungary
325	**Prey** • Monograph • Breeding Biology • Density • Mortality • Cain/Abel • Abnormal Behavior • Egg Measure • Marking • A. chrysaetos • Idaho
327	**Prey** • Nest • Raptors • A. chrysaetos • Utah
332	**Prey** • Field Id • Attitude • Raptors • Popular Monograph • A. audax • Australia
335	**Prey** • Distribution • Nest • Captivity • Vocalization • Raptors • A. chrysaetos • Oregon
336	**Prey** • Nest • A. chrysaetos • Ireland
347	**Prey** • A. verreauxi • Zambia
357	**Prey** • Raptors • A. chrysaetos • Europe
362	**Prey** • Nest • Measure • A. audax • Australia
363	**Prey** • Nest • Clutch Size • A. audax • Australia
365	**Prey** • Lift • Raptors • A. chrysaetos • Norway
366	**Prey** • Hunting Behavior • Raptors • A. chrysaetos • A. heliaca • Spain
377	**Prey** • A. chrysaetos • Czechoslovakia
380	**Prey** • Distribution • Raptors • A. chrysaetos • India
392	**Prey** • Wintering Population • Electrocution • A. chrysaetos • Denmark
401	**Prey** • Nesting Behavior • Photo Account • Popular Monograph • A. chrysaetos • Scotland
411	**Prey** • (Title Only) • A. chrysaetos • Scotland
416	**Prey** • A. chrysaetos • Alabama
427	**Prey** • Aerial Survey • Age Structure • Nest Survey • Banding • Wintering Population • A. chrysaetos • North America
431	**Prey** • Collect • Raptors • A. chrysaetos • Italy
441	**Prey** • Animal Damage • A. chrysaetos • Sweden
454	**Prey** • A. chrysaetos • Washington
466	**Prey** • Distribution • Attitude • Egg Collection • Raptors • A. audax • Australia
468	**Prey** • Nest Chronology • Nest • Egg • A. nipalensis • Russia
471	**Prey** • Attitude • Raptors • A. chrysaetos • United States
475	**Prey** • Status • Distribution • Historical Account • Hunting Behavior • Raptors • A. chrysaetos • Maine
476	**Prey** • Descriptive Summary • Nest • Raptors • A. chrysaetos • Maine
489	**Prey** • Distribution • Nest • Measure • Raptors • A. chrysaetos • Sardinia
495	**Prey** • Wintering Population • Distribution • Hunting Behavior • Age Structure • A. nipalensis • Africa
500	**Prey** • Distribution • Raptors • A. audax • Australia
501	**Prey** • Pellet • Attack • Hunting Behavior • Animal Damage • Habitat • Cache • A. audax • Australia
534	**Prey** • Habitat • Density • Nest Survey • Breeding Biology • Interspecific Competition • Reproductive Success • Etymology • Plumage • Vocalization • Nest Adornment • Raptors • A. verreauxi • A. wahlbergi • Kenya
535	**Prey** • Nest Survey • Breeding Biology • Etymology • Reproductive Success • Interspecific Competition • Raptors • Photo Account • A. wahlbergi • A. verreauxi • Kenya
536	**Prey** • Population Dynamics • Photo Account • Popular Monograph • A. chrysaetos • Scotland
537	**Prey** • Nest • Interspecific Behavior • Habitat • Raptors • A. verreauxi • A. wahlbergi • Kenya
538	**Prey** • Nest Survey • Status • Population Estimate • Reproductive Success • Food Requirements • A. chrysaetos • Scotland
552	**Prey** • Nest Survey • Habitat • Behavior • Status • Reproductive Success • Food Requirements • Animal Damage • Intraspecific Behavior • Predator/Prey • A. chrysaetos • Scotland
554	**Prey** • A. chrysaetos • California
558	**Prey** • Migration • Raptors • A. clanga • A. pomarina • Poland • South Africa
565	**Prey** • Hunting Behavior • Prey Behavior • A. chrysaetos • Alberta
568	**Prey** • Nest • Conservation • Photo Account • Popular Monograph • A. chrysaetos • South Dakota
569	**Prey** • Behavior • A. chrysaetos • California
575	**Prey** • Collect • Raptors • A. heliaca • Cyprus
577	**Prey** • A. chrysaetos • South Dakota
587	**Prey** • Nest • Behavior • Growth • Vocalization • Nest Adornment • Photo Account • Illus. • A. chrysaetos • Switzerland
594	**Prey** • Status • Raptors • A. rapax • Sudan

Master No.	Keywords
606	Prey • Animal Damage • Economics • Attitude • Conservation • Raptors • A. chrysaetos • North America
608	Prey • Interspecific Behavior • Eagles • A. audax • Australia
616	Prey • Nest • Habitat • Egg • Vocalization • Growth • Trapping • Poisoning • Behavior • Lift • Photo Account • A. chrysaetos • Montana
618	Prey • Nest • Attitude • Behavior • Breeding Biology • Flight • Poisoning • Molt • Hunting Behavior • Vocalization • Folklore • Lift • Photo Account • A. chrysaetos • Montana
628	Prey • Hunting Behavior • Nest • Habitat • Photography Account • A. chrysaetos • California
635	Prey • Nest • Egg • Measure • A. audax • Australia
660	Prey • Status • Raptors • A. heliaca • A. clanga • Greece
665	Prey • Raptor • A. chrysaetos • A. clanga • A. pomarina • Hungary
672	Prey • Status • Nest • Raptors • A. chrysaetos • France
673	Prey • Interspecific Competition • Raptors • A. chrysaetos • France
711	Prey • Interspecific Behavior • Lift • Raptors • A. chrysaetos • Scotland
716	Prey • Hunting Behavior • A. chrysaetos • Oregon
725	Prey • Nest • Directional Exposure • A. chrysaetos • California
739	Prey • Lift • A. rapax • Africa
741	Prey • Nest Survey • Pesticides • Reproductive Success • A. chrysaetos • Scotland
755	Prey • Nest • Raptors • A. audax • Australia
756	Prey • Nest Survey • Reproductive Success • Raptors • A. chrysaetos • Colorado
770	Prey • Field Id • Descriptive Summary • Monograph • Distribution • Status • Migration • Plumage • Measure • Etymology • Vocalization • A. pomarina • A. clanga • A. nipalensis • A. heliaca • A. chrysaetos • A. verreauxi • Africa • Europe
773	Prey • Collect • Nest • Raptors • A. heliaca • A. pomarina • India
780	Prey • Collect • Plumage • Raptors • A. heliaca • A. clanga • Hungary
783	Prey • Protection • Conservation • Animal Damage • Hunting Behavior • Prey Behavior • Raptors • A. chrysaetos • Sweden
784	Prey • Conservation • Animal Damage • Attitude • Raptors • A. heliaca • A. audax
789	Prey • Monograph • Habitat • Nest Survey • Photo Account • A. chrysaetos • Colorado
791	Prey • Attitude • Raptors • A. chrysaetos • A. pomarina • France
794	Prey • Animal Damage • Shooting • Bounty • Eagles • A. chrysaetos • California
838	Prey • Historical Account • Status • Population Decline • Nest • A. chrysaetos • Ireland
851	Prey • Clutch Size • Nest • A. chrysaetos • Spain
852	Prey • Clutch Size • Mortality • Nestling • Pesticides • Cain/Abel • A. chrysaetos • Spain
853	Prey • Hunting Behavior • A. chrysaetos • Spain
855	Prey • Lift • A. chrysaetos • India
856	Prey • Nest • Egg • Shooting • A. chrysaetos • Austria
863	Prey • Falconry • Raptors • A. chrysaetos • Kirgizia
871	Prey • Distribution • Etymology • Life History • Habitat • Molt • Measure • Field Id • Migration • Status • Raptors • A. chrysaetos • A. heliaca • A. rapax • A. clanga • A. pomarina • United Soviet Socialists Republic
883	Prey • Descriptive Summary • Etymology • Behavior • Display • A. heliaca • A. rapax • A. pomarina • A. clanga • India
884	Prey • Raptors • A. chrysaetos • Italy
894	Prey • Intraspecific Behavior • Nest • Habitat • Egg • Measure • Photo Account • A. chrysaetos • California
896	Prey • Raptors • A. wahlbergi • A. rapax • South Africa
898	Prey • Hunting Behavior • A. chrysaetos • Alaska
900	Prey • Descriptive Summary • Hunting Behavior • Nest • Interspecific Behavior • Mortality • Vocalization • Raptors • Mt. Mckinley • A. chrysaetos • Alaska
907	Prey • Descriptive Summary • Field Id • Plumage • Measure • Raptors • A. chrysaetos • A. heliaca • A. nipalensis • A. rapax • A. clanga • India
936	Prey • Collect • A. chrysaetos • Kansas
939	Prey • Nest • Habitat • Field Id • Hunting Behavior • Raptors • A. clanga • A. pomarina • A. chrysaetos • Poland
940	Prey • Nest • Habitat Loss • Habitat • Raptors • A. chrysaetos • India
941	Prey • Descriptive Summary • Status • Distribution • Management • Literature Review • A. chrysaetos • Washington
943	Prey • Distribution • Raptors • A. audax • Australia
944	Prey • Nest • Nest Adornment • Behavior • Growth • A. audax • Australia
947	Prey • Roadside Survey • Roost • Nest Survey • Hunting Behavior • Capture • Marking • Interspecific Behavior • Shooting • Electrocution • A. chrysaetos • Utah
949	Prey • Animal Damage • Conservation • Raptors • A. chrysaetos • Scotland
962	Prey • Parental Behavior • A. audax • Australia
963	Prey • Growth • Nest • Plumage • Vocalization • Behavior • Photo Account • A. audax • Australia
964	Prey • Raptors • A. chrysaetos • Spain
970	Prey • Collect • Status • A. rapax • India
1008	Prey • Descriptive Summary • Etymology • Status • Nest Chronology • Raptors • A. chrysaetos • A. heliaca • A. pomarina • A. clanga • Switzerland
1025	Prey • Hunting Behavior • Nest • Density • Habitat • Photo Account • Popular Monograph • A. heliaca • Spain
1031	Prey • Residues • Egg • Raptors • Mercury • A. chrysaetos • Alberta • Saskatchewan
1033	Prey • Habitat • Nest • Flight • Plumage • Raptors • A. wahlbergi • A. rapax • South Africa
1034	Prey • Nest • Egg • Measure • Raptors • A. verreauxi • South Africa
1037	Prey • Nest • Photography • Growth • A. chrysaetos • California
1040	Prey • Monograph • Breeding Biology • Vocalization • Territory • Taxonomy • Migration • Mortality • Molt • Falconry • A. chrysaetos • A. verreauxi • A. audax

Master No.	Keywords
1041	**Prey** • Status • Nest • Habitat • Behavior • Photo Account • A. chrysaetos • Bulgaria
1042	**Prey** • Descriptive Summary • Measure • Hunting Behavior • Animal Damage • Raptors • A. chrysaetos • North America
1052	**Prey** • Interspecific Behavior • Raptors • A. chrysaetos • Washington
1056	**Prey** • Growth • Falconry • Aging • Parental Behavior • Measure • Plumage • Dimorphism • Photo Account • A. audax • Australia
1062	**Prey** • Interspecific Competition • A. chrysaetos • Alberta
1066	**Prey** • Hunting Behavior • A. chrysaetos • California • Nevada
1079	**Prey** • Animal Damage • Eagles • Popular Monograph • A. audax • Australia
1091	**Prey** • Nest • Growth • Photo Account • A. chrysaetos • Switzerland
1103	**Prey** • Status • Hunting Behavior • Poisoning • Raptors • A. chrysaetos • California
1111	**Prey** • A. chrysaetos • Washington
1117	**Prey** • Predator/Prey • Density • A. clanga • Russia
1145	**Prey** • Nest • Parental Behavior • Breeding Biology • Interspecific Behavior • Intraspecific Behavior • Feeding Behavior • Hunting Behavior • Fledging • Display • Raptors • A. verreauxi • Rhodesia
1147	**Prey** • Nest Survey • Photo Account • Popular Monograph • A. verreauxi • Rhodesia
1151	**Prey** • Reproductive Success • Density • Population Dynamics • Nest • Climatology • Territory • Interspecific Competition • Mortality • Clutch Size • Pesticides • Eagles • A. verreauxi • Rhodesia
1154	**Prey** • Nest • Density • Status • Habitat Protection • Habitat • Reproductive Success • A. verreauxi • Rhodesia
1156	**Prey** • Breeding Biology • Habitat • Density • Nest Survey • A. verreauxi • Rhodesia
1163	**Prey** • Nest • Behavior • Clutch Size • A. heliaca • Spain
1167	**Prey** • Nest • Photography • Behavior • Photo Account • A. audax • Australia
1168	**Prey** • Animal Damage • Hunting Behavior • A. audax • Australia
1170	**Prey** • Anatomy • Behavior • Raptors • A. chrysaetos • United Soviet Socialists Republic
1171	**Prey** • Nest • Intraspecific Behavior • A. pomarina • Armenia
1174	**Prey** • Vocalization • Nest • Behavior • Photo Account • Popular Monograph • A. pomarina • West Germany
1178	**Prey** • A. chrysaetos • Africa
1182	**Prey** • Museum Specimen • Collect • Raptors • A. wahlbergi • Cambodia
1189	**Prey** • Nest • Habitat • Raptors • A. chrysaetos • France
1203	**Prey** • Lift • A. chrysaetos • Great Britain
1204	**Prey** • Lift • Animal Damage • A. chrysaetos • Great Britain
1205	**Prey** • Nesting Behavior • Status • Photo Account • A. chrysaetos • Scotland
1208	**Prey** • Lift • A. chrysaetos • Scotland
1212	**Prey** • Descriptive Summary • Taxonomy • Etymology • Nest • A. chrysaetos • Switzerland
1220	**Prey** • Breeding Biology • A. clanga • United Soviet Socialists Republic • Siberia
1222	**Prey** • Animal Damage • Raptors • A. chrysaetos • Kansas
1228	**Prey** • Monograph • Vocalization • Distribution • Plumage • Id Key • A. pomarina • A. chrysaetos • A. nipalensis • Europe
1234	**Prey** • Nest • Raptors • A. pomarina • Belorussia
1235	**Prey** • Breeding Biology • A. pomarina • Belorussia
1236	**Prey** • Breeding Biology • A. pomarina • Belorussia
1237	**Prey** • Predator/Prey • Raptors • A. pomarina • Belorussia
1238	**Prey** • Interspecific Competition • Raptors • A. pomarina • Belorussia
1239	**Prey** • Raptors • A. clanga • Russia
1243	**Prey** • Hunting Behavior • Prey Behavior • A. chrysaetos • Wyoming
1260	**Prey** • Lift • A. chrysaetos • Great Britain
1261	**Prey** • Nest • Behavior • Photo Account • Popular Monograph • A. chrysaetos • Scotland
1273	**Prey** • Mortality • Falconry • Hunting Behavior • Flight • Nest • Mortality • Photo Account • Popular Monograph • A. chrysaetos • Great Britain
1281	**Prey** • Folklore • Behavior • A. chrysaetos • Texas
1296	**Prey** • Trapping • Status • Rarity • A. chrysaetos • Georgia
1298	**Prey** • Hunting Behavior • A. chrysaetos • Montana
1309	**Prey** • Clutch Size • Nest • A. nipalensis • Russia
1345	**Prey** • Rarity • Shooting • Plumage • Measure • A. chrysaetos • West Germany
1357	**Prey** • A. chrysaetos • Washington
1363	**Prey** • Field Id • Migration • Descriptive Summary • Status • Raptors • A. chrysaetos • Wisconsin
1365	**Prey** • Animal Damage • A. chrysaetos • North America
1386	**Prey** • Falconry • Photography • Photo Account • Popular Monograph • A. chrysaetos • Scotland
1387	**Prey** • Time Lapse • Telemetry • A. chrysaetos • Texas
1395	**Prey** • Status • Etymology • Nest • Raptors • A. chrysaetos • A. heliaca • A. rapax • Morocco
1400	**Prey** • Prey Behavior • Raptors • A. chrysaetos • Montana
1404	**Prey** • Hunting Behavior • A. chrysaetos • Manitoba
1406	**Prey** • Nest • A. chrysaetos • Alaska
1413	**Prey** • Lift • A. chrysaetos • Switzerland
1420	**Prey** • Nest • A. chrysaetos • Scotland
1421	**Prey** • Bounty • Popular Monograph • A. audax • Australia
1425	**Prey** • Captivity • A. chrysaetos • Scotland
1427	**Prey** • Behavior • Shooting • Measure • A. chrysaetos • Georgia
1429	**Prey** • Raptors • A. wahlbergi • A. rapax • South Africa
1460	**Prey** • Wintering Population • Raptors • A. chrysaetos • North Dakota
1463	**Prey** • A. chrysaetos • Wyoming
1464	**Prey** • Nest • A. chrysaetos • Spain
1472	**Prey** • A. chrysaetos • Switzerland
1473	**Prey** • Protection • Animal Damage • A. chrysaetos • Switzerland
1499	**Prey** • Banding • A. chrysaetos • Oregon

Master No.	Keywords
1506	**Prey** • Descriptive Summary • Protection • Nest • Raptors • Illus. • A. audax • Australia
1508	**Prey** • Literature Review • Pellet • Shooting • Predator/Prey • Nest Survey • Attitude • Reproductive Success • A. chrysaetos • North America
1513	**Prey** • Nest Failure • Habitat • Behavior • A. chrysaetos • Alaska
1516	**Prey** • Nest • Parasite • A. audax • Australia
1548	**Prey** • Habitat • Nest • A. chrysaetos • Alaska
1556	**Prey** • Status • Raptors • A. audax • New Guinea
1558	**Prey** • Mortality • Wintering Population • Raptors • A. chrysaetos • Alberta
1562	**Prey** • Nest • Photography • Photo Account • A. chrysaetos • Scotland
1597	**Prey** • Collect • Behavior • Habitat • Raptors • A. nipalensis • A. clanga • A. pomarina • India
1598	**Prey** • Measure • Raptors • A. chrysaetos • A. heliaca • A. clanga • A. rapax • India
1599	**Prey** • Taxonomy • A. nipalensis • A. clanga • Burma
1619	**Prey** • Raptors • A. chrysaetos • Scotland
1620	**Prey** • Status • Raptors • A. rapax • A. clanga • India
1625	**Prey** • Status • Shooting • Poisoning • Refuge • Educational Material • Protection • Raptors • A. chrysaetos • Spain
1629	**Prey** • Nest • Parental Behavior • A. chrysaetos • Switzerland
1630	**Prey** • Nest • A. chrysaetos • Switzerland
1631	**Prey** • Nest • Parental Behavior • A. nipalensis • Kazakhstan
1635	**Prey** • Density • Nest • Conservation • Intraspecific Behavior • A. chrysaetos • Belorussia
1636	**Prey** • Collect • A. chrysaetos • West Germany
1639	**Prey** • Taxonomy • Dimorphism • Measure • Plumage • Distribution • Behavior • Management • Habitat • Territory • Mortality • Life History • Economics • Literature Review • A. chrysaetos • Oregon • Washington
1668	**Prey** • Status • Etymology • Nest • Egg • Raptors • A. heliaca • A. nipalensis • A. rapax • A. clanga • A. pomarina • India
1669	**Prey** • Nest • Raptors • A. chrysaetos • Oregon
1679	**Prey** • Raptors • A. chrysaetos
1691	**Prey** • Nest • A. chrysaetos • Texas
1717	**Prey** • Monograph • Hunting Behavior • Animal Damage • Eagles • A. chrysaetos • North America
1720	**Prey** • Status • Nest • Mortality • Raptors • A. chrysaetos • A. clanga • A. heliaca • Yugoslavia
1723	**Prey** • Raptors • A. pomarina • Latvia
1729	**Prey** • Habitat • Mortality • Conservation • A. chrysaetos • Japan
1732	**Prey** • Hunting Behavior • A. chrysaetos • Nevada
1759	**Prey** • Distribution • Raptors • A. rapax • Africa
1763	**Prey** • Nest • Behavior • Photo Account • Popular Monograph • A. chrysaetos • West Germany
1771	**Prey** • Prey Behavior • A. chrysaetos • Washington
1772	**Prey** • Food Requirements • Nest • Predator/Prey • A. chrysaetos • Washington
1773	**Prey** • Winter Survey • Distribution • Population Dynamics • Age Structure • Wintering Population • Eagles • A. chrysaetos • Washington
1774	**Prey** • Wintering Population • Distribution • Raptors • Popular Monograph • A. chrysaetos • Washington
1775	**Prey** • Monograph • Status • Nest Survey • Growth • Reproductive Success • Cain/Abel • Banding • Marking • Mortality • Pesticides • Residues • A. chrysaetos • Idaho
1776	**Prey** • Nest Survey • Refuge • Winter Survey • Density • Electrocution • Shooting • Mortality • Predator/Prey • Raptors • SRBPNA • A. chrysaetos • Idaho
1778	**Prey** • Nest Survey • Nest Chronology • Reproductive Success • Abandonment • Marking • Roadside Survey • Territory • Population Dynamics • Raptors • SRBPNA • A. chrysaetos • Idaho
1779	**Prey** • Reproductive Success • Aerial Survey • Banding • Marking • Population Dynamics • Raptors • SRBPNA • A. chrysaetos • Idaho
1780	**Prey** • Nest Survey • Nest Chronology • Reproductive Success • Abandonment • Marking • Roadside Survey • Territory • Population Dynamics • Raptors • SRBPNA • A. chrysaetos • Idaho
1781	**Prey** • Nest Survey • Reproductive Success • Marking • Habitat • Nest Chronology • Behavior • Banding • Territory • Population Dynamics • Raptors • SRBPNA • A. chrysaetos • Idaho
1782	**Prey** • Marking • Reproductive Success • Aerial Survey • Banding • Population Dynamics • SRBPNA • Raptors • A. chrysaetos • Idaho
1798	**Prey** • Nest • Raptors • A. chrysaetos • Kirgizia
1810	**Prey** • Hunting Behavior • Population Decline • Poisoning • A. chrysaetos • Yugoslavia
1812	**Prey** • Attitude • Behavior • A. chrysaetos • South Dakota
1816	**Prey** • Raptors • A. nipalensis • Kazakhstan
1818	**Prey** • Breeding Biology • Hunting Behavior • Raptors • A. clanga • Siberia
1819	**Prey** • Raptors • A. clanga • Siberia
1848	**Prey** • Hunting Behavior • A. chrysaetos • Massachusetts
1851	**Prey** • Density • A. chrysaetos • Siberia
1853	**Prey** • Interspecific Competition • Raptors • A. chrysaetos • Siberia
1856	**Prey** • Food Requirements • Interspecific Competition • Raptors • A. chrysaetos • A. pomarina
1863	**Prey** • Flight • Hunting Behavior • Flight Speed • Photo Account • Raptors • A. chrysaetos • Scotland
1866	**Prey** • Mortality • A. chrysaetos • Minnesota
1867	**Prey** • Hunting Behavior • Animal Damage • A. audax • Australia
1873	**Prey** • Collect • Raptors • A. clanga • France
1886	**Prey** • Egg Collection • A. verreauxi • Kenya
1900	**Prey** • Mortality • Nest • Dispersal • Pellet • Habitat • Animal Damage • Bounty • Protection • Photo Account • A. audax • Australia
1902	**Prey** • Status • Conservation • Nest • Behavior • A. chrysaetos • Israel
1903	**Prey** • Nest • Habitat • Disturbance • Conservation • Mortality • Hunting Behavior • A. chrysaetos • Israel
1905	**Prey** • Raptors • A. chrysaetos • A. wahlbergi • Africa

Master No.	Keywords
1906	**Prey** • Nest • Behavior • A. heliaca • Spain
1908	**Prey** • Nest • Breeding Biology • Behavior • Clutch Size • A. nipalensis • Bulgaria
1928	**Prey** • Pellet • Nest • Hunting Behavior • Interspecific Behavior • Reproductive Success • Eagles • A. chrysaetos • Oklahoma
1936	**Prey** • A. heliaca • Kazakhstan
1943	**Prey** • Aerial Survey • Time Lapse • Animal Damage • Nest Survey • Habitat • Winter Survey • Capture • Age Structure • Food Requirements • Shooting • A. chrysaetos • Texas
1947	**Prey** • Density • Breeding Biology • (Title Only) • A. chrysaetos • Scotland
1950	**Prey** • Animal Damage • Nest Survey • Abandonment • Management • A. chrysaetos • Scotland
1952	**Prey** • Interspecific Behavior • Raptors • A. chrysaetos • A. heliaca • Rumania
1954	**Prey** • Behavior • Photo Account • Raptors • Popular Monograph • A. chrysaetos • Europe
1955	**Prey** • Nest • Intraspecific Behavior • Raptors • A. chrysaetos • California
1972	**Prey** • Nest • Habitat • A. chrysaetos • Ontario
1985	**Prey** • Habitat • Nest • Behavior • Flight • Photo Account • A. chrysaetos • Scotland
1986	**Prey** • Nest • Clutch Size • Captivity • Nest Survey • Collect • A. chrysaetos • Canada
1994	**Prey** • Distribution • Nest • Hunting Behavior • Raptors • A. verreauxi • South Africa
1997	**Prey** • Abnormal Behavior • Measure • A. chrysaetos • Scotland
1998	**Prey** • Reproductive Success • Growth • Behavior • Photography Account • Popular Monograph • A. chrysaetos • Scotland
1999	**Prey** • A. chrysaetos • Scotland
2024	**Prey** • Rarity • Raptors • A. nipalensis • Uganda
2029	**Prey** • Nest • Status • Habitat • Eagles • A. heliaca • A. chrysaetos • A. clanga • A. pomarina • Yugoslavia
2038	**Prey** • Status • Bounty • Raptors • A. chrysaetos • A. heliaca • India
2039	**Prey** • Nest • Status • Collect • Raptors • A. wahlbergi • Rhodesia
2044	**Prey** • Nest • Raptors • A. verreauxi • South Africa
2066	**Prey** • Rarity • Behavior • A. chrysaetos • Ireland
2079	**Prey** • Distribution • Field Id • Attitude • Raptors • A. chrysaetos • United States
2084	**Prey** • Collect • Trade • A. chrysaetos • California
2086	**Prey** • Monograph • Mortality • Banding • Migration • Breeding Biology • Food Requirements • A. chrysaetos • Montana
2088	**Prey** • Breeding Biology • Banding • Migration • Directional Exposure • Mortality • A. chrysaetos • Montana
2104	**Prey** • (Title Only) • A. audax • Australia
2116	**Prey** • Hunting Behavior • Raptors • A. chrysaetos • A. heliaca • A. nipalensis
2124	**Prey** • Monograph • Breeding Biology • Migration • Hunting Behavior • Attack • Folklore • Falconry • Animal Damage • Popular Monograph • A. chrysaetos
2128	**Prey** • Pesticides • Mortality • Rehabilitation • Wintering Population • Azodrin • Raptors • A. heliaca • A. clanga • A. pomarina • Israel
2138	**Prey** • Nest • Behavior • A. chrysaetos • Switzerland
2143	**Prey** • Hunting Behavior • A. audax • Australia
2144	**Prey** • Nest • Parental Behavior • Vocalization • Nest Adornment • Photo Account • A. chrysaetos • West Germany
2145	**Prey** • Breeding Biology • Habitat • Cain/Abel • Density • Behavior • Nest Survey • Photo Account • A. pomarina • Czechoslovakia
2148	**Prey** • Distribution • Field Id • Management • Taxonomy • Habitat • Nest Survey • Breeding Biology • Photo Account • Eagles • A. pomarina • Europe
2149	**Prey** • Breeding Biology • Cain/Abel • Nest • Taxonomy • Photo Account • Popular Monograph • A. pomarina • West Germany
2152	**Prey** • Habitat • Growth • Parental Behavior • Breeding Biology • Interspecific Behavior • Courtship • Display • Photo Account • Raptors • A. heliaca • Spain
2165	**Prey** • Rarity • A. audax • New Zealand
2168	**Prey** • Literature Review • Owls • A. chrysaetos • Europe
2184	**Prey** • Aerial Survey • Reproductive Success • Management • Survey • Clutch Size • Distribution • Habitat • Nest Survey • Raptors • A. chrysaetos • Arizona
2186	**Prey** • Descriptive Summary • Banding • Raptors • A. wahlbergi • A. rapax • South Africa
2187	**Prey** • Hunting Behavior • A. chrysaetos • Colorado
2188	**Prey** • Density • Breeding Biology • A. nipalensis • Caucasus
2200	**Prey** • Animal Damage • Economics • Nest Survey • Trapping • A. chrysaetos • New Mexico • Texas
2202	**Prey** • Measure • Id Key • Raptors • Photo Account • A. chrysaetos • A. nipalensis • A. clanga • Italy
2204	**Prey** • A. clanga • Italy
2206	**Prey** • Collect • Measure • A. clanga • Italy
2212	**Prey** • Nest • Reproductive Success • Management • Shooting • Density • Raptors • A. chrysaetos • Washington
2217	**Prey** • Nest • Attitude • Photo Account • Popular Monograph • A. chrysaetos • Canada
2220	**Prey** • Distribution • A. verreauxi • Egypt
2224	**Prey** • Nest • Banding • A. chrysaetos • South Dakota
2234	**Prey** • Fat • Shooting • A. chrysaetos • Montana
2239	**Prey** • Survey • Territory • Nest Failure • Nest • Migration • Habitat • Management • Reproductive Success • Roadside Survey • Status • Winter Survey • Uinta Nf • Wasatch Mt. • Raptors • A. chrysaetos • Utah
2240	**Prey** • Nest • Photography • Behavior • Photo Account • A. heliaca • Spain
2242	**Prey** • Nest Survey • Raptors • A. heliaca • Spain
2245	**Prey** • Nest • Vocalization • Etymology • Raptors • A. verreauxi • A. rapax • A. wahlbergi • Rhodesia
2255	**Prey** • Electrocution • Density • Predator/Prey • Nest • A. nipalensis • Russia
2256	**Prey** • Collect • Raptors • A. chrysaetos • Illinois
2257	**Prey** • Prey Behavior • Pellet • Interspecific Behavior • A. chrysaetos • Alaska
2258	**Prey** • A. chrysaetos • Alaska

Master No.	Keywords
2261	**Prey** • Nest Survey • Disturbance • Population Dynamics • Management • Status • Animal Damage • Artificial Nest • Predator/Prey • A. chrysaetos • Utah
2272	**Prey** • Nest • A. chrysaetos • Texas
2281	**Prey** • Trapping • Measure • A. chrysaetos • Maine
2283	**Prey** • Nest Survey • Population Decline • Nest Chronology • A. chrysaetos • France
2293	**Prey** • Conservation • Falconry • Hunting Behavior • Raptors • Popular Monograph • A. chrysaetos • United States
2304	**Prey** • Attitude • Trapping • Raptors • A. audax • Australia
2311	**Prey** • Attitude • Status • Population Estimate • Mortality • Protection • Animal Damage • Disturbance • Raptors • A. chrysaetos • Great Britain
2324	**Prey** • Capture • Animal Damage • Banding • Management • Sexing • Translocation • Marking • Measure • Captive Care • Tech. • Eagles • A. chrysaetos • Montana
2325	**Prey** • Capture • Banding • Marking • Animal Damage • Sexing • Management • Translocation • Survey • Measure • Mortality • Captive Care • Aging • A. chrysaetos • Montana
2326	**Prey** • Capture • Banding • Marking • Animal Damage • Sexing • Management • Translocation • Measure • Mortality • Capt. Care • Aging • Food Requirements • A. chrysaetos • Montana
2327	**Prey** • Mortality • A. chrysaetos • Norway
2332	**Prey** • Behavior • Conservation • A. audax • Australia
2333	**Prey** • Status • Nest • Raptors • A. heliaca • Spain
2340	**Prey** • Descriptive Summary • Plumage • Distribution • Nest • Egg • Photo Account • Raptors • A. audax • Australia
2345	**Prey** • Shooting • Density • Nest • Reproductive Success • Eagles • A. chrysaetos • Italy
2352	**Prey** • Collect • Plumage • A. nipalensis • A. heliaca • India
2369	**Prey** • Folklore • Lift • Animal Damage • Conservation • Electrocution • Eagles • Popular Monograph • A. chrysaetos • North America
2370	**Prey** • Status • Nest • Nest Survey • Monograph • Habitat • Mortality • Breeding Biology • Nest Chronology • Courtship • Density • Management • A. chrysaetos • Colorado
2371	**Prey** • Popular Monograph • Story • Breeding Biology • Behavior • Habitat • Nest Survey • Hunting Behavior • Raptors • Illus. • A. chrysaetos • Colorado
2373	**Prey** • Literature Review • Animal Damage • Food Requirements • A. chrysaetos • Colorado
2385	**Prey** • Collect • Eagles • A. chrysaetos • Italy
2390	**Prey** • Raptors • A. chrysaetos • Siberia
2393	**Prey** • Distribution • Beneficial Effect • Raptors • A. heliaca • A. nipalensis • Kazakhstan
2395	**Prey** • Status • Symbolism • Nest • Raptors • A. chrysaetos • South Dakota
2396	**Prey** • Collect • Raptors • A. clanga • Kenya
2407	**Prey** • Predator/Prey • Photo Account • A. pomarina • Czechoslovakia
2411	**Prey** • Interspecific Behavior • Nest • Behavior • Photography • Nest Adornment • Talon Grappling • Cain/Abel • Photo Account • Popular Monograph • A. chrysaetos • Scotland
2416	**Prey** • Breeding Biology • A. clanga • Siberia
2425	**Prey** • Capture • Lift • Historical Account • A. chrysaetos • Scotland
2439	**Prey** • Nest • Raptors • A. audax • Australia
2445	**Prey** • Flight • Agonistic Behavior • Aircraft • Raptors • A. rapax • Kenya • Tanzania
2451	**Prey** • Nest • Raptors • A. chrysaetos • Siberia
2458	**Prey** • Piracy • Eagles • A. chrysaetos • Sweden
2464	**Prey** • Management • Habitat • Directional Exposure • Disturbance • Mortality • Nest Survey • Density • Nest Chronology • Reproductive Success • Prog. Rep. • A. chrysaetos • Wyoming
2465	**Prey** • Habitat • Mining • Nest Survey • Disturbance • Marking • Time Lapse • Raptors • Proj. Out. • A. chrysaetos • Wyoming • Montana
2471	**Prey** • Raptors • A. wahlbergi • A. rapax • South Africa
2477	**Prey** • Lift • A. chrysaetos • Great Britain
2484	**Prey** • Hunting Behavior • A. chrysaetos • West Germany
2488	**Prey** • Abandonment • Nest • Raptors • A. chrysaetos • Utah
2489	**Prey** • Hunting Behavior • Interspecific Behavior • A. chrysaetos • Utah
2490	**Prey** • A. chrysaetos • Scotland
2510	**Prey** • Wintering Population • Raptors • A. chrysaetos • A. clanga • A. heliaca • Albania
2513	**Prey** • Rarity • Measure • A. clanga • Great Britain
2517	**Prey** • Nest • A. pomarina • West Germany
2526	**Prey** • Nest • Habitat • A. clanga • Russia
2531	**Prey** • Nest • Behavior • A. verreauxi • South Africa
2532	**Prey** • Animal Damage • Raptors • A. verreauxi • South Africa
2537	**Prey** • Nest • Status • A. clanga • Russia
2539	**Prey** • Status • Nest • Raptors • A. chrysaetos • A. clanga • Russia
2544	**Prey** • Collect • Habitat • Behavior • A. clanga • West Germany
2546	**Prey** • A. chrysaetos • Russia
2554	**Prey** • Conservation • Nest • Behavior • Interspecific Behavior • Trapping • Animal Damage • Rehabilitation • Photo Account • Popular Monograph • A. chrysaetos • Wyoming
2559	**Prey** • A. chrysaetos • Scotland
2568	**Prey** • Behavior • A. chrysaetos • North America
2569	**Prey** • Nest • Behavior • Disturbance • A. chrysaetos • Colorado
2572	**Prey** • Animal Damage • A. chrysaetos • New York
2583	**Prey** • Collect • Habitat • Status • Raptors • A. heliaca • A. nipalensis • A. clanga • A. rapax • A. pomarina • India
2591	**Prey** • Roadside Survey • Wintering Population • Animal Damage • Age Structure • Raptors • A. chrysaetos • New Mexico
2594	**Prey** • Monograph • Pesticides • Density • Nest Survey • Mortality • Reproductive Success • A. chrysaetos • Montana
2597	**Prey** • Nest • Illus. • A. chrysaetos • Switzerland

Master No.	Keywords
2601	**Prey** • Banding • Nest • Photo Account • A. chrysaetos • Switzerland
2602	**Prey** • Nest • Photography • Habitat • Photo Account • A. chrysaetos • Switzerland
2611	**Prey** • Banding • Conservation • Nest • Photo Account • Popular Monograph • A. audax • Australia
2612	**Prey** • Breeding Biology • Habitat • Dispersal • A. audax • Australia
2637	**Prey** • Descriptive Summary • Field Id • Distribution • Etymology • Raptors • A. verreauxi • A. rapax • A. nipalensis • A. pomarina • A. wahlbergi • South Africa
2643	**Prey** • Raptors • A. rapax • Southwest Africa
2670	**Prey** • Cain/Abel • Parental Behavior • Growth • Breeding Biology • A. verreauxi • Africa
2672	**Prey** • Animal Damage • Hunting Behavior • Feeding Behavior • A. audax • Australia
2675	**Prey** • Collect • Wintering Population • Etymology • Status • Raptors • A. chrysaetos • A. clanga • Yugoslavia
2688	**Prey** • Hunting Behavior • Animal Damage • Eagles • A. chrysaetos • Michigan
2689	**Prey** • A. chrysaetos • Russia
2694	**Prey** • Population Decline • Raptors • A. nipalensis • Kazakhstan
2714	**Prey** • Popular Monograph • Shooting • Interspecific Behavior • Hunting Behavior • Eagles • A. chrysaetos • North America
2721	**Prey** • Measure • A. chrysaetos • California
2732	**Prey** • Descriptive Summary • Distribution • Habitat • Raptors • A. chrysaetos • A. nipalensis • Tibet
2741	**Prey** • Collect • A. chrysaetos • Italy
2753	**Prey** • Monograph • Habitat • Nest Survey • Behavior • Pellet • Abandonment • A. chrysaetos • Wyoming
2760	**Prey** • Nest • Vocalization • Behavior • A. chrysaetos • West Germany
2761	**Prey** • Breeding Biology • Directional Exposure • Vocalization • A. chrysaetos • West Germany
2763	**Prey** • Behavior • Nest Chronology • A. chrysaetos • West Germany
2779	**Prey** • Status • Nest • A. chrysaetos • Siberia
2801	**Prey** • Nest • Forestry • Mortality • Habitat • A. chrysaetos • Washington
2816	**Prey** • Collect • Nest • Status • Raptors • A. wahlbergi • South Africa
2819	**Prey** • Hunting Behavior • Raptors • A. nipalensis • Kazakhstan
2830	**Prey** • Breeding Biology • Habitat • Cain/Abel • Nest Adornment • Growth • Vocalization • Plumage • Hunting Behavior • Illus. • A. pomarina • West Germany
2845	**Prey** • Status • Raptors • Yellowstone • A. chrysaetos • Wyoming
2846	**Prey** • Protection • Distribution • Nest Survey • Photo Account • A. pomarina • Czechoslovakia
2849	**Prey** • Nest • A. pomarina • Czechoslovakia
2850	**Prey** • Status • Nest Survey • Photo Account • A. heliaca • Czechoslovakia
2852	**Prey** • Roadside Survey • Monograph • Density • Habitat • Interspecific Competition • Taxonomy • Vocalization • Hunting Behavior • Nest Chronology • Photo Account • Raptors • A. wahlbergi • A. rapax • A. nipalensis • A. heliaca • Kenya
2854	**Prey** • Population Dynamics • Nest • Density • Breeding Biology • Distribution • Habitat • Directional Exposure • Reproductive Success • Mortality • Abandonment • Disturbance • Interspecific Behavior • Territory • A. chrysaetos • Utah
2856	**Prey** • Breeding Biology • Nest Survey • Habitat • Hunting Behavior • Territory • Interspecific Behavior • Raptors • A. chrysaetos • Utah
2870	**Prey** • Banding • Survey • Marking • Nest • Capture • Food Requirements • Raptors • A. chrysaetos • A. wahlbergi • South Africa
2882	**Prey** • Hunting Behavior • A. heliaca • United Soviet Socialists Republic
2883	**Prey** • Habitat • Raptors • A. heliaca • Kazakhstan
2885	**Prey** • Hunting Behavior • A. chrysaetos • Oregon
2886	**Prey** • Nest • Behavior • Raptors • A. chrysaetos • Turkmenia
2895	**Prey** • Status • Raptors • A. chrysaetos • A. clanga • China
2898	**Prey** • Prey Behavior • Nest • A. chrysaetos • Colorado
2906	**Prey** • Attitude • Mortality • Animal Damage • Raptors • A. chrysaetos • Texas
2913	**Prey** • Shooting • Survey • Animal Damage • Plumage • A. chrysaetos • Texas
2920	**Prey** • Animal Damage • Age Structure • A. chrysaetos • North America
2938	**Prey** • Nest • Photo Account • A. chrysaetos • Kansas
2941	**Prey** • Rarity • Eagles • A. pomarina • A. nipalensis • West Germany
2949	**Prey** • Nest • Status • Vocalization • Behavior • Photo Account • A. chrysaetos • Switzerland
2951	**Prey** • Nest • Photo Account • A. chrysaetos • Switzerland
2960	**Prey** • (Title Only) • A. chrysaetos • Scotland
2962	**Prey** • Piracy • Density • Raptors • A. chrysaetos • Washington
2966	**Prey** • Behavior • Incubation • Cain/Abel • Reproductive Success • A. rapax • South Africa
2969	**Prey** • Nest Survey • Habitat • Breeding Biology • Reproductive Success • Growth • Field Id • Vocalization • Migration • Behavior • Eagles • Photo Account • A. wahlbergi • A. rapax • Rhodesia
2972	**Prey** • Plumage • Photography • Story • Breeding Biology • Photo Account • Popular Monograph • A. verreauxi • A. wahlbergi • A. rapax • Rhodesia • South Africa
2973	**Prey** • Life History • Plumage • Cain/Abel • Banding • Reproductive Success • Interspecific Behavior • Piracy • Photo Account • Raptors • A. chrysaetos • Rhodesia
2976	**Prey** • Reproductive Success • A. wahlbergi • Rhodesia
2984	**Prey** • A. nipalensis • Africa
2985	**Prey** • Shooting • Measure • Photo Account • Eagles • A. chrysaetos • New York

Master No.	Keywords
3000	**Prey** • Egg • Museum Specimen • Raptors • A. chrysaetos • A. nipalensis • China
3008	**Prey** • Raptors • A. heliaca • China
3012	**Prey** • Breeding Biology • Raptors • A. nipalensis • Turkmenia
3013	**Prey** • Hunting Behavior • Raptors • A. clanga • A. chrysaetos • A. heliaca • Turkmenia
3014	**Prey** • A. chrysaetos • Finland
3015	**Prey** • Nest • A. chrysaetos • Finland
3032	**Prey** • A. chrysaetos • Switzerland
3036	**Prey** • Descriptive Summary • Distribution • Status • Measure • Raptors • A. chrysaetos • Oklahoma
3047	**Prey** • Nest Failure • Raptors • A. chrysaetos • Scotland
3054	**Prey** • Status • Raptors • A. rapax • A. wahlbergi • Rhodesia
3058	**Prey** • Habitat • Prey Behavior • Nest • Animal Damage • Shooting • Egg • Behavior • A. verreauxi • South Africa
3063	**Prey** • Nest • A. pomarina • West Germany
3064	**Prey** • Nest • Status • A. pomarina • Australia
3079	**Prey** • Nest Survey • Clutch Size • Reproductive Success • Display • Hunting Behavior • Density • Migration • A. wahlbergi • South Africa
3080	**Prey** • Cain/Abel • Distribution • Educational Material • Field Id • Habitat • Hunting Behavior • Artificial Nest • Nest • Incubation • Wintering Population • Conservation • Photo Account • Eagles • A. chrysaetos • A. wahlbergi • A. verreauxi • A. nipalensis • A. pomarina • South Africa
3095	**Prey** • Animal Damage • Nest • Behavior • A. verreauxi • South Africa
3118	**Prey** • Distribution • Habitat • Descriptive Summary • Density • Interspecific Competition • Eagles • A. chrysaetos • France
3131	**Prey** • Habitat • Niche • Raptors • A. wahlbergi • A. rapax • Ivory Coast
3133	**Prey** • Wintering Population • Habitat • Density • Raptors • A. wahlbergi • A. rapax • A. nipalensis • A. pomarina • Cambodia
3146	**Prey** • Status • Population Dynamics • Nest Survey • A. chrysaetos • Idaho
3152	**Prey** • Feeding Behavior • A. heliaca • Wyoming
3170	**Prey** • Nest • Photo Account • Popular Monograph • A. nipalensis • West Germany
3177	**Prey** • A. pomarina • A. wahlbergi • Botswana
3183	**Prey** • Nest • Photo Account • Popular Monograph • A. chrysaetos • Montana
3187	**Prey** • Nest Survey • Banding • Raptors • A. wahlbergi • Rhodesia
3199	**Prey** • Status • Raptors • A. nipalensis • Kazakhstan • Russia
3207	**Prey** • Refuge • Habitat • Nest Survey • Density • Predator/Prey • Disturbance • Telemetry • Territory • Reproductive Success • Raptors • SRBPNA • Photo Account • A. chrysaetos • Idaho
3216	**Prey** • Nest • Raptors • A. pomarina • West Germany
3217	**Prey** • Nest • A. pomarina • West Germany
3218	**Prey** • Descriptive Summary • Hunting Behavior • Habitat • Raptors • A. chrysaetos • A. heliaca • A. nipalensis • A. clanga • A. pomarina • West Germany
3225	**Prey** • Habitat • Nest • Raptors • A. heliaca • Spain
3231	**Prey** • Wintering Population • Raptors • A. chrysaetos • Washington
3235	**Prey** • Growth • Nest • Photography • Behavior • Hunting Behavior • Vocalization • Nest Adornment • Egg • Raptors • Popular Monograph • A. rapax • Kenya
3240	**Prey** • Habitat • Raptors • A. heliaca • A. nipalensis • Kazakhstan
3242	**Prey** • Raptors • A. heliaca • A. nipalensis • Kazakhstan
3243	**Prey** • Hunting Behavior • Habitat • Raptors • A. chrysaetos • A. heliaca • A. nipalensis • United Soviet Socialists Republic
3244	**Prey** • A. chrysaetos • Kazakhstan
3245	**Prey** • Distribution • Habitat • Raptors • A. nipalensis • A. heliaca • A. chrysaetos • United Soviet Socialists Republic
3258	**Prey** • Collect • Rarity • A. pomarina • France
3259	**Prey** • Nest Survey • Habitat • Territory • Intraspecific Behavior • Density • Nest Chronology • Directional Exposure • Raptors • A. verreauxi • Rhodesia
3260	**Prey** • Nest • Raptors • A. rapax • Rhodesia
3264	**Prey** • A. chrysaetos • Connecticut
3278	**Prey** • Nest • A. wahlbergi • Ethiopia
3283	**Prey** • Breeding Biology • Conservation • Mortality • Distribution • Photo Account • A. chrysaetos • Czechoslovakia
3284	**Prey** • Hunting Behavior • A. chrysaetos • United States
3292	**Prey** • Distribution • Measure • Raptors • A. gurneyi • New Guinea
3305	**Prey** • Status • Habitat • Raptors • A. chrysaetos • Scotland
3316	**Prey** • Animal Damage • Food Requirements • Management • A. chrysaetos • Scotland
3327	**Prey** • Monograph • Breeding Biology • A. pomarina • A. clanga
3335	**Prey** • Descriptive Summary • Lift • Raptors • A. chrysaetos • North America
3345	**Prey** • Hunting Behavior • Eagles • A. chrysaetos • A. nipalensis • A. rapax • India
3357	**Prey** • Status • Piracy • Collect • Raptors • A. heliaca • A. nipalensis • A. clanga • A. rapax • India
3360	**Prey** • Nest Survey • Habitat • Directional Exposure • Eagles • A. chrysaetos • Saskatchewan
3385	**Prey** • Flight Speed • Status • Raptors • A. wahlbergi • Rhodesia
3387	**Prey** • Status • Raptors • A. rapax • A. verreauxi • South Africa
3391	**Prey** • Rarity • A. clanga • Great Britain
3399	**Prey** • Hunting Behavior • A. audax • Australia
3401	**Prey** • Shooting • A. chrysaetos • West Germany
3413	**Prey** • Bounty • Lift • Hunting Behavior • A. chrysaetos • Montana
3416	**Prey** • Raptors • A. chrysaetos • Great Britain
3424	**Prey** • Behavior • A. chrysaetos • California
3439	**Prey** • Nest • Clutch Size • Nest Chronology • A. audax • Australia

Master No.	Keywords
783	**Protection** • Conservation • Prey • Animal Damage • Hunting Behavior • Prey Behavior • Raptors • A. chrysaetos • Sweden
792	**Protection** • Distribution • Eagles • A. chrysaetos • Sweden
802	**Protection** • Animal Damage • Beneficial Effect • Raptors • A. clanga • Siberia
808	**Protection** • Conservation • A. chrysaetos • Switzerland
878	**Protection** • Status • A. chrysaetos • France
903	**Protection** • Status • Conservation • Raptors • A. chrysaetos • A. clanga • Belorussia
1046	**Protection** • Falconry • Photo Account • Raptors • A. chrysaetos • A. heliaca
1106	**Protection** • Status • Shooting • Raptors • A. chrysaetos • United States
1107	**Protection** • A. chrysaetos • United States
1115	**Protection** • Status • A. chrysaetos • A. nipalensis • A. heliaca • United Soviet Socialists Republic
1123	**Protection** • Density • A. clanga • A. chrysaetos • United Soviet Socialists Republic
1125	**Protection** • Density • Popular Monograph • A. chrysaetos • United Soviet Socialists Republic
1131	**Protection** • Folklore • Hunting Behavior • Trapping • Animal Damage • Eagles • A. chrysaetos • Tennessee
1162	**Protection** • Status • Nest Survey • Conservation • Shooting • Habitat • Raptors • A. chrysaetos • A. heliaca • Spain
1177	**Protection** • Status • Nest Survey • Distribution • Reproductive Success • Photo Account • A. pomarina • East Germany
1187	**Protection** • Habitat • Nest • Vocalization • Raptors • A. chrysaetos • A. heliaca • Yugoslavia
1259	**Protection** • Poetry • Captivity • Attitude • Flight • Folklore • Raptors • A. chrysaetos • Great Britain
1346	**Protection** • Raptors • A. chrysaetos • United States
1366	**Protection** • Status • Mortality • Management • Disturbance • Eagles • A. chrysaetos • North America
1468	**Protection** • Status • Habitat Loss • Raptors • A. chrysaetos • Switzerland
1473	**Protection** • Animal Damage • Prey • A. chrysaetos • Switzerland
1478	**Protection** • A. chrysaetos • Switzerland
1506	**Protection** • Descriptive Summary • Nest • Prey • Raptors • Illus. • A. audax • Australia
1507	**Protection** • Conservation • Economics • Mortality • Reward • Pesticides • Poisoning • Eagles • A. chrysaetos • Wyoming • Colorado
1625	**Protection** • Status • Shooting • Poisoning • Prey • Refuge • Educational Material • Raptors • A. chrysaetos • Spain
1650	**Protection** • A. chrysaetos • United States
1742	**Protection** • Eagles • A. chrysaetos • Washington
1799	**Protection** • Nest • Nest Failure • Raptors • A. nipalensis • Caucasus
1820	**Protection** • Population Decline • Conservation • Raptors • A. chrysaetos • A. clanga • A. pomarina • West Germany
1822	**Protection** • Raptors • A. chrysaetos • A. clanga • Estonia
1899	**Protection** • Shooting • A. chrysaetos • Texas
1900	**Protection** • Mortality • Prey • Nest • Dispersal • Pellet • Habitat • Animal Damage • Bounty • Photo Account • A. audax • Australia
1901	**Protection** • Popular Monograph • A. chrysaetos • Israel
1946	**Protection** • Mortality • Management • Photo Account • A. chrysaetos • Scotland
1971	**Protection** • Conservation • Animal Damage • Eagles • A. chrysaetos • North America
2041	**Protection** • Shooting • A. chrysaetos • Wyoming
2043	**Protection** • Symbolism • Eagles • Feathers • A. chrysaetos • North America
2049	**Protection** • Status • Shooting • Pesticides • Migration • Poisoning • A. chrysaetos • Siberia
2089	**Protection** • Bounty • Raptors • A. audax • Australia
2096	**Protection** • Animal Damage • (Title Only) • A. chrysaetos • North America
2125	**Protection** • Poisoning • Pesticides • Status • Captivity • Density • Food Requirements • Habitat • Predator/Prey • Raptors • Photography Account • PCB • A. clanga • A. heliaca • A. nipalensis • A. verreauxi • A. pomarina • Israel
2169	**Protection** • Hunting Behavior • Descriptive Summary • Illus. • Popular Monograph • A. chrysaetos • Wyoming
2215	**Protection** • Status • Rehabilitation • Mortality • Raptors • A. audax • Australia
2233	**Protection** • Shooting • Attitude • Punishment • Eagles • Popular Monograph • A. chrysaetos • North America
2236	**Protection** • Raptors • A. chrysaetos • North America
2280	**Protection** • Trapping • Economics • Raptors • A. chrysaetos • Nevada
2311	**Protection** • Attitude • Status • Population Estimate • Mortality • Prey • Animal Damage • Disturbance • Raptors • A. chrysaetos • Great Britain
2315	**Protection** • Status • Animal Damage • A. chrysaetos • Austria
2349	**Protection** • Feeding Behavior • Animal Damage • Capture • Predator/Prey • Management • Economics • A. chrysaetos • Montana
2466	**Protection** • Raptors • A. chrysaetos • United States
2505	**Protection** • Field Id • Photo Account • A. chrysaetos • A. clanga • A. heliaca • A. nipalensis • A. pomarina • Europe
2514	**Protection** • A. chrysaetos • France
2558	**Protection** • Nest Survey • Reproductive Success • Habitat • Clutch Size • Eagles • A. chrysaetos • Estonia
2615	**Protection** • Raptors • A. chrysaetos • Switzerland
2627	**Protection** • Status • Pesticides • Raptors • A. chrysaetos • Italy
2645	**Protection** • Conservation • Shooting • Poisoning • Egg Collection • Nest Robbing • Trade • Trapping • Raptors • A. chrysaetos • Great Britain
2656	**Protection** • Winter Survey • Conservation • Shooting • Poisoning • Reward • Photo Account • Eagles • Popular Monograph • A. chrysaetos • Wyoming
2691	**Protection** • Nest • A. chrysaetos • West Germany

Master No.	Keywords
2726	**Protection** • Banding • Eagles • A. chrysaetos • Finland
2736	**Protection** • Status • Density • Conservation • Survey • Raptors • A. chrysaetos • Sardinia
2739	**Protection** • Raptors • A. chrysaetos • Hungary
2740	**Protection** • Raptors • A. chrysaetos • Hungary
2767	**Protection** • A. chrysaetos • Switzerland
2787	**Protection** • Status • Conservation • Habitat Loss • Mortality • Raptors • A. chrysaetos • A. clanga • Sweden
2797	**Protection** • Bounty • Raptors • A. audax • Australia
2798	**Protection** • Raptors • A. audax • Australia
2804	**Protection** • Bounty • Shooting • Attitude • Economics • A. audax • Australia
2827	**Protection** • Conservation • Raptors • A. chrysaetos • New Mexico
2846	**Protection** • Distribution • Prey • Nest Survey • Photo Account • A. pomarina • Czechoslovakia
2990	**Protection** • Mortality • Conservation • Attitude • Photo Account • Popular Monograph • A. chrysaetos • North America
3112	**Protection** • Status • Shooting • Raptors • A. chrysaetos • A. pomarina • West Germany
3155	**Protection** • Conservation • Mortality • A. chrysaetos • North America
3192	**Protection** • Poisoning • Shooting • Electrocution • Attitude • Popular Monograph • A. chrysaetos • Wyoming
3200	**Protection** • Attitude • Punishment • Reward • Aircraft • Raptors • A. chrysaetos • United States
3202	**Protection** • Disturbance • Aircraft • Mitigation • Environmental Impact • A. chrysaetos • California
3265	**Protection** • Raptors • A. pomarina • A. clanga • A. chrysaetos • A. heliaca • Hungary
3277	**Protection** • Mortality • Nest • Popular Monograph • A. heliaca • South Africa
3339	**Protection** • Conservation • A. chrysaetos • North America
3365	**Protection** • Conservation • Anatomy • Bounty • Raptors • Illus. • A. chrysaetos • Finland
3372	**Protection** • Population Decline • Mortality • Conservation • Animal Damage • Bounty • A. chrysaetos • Norway
3374	**Protection** • Status • Distribution • Conservation • Raptors • A. chrysaetos • Norway
3375	**Protection** • Symbolism • Shooting • Economics • Attitude • Conservation • A. chrysaetos • United States
3378	**Protection** • Economics • Animal Damage • Popular Monograph • A. chrysaetos • North America
3444	**Protection** • Breeding Biology • Nest Survey • Habitat • Prey • Distribution • Reproductive Success • Photo Account • A. chrysaetos • Estonia
151	**Punishment** • Poisoning • Mortality • Wintering Population • Conservation • Attitude • A. chrysaetos • Wyoming
152	**Punishment** • Raptors • A. chrysaetos • Turkey • West Germany
170	**Punishment** • Shooting • Attitude • Eagles • A. chrysaetos • Mississippi
1653	**Punishment** • Conservation • Disturbance • Nest Robbing • Popular Monograph • A. chrysaetos • Scotland
2233	**Punishment** • Shooting • Attitude • Protection • Eagles • Popular Monograph • A. chrysaetos • North America
2555	**Punishment** • Aircraft • Shooting • Attitude • Eagle • Popular Monograph • A. chrysaetos • Texas
2758	**Punishment** • Attitude • Shooting • Conservation • Story • Popular Monograph • A. chrysaetos • Texas
2759	**Punishment** • Animal Damage • Shooting • Conservation • Attitude • Reward • Translocation • Popular Monograph • A. chrysaetos • Texas
2939	**Punishment** • Poisoning • Eagles • A. chrysaetos • Wyoming
3200	**Punishment** • Protection • Attitude • Reward • Aircraft • Raptors • A. chrysaetos • United States
3310	**Punishment** • Shooting • Poisoning • Bounty • Conservation • Story • Eagles • Popular Monograph • A. chrysaetos • Wyoming
2322	**Questionnaire** • Animal Damage • Economics • A. chrysaetos • Utah
52	**Rarity** • Raptors • A. clanga • A. pomarina • Denmark
70	**Rarity** • A. clanga • Austria
72	**Rarity** • Prey • A. chrysaetos • Great Britain
92	**Rarity** • A. clanga • Belgium
205	**Rarity** • Attitude • Rehabilitation • A. clanga • Great Britain
212	**Rarity** • A. clanga • West Germany
218	**Rarity** • Descriptive Summary • Prey • Distribution • Etymology • Raptors • A. chrysaetos • A. heliaca • Japan
227	**Rarity** • Nest • Raptors • A. clanga • Finland
232	**Rarity** • A. clanga • Finland
258	**Rarity** • Shooting • A. clanga • West Germany
259	**Rarity** • A. heliaca • Morocco
266	**Rarity** • Collect • A. clanga • West Germany
272	**Rarity** • A. chrysaetos • Great Britain
283	**Rarity** • A. chrysaetos • Great Britain
298	**Rarity** • Habitat • A. pomarina • A. heliaca • Austria
299	**Rarity** • Status • Eagles • A. clanga • A. pomarina • A. heliaca • Austria
307	**Rarity** • Collect • Raptors • A. clanga • Cyprus
320	**Rarity** • A. clanga • Kenya
353	**Rarity** • (Title Only) • A. heliaca • Sweden
359	**Rarity** • Status • Raptors • A. chrysaetos • A. heliaca • A. clanga • Spain
360	**Rarity** • A. chrysaetos • West Germany
410	**Rarity** • A. pomarina • A. heliaca • Malawi
413	**Rarity** • Plumage • A. heliaca • A. rapax • Morocco
419	**Rarity** • A. clanga • West Germany
432	**Rarity** • Raptors • A. heliaca • Hungary
448	**Rarity** • Collect • A. clanga • Czechoslovakia
478	**Rarity** • A. chrysaetos • Great Britain
491	**Rarity** • Raptors • A. pomarina • Rhodesia • Malawi
493	**Rarity** • Raptors • A. pomarina • Mozambique
496	**Rarity** • Trapping • A. nipalensis • South Africa
529	**Rarity** • A. clanga • Netherlands
530	**Rarity** • A. chrysaetos • Netherlands
564	**Rarity** • Raptors • A. heliaca • A. clanga • A. pomarina • Denmark

Master No.	Keywords
583	**Rarity** • Status • Raptors • A. chrysaetos • A. pomarina • Libya
643	**Rarity** • A. clanga • France
645	**Rarity** • Interspecific Behavior • A. clanga • Czechoslovakia
661	**Rarity** • Raptors • A. chrysaetos • Azores
683	**Rarity** • Field Id • A. heliaca • Denmark
684	**Rarity** • A. pomarina • Denmark
691	**Rarity** • A. clanga • Great Britain
709	**Rarity** • A. clanga • Great Britain
715	**Rarity** • A. clanga • France
740	**Rarity** • A. chrysaetos • Great Britain
774	**Rarity** • Raptors • A. chrysaetos • A. clanga • A. pomarina • France
775	**Rarity** • A. chrysaetos • Australia
776	**Rarity** • Eagles • A. clanga • Austria
777	**Rarity** • Status • A. clanga • Austria
778	**Rarity** • A. nipalensis • Hungary
825	**Rarity** • A. pomarina • Ireland
845	**Rarity** • Collect • A. chrysaetos • A. clanga • Thailand
849	**Rarity** • A. heliaca • A. rapax • A. clanga • Viet Nam • Laos
862	**Rarity** • Raptors • A. pomarina • Corsica
879	**Rarity** • Etymology • Raptors • A. heliaca • A. chrysaetos • A. pomarina • Malawi
918	**Rarity** • A. chrysaetos • A. clanga • Great Britain
922	**Rarity** • A. chrysaetos • Belgium
937	**Rarity** • Raptors • A. chrysaetos • A. heliaca • A. pomarina • A. clanga • Denmark
967	**Rarity** • A. clanga • West Germany
979	**Rarity** • Habitat • Raptors • A. chrysaetos • France
981	**Rarity** • Museum Specimen • A. pomarina • France
984	**Rarity** • A. clanga • Finland
1011	**Rarity** • A. clanga • Italy
1013	**Rarity** • Captivity • Poisoning • Measure • Plumage • Photo Account • A. heliaca • Italy
1053	**Rarity** • Collect • A. heliaca • Czechoslovakia
1076	**Rarity** • Raptors • A. clanga • Algeria
1166	**Rarity** • Raptors • A. pomarina • Malawi
1180	**Rarity** • (Title Only) • A. heliaca • Sweden
1183	**Rarity** • A. chrysaetos • Switzerland
1195	**Rarity** • A. chrysaetos • Switzerland
1211	**Rarity** • Collect • Measure • Plumage • Photo Account • A. pomarina • Italy
1214	**Rarity** • Raptor • A. clanga • France
1215	**Rarity** • Raptors • A. heliaca • France
1218	**Rarity** • A. chrysaetos • Ireland
1229	**Rarity** • Collect • A. clanga • Great Britain
1296	**Rarity** • Trapping • Prey • Status • A. chrysaetos • Georgia
1297	**Rarity** • Photo Account • A. pomarina • East Germany
1302	**Rarity** • A. clanga • France
1303	**Rarity** • Nest • A. clanga • Finland
1329	**Rarity** • Measure • Plumage • A. clanga • A. pomarina • Great Britain
1345	**Rarity** • Shooting • Plumage • Prey • Measure • A. chrysaetos • West Germany
1367	**Rarity** • A. nipalensis • Hungary
1375	**Rarity** • Raptors • A. clanga • A. pomarina • Denmark
1376	**Rarity** • Raptors • A. heliaca • A. pomarina • A. clanga • Denmark
1377	**Rarity** • Raptors • A. chrysaetos • A. clanga • A. pomarina • Denmark
1398	**Rarity** • Popular Monograph • (Title Only) • A. chrysaetos • Great Britain
1428	**Rarity** • A. heliaca • A. nipalensis • Sweden
1430	**Rarity** • Collect • A. clanga • A. pomarina • Switzerland
1446	**Rarity** • Shooting • A. clanga • France
1456	**Rarity** • A. clanga • West Germany
1459	**Rarity** • A. clanga • France
1474	**Rarity** • A. chrysaetos • Switzerland
1475	**Rarity** • Collect • A. heliaca • Switzerland
1476	**Rarity** • Collect • A. pomarina • Switzerland
1512	**Rarity** • A. clanga • Finland
1519	**Rarity** • Shooting • A. chrysaetos • A. clanga • Australia
1520	**Rarity** • Field Id • Eagles • A. chrysaetos • A. clanga • Switzerland
1549	**Rarity** • Literature Review • A. chrysaetos • Ontario
1560	**Rarity** • A. pomarina • Hungary
1567	**Rarity** • A. verreauxi • Israel
1568	**Rarity** • A. verreauxi • Israel
1569	**Rarity** • Raptors • A. clanga • A. pomarina • France
1595	**Rarity** • Status • Raptors • A. chrysaetos • A. clanga • A. heliaca • France
1596	**Rarity** • Raptors • A. pomarina • Luxembourg
1610	**Rarity** • A. chrysaetos • Netherlands
1611	**Rarity** • A. verreauxi • Israel
1645	**Rarity** • Raptors • A. chrysaetos • A. clanga • A. pomarina • Denmark
1646	**Rarity** • A. chrysaetos • Sweden
1655	**Rarity** • Raptors • A. pomarina • Poland
1664	**Rarity** • A. pomarina • South Africa
1694	**Rarity** • A. chrysaetos • West Germany
1707	**Rarity** • A. heliaca • Italy
1711	**Rarity** • A. pomarina • Netherlands
1712	**Rarity** • Etymology • Raptors • A. chrysaetos • A. clanga • Netherlands
1713	**Rarity** • Raptors • A. clanga • Netherlands
1718	**Rarity** • A. chrysaetos • West Germany
1787	**Rarity** • Raptors • A. chrysaetos • A. pomarina • A. clanga • Netherlands
1823	**Rarity** • A. rapax • A. clanga • West Germany
1832	**Rarity** • A. clanga • West Germany
1833	**Rarity** • Raptors • A. clanga • A. pomarina • A. heliaca • Austria
1834	**Rarity** • A. clanga • West Germany
1841	**Rarity** • Collect • Measure • Plumage • A. heliaca • France
1875	**Rarity** • A. clanga • Great Britain
1876	**Rarity** • Capture • A. clanga • Great Britain
1880	**Rarity** • Raptors • A. nipalensis • Algeria
1907	**Rarity** • Museum Specimen • Raptors • A. clanga • France
1959	**Rarity** • A. pomarina • Denmark
1961	**Rarity** • A. heliaca • Morocco
1969	**Rarity** • A. heliaca • West Germany
1974	**Rarity** • Raptors • A. clanga • Poland
1996	**Rarity** • Nest Survey • Raptors • A. chrysaetos • A. clanga • Scotland
2007	**Rarity** • A. chrysaetos • France
2020	**Rarity** • Museum Specimen • Measure • Plumage • Photo Account • A. heliaca • Italy
2024	**Rarity** • Prey • Raptors • A. nipalensis • Uganda
2027	**Rarity** • A. pomarina • A. clanga • A. chrysaetos • Hungary
2031	**Rarity** • A. chrysaetos • France

Master No.	Keywords
2033	**Rarity** • Status • (Title Only) • A. clanga • Sweden
2034	**Rarity** • Field Id • Eagles • A. heliaca • Sweden
2050	**Rarity** • Raptors • A. heliaca • Italy
2066	**Rarity** • Behavior • Prey • A. chrysaetos • Ireland
2070	**Rarity** • A. nipalensis • France
2073	**Rarity** • Raptors • A. pomarina • Morocco
2074	**Rarity** • Raptors • A. chrysaetos • A. nipalensis • France
2076	**Rarity** • Collect • A. heliaca • Austria
2137	**Rarity** • Photo Account • A. pomarina • France
2162	**Rarity** • A. chrysaetos • France
2164	**Rarity** • A. chrysaetos • Switzerland
2165	**Rarity** • Prey • A. audax • New Zealand
2167	**Rarity** • A. nipalensis • Hungary
2182	**Rarity** • A. chrysaetos • Ireland
2185	**Rarity** • A. chrysaetos • France
2197	**Rarity** • A. heliaca • West Germany
2201	**Rarity** • Captivity • A. pomarina • Italy
2203	**Rarity** • Collect • Photo Account • A. nipalensis • Italy
2216	**Rarity** • A. chrysaetos • Great Britain
2227	**Rarity** • Collect • A. clanga • France
2305	**Rarity** • A. clanga • Ireland
2358	**Rarity** • A. rapax • Italy
2364	**Rarity** • Raptors • A. heliaca • Formosa
2365	**Rarity** • Measure • A. clanga • Great Britain
2366	**Rarity** • Raptors • A. chrysaetos • A. pomarina • Poland
2418	**Rarity** • Taxonomy • Raptors • A. nipalensis • A. rapax • A. clanga • A. pomarina • A. heliaca • Great Britain • Europe
2423	**Rarity** • (Title Only) • A. rapax • Sweden
2435	**Rarity** • Wintering Population • Raptors • A. clanga • A. pomarina • A. heliaca • A. nipalensis • Kenya
2447	**Rarity** • Photo Account • A. heliaca • Italy
2487	**Rarity** • Shooting • Raptors • A. pomarina • Czechoslovakia
2508	**Rarity** • A. pomarina • France
2513	**Rarity** • Prey • Measure • A. clanga • Great Britain
2607	**Rarity** • Raptors • A. clanga • China
2617	**Rarity** • A. clanga • France
2622	**Rarity** • Collect • Plumage • A. clanga • West Germany
2623	**Rarity** • Photo Account • A. chrysaetos • West Germany
2624	**Rarity** • A. clanga • Great Britain
2632	**Rarity** • A. chrysaetos • Great Britain
2650	**Rarity** • A. clanga • France
2651	**Rarity** • Collect • A. clanga • Great Britain
2652	**Rarity** • Measure • Plumage • A. clanga • Great Britain
2687	**Rarity** • Raptors • A. chrysaetos • West Virginia
2692	**Rarity** • A. chrysaetos • Finland
2693	**Rarity** • A. clanga • West Germany
2777	**Rarity** • Raptors • A. nipalensis • Nepal
2828	**Rarity** • Capture • A. chrysaetos • West Germany
2841	**Rarity** • A. chrysaetos • Finland
2861	**Rarity** • Eagle • A. heliaca • Corsica
2941	**Rarity** • Prey • Eagles • A. pomarina • A. nipalensis • West Germany
2963	**Rarity** • Animal Damage • Measure • Captivity • Behavior • A. chrysaetos • Georgia
2964	**Rarity** • A. chrysaetos • Great Britain
3041	**Rarity** • Photo Account • A. nipalensis • Sweden
3042	**Rarity** • Eagles • A. heliaca • A. chrysaetos • A. pomarina • Sweden
3043	**Rarity** • Raptors • A. clanga • A. pomarina • A. nipalensis • Sweden
3045	**Rarity** • Taxonomy • Museum Specimen • Measure • A. rapax • Rumania
3056	**Rarity** • Status • A. clanga • A. pomarina • Sweden
3077	**Rarity** • A. chrysaetos • West Germany
3098	**Rarity** • Raptors • A. chrysaetos • A. pomarina • Netherlands
3100	**Rarity** • A. chrysaetos • Netherlands
3158	**Rarity** • A. nipalensis • West Germany
3178	**Rarity** • A. chrysaetos • Italy
3185	**Rarity** • A. chrysaetos • Tennessee
3210	**Rarity** • Migration • Raptors • A. clanga • Sweden
3212	**Rarity** • Field Id • (Title Only) • A. heliaca • A. nipalensis • Sweden
3213	**Rarity** • A. verreauxi • Egypt
3221	**Rarity** • Migration • A. chrysaetos • Tunisia
3222	**Rarity** • A. pomarina • France
3224	**Rarity** • A. chrysaetos • Italy
3254	**Rarity** • Nest • A. chrysaetos • Scotland
3258	**Rarity** • Collect • Prey • A. pomarina • France
3282	**Rarity** • A. nipalensis • Netherlands
3320	**Rarity** • A. chrysaetos • West Germany
3361	**Rarity** • A. clanga • Great Britain
3377	**Rarity** • Trapping • Raptors • A. chrysaetos • Ireland
3391	**Rarity** • Prey • A. clanga • Great Britain
3392	**Rarity** • Collect • A. clanga • Great Britain
3393	**Rarity** • A. clanga • Great Britain
3394	**Rarity** • A. clanga • Great Britain
3396	**Rarity** • A. chrysaetos • Ireland
3428	**Rarity** • A. chrysaetos • West Germany
3429	**Rarity** • A. pomarina • West Germany
3436	**Rarity** • Supplemental Feeding • Behavior • Plumage • Photo Account • A. heliaca • Spain
297	**Refuge** • Photo Account • Raptors • SRBPNA • Popular Monograph • A. chrysaetos • Idaho
474	**Refuge** • Migration • Field Id • Conservation • Raptors • A. chrysaetos • Pennsylvania
479	**Refuge** • Migration • Shooting • Protection • Conservation • Raptors • Hawk Mt. • A. chrysaetos • Pennsylvania
526	**Refuge** • Migration • Hawk Mt. • Raptors • A. chrysaetos • Pennsylvania
705	**Refuge** • Art • Conservation • Population Decline • A. heliaca • Spain
999	**Refuge** • Conservation • Management • Wintering Population • Raptors • A. chrysaetos • Great Britain
1024	**Refuge** • Raptors • Photo Account • Popular Monograph • A. heliaca • Spain
1355	**Refuge** • Popular Monograph • Habitat • A. chrysaetos • Idaho • Switzerland
1625	**Refuge** • Status • Shooting • Poisoning • Prey • Educational Material • Protection • Raptors • A. chrysaetos • Spain
1776	**Refuge** • Nest Survey • Winter Survey • Density • Electrocution • Shooting • Mortality • Prey • Predator/Prey • Raptors • SRBPNA • A. chrysaetos • Idaho
1973	**Refuge** • Poisoning • Winter Survey • Roost • Photo Account • Popular Monograph • A. chrysaetos • Wyoming

Master No.	Keywords
2109	**Refuge** • Conservation • Educational Material • Raptors • A. chrysaetos • Idaho
2241	**Refuge** • Photography • Behavior • Nest Survey • Reproductive Success • Mortality • Photo Account • Raptors • Popular Monograph • A. heliaca • Spain
2278	**Refuge** • Migration • Conservation • Survey • Shooting • Hawk Mt. • A. chrysaetos • Pennsylvania
2312	**Refuge** • Habitat • Raptors • A. chrysaetos • A. heliaca • Europe
2374	**Refuge** • Management • Conservation • Habitat Protection • Theory • Tech. • Raptors • A. chrysaetos • United States
2644	**Refuge** • Conservation • Nest • Measure • SRBPNA • Photo Account • Popular Monograph • A. chrysaetos • Idaho
3204	**Refuge** • Habitat • Conservation • Raptors • SRBPNA • A. chrysaetos • Idaho
3207	**Refuge** • Habitat • Nest Survey • Density • Prey • Predator/Prey • Disturbance • Telemetry • Territory • Reproductive Success • Raptors • SRBPNA • Photo Account • A. chrysaetos • Idaho
154	**Rehabilitation** • Banding • Captivity • Animal Damage • Popular Monograph • A. chrysaetos • North America
166	**Rehabilitation** • Raptors • A. chrysaetos • Minnesota
172	**Rehabilitation** • Mortality • Protection • Raptors • A. chrysaetos • Utah
181	**Rehabilitation** • Marking • Eagles • A. chrysaetos • Washington
205	**Rehabilitation** • Rarity • Attitude • A. clanga • Great Britain
601	**Rehabilitation** • A. chrysaetos • Arkansas
631	**Rehabilitation** • Conservation • Story • Trapping • Falconry • Trade • Wintering Population • Captivity • Raptors • Popular Monograph • A. chrysaetos • New York
668	**Rehabilitation** • A. chrysaetos • Switzerland
737	**Rehabilitation** • A. chrysaetos • Great Britain
827	**Rehabilitation** • Hacking • Raptors • A. chrysaetos
974	**Rehabilitation** • Shooting • Marking • Behavior • Eagles • A. chrysaetos • United States
1075	**Rehabilitation** • Telemetry • Raptors • A. chrysaetos • Washington
1096	**Rehabilitation** • A. chrysaetos • Minnesota
1097	**Rehabilitation** • Shooting • Raptors • A. chrysaetos • Minnesota
1837	**Rehabilitation** • Photo Account • A. chrysaetos • Northwest Territory
1940	**Rehabilitation** • Nest Survey • Raptors • A. chrysaetos • Nebraska
2019	**Rehabilitation** • Management • Educational Material • Falconry • Survey • Raptors • A. chrysaetos • California
2126	**Rehabilitation** • Status • Nest Survey • Wintering Population • Mortality • Raptors • A. chrysaetos • A. heliaca • A. pomarina • A. nipalensis • A. clanga • Israel
2128	**Rehabilitation** • Pesticides • Mortality • Prey • Wintering Population • Azodrin • Raptors • A. heliaca • A. clanga • A. pomarina • Israel
2129	**Rehabilitation** • Poisoning • Raptors • A. clanga • A. pomarina • A. heliaca • Israel
2215	**Rehabilitation** • Status • Protection • Mortality • Raptors • A. audax • Australia
2237	**Rehabilitation** • Shooting • Conservation • Disease • Photo Account • Raptors • A. chrysaetos • California
2285	**Rehabilitation** • Shooting • Raptors • A. chrysaetos • New Mexico
2286	**Rehabilitation** • Conservation • Photo Account • Raptors • Popular Monograph • A. chrysaetos • New Mexico
2301	**Rehabilitation** • A. chrysaetos • California
2363	**Rehabilitation** • Shooting • Raptors • Popular Monograph • A. chrysaetos • Texas
2459	**Rehabilitation** • A. chrysaetos • New York
2553	**Rehabilitation** • Story • Shooting • Nest • Conservation • Popular Monograph • A. chrysaetos • Wyoming
2554	**Rehabilitation** • Conservation • Nest • Behavior • Interspecific Behavior • Prey • Trapping • Animal Damage • Photo Account • Popular Monograph • A. chrysaetos • Wyoming
2576	**Rehabilitation** • Shooting • Trapping • Disease • Anaesthesia • Blood Sample • Raptors • A. chrysaetos • United States
2711	**Rehabilitation** • Trapping • Shooting • Banding • A. chrysaetos • Tennessee
2831	**Rehabilitation** • A. chrysaetos • Italy
2872	**Rehabilitation** • Raptors • A. chrysaetos • United States
2905	**Rehabilitation** • Falconry • Raptors • A. chrysaetos • California
2937	**Rehabilitation** • Disease • Physiology • Riboflavin • A. chrysaetos • Idaho
3007	**Rehabilitation** • Trapping • A. chrysaetos • North Carolina
3182	**Rehabilitation** • Story • Hunting Behavior • Falconry • Measure • Popular Monograph • A. chrysaetos • Texas
3236	**Rehabilitation** • Falconry • Shooting • Wintering Population • Photo Account • A. chrysaetos • Missouri
3389	**Rehabilitation** • A. chrysaetos • West Germany
3390	**Rehabilitation** • Raptors • A. chrysaetos • California
8	**Reproductive Success** • Density • Prey • Nest Failure • A. nipalensis • Russia
53	**Reproductive Success** • Nest Survey • A. chrysaetos • Colorado
198	**Reproductive Success** • Prey • Monograph • Nest Survey • Habitat • Literature Review • Disturbance • Pellet • Animal Damage • A. chrysaetos • Utah
233	**Reproductive Success** • Attitude • Density • Wintering Population • Food Requirements • Habitat • Management • Aerial Survey • Nest Survey • Raptors • A. chrysaetos • Montana
324	**Reproductive Success** • Food Requirements • Territory • Abandonment • Mortality • Cain/Abel • Density • Dispersal • Breeding Biology • Nest Survey • A. chrysaetos • Idaho
373	**Reproductive Success** • Territory • Sub-Adult • Shooting • Photo Account • A. chrysaetos • France
428	**Reproductive Success** • Wintering Population • Status • Aerial Survey • Proj. Out. • A. chrysaetos • Colorado • Idaho • Montana • Utah • New Mexico • Wyoming

Master No.	Keywords
521	**Reproductive Success** • Clutch Size • Raptors • A. chrysaetos • Morocco
534	**Reproductive Success** • Habitat • Density • Nest Survey • Breeding Biology • Interspecific Competition • Prey • Etymology • Plumage • Vocalization • Nest Adornment • Raptors • A. verreauxi • A. wahlbergi • Kenya
535	**Reproductive Success** • Nest Survey • Breeding Biology • Etymology • Prey • Interspecific Competition • Raptors • Photo Account • A. wahlbergi • A. verreauxi • Kenya
538	**Reproductive Success** • Nest Survey • Status • Population Estimate • Prey • Food Requirements • A. chrysaetos • Scotland
540	**Reproductive Success** • Nest Failure • Food Requirements • Home Range • Breeding Biology • Parental Behavior • Eagles • Theory • A. wahlbergi • A. verreauxi • Africa
552	**Reproductive Success** • Nest Survey • Prey • Habitat • Behavior • Status • Food Requirements • Animal Damage • Intraspecific Behavior • Predator/Prey • A. chrysaetos • Scotland
553	**Reproductive Success** • Cain/Abel • Behavior • Eagles • A. verreauxi • A. rapax • A. wahlbergi • Africa
614	**Reproductive Success** • Monograph • Nest Survey • Habitat • Directional Exposure • Behavior • A. chrysaetos • Utah
741	**Reproductive Success** • Nest Survey • Prey • Pesticides • A. chrysaetos • Scotland
756	**Reproductive Success** • Nest Survey • Prey • Raptors • A. chrysaetos • Colorado
757	**Reproductive Success** • Nest Survey • Aerial Survey • Density • Raptors • A. chrysaetos • Colorado
758	**Reproductive Success** • Nest Survey • Density • Winter Survey • Wintering Population • Status • Aerial Survey • A. chrysaetos • Colorado
859	**Reproductive Success** • Nest • A. chrysaetos • Austria
872	**Reproductive Success** • Mortality • A. chrysaetos • Scotland
873	**Reproductive Success** • Raptors • A. chrysaetos • Scotland
995	**Reproductive Success** • Nest Survey • Disturbance • A. chrysaetos • Scotland
996	**Reproductive Success** • Survey • Pesticides • Status • Disturbance • A. chrysaetos • Scotland
1007	**Reproductive Success** • Nest Survey • Territory • A. chrysaetos • Italy
1141	**Reproductive Success** • Nest Survey • Habitat • Density • Disturbance • Clutch Size • Raptors • A. verreauxi • Rhodesia
1144	**Reproductive Success** • Nest Survey • Status • Behavior • Fledging • Mortality • Banding • Marking • Cain/Abel • Breeding Biology • Interspecific Behavior • Hunting Behavior • Display • Courtship • Raptors • A. verreauxi • Rhodesia
1150	**Reproductive Success** • Density • Territory • Dispersal • Nest • Survey • Intraspecific Behavior • Marking • Display • A. verreauxi • Rhodesia
1151	**Reproductive Success** • Density • Population Dynamics • Prey • Nest • Climatology • Territory • Interspecific Competition • Mortality • Clutch Size • Pesticides • Eagles • A. verreauxi • Rhodesia
1154	**Reproductive Success** • Nest • Density • Prey • Status • Habitat Protection • Habitat • A. verreauxi • Rhodesia
1157	**Reproductive Success** • Nest Survey • A. verreauxi • Rhodesia
1158	**Reproductive Success** • Nest Survey • Cain/Abel • Raptors • A. verreauxi • Rhodesia
1177	**Reproductive Success** • Status • Nest Survey • Distribution • Protection • Photo Account • A. pomarina • East Germany
1223	**Reproductive Success** • Raptors • A. chrysaetos • Switzerland
1371	**Reproductive Success** • A. chrysaetos • Switzerland
1469	**Reproductive Success** • Roadside Survey • Raptors • A. chrysaetos • Nevada
1470	**Reproductive Success** • Roadside Survey • Raptors • A. chrysaetos • Nevada
1471	**Reproductive Success** • Roadside Survey • Raptors • A. chrysaetos • Nevada
1508	**Reproductive Success** • Literature Review • Prey • Pellet • Shooting • Predator/Prey • Nest Survey • Attitude • A. chrysaetos • North America
1775	**Reproductive Success** • Monograph • Status • Nest Survey • Growth • Cain/Abel • Banding • Marking • Mortality • Pesticides • Prey • Residues • A. chrysaetos • Idaho
1778	**Reproductive Success** • Nest Survey • Nest Chronology • Prey • Abandonment • Marking • Roadside Survey • Territory • Population Dynamics • Raptors • SRBPNA • A. chrysaetos • Idaho
1779	**Reproductive Success** • Prey • Aerial Survey • Banding • Marking • Population Dynamics • Raptors • SRBPNA • A. chrysaetos • Idaho
1780	**Reproductive Success** • Nest Survey • Nest Chronology • Prey • Abandonment • Marking • Roadside Survey • Territory • Population Dynamics • Raptors • SRBPNA • A. chrysaetos • Idaho
1781	**Reproductive Success** • Nest Survey • Marking • Habitat • Nest Chronology • Behavior • Prey • Banding • Territory • Population Dynamics • Raptors • SRBPNA • A. chrysaetos • Idaho
1782	**Reproductive Success** • Marking • Prey • Aerial Survey • Banding • Population Dynamics • SRBPNA • Raptors • A. chrysaetos • Idaho
1928	**Reproductive Success** • Prey • Pellet • Nest • Hunting Behavior • Interspecific Behavior • Eagles • A. chrysaetos • Oklahoma
1942	**Reproductive Success** • Food Habits • Nest • Time Lapse • Behavior • Economics • Animal Damage • Status • A. chrysaetos • Texas
1945	**Reproductive Success** • Mining • Telemetry • Disturbance • Nest Survey • Raptors • A. chrysaetos • Montana • Wyoming
1949	**Reproductive Success** • Pesticides • Nest Failure • Egg • Disturbance • A. chrysaetos • Scotland
1951	**Reproductive Success** • Pesticides • A. chrysaetos • Scotland
1998	**Reproductive Success** • Growth • Prey • Behavior • Photography Account • Popular Monograph • A. chrysaetos • Scotland

Master No.	Keywords
2154	**Reproductive Success** • Cain/Abel • Mortality • Management • Experiment • A. chrysaetos • A. heliaca
2157	**Reproductive Success** • Conservation • Fostering • Cain/Abel • Photo Account • A. heliaca • Spain
2184	**Reproductive Success** • Aerial Survey • Management • Survey • Clutch Size • Distribution • Habitat • Nest Survey • Prey • Raptors • A. chrysaetos • Arizona
2212	**Reproductive Success** • Nest • Management • Shooting • Prey • Density • Raptors • A. chrysaetos • Washington
2239	**Reproductive Success** • Survey • Territory • Nest Failure • Nest • Migration • Habitat • Management • Prey • Roadside Survey • Status • Winter Survey • Uinta Nf • Wasatch Mt. • Raptors • A. chrysaetos • Utah
2241	**Reproductive Success** • Refuge • Photography • Behavior • Nest Survey • Mortality • Photo Account • Raptors • Popular Monograph • A. heliaca • Spain
2269	**Reproductive Success** • Nest Survey • Directional Exposure • Density • Raptors • A. chrysaetos • Utah
2345	**Reproductive Success** • Prey • Shooting • Density • Nest • Eagles • A. chrysaetos • Italy
2347	**Reproductive Success** • Nest Survey • Raptors • A. chrysaetos • South Dakota
2398	**Reproductive Success** • A. chrysaetos • Switzerland
2464	**Reproductive Success** • Management • Habitat • Directional Exposure • Disturbance • Mortality • Nest Survey • Density • Nest Chronology • Prey • Prog. Rep. • A. chrysaetos • Wyoming
2515	**Reproductive Success** • Nest • Photo Account • A. chrysaetos • Switzerland
2522	**Reproductive Success** • Nest Survey • Raptors • A. chrysaetos • Idaho
2540	**Reproductive Success** • Nest Survey • Raptors • A. chrysaetos • South Dakota
2558	**Reproductive Success** • Nest Survey • Habitat • Protection • Clutch Size • Eagles • A. chrysaetos • Estonia
2594	**Reproductive Success** • Monograph • Pesticides • Density • Prey • Nest Survey • Mortality • A. chrysaetos • Montana
2712	**Reproductive Success** • Nest Survey • Disturbance • A. chrysaetos • Scotland
2723	**Reproductive Success** • Nest Survey • Banding • Mortality • Status • Raptors • A. chrysaetos • Finland
2854	**Reproductive Success** • Population Dynamics • Nest • Density • Prey • Breeding Biology • Distribution • Habitat • Directional Exposure • Mortality • Abandonment • Disturbance • Interspecific Behavior • Territory • A. chrysaetos • Utah
2966	**Reproductive Success** • Prey • Behavior • Incubation • Cain/Abel • A. rapax • South Africa
2969	**Reproductive Success** • Nest Survey • Habitat • Prey • Breeding Biology • Growth • Field Id • Vocalization • Migration • Behavior • Eagles • Photo Account • A. wahlbergi • A. rapax • Rhodesia
2973	**Reproductive Success** • Life History • Prey • Plumage • Cain/Abel • Banding • Interspecific Behavior • Piracy • Photo Account • Raptors • A. chrysaetos • Rhodesia
2975	**Reproductive Success** • Prey Behavior • Incubation • Cain/Abel • A. rapax • South Africa
2976	**Reproductive Success** • Prey • A. wahlbergi • Rhodesia
3039	**Reproductive Success** • Nest Survey • Mortality • A. heliaca • A. pomarina • Czechoslovakia
3079	**Reproductive Success** • Prey • Nest Survey • Clutch Size • Display • Hunting Behavior • Density • Migration • A. wahlbergi • South Africa
3110	**Reproductive Success** • Nest Survey • Distribution • Status • Density • Habitat Loss • Density • Raptors • Photo Account • A. chrysaetos • California
3207	**Reproductive Success** • Refuge • Habitat • Nest Survey • Density • Prey • Predator/Prey • Disturbance • Telemetry • Territory • Raptors • SRBPNA • Photo Account • A. chrysaetos • Idaho
3226	**Reproductive Success** • Density • Sub-Adult • Plumage • Museum Specimen • Population Decline • Illus. • A. heliaca • Spain
3270	**Reproductive Success** • Clutch Size • A. heliaca • Spain
3304	**Reproductive Success** • Nest Survey • Density • Directional Exposure • Territory • A. chrysaetos • Scotland
3444	**Reproductive Success** • Breeding Biology • Nest Survey • Habitat • Protection • Prey • Distribution • Photo Account • A. chrysaetos • Estonia
313	**Residues** • Poisoning • Toxicology • Autopsy • Captivity • Thal. Sulfate • A. chrysaetos
1031	**Residues** • Egg • Prey • Raptors • Mercury • A. chrysaetos • Alberta • Saskatchewan
1775	**Residues** • Monograph • Status • Nest Survey • Growth • Reproductive Success • Cain/Abel • Banding • Marking • Mortality • Pesticides • Prey • A. chrysaetos • Idaho
2582	**Residues** • PCB • Eagles • A. chrysaetos • North America
153	**Reward** • Shooting • Conservation • Protection • Aircraft • Attitude • Economics • A. chrysaetos • United States
161	**Reward** • Shooting • Mortality • Protection • A. chrysaetos • Idaho
173	**Reward** • Field Id • Measure • Protection • Winter Survey • Eagles • A. chrysaetos • Arkansas
1507	**Reward** • Conservation • Protection • Economics • Mortality • Pesticides • Poisoning • Eagles • A. chrysaetos • Wyoming • Colorado
2656	**Reward** • Winter Survey • Conservation • Shooting • Poisoning • Protection • Photo Account • Eagles • Popular Monograph • A. chrysaetos • Wyoming
2759	**Reward** • Animal Damage • Shooting • Conservation • Punishment • Attitude • Translocation • Popular Monograph • A. chrysaetos • Texas
3200	**Reward** • Protection • Attitude • Punishment • Aircraft • Raptors • A. chrysaetos • United States
30	**Roadside Survey** • Habitat • Status • A. chrysaetos • Texas

Master No.	Keywords
168	**Roadside Survey** • Raptors • A. chrysaetos • Oregon
423	**Roadside Survey** • Aerial Survey • Wintering Population • Nest Survey • Banding • Electrocution • A. chrysaetos • North America
762	**Roadside Survey** • Winter Survey • Raptors • A. chrysaetos • Idaho
763	**Roadside Survey** • Wintering Population • Nest • Survey • Raptors • A. chrysaetos • Idaho
947	**Roadside Survey** • Roost • Nest Survey • Prey • Hunting Behavior • Capture • Marking • Interspecific Behavior • Shooting • Electrocution • A. chrysaetos • Utah
968	**Roadside Survey** • Winter Survey • Raptors • A. chrysaetos • Colorado
1049	**Roadside Survey** • Raptors • A. chrysaetos • Kansas
1469	**Roadside Survey** • Reproductive Success • Raptors • A. chrysaetos • Nevada
1470	**Roadside Survey** • Reproductive Success • Raptors • A. chrysaetos • Nevada
1471	**Roadside Survey** • Reproductive Success • Raptors • A. chrysaetos • Nevada
1495	**Roadside Survey** • Raptors • A. chrysaetos • Montana
1675	**Roadside Survey** • Winter Survey • Raptors • A. chrysaetos • Colorado
1752	**Roadside Survey** • Distribution • Density • Raptors • A. chrysaetos • California
1756	**Roadside Survey** • Distribution • Raptors • A. chrysaetos • California
1778	**Roadside Survey** • Nest Survey • Nest Chronology • Prey • Reproductive Success • Abandonment • Marking • Territory • Population Dynamics • Raptors • SRBPNA • A. chrysaetos • Idaho
1780	**Roadside Survey** • Nest Survey • Nest Chronology • Prey • Reproductive Success • Abandonment • Marking • Territory • Population Dynamics • Raptors • SRBPNA • A. chrysaetos • Idaho
1938	**Roadside Survey** • Distribution • Raptors • A. chrysaetos • Nebraska
1939	**Roadside Survey** • Distribution • Raptors • A. chrysaetos • Nebraska
2058	**Roadside Survey** • Wintering Population • Raptors • A. chrysaetos • Nebraska
2147	**Roadside Survey** • Raptors • A. chrysaetos • Spain
2231	**Roadside Survey** • Raptors • A. audax • Australia
2239	**Roadside Survey** • Survey • Territory • Nest Failure • Nest • Migration • Habitat • Management • Reproductive Success • Prey • Status • Winter Survey • Uinta Nf • Wasatch Mt. • Raptors • A. chrysaetos • Utah
2549	**Roadside Survey** • Raptors • A. chrysaetos • New Mexico
2591	**Roadside Survey** • Wintering Population • Animal Damage • Age Structure • Prey • Raptors • A. chrysaetos • New Mexico
2826	**Roadside Survey** • Raptors • A. chrysaetos • New Hampshire
2852	**Roadside Survey** • Prey • Monograph • Density • Habitat • Interspecific Competition • Taxonomy • Vocalization • Hunting Behavior • Nest Chronology • Photo Account • Raptors • A. wahlbergi • A. rapax • A. nipalensis • A. heliaca • Kenya
2932	**Roadside Survey** • Wintering Population • Habitat • Raptors • A. chrysaetos • Colorado
2961	**Roadside Survey** • Winter Survey • Nest • Raptors • A. chrysaetos • Kansas
3121	**Roadside Survey** • Density • Raptors • A. wahlbergi • A. rapax • Ivory Coast
3125	**Roadside Survey** • Raptors • A. wahlbergi • A. rapax • Ivory Coast
3128	**Roadside Survey** • Status • Distribution • Habitat • A. chrysaetos • A. pomarina • A. clanga • A. wahlbergi • A. rapax • A. verreauxi • Ivory Coast • Upper Volta • Mali
3130	**Roadside Survey** • Habitat • Predator/Prey • A. rapax • A. nipalensis • A. pomarina • A. wahlbergi • Uganda
3134	**Roadside Survey** • Habitat • Raptors • A. rapax • A. clanga • A. nipalensis • A. pomarina • A. wahlbergi • A. verreauxi • Uganda
3348	**Roadside Survey** • Raptors • A. chrysaetos • Utah • Colorado • Kansas
3402	**Roadside Survey** • Raptors • A. chrysaetos • Utah
947	**Roost** • Roadside Survey • Nest Survey • Prey • Hunting Behavior • Capture • Marking • Interspecific Behavior • Shooting • Electrocution • A. chrysaetos • Utah
1269	**Roost** • Popular Monograph • A. chrysaetos • Scotland
1615	**Roost** • Wintering Population • Status • Raptors • A. chrysaetos • Kansas
1719	**Roost** • Raptors • A. heliaca • Yugoslavia
1893	**Roost** • Raptors • A. wahlbergi • A. verreauxi • Rhodesia
1973	**Roost** • Poisoning • Winter Survey • Refuge • Photo Account • Popular Monograph • A. chrysaetos • Wyoming
1334	**Senses** • Anatomy • A. wahlbergi
2382	**Senses** • Captivity • Raptors • A. audax • Australia
2541	**Senses** • Anatomy • Raptors • A. chrysaetos
3097	**Senses** • Trapping • Raptors • A. chrysaetos • Oregon
85	**Sexing** • Shooting • A. chrysaetos • Great Britain
735	**Sexing** • Anaesthesia • Raptors • A. rapax • Kenya
1358	**Sexing** • Raptors • A. chrysaetos
1505	**Sexing** • Dimorphism • Taxonomy • Museum Specimen • Measure • Raptors • A. chrysaetos • North America
2324	**Sexing** • Capture • Animal Damage • Banding • Management • Prey • Translocation • Marking • Measure • Captive Care • Tech. • Eagles • A. chrysaetos • Montana
2325	**Sexing** • Capture • Banding • Marking • Animal Damage • Management • Translocation • Prey • Survey • Measure • Mortality • Captive Care • Aging • A. chrysaetos • Montana
2326	**Sexing** • Capture • Banding • Marking • Animal Damage • Management • Translocation • Prey • Measure • Mortality • Captive Care • Aging • Food Requirements • A. chrysaetos • Montana
3033	**Sexing** • Agonistic Behavior • Museum Specimen • A. chrysaetos • Switzerland
714	**Sex Ratio** • Food Requirements • Breeding Biology • Predator/Prey • Cain/Abel • Directional Exposure • Growth • Behavior • A. chrysaetos • Idaho

Master No.	Keywords
7	**Shooting** • Status • Trapping • Eagles • A. chrysaetos • Virginia
71	**Shooting** • Prey • A. chrysaetos • Scotland
84	**Shooting** • Measure • A. chrysaetos • Great Britain
85	**Shooting** • Sexing • A. chrysaetos • Great Britain
99	**Shooting** • A. chrysaetos • Switzerland
100	**Shooting** • A. chrysaetos • Switzerland
116	**Shooting** • A. chrysaetos • Italy
121	**Shooting** • Eagles • A. chrysaetos • Texas
125	**Shooting** • Folklore • Protection • A. chrysaetos • Switzerland
129	**Shooting** • Prey • A. chrysaetos • Virginia
130	**Shooting** • A. chrysaetos • Virginia
138	**Shooting** • Mortality • A. audax • Australia
149	**Shooting** • Attitude • Aircraft • Eagles • Photo Account • Popular Monograph • A. chrysaetos • Wyoming • Colorado
153	**Shooting** • Conservation • Protection • Aircraft • Reward • Attitude • Economics • A. chrysaetos • United States
161	**Shooting** • Mortality • Protection • Reward • A. chrysaetos • Idaho
165	**Shooting** • A. chrysaetos • Texas
170	**Shooting** • Attitude • Punishment • Eagles • A. chrysaetos • Mississippi
258	**Shooting** • Rarity • A. clanga • West Germany
267	**Shooting** • Attitude • Conservation • Raptors • A. chrysaetos • Scotland
279	**Shooting** • Aircraft • A. chrysaetos • California
281	**Shooting** • A. chrysaetos • Illinois
290	**Shooting** • Status • Nest • Conservation • A. chrysaetos • France
294	**Shooting** • Animal Damage • Poisoning • Captivity • Lift • A. audax • Australia
309	**Shooting** • A. chrysaetos • New York
319	**Shooting** • Folklore • Animal Damage • Attitude • Aircraft • Popular Monograph • A. chrysaetos • Texas
333	**Shooting** • Museum Specimen • A. chrysaetos • Pennsylvania
356	**Shooting** • A. pomarina • Hungary
370	**Shooting** • A. chrysaetos • France
373	**Shooting** • Territory • Reproductive Success • Sub-Adult • Photo Account • A. chrysaetos • France
386	**Shooting** • Bounty • Disturbance • Habitat Loss • Protection • Status • Pesticides • Migration • Distribution • Population Decline • Raptors • A. chrysaetos • A. heliaca • A. pomarina • A. clanga • Europe
417	**Shooting** • A. chrysaetos • New Brunswick
420	**Shooting** • Raptors • A. audax • Australia
436	**Shooting** • A. chrysaetos • Missouri
456	**Shooting** • Status • Raptors • A. chrysaetos • A. heliaca • Spain
457	**Shooting** • Conservation • Attitude • Protection • Status • A. chrysaetos • Austria
479	**Shooting** • Migration • Protection • Refuge • Conservation • Raptors • Hawk Mt. • A. chrysaetos • Pennsylvania
527	**Shooting** • Migration • Conservation • Population Decline • Historical Account • Hawk Mt. • Raptors • Popular Monograph • A. chrysaetos • Pennsylvania
557	**Shooting** • Attitude • Photo Account • Popular Monograph • A. chrysaetos • Montana
572	**Shooting** • Economics • Protection • Eagles • A. chrysaetos • United States
573	**Shooting** • Conservation • Protection • Eagles • A. chrysaetos • United States
574	**Shooting** • Status • Nest • Egg • Raptors • A. heliaca • Cyprus
578	**Shooting** • Hunting Behavior • Attitude • Animal Damage • A. chrysaetos • Texas
584	**Shooting** • Banding • A. chrysaetos • Switzerland
599	**Shooting** • Nest • A. chrysaetos • Great Britain
632	**Shooting** • Population Decline • A. chrysaetos • Ireland
634	**Shooting** • Aircraft • Attitude • Animal Damage • A. audax • Australia
639	**Shooting** • Nest • A. chrysaetos • Italy
650	**Shooting** • A. chrysaetos • North Carolina • South Carolina
674	**Shooting** • Status • Protection • Raptors • A. chrysaetos • Italy
717	**Shooting** • Protection • A. pomarina • A. clanga • A. chrysaetos • A. heliaca • Europe
753	**Shooting** • Behavior • Nest • Photography • Story • Banding • Photo Account • Raptors • A. verreauxi • South Africa
786	**Shooting** • Poisoning • Raptors • A. chrysaetos • Yugoslavia
794	**Shooting** • Animal Damage • Bounty • Prey • Eagles • A. chrysaetos • California
812	**Shooting** • Nest • A. chrysaetos • Switzerland
819	**Shooting** • Mortality • Animal Damage • Poisoning • Eagles • A. chrysaetos • Ohio
820	**Shooting** • Animal Damage • A. chrysaetos • Ohio
856	**Shooting** • Nest • Egg • Prey • A. chrysaetos • Austria
891	**Shooting** • A. clanga • Italy
909	**Shooting** • Captivity • A. chrysaetos • Great Britain
947	**Shooting** • Roadside Survey • Roost • Nest Survey • Prey • Hunting Behavior • Capture • Marking • Interspecific Behavior • Electrocution • A. chrysaetos • Utah
974	**Shooting** • Rehabilitation • Marking • Behavior • Eagles • A. chrysaetos • United States
1010	**Shooting** • Eagles • A. clanga • Italy
1015	**Shooting** • A. clanga • Italy
1095	**Shooting** • Museum Specimen • Plumage • Measure • Taxonomy • A. clanga • Albania
1097	**Shooting** • Rehabilitation • Raptors • A. chrysaetos • Minnesota
1106	**Shooting** • Protection • Status • Raptors • A. chrysaetos • United States
1162	**Shooting** • Status • Nest Survey • Protection • Conservation • Habitat • Raptors • A. chrysaetos • A. heliaca • Spain
1197	**Shooting** • Measure • A. clanga • Italy
1345	**Shooting** • Rarity • Plumage • Prey • Measure • A. chrysaetos • West Germany
1372	**Shooting** • Attitude • A. chrysaetos • Switzerland
1427	**Shooting** • Prey • Behavior • Measure • A. chrysaetos • Georgia
1446	**Shooting** • Rarity • A. clanga • France
1449	**Shooting** • Raptors • A. chrysaetos • A. heliaca • Czechoslovakia
1457	**Shooting** • Measure • A. chrysaetos • Ohio
1479	**Shooting** • A. chrysaetos • Switzerland
1481	**Shooting** • A. chrysaetos • Switzerland
1484	**Shooting** • A. chrysaetos • Switzerland

Master No.	Keywords
1486	**Shooting** • A. chrysaetos • Switzerland
1487	**Shooting** • Attitude • A. chrysaetos • Switzerland
1489	**Shooting** • Historical Account • A. chrysaetos • Switzerland
1508	**Shooting** • Literature Review • Prey • Pellet • Predator/Prey • Nest Survey • Attitude • Reproductive Success • A. chrysaetos • North America
1519	**Shooting** • Rarity • A. chrysaetos • A. clanga • Australia
1605	**Shooting** • A. chrysaetos • North Carolina
1625	**Shooting** • Status • Poisoning • Prey • Refuge • Educational Material • Protection • Raptors • A. chrysaetos • Spain
1681	**Shooting** • Bounty • Mortality • A. chrysaetos • Texas
1702	**Shooting** • Measure • A. chrysaetos • Great Britain
1762	**Shooting** • Migration • Raptors • Photo Account • A. pomarina • Turkey • Greece
1768	**Shooting** • A. chrysaetos • West Germany
1776	**Shooting** • Nest Survey • Refuge • Winter Survey • Density • Electrocution • Mortality • Prey • Predator/Prey • Raptors • SRBPNA • A. chrysaetos • Idaho
1793	**Shooting** • A. chrysaetos • West Germany
1794	**Shooting** • Hunting Behavior • A. chrysaetos • West Germany
1854	**Shooting** • Nest • Captivity • A. chrysaetos • Texas
1899	**Shooting** • Protection • A. chrysaetos • Texas
1922	**Shooting** • Conservation • Status • Raptors • A. chrysaetos • California
1943	**Shooting** • Aerial Survey • Time Lapse • Animal Damage • Nest Survey • Prey • Habitat • Winter Survey • Capture • Age Structure • Food Requirements • A. chrysaetos • Texas
1970	**Shooting** • Captivity • A. chrysaetos • Montana
2041	**Shooting** • Protection • A. chrysaetos • Wyoming
2049	**Shooting** • Status • Pesticides • Migration • Protection • Poisoning • A. chrysaetos • Siberia
2103	**Shooting** • Conservation • Folklore • Habitat • Animal Damage • A. chrysaetos • United States
2108	**Shooting** • Nest • A. chrysaetos • New York
2212	**Shooting** • Nest • Reproductive Success • Management • Prey • Density • Raptors • A. chrysaetos • Washington
2233	**Shooting** • Attitude • Protection • Punishment • Eagles • Popular Monograph • A. chrysaetos • North America
2234	**Shooting** • Prey • Fat • A. chrysaetos • Montana
2237	**Shooting** • Rehabilitation • Conservation • Disease • Photo Account • Raptors • A. chrysaetos • California
2273	**Shooting** • Animal Damage • Wintering Population • Measure • Eagles • A. chrysaetos • Illinois
2278	**Shooting** • Migration • Refuge • Conservation • Survey • Hawk Mt. • A. chrysaetos • Pennsylvania
2285	**Shooting** • Rehabilitation • Raptors • A. chrysaetos • New Mexico
2298	**Shooting** • Electrocution • Mortality • Artificial Nest. • Eagles • A. chrysaetos • Idaho
2299	**Shooting** • Electrocution • Artificial Nest • Conservation • Management • Eagles • Popular Monograph • A. chrysaetos • Idaho
2345	**Shooting** • Prey • Density • Nest • Reproductive Success • Eagles • A. chrysaetos • Italy
2353	**Shooting** • Descriptive Summary • Economics • Eagles • A. chrysaetos • North America
2363	**Shooting** • Rehabilitation • Raptors • Popular Monograph • A. chrysaetos • Texas
2386	**Shooting** • Measure • A. chrysaetos • Italy
2457	**Shooting** • A. chrysaetos • France
2487	**Shooting** • Rarity • Raptors • A. pomarina • Czechoslovakia
2499	**Shooting** • Raptors • A. chrysaetos • Pennsylvania
2553	**Shooting** • Story • Rehabilitation • Nest • Conservation • Popular Monograph • A. chrysaetos • Wyoming
2555	**Shooting** • Aircraft • Punishment • Attitude • Eagle • Popular Monograph • A. chrysaetos • Texas
2566	**Shooting** • Nest • A. chrysaetos • Pennsylvania
2576	**Shooting** • Rehabilitation • Trapping • Disease • Anaesthesia • Blood Sample • Raptors • A. chrysaetos • United States
2587	**Shooting** • Nest • A. chrysaetos • Switzerland
2613	**Shooting** • A. chrysaetos • Tennessee
2629	**Shooting** • Poetry • A. chrysaetos • Ontario
2645	**Shooting** • Conservation • Poisoning • Egg Collection • Nest Robbing • Protection • Trade • Trapping • Raptors • A. chrysaetos • Great Britain
2656	**Shooting** • Winter Survey • Conservation • Poisoning • Protection • Reward • Photo Account • Eagles • Popular Monograph • A. chrysaetos • Wyoming
2668	**Shooting** • A. chrysaetos • Michigan
2682	**Shooting** • Migration • Population Decline • Raptors • A. chrysaetos • Sweden
2711	**Shooting** • Trapping • Rehabilitation • Banding • A. chrysaetos • Tennessee
2714	**Shooting** • Popular Monograph • Prey • Interspecific Behavior • Hunting Behavior • Eagles • A. chrysaetos • North America
2722	**Shooting** • Raptors • A. chrysaetos • Ontario
2756	**Shooting** • Status • Trapping • Collect • Photo Account • A. chrysaetos • Wisconsin
2758	**Shooting** • Attitude • Conservation • Story • Punishment • Popular Monograph • A. chrysaetos • Texas
2759	**Shooting** • Animal Damage • Conservation • Punishment • Attitude • Reward • Translocation • Popular Monograph • A. chrysaetos • Texas
2794	**Shooting** • Measure • Plumage • A. chrysaetos • Pennsylvania
2803	**Shooting** • Attitude • Aircraft • Animal Damage • Economics • Popular Monograph • A. chrysaetos • Texas
2804	**Shooting** • Bounty • Attitude • Protection • Economics • A. audax • Australia
2806	**Shooting** • Hunting Behavior • Predator/Prey • A. chrysaetos • Montana
2835	**Shooting** • Attitude • A. chrysaetos • Kansas
2913	**Shooting** • Survey • Prey • Animal Damage • Plumage • A. chrysaetos • Texas
2915	**Shooting** • Banding • Migration • A. chrysaetos • North America
2916	**Shooting** • Migration • A. chrysaetos • North America
2985	**Shooting** • Prey • Measure • Photo Account • Eagles • A. chrysaetos • New York
3011	**Shooting** • Nest • A. heliaca • Spain

Master No.	Keywords
3017	**Shooting** • Status • Migration • Conservation • Raptors • A. pomarina • Malawi
3026	**Shooting** • Measure • A. chrysaetos • Virginia
3034	**Shooting** • Trapping • Measure • A. chrysaetos • Pennsylvania
3058	**Shooting** • Habitat • Prey • Prey Behavior • Nest • Animal Damage • Egg • Behavior • A. verreauxi • South Africa
3112	**Shooting** • Status • Protection • Raptors • A. chrysaetos • A. pomarina • West Germany
3163	**Shooting** • Aircraft • Attitude • A. chrysaetos • Texas
3189	**Shooting** • Attitude • Popular Monograph • A. chrysaetos • Texas
3191	**Shooting** • Conservation • Poisoning • Electrocution • Attitude • Eagles • A. chrysaetos • United States
3192	**Shooting** • Protection • Poisoning • Electrocution • Attitude • Popular Monograph • A. chrysaetos • Wyoming
3236	**Shooting** • Rehabilitation • Falconry • Wintering Population • Photo Account • A. chrysaetos • Missouri
3263	**Shooting** • Status • Museum Specimen • Animal Damage • Raptors • A. verreauxi • A. nipalensis • A. rapax • South Africa
3287	**Shooting** • Captivity • A. chrysaetos • Iowa
3310	**Shooting** • Poisoning • Bounty • Conservation • Punishment • Story • Eagles • Popular Monograph • A. chrysaetos • Wyoming
3375	**Shooting** • Protection • Symbolism • Economics • Attitude • Conservation • A. chrysaetos • United States
3382	**Shooting** • Hunting Behavior • Raptors • A. chrysaetos • Switzerland
3401	**Shooting** • Prey • A. chrysaetos • West Germany
3417	**Shooting** • Measure • A. chrysaetos • New York
3440	**Shooting** • Animal Damage • A. chrysaetos • South Dakota • Iowa • Minnesota • Nebraska
3	**Status** • Migration • Raptors • A. nipalensis • Turkey
4	**Status** • Nest • Raptors • A. rapax • India
7	**Status** • Shooting • Trapping • Eagles • A. chrysaetos • Virginia
12	**Status** • Nest • Egg • A. chrysaetos • Lebanon • Syria
13	**Status** • Raptors • A. chrysaetos • A. heliaca • Egypt
19	**Status** • Raptors • A. chrysaetos • Washington
23	**Status** • Descriptive Summary • Field Id • Raptors • A. rapax • A. clanga • A. nipalensis • A. heliaca • India
26	**Status** • Id Key • Descriptive Summary • Etymology • Distribution • Field Id • Raptors • Illus. • A. chrysaetos • A. heliaca • A. rapax • A. nipalensis • A. clanga • A. pomarina • India
30	**Status** • Roadside Survey • Habitat • A. chrysaetos • Texas
32	**Status** • Nest • A. chrysaetos • New York
36	**Status** • Collect • Raptors • A. chrysaetos • Massachusetts
37	**Status** • Collect • Raptors • A. chrysaetos • Siberia
43	**Status** • Raptors • A. clanga • A. pomarina • A. heliaca • A. chrysaetos • Rumania • Bulgaria

Master No.	Keywords
45	**Status** • Raptors • A. heliaca • A. pomarina • A. clanga • A. nipalensis • A. chrysaetos • Israel
47	**Status** • Population Decline • Historical Account • Management • Distribution • Raptors • A. chrysaetos • Tennessee
65	**Status** • Museum Specimen • Wintering Population • Migration • Descriptive Summary • A. chrysaetos • Iowa
86	**Status** • Historical Account • Flight • Attitude • Prey • Photo Account • A. chrysaetos • Great Britain
106	**Status** • Raptors • A. wahlbergi • A. rapax • Kenya
132	**Status** • Raptors • A. chrysaetos • A. pomarina • A. clanga • Netherlands
140	**Status** • Raptors • A. chrysaetos • A. pomarina • A. clanga • Belgium
160	**Status** • Raptors • A. chrysaetos • Tennessee
177	**Status** • Raptors • A. chrysaetos • Kazakhstan
178	**Status** • Raptors • A. chrysaetos • A. heliaca • A. nipalensis • United Soviet Socialists Republic
179	**Status** • Raptors • A. nipalensis • A. heliaca • A. clanga • A. pomarina • Moldavia
188	**Status** • Habitat • Raptors • A. heliaca • A. pomarina • A. clanga • A. chrysaetos • Yugoslavia
192	**Status** • Nest • Prey • Raptors • A. chrysaetos • A. heliaca • Spain
208	**Status** • Raptors • A. audax • Australia
216	**Status** • Etymology • Raptors • A. chrysaetos • A. heliaca • A. clanga • Korea
219	**Status** • Distribution • A. chrysaetos • France
220	**Status** • Distribution • Raptors • A. heliaca • A. chrysaetos • A. clanga • A. pomarina • Moldavia
229	**Status** • Distribution • Nest • Prey • A. audax • Australia
249	**Status** • Prey • Nest • A. chrysaetos • Algeria • Morocco • Tunisia
252	**Status** • Raptors • A. chrysaetos • A. pomarina • Czechoslovakia
262	**Status** • Descriptive Summary • Etymology • Nest • Behavior • Plumage • A. heliaca • A. nipalensis • A. clanga • A. chrysaetos • Cyprus
263	**Status** • Descriptive Summary • Field Id • Raptors • A. heliaca • A. clanga • A. pomarina • Cyprus
268	**Status** • Wintering Population • Historical Account • A. chrysaetos • Kentucky
282	**Status** • Raptors • A. chrysaetos • Delaware
289	**Status** • Nest • Raptors • A. chrysaetos • France
290	**Status** • Shooting • Nest • Conservation • A. chrysaetos • France
299	**Status** • Rarity • Eagles • A. clanga • A. pomarina • A. heliaca • Austria
300	**Status** • Disturbance • Raptors • A. heliaca • A. pomarina • A. chrysaetos • Austria
301	**Status** • Nest • Raptors • A. chrysaetos • Czechoslovakia
303	**Status** • Raptors • Photo Account • A. chrysaetos • A. heliaca • A. pomarina • Bulgaria
308	**Status** • Distribution • Raptors • A. audax • Australia
312	**Status** • Raptors • A. chrysaetos • A. heliaca • A. nipalensis • A. clanga • A. pomarina • Turkey

Master No.	Keywords
318	**Status** • Protection • Raptors • A. chrysaetos • A. heliaca • A. clanga • A. nipalensis • A. pomarina • Hungary
330	**Status** • Wintering Population • Raptors • A. chrysaetos • Saskatchewan
331	**Status** • Raptors • A. chrysaetos • A. heliaca • Rumania
334	**Status** • Distribution • Raptors • A. clanga • Russia
342	**Status** • Raptors • A. verreauxi • Malawi
354	**Status** • Raptors • A. chrysaetos • Poland
358	**Status** • Management • Population Estimate • Protection • Artificial Nest • Raptors • A. chrysaetos • A. clanga • Finland
359	**Status** • Rarity • Raptors • A. chrysaetos • A. heliaca • A. clanga • Spain
361	**Status** • A. chrysaetos • West Germany
369	**Status** • Nest • Raptors • A. chrysaetos • France
379	**Status** • Collect • Raptors • A. chrysaetos • A. clanga • Italy
381	**Status** • Hunting Behavior • Raptors • A. chrysaetos • India
386	**Status** • Bounty • Disturbance • Habitat Loss • Protection • Shooting • Pesticides • Migration • Distribution • Population Decline • Raptors • A. chrysaetos • A. heliaca • A. pomarina • A. clanga • Europe
396	**Status** • Nest • Raptors • A. chrysaetos • Alaska
405	**Status** • Etymology • A. wahlbergi • A. rapax • Africa
409	**Status** • Raptors • A. chrysaetos • A. pomarina • West Germany
428	**Status** • Wintering Population • Reproductive Success • Aerial Survey • Proj. Out. • A. chrysaetos • Colorado • Idaho • Montana • Utah • New Mexico • Wyoming
429	**Status** • Raptors • A. chrysaetos • A. heliaca • A. nipalensis • A. pomarina • A. clanga • Bulgaria
430	**Status** • Etymology • Raptors • A. chrysaetos • A. pomarina • A. heliaca • Bulgaria
440	**Status** • Id Key • Raptors • Illus. • A. chrysaetos • North America
445	**Status** • Distribution • A. verreauxi • South Africa
446	**Status** • Raptors • A. rapax • A. wahlbergi • A. verreauxi • Ethiopia
450	**Status** • Collect • A. chrysaetos • France
456	**Status** • Shooting • Raptors • A. chrysaetos • A. heliaca • Spain
457	**Status** • Conservation • Shooting • Attitude • Protection • A. chrysaetos • Austria
469	**Status** • Poisoning • Raptors • A. audax • Australia
475	**Status** • Distribution • Prey • Historical Account • Hunting Behavior • Raptors • A. chrysaetos • Maine
477	**Status** • Habitat • Collect • Raptors • A. chrysaetos • A. clanga • Italy
482	**Status** • Raptors • A. wahlbergi • A. rapax • Zambia
502	**Status** • Distribution • Raptors • A. audax • Australia
516	**Status** • Raptors • A. heliaca • A. clanga • A. rapax • India
519	**Status** • Raptors • A. chrysaetos • A. rapax • Morocco
520	**Status** • Raptors • A. chrysaetos • A. rapax • Morocco
538	**Status** • Nest Survey • Population Estimate • Reproductive Success • Prey • Food Requirements • A. chrysaetos • Scotland
543	**Status** • Wintering Population • Habitat Loss • Raptors • AF • Africa
544	**Status** • Taxonomy • Field Id • Flight • Habitat • Terminology • Distribution • Mortality • Behavior • Life History • Longevity • Territory • Photo Account • Raptors • A. chrysaetos • A. clanga • Great Britain • Ireland • Scotland
546	**Status** • Hunting Behavior • Anatomy • A. chrysaetos • A. heliaca • A. audax • A. gurneyi • A. rapax • A. nipalensis • A. clanga • A. verreauxi • A. wahlbergi • A. pomarina
548	**Status** • Poisoning • Habitat Loss • Conservation • A. pomarina • A. clanga • A. heliaca • East Africa • Ethiopia
550	**Status** • Nest • A. wahlbergi • Africa
552	**Status** • Nest Survey • Prey • Habitat • Behavior • Reproductive Success • Food Requirements • Animal Damage • Intraspecific Behavior • Predator/Prey • A. chrysaetos • Scotland
555	**Status** • Raptors • A. audax • Tasmania
561	**Status** • A. chrysaetos • Switzerland
562	**Status** • Raptors • A. rapax • A. wahlbergi • Ivory Coast
563	**Status** • Raptors • A. chrysaetos • Nebraska
571	**Status** • Baja • A. chrysaetos • Mexico
574	**Status** • Nest • Shooting • Egg • Raptors • A. heliaca • Cyprus
582	**Status** • Raptors • A. chrysaetos • A. clanga • A. heliaca • A. pomarina • Libya
583	**Status** • Rarity • Raptors • A. chrysaetos • A. pomarina • Libya
593	**Status** • Collect • Raptors • A. rapax • Sudan
594	**Status** • Prey • Raptors • A. rapax • Sudan
595	**Status** • Collect • Behavior • Raptors • A. rapax • A. clanga • India
596	**Status** • Raptors • A. chrysaetos • A. heliaca • A. clanga • A. rapax • India
602	**Status** • Poisoning • Raptors • A. chrysaetos • Great Britain
625	**Status** • Raptors • A. chrysaetos • British Columbia
627	**Status** • Raptors • A. chrysaetos • Italy
638	**Status** • Raptors • A. chrysaetos • A. clanga • Italy
641	**Status** • Raptors • A. wahlbergi • Gambia
653	**Status** • Disturbance • A. heliaca • Spain
657	**Status** • Raptors • A. chrysaetos • A. rapax • Algeria
660	**Status** • Prey • Raptors • A. heliaca • A. clanga • Greece
664	**Status** • Distribution • A. chrysaetos • A. heliaca • A. nipalensis • A. clanga • China
672	**Status** • Nest • Prey • Raptors • A. chrysaetos • France
674	**Status** • Shooting • Protection • Raptors • A. chrysaetos • Italy
676	**Status** • A. audax • Australia
678	**Status** • Descriptive Summary • Etymology • Raptors • A. clanga • A. heliaca • A. chrysaetos • Japan
688	**Status** • Museum Specimen • A. clanga • West Germany
689	**Status** • (Title Only) • A. pomarina • West Germany

Permuted List of Keywords (Cont.)

Master No.	Keywords
696	**Status** • Distribution • Habitat • Raptors • A. audax • Australia
697	**Status** • Raptors • A. heliaca • Yugoslavia
720	**Status** • Mortality • Conservation • Raptors • A. audax • Australia
727	**Status** • Nest • Raptors • A. chrysaetos • Michigan
758	**Status** • Nest Survey • Density • Reproductive Success • Winter Survey • Wintering Population • Aerial Survey • A. chrysaetos • Colorado
761	**Status** • Conservation • Field Id • Raptors • A. chrysaetos • Colorado
769	**Status** • Conservation • A. heliaca • A. pomarina • Rumania
770	**Status** • Field Id • Descriptive Summary • Monograph • Distribution • Migration • Prey • Plumage • Measure • Etymology • Vocalization • A. pomarina • A. clanga • A. nipalensis • A. heliaca • A. chrysaetos • A. verreauxi • Africa • Europe
772	**Status** • Raptors • A. audax • Australia
777	**Status** • Rarity • A. clanga • Austria
788	**Status** • Raptors • A. chrysaetos • A. clanga • Poland
799	**Status** • Captivity • Plumage • Raptors • A. chrysaetos • A. heliaca • A. clanga • Rumania
804	**Status** • Raptors • A. chrysaetos • A. clanga • A. pomarina • Denmark
809	**Status** • A. chrysaetos • Switzerland
811	**Status** • A. chrysaetos • Switzerland
814	**Status** • A. clanga • France
815	**Status** • Egg Collection • Collect • Raptors • A. clanga • A. rapax • A. heliaca • India
817	**Status** • A. nipalensis • A. rapax • India
835	**Status** • Raptors • A. rapax • Southwest Africa
838	**Status** • Historical Account • Population Decline • Nest • Prey • A. chrysaetos • Ireland
843	**Status** • Raptors • A. pomarina • A. clanga • A. heliaca • A. chrysaetos • United Soviet Socialists Republic
844	**Status** • Raptors • A. chrysaetos • China
848	**Status** • Raptors • A. clanga • Cambodia
854	**Status** • Raptors • A. chrysaetos • A. heliaca • A. clanga • France
858	**Status** • Nest Survey • Conservation • A. chrysaetos • Austria
861	**Status** • A. pomarina • A. clanga • West Germany
871	**Status** • Distribution • Etymology • Life History • Habitat • Molt • Prey • Measure • Field Id • Migration • Raptors • A. chrysaetos • A. heliaca • A. rapax • A. clanga • A. pomarina • United Soviet Socialists Republic
875	**Status** • Migration • A. chrysaetos • Europe
878	**Status** • Protection • A. chrysaetos • France
880	**Status** • Etymology • Raptors • A. chrysaetos • A. pomarina • A. heliaca • Malawi
885	**Status** • Nest • Raptors • A. chrysaetos • Italy
886	**Status** • Nest Survey • Raptors • A. chrysaetos • Italy
889	**Status** • Mortality • Raptors • A. chrysaetos • Sardinia
903	**Status** • Protection • Conservation • Raptors • A. chrysaetos • A. clanga • Belorussia
917	**Status** • Raptors • A. chrysaetos • A. heliaca • A. nipalensis • A. clanga • Turkmenia
923	**Status** • A. chrysaetos • A. heliaca • A. pomarina • A. clanga • Poland
935	**Status** • Trapping • A. chrysaetos • Prince Edward Island
938	**Status** • Raptors • A. pomarina • Poland
941	**Status** • Descriptive Summary • Distribution • Management • Literature Review • Prey • A. chrysaetos • Washington
952	**Status** • Migration • Raptors • A. wahlbergi • Nigeria
965	**Status** • Nest • Egg • Taxonomy • Raptors • A. chrysaetos • A. heliaca • A. nipalensis • A. clanga • Yugoslavia • Bulgaria
970	**Status** • Collect • Prey • A. rapax • India
977	**Status** • Distribution • Raptors • A. chrysaetos • Switzerland
982	**Status** • A. chrysaetos • West Germany
987	**Status** • Descriptive Summary • Field Id • Distribution • Nest • Taxonomy • Egg • Illus. • A. chrysaetos • A. heliaca • A. nipalensis • A. rapax • A. clanga • A. pomarina • A. verreauxi • Africa
991	**Status** • Raptors • A. chrysaetos • Scotland
996	**Status** • Survey • Pesticides • Reproductive Success • Disturbance • A. chrysaetos • Scotland
997	**Status** • Distribution • A. chrysaetos • Scotland
1001	**Status** • Collect • Egg • Raptors • A. audax • Australia
1004	**Status** • Migration • Raptors • A. chrysaetos • New Jersey
1005	**Status** • Raptors • A. pomarina • Hungary
1008	**Status** • Descriptive Summary • Etymology • Nest Chronology • Prey • Raptors • A. chrysaetos • A. heliaca • A. pomarina • A. clanga • Switzerland
1020	**Status** • Habitat • Population Decline • Mortality • Raptors • A. chrysaetos • Europe
1041	**Status** • Prey • Nest • Habitat • Behavior • Photo Account • A. chrysaetos • Bulgaria
1043	**Status** • Behavior • Raptors • A. audax • Australia
1057	**Status** • Raptors • A. chrysaetos • Ontario
1060	**Status** • Field Id • Conservation • Raptors • A. chrysaetos • Colorado
1061	**Status** • Raptors • A. chrysaetos • A. pomarina • A. clanga • West Germany
1064	**Status** • Descriptive Summary • Hunting Behavior • Folklore • Raptors • A. chrysaetos
1068	**Status** • Raptors • A. audax • Australia
1074	**Status** • A. chrysaetos • Italy
1082	**Status** • Collect • A. chrysaetos • West Germany
1093	**Status** • Raptors • A. wahlbergi • A. rapax • Nigeria
1100	**Status** • Raptor • A. chrysaetos • Canada
1101	**Status** • Habitat Loss • Pesticides • Disturbance • Raptors • A. chrysaetos • Canada
1103	**Status** • Hunting Behavior • Prey • Poisoning • Raptors • A. chrysaetos • California
1106	**Status** • Protection • Shooting • Raptors • A. chrysaetos • United States
1112	**Status** • Raptors • A. audax • Australia
1115	**Status** • Protection • A. chrysaetos • A. nipalensis • A. heliaca • United Soviet Socialists Republic
1116	**Status** • Nest Survey • Raptors • A. clanga • Russia
1119	**Status** • Density • Raptors • A. clanga • Russia
1120	**Status** • Raptors • A. clanga • United Soviet Socialists Republic

Master No.	Keywords
1128	**Status** • Distribution • Raptors • A. chrysaetos • A. heliaca • A. nipalensis • United Soviet Socialists Republic
1129	**Status** • Distribution • Raptors • A. nipalensis • A. clanga • A. pomarina • A. heliaca • A. chrysaetos • United Soviet Socialists Republic
1135	**Status** • Raptors • A. heliaca • A. nipalensis • A. clanga • A. pomarina • Moldavia
1136	**Status** • Population Decline • A. heliaca • A. chrysaetos • A. nipalensis • A. clanga • A. pomarina • Moldavia
1144	**Status** • Nest Survey • Behavior • Fledging • Mortality • Banding • Marking • Cain/Abel • Reproductive Success • Breeding Biology • Interspecific Behavior • Hunting Behavior • Display • Courtship • Raptors • A. verreauxi • Rhodesia
1154	**Status** • Nest • Density • Prey • Habitat Protection • Habitat • Reproductive Success • A. verreauxi • Rhodesia
1162	**Status** • Nest Survey • Protection • Conservation • Shooting • Habitat • Raptors • A. chrysaetos • A. heliaca • Spain
1164	**Status** • Pesticides • Conservation • Disturbance • Population Estimate • Raptors • A. heliaca • A. chrysaetos • Spain
1169	**Status** • A. heliaca • A. nipalensis • A. clanga • Hungary
1173	**Status** • Etymology • Raptors • A. chrysaetos • A. heliaca • A. clanga • Russia
1177	**Status** • Nest Survey • Distribution • Reproductive Success • Protection • Photo Account • A. pomarina • East Germany
1190	**Status** • Raptors • A. pomarina • A. chrysaetos • West Germany
1191	**Status** • A. chrysaetos • Switzerland
1194	**Status** • Raptors • A. pomarina • West Germany
1201	**Status** • Museum Specimen • A. chrysaetos • A. pomarina • Italy
1205	**Status** • Nesting Behavior • Prey • Photo Account • A. chrysaetos • Scotland
1207	**Status** • Raptors • A. audax • Australia
1213	**Status** • Raptors • A. chrysaetos • A. heliaca • A. clanga • A. pomarina • Yugoslavia
1226	**Status** • A. chrysaetos • Switzerland
1227	**Status** • Nest Survey • A. chrysaetos • Switzerland
1231	**Status** • Descriptive Summary • Egg • Incubation • Raptors • A. pomarina • A. clanga • A. chrysaetos • West Germany
1232	**Status** • Nest • Clutch Size • Raptors • A. nipalensis • Ukraine
1241	**Status** • Field Id • Raptors • A. chrysaetos • Kansas
1276	**Status** • Descriptive Summary • Habitat • Raptors • A. chrysaetos • A. heliaca • A. clanga • Korea
1280	**Status** • Raptors • A. clanga • A. chrysaetos • A. rapax • Tunisia
1296	**Status** • Trapping • Prey • Rarity • A. chrysaetos • Georgia
1300	**Status** • Nest • A. chrysaetos • California
1319	**Status** • Piracy • Raptors • A. rapax • South Africa
1340	**Status** • Raptors • A. rapax • A. wahlbergi • South Africa
1356	**Status** • Raptors • A. chrysaetos • West Virginia
1363	**Status** • Field Id • Migration • Descriptive Summary • Prey • Raptors • A. chrysaetos • Wisconsin
1366	**Status** • Mortality • Protection • Management • Disturbance • Eagles • A. chrysaetos • North America
1381	**Status** • Raptors • A. rapax • South Africa
1390	**Status** • Raptors • A. pomarina • A. clanga • A. chrysaetos • West Germany
1395	**Status** • Prey • Etymology • Nest • Raptors • A. chrysaetos • A. heliaca • A. rapax • Morocco
1405	**Status** • Habitat • Raptors • A. audax • Australia
1422	**Status** • Raptors • A. wahlbergi • A. rapax • Rhodesia
1424	**Status** • Mortality • Nest Chronology • A. chrysaetos • Utah
1452	**Status** • Raptors • A. chrysaetos • A. pomarina • A. clanga • Czechoslovakia
1465	**Status** • Raptors • A. audax • Australia
1466	**Status** • Raptors • A. audax • Australia
1468	**Status** • Protection • Habitat Loss • Raptors • A. chrysaetos • Switzerland
1490	**Status** • Raptors • A. chrysaetos • A. pomarina • West Germany
1496	**Status** • Measure • Egg • Pesticides • Raptors • A. chrysaetos • California
1510	**Status** • Nest • A. chrysaetos • New York
1515	**Status** • Nest Chronology • Raptors • A. audax • Australia
1523	**Status** • A. chrysaetos • Yukon Territory
1525	**Status** • A. rapax • Southwest Africa
1526	**Status** • Descriptive Summary • Distribution • Measure • A. verreauxi • A. nipalensis • A. rapax • A. pomarina • A. wahlbergi • Southwest Africa
1530	**Status** • (Title Only) • A. pomarina • Europe
1531	**Status** • (Title Only) • A. clanga • Europe
1547	**Status** • Hunting Behavior • A. wahlbergi • Gold Coast
1552	**Status** • Taxonomy • Raptors • A. chrysaetos • A. heliaca • A. nipalensis • A. clanga • United Soviet Socialists Republic
1554	**Status** • Raptors • A. chrysaetos • Bulgaria
1556	**Status** • Prey • Raptors • A. audax • New Guinea
1557	**Status** • Raptors • A. audax • Australia
1559	**Status** • Eagles • A. chrysaetos • Finland
1589	**Status** • Population Decline • Distribution • Conservation • Raptors • Popular Monograph • A. clanga • A. heliaca • Europe
1590	**Status** • A. chrysaetos • Syria • Jordan
1595	**Status** • Rarity • Raptors • A. chrysaetos • A. clanga • A. heliaca • France
1613	**Status** • Nest • A. chrysaetos • A. clanga • A. heliaca • United Soviet Socialists Republic
1615	**Status** • Wintering Population • Roost • Raptors • A. chrysaetos • Kansas
1620	**Status** • Prey • Raptors • A. rapax • A. clanga • India
1621	**Status** • Raptors • A. clanga • Spain
1622	**Status** • Raptors • A. chrysaetos • Spain
1625	**Status** • Shooting • Poisoning • Prey • Refuge • Educational Material • Protection • Raptors • A. chrysaetos • Spain
1628	**Status** • Taxonomy • Raptors • A. rapax • A. nipalensis • A. verreauxi • Zambia

Master No.	Keywords
1637	**Status** • Raptors • A. pomarina • Poland
1641	**Status** • Descriptive Summary • Id Key • Distribution • A. verreauxi • A. nipalensis • A. rapax • A. pomarina • A. wahlbergi • Kenya • Uganda
1649	**Status** • Etymology • A. chrysaetos • Japan
1651	**Status** • Raptors • A. chrysaetos • Ontario
1657	**Status** • Raptors • A. audax • Australia
1663	**Status** • (Title Only) • A. nipalensis • Southwest Africa
1665	**Status** • Taxonomy • Raptors • A. chrysaetos • A. heliaca • A. clanga • A. rapax • A. pomarina • India
1666	**Status** • Etymology • Nest • Raptors • A. rapax • India
1668	**Status** • Etymology • Prey • Nest • Egg • Raptors • A. heliaca • A. nipalensis • A. rapax • A. clanga • A. pomarina • India
1672	**Status** • Raptors • A. chrysaetos • Denmark
1674	**Status** • Raptors • A. chrysaetos • New Brunswick
1690	**Status** • A. wahlbergi • Rhodesia
1695	**Status** • Raptors • A. chrysaetos • A. heliaca • Spain
1697	**Status** • Plumage • Distribution • Taxonomy • Etymology • Field Id • Literature Review • Eagles • A. pomarina • A. clanga • France • Belgium • Switzerland • Italy
1704	**Status** • A. chrysaetos • Ireland
1715	**Status** • Raptors • A. pomarina • West Germany
1720	**Status** • Nest • Prey • Mortality • Raptors • A. chrysaetos • A. clanga • A. heliaca • Yugoslavia
1724	**Status** • Density • Raptors • A. pomarina • A. clanga • A. chrysaetos • Latvia
1735	**Status** • Distribution • Raptors • A. verreauxi • A. wahlbergi • A. pomarina • A. rapax • South Africa
1737	**Status** • Nest • Distribution • Raptors • A. nipalensis • A. rapax • A. wahlbergi • A. verreauxi • South Africa
1745	**Status** • Raptors • A. clanga • Hungary
1749	**Status** • Raptors • A. chrysaetos • A. heliaca • A. clanga • United Soviet Socialists Republic
1753	**Status** • Raptors • A. audax • Australia
1754	**Status** • Distribution • Nest • Wintering Population • Habitat • Raptors • A. chrysaetos • Colorado
1755	**Status** • Nest • A. pomarina • Hungary
1775	**Status** • Monograph • Nest Survey • Growth • Reproductive Success • Cain/Abel • Banding • Marking • Mortality • Pesticides • Prey • Residues • A. chrysaetos • Idaho
1783	**Status** • Animal Damage • Raptors • A. chrysaetos • India
1785	**Status** • Descriptive Summary • Etymology • Collect • Raptors • A. chrysaetos • A. heliaca • A. pomarina • A. clanga • Egypt
1786	**Status** • Raptors • A. chrysaetos • A. pomarina • A. clanga • West Germany
1792	**Status** • Distribution • Raptors • A. chrysaetos • A. clanga • Ukraine
1795	**Status** • Nest • Raptors • A. chrysaetos • Alabama
1802	**Status** • Raptors • A. chrysaetos • Italy
1803	**Status** • Raptors • A. chrysaetos • Italy
1804	**Status** • A. chrysaetos • West Germany
1807	**Status** • Poisoning • Raptors • A. chrysaetos • Yugoslavia
1811	**Status** • Raptors • A. chrysaetos • Yugoslavia
1813	**Status** • Nest • Eagles • A. pomarina • A. clanga • West Germany
1817	**Status** • Raptors • A. pomarina • Czechoslovakia
1821	**Status** • Nest • Etymology • Raptors • A. chrysaetos • A. pomarina • A. clanga • Poland
1825	**Status** • Conservation • Raptors • A. chrysaetos • A. nipalensis • A. clanga • A. pomarina • A. heliaca • Turkey
1826	**Status** • Raptors • A. chrysaetos • A. heliaca • A. nipalensis • A. clanga • A. pomarina • Syria
1827	**Status** • A. chrysaetos • A. pomarina • A. nipalensis • A. heliaca • Turkey • Syria • Lebanon
1828	**Status** • Museum Specimen • Raptors • A. chrysaetos • A. heliaca • A. pomarina • A. clanga • Turkey
1840	**Status** • Collect • Raptors • A. chrysaetos • A. clanga • France
1842	**Status** • Raptors • A. chrysaetos • A. heliaca • A. clanga • France
1857	**Status** • Raptors • A. heliaca • A. nipalensis • A. verreauxi • A. rapax • A. wahlbergi • Kenya
1860	**Status** • Raptors • A. pomarina • A. chrysaetos • A. clanga • A. heliaca • Greece
1874	**Status** • Taxonomy • Plumage • Photo Account • A. chrysaetos • A. clanga • A. nipalensis • France
1877	**Status** • Nest • Raptors • A. audax • Australia
1882	**Status** • Raptors • A. chrysaetos • A. clanga • A. pomarina • A. nipalensis • Egypt
1895	**Status** • Raptors • A. chrysaetos • A. clanga • France
1898	**Status** • Egg Collection • Nest • A. chrysaetos • North Carolina
1902	**Status** • Conservation • Prey • Nest • Behavior • A. chrysaetos • Israel
1904	**Status** • Poisoning • Conservation • A. chrysaetos • Israel
1914	**Status** • Distribution • Habitat • Nest • Clutch Size • Descriptive Summary • Raptors • A. chrysaetos • New Mexico
1920	**Status** • Raptors • A. nipalensis • Kazakhstan
1922	**Status** • Conservation • Shooting • Raptors • A. chrysaetos • California
1924	**Status** • Etymology • A. chrysaetos • A. pomarina • A. clanga • A. heliaca • Yugoslavia
1937	**Status** • Nest • Raptors • A. chrysaetos • Siberia
1942	**Status** • Food Habits • Nest • Time Lapse • Behavior • Economics • Animal Damage • Reproductive Success • A. chrysaetos • Texas
1956	**Status** • Raptors • A. chrysaetos • Tunisia
1960	**Status** • Nest Chronology • Raptors • A. audax • Australia
1967	**Status** • Falconry • Raptors • A. chrysaetos • China
1982	**Status** • Museum Specimen • Raptors • A. wahlbergi • A. rapax • A. nipalensis • A. verreauxi • Sudan
1983	**Status** • Population Decline • A. nipalensis • Ukraine
1991	**Status** • Raptors • A. chrysaetos • Southwest Africa
2003	**Status** • Habitat • Raptors • A. audax • Australia
2004	**Status** • Attitude • Raptors • A. audax • Australia
2006	**Status** • Descriptive Summary • Field Id • Egg • Distribution • A. nipalensis • A. rapax • A. pomarina • A. wahlbergi • Africa
2009	**Status** • Nest • A. heliaca • Yugoslavia
2010	**Status** • Nest • A. pomarina • Yugoslavia

Master No.	Keywords
2011	**Status** • A. rapax • A. verreauxi • Chad
2012	**Status** • Population Decline • Attitude • Raptors • A. chrysaetos • A. clanga • Russia
2018	**Status** • Field Id • Educational Material • Raptors • A. chrysaetos • California
2022	**Status** • A. chrysaetos • Ireland
2029	**Status** • Nest • Habitat • Prey • Eagles • A. heliaca • A. chrysaetos • A. clanga • A. pomarina • Yugoslavia
2033	**Status** • Rarity • (Title Only) • A. clanga • Sweden
2038	**Status** • Bounty • Prey • Raptors • A. chrysaetos • A. heliaca • India
2039	**Status** • Nest • Collect • Prey • Raptors • A. wahlbergi • Rhodesia
2042	**Status** • Museum Specimen • Raptors • A. heliaca • A. nipalensis • A. clanga • China
2049	**Status** • Shooting • Pesticides • Migration • Protection • Poisoning • A. chrysaetos • Siberia
2052	**Status** • Raptors • A. audax • Australia
2061	**Status** • Raptors • A. pomarina • A. clanga • Poland
2064	**Status** • Raptors • A. heliaca • Bulgaria
2067	**Status** • Collect • Raptors • A. chrysaetos • France
2069	**Status** • Raptors • A. chrysaetos • A. nipalensis • A. clanga • France
2072	**Status** • Museum Specimen • Raptors • A. pomarina • A. chrysaetos • France
2075	**Status** • Raptors • A. rapax • A. pomarina • Africa
2083	**Status** • Nest • A. chrysaetos • Wyoming
2090	**Status** • Nest • Raptors • A. audax • Australia
2095	**Status** • Raptors • A. heliaca • A. clanga • A. chrysaetos • Turkey
2097	**Status** • Distribution • Habitat • Nest • Raptors • A. chrysaetos • North Dakota
2110	**Status** • Collect • Hunting Behavior • Raptors • A. chrysaetos • A. nipalensis • Pakistan
2112	**Status** • Collect • Raptors • A. nipalensis • A. heliaca • Iraq
2125	**Status** • Poisoning • Pesticides • Captivity • Density • Food Requirements • Habitat • Protection • Predator/Prey • Raptors • Photography Account • PCB • A. clanga • A. heliaca • A. nipalensis • A. verreauxi • A. pomarina • Israel
2126	**Status** • Nest Survey • Wintering Population • Mortality • Rehabilitation • Raptors • A. chrysaetos • A. heliaca • A. pomarina • A. nipalensis • A. clanga • Israel
2131	**Status** • Distribution • Plumage • Raptors • A. chrysaetos • A. heliaca • A. clanga • A. pomarina • Russia
2136	**Status** • Descriptive Summary • Distribution • Population Decline • Raptors • A. chrysaetos • A. clanga • A. pomarina • Finland
2139	**Status** • A. chrysaetos • New York
2140	**Status** • Trapping • A. chrysaetos • Quebec
2153	**Status** • Habitat Loss • Raptors • Photo Account • A. heliaca • Spain
2159	**Status** • Habitat Loss • Conservation • Forestry • Habitat Protection • Raptors • A. heliaca • Spain
2166	**Status** • Raptors • A. chrysaetos • A. clanga • Ukraine
2170	**Status** • A. nipalensis • A. rapax • A. wahlbergi • Rhodesia
2183	**Status** • Nest Survey • Population Decline • Habitat • Raptors • A. heliaca • Morocco
2192	**Status** • Raptors • A. chrysaetos • Syria • Iran
2198	**Status** • A. heliaca • West Germany
2199	**Status** • Raptors • A. chrysaetos • Norway
2205	**Status** • Raptors • A. chrysaetos • A. heliaca • A. nipalensis • A. clanga • Italy
2208	**Status** • Etymology • Raptors • A. chrysaetos • Italy
2209	**Status** • Raptors • A. chrysaetos • A. clanga • A. pomarina • A. rapax • A. nipalensis • A. heliaca
2215	**Status** • Protection • Rehabilitation • Mortality • Raptors • A. audax • Australia
2225	**Status** • Nest • Raptors • A. audax • Australia
2239	**Status** • Survey • Territory • Nest Failure • Nest • Migration • Habitat • Management • Reproductive Success • Prey • Roadside Survey • Winter Survey • Uinta Nf • Wasatch Mt. • Raptors • A. chrysaetos • Utah
2243	**Status** • Nest • Display • Raptors • A. pomarina • A. clanga • A. chrysaetos • A. heliaca • Bulgaria
2252	**Status** • Habitat • Raptors • A. clanga • India
2260	**Status** • Population Estimate • Eagles • A. chrysaetos • North America
2261	**Status** • Nest Survey • Disturbance • Prey • Population Dynamics • Management • Animal Damage • Artificial Nest • Predator/Prey • A. chrysaetos • Utah
2276	**Status** • Raptors • A. chrysaetos • Norway
2279	**Status** • Raptors • A. pomarina • A. clanga • A. heliaca • Hungary
2284	**Status** • Raptors • A. chrysaetos • Morocco
2291	**Status** • Behavior • A. chrysaetos • Wisconsin
2307	**Status** • Density • Conservation • Mortality • Pesticides • Raptors • A. chrysaetos • Scotland
2310	**Status** • Raptors • A. verreauxi • Rhodesia
2311	**Status** • Attitude • Population Estimate • Mortality • Protection • Prey • Animal Damage • Disturbance • Raptors • A. chrysaetos • Great Britain
2313	**Status** • Capture • Raptors • A. clanga • A. heliaca • Egypt
2315	**Status** • Protection • Animal Damage • A. chrysaetos • Austria
2316	**Status** • Population Decline • Banding • Nest Survey • Disturbance • Attitude • A. chrysaetos • Europe
2317	**Status** • A. chrysaetos • Europe
2328	**Status** • A. chrysaetos • A. clanga • Sweden
2333	**Status** • Nest • Prey • Raptors • A. heliaca • Spain
2342	**Status** • Etymology • Habitat • Interspecific Behavior • Collect • Nest • Raptors • A. verreauxi • A. nipalensis • A. rapax • Africa
2346	**Status** • Raptors • A. chrysaetos • Norway
2362	**Status** • Raptors • A. chrysaetos • A. heliaca • A. pomarina • Austria
2370	**Status** • Nest • Nest Survey • Monograph • Habitat • Mortality • Breeding Biology • Nest Chronology • Courtship • Density • Prey • Management • A. chrysaetos • Colorado
2372	**Status** • Nest Survey • Density • Habitat • Raptors • A. chrysaetos • Colorado
2377	**Status** • Conservation • Electrocution • Management • Literature Review • Behavior • Eagles • A. chrysaetos • North America

Master No.	Keywords
2378	**Status** • Nest • A. heliaca • Portugal
2388	**Status** • Collect • Eagles • A. chrysaetos • A. clanga • A. heliaca • Italy
2395	**Status** • Symbolism • Prey • Nest • Raptors • A. chrysaetos • South Dakota
2397	**Status** • Hunting Behavior • Popular Monograph • A. chrysaetos • Wyoming
2400	**Status** • Animal Damage • Predator/Prey • Aerial Survey • Habitat • Density • Population Dynamics • Food Habits • Nest • Interspecific Competition • Literature Review • A. chrysaetos • Texas • New Mexico
2402	**Status** • Raptors • A. verreauxi • South Africa
2404	**Status** • Nest • Nest Chronology • Clutch Size • Raptors • A. wahlbergi • A. rapax • Malawi
2405	**Status** • Nest • Raptors • A. rapax • A. verreauxi • Uganda
2408	**Status** • Distribution • Nest Survey • Photo Account • A. chrysaetos • Czechoslovakia
2422	**Status** • Distribution • Pesticides • Raptors • A. chrysaetos • Great Britain
2424	**Status** • Raptors • A. chrysaetos • A. heliaca • A. clanga • Iran
2427	**Status** • A. chrysaetos • Hungary
2431	**Status** • Raptors • A. chrysaetos • A. rapax • Tunisia • Algeria
2434	**Status** • Display • Flight • A. chrysaetos • Scotland
2440	**Status** • A. chrysaetos • A. clanga • West Germany
2441	**Status** • Raptors • A. heliaca • Yugoslavia
2442	**Status** • Raptors • A. clanga • A. pomarina • A. heliaca • Yugoslavia
2461	**Status** • Distribution • Raptors • A. chrysaetos • Arizona
2462	**Status** • Raptors • A. chrysaetos • Arizona
2463	**Status** • Collect • Vocalization • Behavior • Raptors • A. rapax • Somalia
2468	**Status** • Museum Specimen • Raptors • A. chrysaetos • A. clanga • A. pomarina • Italy
2469	**Status** • Taxonomy • Nest • Raptors • A. chrysaetos • A. pomarina • A. clanga • Yugoslavia
2481	**Status** • Raptors • A. chrysaetos • A. heliaca • Morocco
2491	**Status** • Distribution • Density • Habitat • Survey • Raptors • A. chrysaetos • New Mexico
2494	**Status** • Nest • Raptors • A. chrysaetos • Italy
2495	**Status** • Raptors • A. chrysaetos • A. heliaca • A. nipalensis • A. pomarina • Kazakhstan
2501	**Status** • A. chrysaetos • Pennsylvania
2502	**Status** • Raptors • A. chrysaetos • A. heliaca • A. clanga • Russia
2519	**Status** • Disturbance • A. chrysaetos • Great Britain
2527	**Status** • Wintering Population • Raptors • A. chrysaetos • Russia
2535	**Status** • Field Id • Distribution • Behavior • Raptors • A. verreauxi • A. rapax • A. nipalensis • A. pomarina • A. wahlbergi • South Africa
2537	**Status** • Nest • Prey • A. clanga • Russia
2539	**Status** • Nest • Prey • Raptors • A. chrysaetos • A. clanga • Russia
2548	**Status** • Raptors • A. clanga • A. pomarina • Greece
2556	**Status** • Raptors • A. chrysaetos • Estonia
2573	**Status** • Trade • Symbolism • Raptors • A. chrysaetos • South Dakota
2574	**Status** • Raptors • A. chrysaetos • A. pomarina • A. clanga • Italy
2575	**Status** • Collect • A. chrysaetos • France
2577	**Status** • Habitat • Raptors • A. chrysaetos • A. heliaca • A. nipalensis • A. clanga • Afghanistan
2583	**Status** • Collect • Prey • Habitat • Raptors • A. heliaca • A. nipalensis • A. clanga • A. rapax • A. pomarina • India
2590	**Status** • Management • A. chrysaetos • Idaho
2610	**Status** • (Title Only) • A. pomarina • West Germany
2620	**Status** • Raptors • A. nipalensis • Burma
2627	**Status** • Pesticides • Protection • Raptors • A. chrysaetos • Italy
2631	**Status** • Photo Account • A. chrysaetos • West Germany
2647	**Status** • Raptors • A. chrysaetos • A. clanga • A. pomarina • A. heliaca • Switzerland
2648	**Status** • Museum Specimen • A. chrysaetos • A. clanga • France
2675	**Status** • Collect • Prey • Wintering Population • Etymology • Raptors • A. chrysaetos • A. clanga • Yugoslavia
2683	**Status** • Migration • Historical Account • Raptors • A. chrysaetos • Sweden
2686	**Status** • (Title Only) • A. pomarina • West Germany
2703	**Status** • Raptors • A. heliaca • Hungary
2707	**Status** • Raptors • A. chrysaetos • A. rapax • A. wahlbergi • Chad
2708	**Status** • A. rapax • Congo
2715	**Status** • Nest Chronology • Raptors • A. chrysaetos • Montana
2716	**Status** • Raptors • A. chrysaetos • A. heliaca • Spain
2719	**Status** • Raptors • A. chrysaetos • A. heliaca • A. clanga • Spain • France
2720	**Status** • Museum Specimen • Behavior • Raptors • A. chrysaetos • A. clanga • Switzerland
2723	**Status** • Nest Survey • Banding • Mortality • Reproductive Success • Raptors • A. chrysaetos • Finland
2727	**Status** • Raptors • A. nipalensis • Caucasus
2736	**Status** • Density • Conservation • Protection • Survey • Raptors • A. chrysaetos • Sardinia
2744	**Status** • A. pomarina • East Germany
2756	**Status** • Shooting • Trapping • Collect • Photo Account • A. chrysaetos • Wisconsin
2779	**Status** • Nest • Prey • A. chrysaetos • Siberia
2786	**Status** • Photo Account • Raptors • A. chrysaetos • Sweden
2787	**Status** • Conservation • Protection • Habitat Loss • Mortality • Raptors • A. chrysaetos • A. clanga • Sweden
2795	**Status** • Nest • Egg • Raptors • A. rapax • Nigeria
2802	**Status** • Animal Damage • Raptors • A. wahlbergi • Uganda
2807	**Status** • Raptors • A. verreauxi • A. rapax • A. wahlbergi • Kenya
2809	**Status** • Collect • Raptors • A. rapax • Uganda
2811	**Status** • Distribution • Habitat • A. chrysaetos • Great Britain • Ireland • Scotland
2813	**Status** • Nest • Raptors • A. chrysaetos • Alaska
2816	**Status** • Collect • Nest • Prey • Raptors • A. wahlbergi • South Africa
2839	**Status** • Nest Survey • Habitat Loss • Raptors • Photo Account • A. chrysaetos • Nevada
2843	**Status** • Raptors • A. chrysaetos • Wyoming

Master No.	Keywords
2845	**Status** • Prey • Raptors • Yellowstone • A. chrysaetos • Wyoming
2850	**Status** • Nest Survey • Prey • Photo Account • A. heliaca • Czechoslovakia
2853	**Status** • Museum Specimen • Raptors • A. chrysaetos • A. heliaca • Portugal
2857	**Status** • Raptors • A. chrysaetos • A. heliaca • A. nipalensis • A. clanga • Afghanistan
2864	**Status** • Raptors • A. verreauxi • A. heliaca • A. nipalensis • A. rapax • A. clanga • Ethiopia
2866	**Status** • Raptors • A. chrysaetos • A. rapax • Morocco
2879	**Status** • Etymology • Descriptive Summary • Distribution • Raptors • A. chrysaetos • Canada
2895	**Status** • Prey • Raptors • A. chrysaetos • A. clanga • China
2896	**Status** • Nest Failure • Raptors • A. nipalensis • United Soviet Socialists Republic
2912	**Status** • Mortality • A. chrysaetos • North America
2935	**Status** • Raptors • A. chrysaetos • Alberta
2940	**Status** • Raptors • A. chrysaetos • A. clanga • Massachusettsn
2945	**Status** • Raptors • A. chrysaetos • West Germany
2948	**Status** • Distribution • A. chrysaetos • A. heliaca • A. pomarina • Switzerland
2949	**Status** • Nest • Vocalization • Prey • Behavior • Photo Account • A. chrysaetos • Switzerland
2953	**Status** • Nest Survey • A. chrysaetos • Australia
2955	**Status** • Distribution • (Title Only) • A. chrysaetos • Switzerland
2959	**Status** • (Title Only) • A. chrysaetos • Scotland
2965	**Status** • A. chrysaetos • North Carolina
2968	**Status** • Migration • Raptors • A. chrysaetos • Maryland
2988	**Status** • A. chrysaetos • Ireland
2995	**Status** • Raptors • A. heliaca • A. chrysaetos • A. rapax
2996	**Status** • Raptors • A. chrysaetos • Crete
2998	**Status** • Nest • Raptors • A. chrysaetos • A. clanga • Crete
3009	**Status** • Raptors • A. heliaca • A. clanga • China
3016	**Status** • Raptors • A. audax • Australia
3017	**Status** • Shooting • Migration • Conservation • Raptors • A. pomarina • Malawi
3029	**Status** • Raptors • A. chrysaetos • A. heliaca • A. nipalensis • A. clanga • Kirgizia
3030	**Status** • Raptors • A. chrysaetos • A. heliaca • A. nipalensis • A. clanga • Kirgizia
3036	**Status** • Descriptive Summary • Distribution • Prey • Measure • Raptors • A. chrysaetos • Oklahoma
3049	**Status** • Raptors • A. chrysaetos • Arizona
3052	**Status** • Collect • Raptors • A. heliaca • Afghanistan
3053	**Status** • Collect • Raptors • A. clanga • A. rapax • India
3054	**Status** • Prey • Raptors • A. rapax • A. wahlbergi • Rhodesia
3056	**Status** • Rarity • A. clanga • A. pomarina • Sweden
3057	**Status** • Nest • (Title Only) • A. chrysaetos • Sweden
3064	**Status** • Nest • Prey • A. pomarina • Australia
3074	**Status** • Raptors • A. chrysaetos • A. pomarina • A. heliaca • A. clanga • Poland
3078	**Status** • Nest • Migration • Photo Account • A. heliaca • Hungary
3081	**Status** • Raptors • A. chrysaetos • Oklahoma
3087	**Status** • Raptors • A. heliaca • A. clanga • Egypt
3096	**Status** • Raptors • A. chrysaetos • Nebraska
3103	**Status** • Distribution • Field Id • Conservation • Photo Account • Raptors • A. chrysaetos • France
3106	**Status** • Distribution • Raptors • A. audax • Australia
3110	**Status** • Nest Survey • Distribution • Density • Habitat Loss • Density • Reproductive Success • Raptors • Photo Account • A. chrysaetos • California
3111	**Status** • Distribution • Raptors • A. chrysaetos • A. rapax • A. heliaca • Morocco
3112	**Status** • Protection • Shooting • Raptors • A. chrysaetos • A. pomarina • West Germany
3122	**Status** • Habitat • Density • Raptors • A. rapax • A. wahlbergi • A. nipalensis • A. pomarina • Chad
3123	**Status** • Nest • Hunting Behavior • Raptors • A. wahlbergi • A. rapax • Ivory Coast
3128	**Status** • Roadside Survey • Distribution • Habitat • A. chrysaetos • A. pomarina • A. clanga • A. wahlbergi • A. rapax • A. verreauxi • Ivory Coast • Upper Volta • Mali
3132	**Status** • Raptors • A. chrysaetos • Afghanistan
3137	**Status** • Raptors • A. rapax • A. verreauxi • A. chrysaetos • A. nipalensis • A. pomarina • A. clanga • Yemen
3145	**Status** • Raptors • A. chrysaetos • A. heliaca • A. clanga • Albania • Yugoslavia
3146	**Status** • Population Dynamics • Nest Survey • Prey • A. chrysaetos • Idaho
3147	**Status** • Etymology • Raptors • A. chrysaetos • A. heliaca • A. nipalensis • A. rapax • A. clanga • India
3150	**Status** • Raptors • A. clanga • A. pomarina • Albania
3159	**Status** • Raptors • A. pomarina • A. clanga • Poland
3166	**Status** • Descriptive Summary • Raptors • A. clanga • A. pomarina • A. heliaca • A. chrysaetos • Poland
3171	**Status** • Density • A. chrysaetos • Austria
3172	**Status** • Collect • Raptors • A. wahlbergi • Rhodesia
3174	**Status** • Migration • Raptors • A. pomarina • A. nipalensis • A. wahlbergi • Rhodesia
3193	**Status** • Etymology • Raptors • A. chrysaetos • Labrador • Quebec
3194	**Status** • Animal Damage • A. chrysaetos • Tennessee
3197	**Status** • Raptors • A. rapax • Ethiopia
3199	**Status** • Prey • Raptors • A. nipalensis • Kazakhstan • Russia
3205	**Status** • Management • Animal Damage • Distribution • Wintering Population • A. chrysaetos • United States
3209	**Status** • Habitat • Raptors • A. heliaca • Hungary
3220	**Status** • Distribution • Habitat Loss • Raptors • A. chrysaetos • A. heliaca • A. clanga • A. pomarina • Greece
3223	**Status** • A. chrysaetos • Spain
3228	**Status** • Raptors • A. chrysaetos • A. pomarina • Lithuania
3232	**Status** • Nest • Raptors • A. heliaca • A. nipalensis • A. chrysaetos • A. pomarina • Rumania

Master No.	Keywords
3237	**Status** • Migration • Raptors • A. chrysaetos • A. pomarina • A. wahlbergi • A. rapax • Rwanda
3241	**Status** • Survey • Habitat • Raptors • A. heliaca • A. nipalensis • United Soviet Socialists Republic
3246	**Status** • Raptors • A. heliaca • A. nipalensis • Afghanistan
3251	**Status** • Nest • Raptors • A. chrysaetos • A. heliaca • A. rapax • Morocco
3256	**Status** • Raptors • A. nipalensis • A. clanga • A. chrysaetos • Mongolia
3263	**Status** • Museum Specimen • Shooting • Animal Damage • Raptors • A. verreauxi • A. nipalensis • A. rapax • South Africa
3266	**Status** • A. pomarina • France
3267	**Status** • Raptors • A. clanga • A. chrysaetos • Afghanistan
3268	**Status** • Raptors • A. rapax • A. pomarina • A. clanga • A. wahlbergi • A. verreauxi • Chad
3269	**Status** • Measure • Raptors • A. pomarina • Chad
3273	**Status** • Density • A. nipalensis • Caucasus
3275	**Status** • Raptors • A. chrysaetos • South Dakota
3276	**Status** • Raptors • A. chrysaetos • South Dakota
3297	**Status** • Raptors • A. heliaca • A. chrysaetos • Bulgaria
3305	**Status** • Habitat • Prey • Raptors • A. chrysaetos • Scotland
3311	**Status** • Raptors • A. chrysaetos • Washington
3325	**Status** • Breeding Biology • A. pomarina • Europe
3330	**Status** • Habitat • Management • A. chrysaetos • Washington
3331	**Status** • Raptors • A. chrysaetos • A. clanga • Spain
3341	**Status** • Egg Measure • Raptors • A. heliaca • A. nipalensis • A. rapax • India
3342	**Status** • Etymology • Raptors • A. chrysaetos • A. heliaca • A. nipalensis • A. rapax • India
3343	**Status** • Nest • Raptors • A. heliaca • A. nipalensis • A. clanga • A. rapax • India
3347	**Status** • Nest • Raptors • A. chrysaetos • Nebraska
3352	**Status** • Habitat • Raptors • A. verreauxi • A. rapax • A. wahlbergi • Rhodesia
3357	**Status** • Prey • Piracy • Collect • Raptors • A. heliaca • A. nipalensis • A. clanga • A. rapax • India
3358	**Status** • Raptors • A. chrysaetos • India
3362	**Status** • Nest • A. audax • Australia
3366	**Status** • Museum Specimen • Plumage • Raptors • A. nipalensis • A. heliaca • Hungary
3373	**Status** • Bounty • Conservation • Mortality • Eagles • A. chrysaetos • Norway
3374	**Status** • Distribution • Protection • Conservation • Raptors • A. chrysaetos • Norway
3381	**Status** • Nest • Raptors • A. audax • Australia
3384	**Status** • Nest • Raptors • A. wahlbergi • Rhodesia
3385	**Status** • Prey • Flight Speed • Raptors • A. wahlbergi • Rhodesia
3386	**Status** • Raptors • A. wahlbergi • Rhodesia
3387	**Status** • Prey • Raptors • A. rapax • A. verreauxi • South Africa
3395	**Status** • Interspecific Behavior • Raptors • A. chrysaetos • A. heliaca • Spain
3403	**Status** • Raptors • A. chrysaetos • Utah
3410	**Status** • Collect • Raptors • A. chrysaetos • Michigan
3421	**Status** • Raptors • A. heliaca • A. clanga • Malawi
3423	**Status** • Raptors • A. chrysaetos • Malawi
3430	**Status** • Raptors • A. chrysaetos • A. heliaca • A. clanga • A. pomarina • West Germany
3431	**Status** • (Title Only) • A. chrysaetos • West Germany
3442	**Status** • Etymology • Raptors • A. chrysaetos • A. clanga • Italy
3450	**Status** • Raptors • A. chrysaetos • A. heliaca • A. clanga • A. pomarina • Poland
3453	**Status** • Raptors • A. chrysaetos • Georgia Ssr
3454	**Status** • Raptors • A. chrysaetos • A. heliaca • A. nipalensis • A. pomarina • Ukraine
3456	**Status** • Nest • Raptors • A. chrysaetos • Italy
559	**Story** • Falconry • Measure • Photo Account • A. chrysaetos • Kirgizia • Great Britain
631	**Story** • Conservation • Trapping • Falconry • Rehabilitation • Trade • Wintering Population • Captivity • Raptors • Popular Monograph • A. chrysaetos • New York
753	**Story** • Behavior • Nest • Shooting • Photography • Banding • Photo Account • Raptors • A. verreauxi • South Africa
1311	**Story** • Conservation • Children's • Photo Account • SRBPNA • Popular Monograph • A. chrysaetos • Idaho
1361	**Story** • Falconry • Captive Breeding • Artificial Insemination • Photography Account • Popular Monograph • A. chrysaetos • United States
1403	**Story** • Migration • Raptors • Popular Monograph • A. chrysaetos • United States
1770	**Story** • Falconry • Behavior • Eagles • Photography Account • A. chrysaetos • Great Britain
2270	**Story** • Children's • A. chrysaetos • Colorado
2371	**Story** • Popular Monograph • Breeding Biology • Behavior • Habitat • Nest Survey • Hunting Behavior • Prey • Raptors • Illus. • A. chrysaetos • Colorado
2553	**Story** • Shooting • Rehabilitation • Nest • Conservation • Popular Monograph • A. chrysaetos • Wyoming
2758	**Story** • Attitude • Shooting • Conservation • Punishment • Popular Monograph • A. chrysaetos • Texas
2972	**Story** • Plumage • Photography • Breeding Biology • Prey • Photo Account • Popular Monograph • A. verreauxi • A. wahlbergi • A. rapax • Rhodesia • South Africa
3018	**Story** • Falconry • Behavior • Illus. • A. chrysaetos • Great Britain
3182	**Story** • Hunting Behavior • Rehabilitation • Falconry • Measure • Popular Monograph • A. chrysaetos • Texas
3310	**Story** • Shooting • Poisoning • Bounty • Conservation • Punishment • Eagles • Popular Monograph • A. chrysaetos • Wyoming
293	**Sub-Adult** • A. chrysaetos • Scotland
373	**Sub-Adult** • Territory • Reproductive Success • Shooting • Photo Account • A. chrysaetos • France
585	**Sub-Adult** • Nest • Collect • A. heliaca • Spain
3226	**Sub-Adult** • Density • Reproductive Success • Plumage • Museum Specimen • Population Decline • Illus. • A. heliaca • Spain
3314	**Sub-Adult** • Raptors • A. chrysaetos • Scotland
3368	**Sub-Adult** • Nest • Vocalization • A. chrysaetos • Norway

Master No.	Keywords
1448	**Supplemental Feeding** • Eagles • A. chrysaetos • Sweden
2705	**Supplemental Feeding** • Behavior • Raptors • Photo Account • A. chrysaetos • Corsica
3160	**Supplemental Feeding** • Wintering Population • Survey • Territory • Field Id • Age Structure • Illus. • A. chrysaetos • Sweden
3436	**Supplemental Feeding** • Behavior • Rarity • Plumage • Photo Account • A. heliaca • Spain
286	**Survey** • Trapping • (Title Only) • A. chrysaetos • South Dakota
434	**Survey** • Animal Damage • Economics • Attitude • Management • A. chrysaetos • North America
533	**Survey** • Nest • Photo Account • Popular Monograph • A. chrysaetos • Scotland
763	**Survey** • Roadside Survey • Wintering Population • Nest • Raptors • A. chrysaetos • Idaho
823	**Survey** • Distribution • Wintering Population • A. chrysaetos • New Mexico
830	**Survey** • Distribution • Raptors • A. audax • Australia
996	**Survey** • Pesticides • Reproductive Success • Status • Disturbance • A. chrysaetos • Scotland
1098	**Survey** • Banding • Disturbance • Nest • A. chrysaetos • Canada
1150	**Survey** • Density • Territory • Dispersal • Nest • Intraspecific Behavior • Marking • Reproductive Success • Display • A. verreauxi • Rhodesia
2014	**Survey** • Field Id • Raptors • A. chrysaetos • California
2015	**Survey** • Raptors • A. chrysaetos • California
2016	**Survey** • Raptors • A. chrysaetos • California
2017	**Survey** • Raptors • A. chrysaetos • California
2019	**Survey** • Management • Educational Material • Falconry • Rehabilitation • Raptors • A. chrysaetos • California
2184	**Survey** • Aerial Survey • Reproductive Success • Management • Clutch Size • Distribution • Habitat • Nest Survey • Prey • Raptors • A. chrysaetos • Arizona
2239	**Survey** • Territory • Nest Failure • Nest • Migration • Habitat • Management • Reproductive Success • Prey • Roadside Survey • Status • Winter Survey • Uinta Nf • Wasatch Mt. • Raptors • A. chrysaetos • Utah
2278	**Survey** • Migration • Refuge • Conservation • Shooting • Hawk Mt. • A. chrysaetos • Pennsylvania
2325	**Survey** • Capture • Banding • Marking • Animal Damage • Sexing • Management • Translocation • Prey • Measure • Mortality • Captive Care • Aging • A. chrysaetos • Montana
2360	**Survey** • Nest • Egg • Photo Account • A. chrysaetos • Europe
2467	**Survey** • Raptors • A. audax • Australia
2474	**Survey** • Historical Account • Trapping • A. chrysaetos • California
2491	**Survey** • Distribution • Density • Habitat • Status • Raptors • A. chrysaetos • New Mexico
2736	**Survey** • Status • Density • Conservation • Protection • Raptors • A. chrysaetos • Sardinia
2870	**Survey** • Banding • Marking • Nest • Capture • Prey • Food Requirements • Raptors • A. chrysaetos • A. wahlbergi • South Africa
2874	**Survey** • Management • Raptors • A. chrysaetos • Idaho
2913	**Survey** • Shooting • Prey • Animal Damage • Plumage • A. chrysaetos • Texas
2928	**Survey** • Mortality • Electrocution • Habitat Loss • Nest • Raptors • A. chrysaetos • Colorado
3160	**Survey** • Wintering Population • Territory • Field Id • Age Structure • Supplemental Feeding • Illus. • A. chrysaetos • Sweden
3241	**Survey** • Habitat • Status • Raptors • A. heliaca • A. nipalensis • United Soviet Socialists Republic
3350	**Survey** • Nest • Raptors • A. chrysaetos • Alaska
73	**Symbolism** • Folklore • Historical Account • Poetry • Eagles • A. chrysaetos
109	**Symbolism** • History • A. chrysaetos • United States
217	**Symbolism** • Life History • Raptors • A. chrysaetos
404	**Symbolism** • Descriptive Summary • Etymology • Raptors • A. chrysaetos • North America
729	**Symbolism** • Etymology • A. chrysaetos • Minnesota
966	**Symbolism** • Historical Account • Eagles • A. chrysaetos
1538	**Symbolism** • Folklore • Photo Account • A. chrysaetos • Canada
1661	**Symbolism** • Nest • Capture • A. chrysaetos • New Mexico
2013	**Symbolism** • Field Id • Conservation • Popular Monograph • A. chrysaetos • Colorado
2043	**Symbolism** • Protection • Eagles • Feathers • A. chrysaetos • North America
2178	**Symbolism** • Anthropology • A. chrysaetos • Oregon
2395	**Symbolism** • Status • Prey • Nest • Raptors • A. chrysaetos • South Dakota
2420	**Symbolism** • Capture • Historical Account • Anthropology • Raptors • Photo Account • A. chrysaetos • South Dakota
2573	**Symbolism** • Trade • Status • Raptors • A. chrysaetos • South Dakota
2654	**Symbolism** • Educational Material • Life History • Eagles • Children's Book • Popular Monograph • A. chrysaetos
2730	**Symbolism** • Trade • Eagles • A. chrysaetos • Pennsylvania
3375	**Symbolism** • Protection • Shooting • Economics • Attitude • Conservation • A. chrysaetos • United States
3379	**Symbolism** • Capture • Captivity • Mortality • Anthropology • A. chrysaetos • North America
3412	**Symbolism** • Nest • Distribution • Raptors • A. chrysaetos • Utah
1337	**Synonomy** • Collect • A. wahlbergi • South Africa
2141	**Synonomy** • A. chrysaetos
2604	**Synonomy** • A. chrysaetos
127	**Talon Grappling** • Intraspecific Behavior • A. chrysaetos • Scotland • Switzerland
487	**Talon Grappling** • Intraspecific Behavior • A. verreauxi • South Africa
1382	**Talon Grappling** • A. verreauxi • Rhodesia
1455	**Talon Grappling** • Capture • A. chrysaetos • Great Britain
2391	**Talon Grappling** • Intraspecific Behavior • A. chrysaetos • Scotland

Permuted List of Keywords (Cont.)

Master No.	Keywords
2411	**Talon Grappling** • Interspecific Behavior • Nest • Behavior • Photography • Prey • Nest Adornment • Cain/Abel • Photo Account • Popular Monograph • A. chrysaetos • Scotland
2593	**Talon Grappling** • Behavior • Interspecific Behavior • Raptors • A. chrysaetos • Washington
2680	**Talon Grappling** • Nest • Raptors • A. pomarina • Yugoslavia
2974	**Talon Grappling** • Interspecific Behavior • Raptors • A. verreauxi • Rhodesia
3037	**Talon Grappling** • Behavior • A. chrysaetos • Oklahoma
2323	**Taxidermy** • A. chrysaetos • North America
16	**Taxonomy** • Etymology • A. pomarina • West Germany
38	**Taxonomy** • Raptors • A. chrysaetos • North America
48	**Taxonomy** • Distribution • Raptors • A. audax • A. gurneyi • Australia
55	**Taxonomy** • Plumage • A. pomarina • A. nipalensis • A. heliaca • India
59	**Taxonomy** • Plumage • Measure • A. nipalensis • A. heliaca • India
67	**Taxonomy** • Distribution • Egg • Plumage • Prey • Raptors • A. verreauxi • A. rapax • Southwest Africa
68	**Taxonomy** • Museum Specimen • Plumage • Distribution • A. nipalensis • A. clanga • A. pomarina • France
74	**Taxonomy** • Plumage • A. nipalensis • A. pomarina • A. clanga • West Germany
88	**Taxonomy** • A. clanga • A. pomarina
95	**Taxonomy** • A. clanga • Great Britain
96	**Taxonomy** • A. chrysaetos • Great Britain
105	**Taxonomy** • A. chrysaetos • Great Britain
112	**Taxonomy** • A. pomarina • A. nipalensis
251	**Taxonomy** • Descriptive Summary • Nest • Egg • Measure • Plumage • Photo Account • Raptors • A. chrysaetos • A. rapax • Algeria
384	**Taxonomy** • Capture • Aging • Field Id • Photo Account • A. nipalensis • A. rapax • Southwest Africa
407	**Taxonomy** • Raptors • A. heliaca • A. nipalensis • A. clanga
408	**Taxonomy** • Descriptive Summary • Plumage • Measure • A. chrysaetos • A. heliaca • A. clanga • A. pomarina
437	**Taxonomy** • A. chrysaetos • North America
498	**Taxonomy** • Field Id • Molt • Measure • Museum Specimen • Behavior • Wintering Population • Interspecific Competition • Conservation • Eagles • Photo Account • A. nipalensis • A. pomarina • Africa
505	**Taxonomy** • Plumage • A. heliaca • A. clanga • India
506	**Taxonomy** • Plumage • Raptors • A. clanga • India
507	**Taxonomy** • Plumage • A. heliaca • A. nipalensis • India
509	**Taxonomy** • Plumage • Etymology • A. nipalensis • A. clanga • A. rapax • A. pomarina • India
510	**Taxonomy** • Plumage • Etymology • A. heliaca • A. nipalensis • A. clanga • A. pomarina • India
511	**Taxonomy** • Plumage • Etymology • A. rapax • India
512	**Taxonomy** • Plumage • A. rapax • A. nipalensis • A. pomarina • India
513	**Taxonomy** • Plumage • Etymology • A. clanga • A. pomarina • India
514	**Taxonomy** • Plumage • Etymology • Raptors • A. clanga • A. pomarina • A. rapax
515	**Taxonomy** • A. pomarina • A. clanga
517	**Taxonomy** • A. heliaca • A. pomarina
518	**Taxonomy** • Plumage • A. clanga • A. rapax
544	**Taxonomy** • Status • Field Id • Flight • Habitat • Terminology • Distribution • Mortality • Behavior • Life History • Longevity • Territory • Photo Account • Raptors • A. chrysaetos • A. clanga • Great Britain • Ireland • Scotland
551	**Taxonomy** • Monograph • Conservation • Courtship • Territory • A. chrysaetos • A. heliaca • A. clanga • A. rapax • A. verreauxi • A. wahlbergi • A. audax • A. gurneyi • A. nipalensis • A. pomarina
598	**Taxonomy** • A. chrysaetos • United Soviet Socialists Republic
721	**Taxonomy** • (Title Only) • A. audax • Australia
864	**Taxonomy** • Descriptive Summary • Distribution • Raptors • A. chrysaetos • A. heliaca • A. nipalensis • A. pomarina • A. clanga • Europe • Asia
866	**Taxonomy** • A. chrysaetos
882	**Taxonomy** • Etymology • A. chrysaetos • A. clanga • A. rapax • A. audax • A. verreauxi • A. gurneyi • A. wahlbergi • A. pomarina • A. clanga • A. heliaca
913	**Taxonomy** • A. nipalensis • Zambia
915	**Taxonomy** • Plumage • A. nipalensis • A. heliaca • A. clanga • A. pomarina • A. rapax • India
916	**Taxonomy** • Plumage • Measure • A. nipalensis • A. clanga • A. pomarina
965	**Taxonomy** • Status • Nest • Egg • Raptors • A. chrysaetos • A. heliaca • A. nipalensis • A. clanga • Yugoslavia • Bulgaria
987	**Taxonomy** • Descriptive Summary • Field Id • Distribution • Nest • Status • Egg • Illus. • A. chrysaetos • A. heliaca • A. nipalensis • A. rapax • A. clanga • A. pomarina • A. verreauxi • Africa
989	**Taxonomy** • Plumage • Measure • Museum Specimen • A. clanga • A. pomarina • A. rapax • A. nipalensis • France
1040	**Taxonomy** • Monograph • Breeding Biology • Prey • Vocalization • Territory • Migration • Mortality • Molt • Falconry • A. chrysaetos • A. verreauxi • A. audax
1058	**Taxonomy** • Museum Specimen • Collect • Behavior • Raptors • A. pomarina • A. clanga • A. nipalensis • Nepal
1081	**Taxonomy** • Literature Review • Collect • A. rapax • Africa
1095	**Taxonomy** • Museum Specimen • Shooting • Plumage • Measure • A. clanga • Albania
1209	**Taxonomy** • Plumage • Captivity • Behavior • Eagles • A. chrysaetos • Nova Scotia
1212	**Taxonomy** • Descriptive Summary • Etymology • Nest • Prey • A. chrysaetos • Switzerland
1289	**Taxonomy** • Museum Specimen • Type Spec. • Raptors • A. heliaca • A. rapax • A. nipalensis • A. clanga • A. pomarina • A. audax

Master No.	Keywords
1314	**Taxonomy** • Plumage • Measure • A. clanga • Europe
1322	**Taxonomy** • Plumage • A. heliaca • France
1323	**Taxonomy** • Raptors • A. heliaca • A. nipalensis • Egypt
1324	**Taxonomy** • A. clanga
1325	**Taxonomy** • Anatomy • A. heliaca • A. nipalensis
1326	**Taxonomy** • Plumage • Museum Specimen • A. nipalensis • Turkey
1327	**Taxonomy** • Plumage • A. heliaca • A. nipalensis
1328	**Taxonomy** • Museum Specimen • Plumage • Captivity • Eagles • A. chrysaetos • A. heliaca • A. rapax • A. clanga • A. nipalensis
1347	**Taxonomy** • Plumage • Anatomy • Behavior • Raptors • Theory • A. chrysaetos
1394	**Taxonomy** • Museum Specimen • Raptors • A. heliaca • A. rapax • A. clanga • A. pomarina
1396	**Taxonomy** • Plumage • Measure • Museum Specimen • A. rapax • A. nipalensis • Great Britain
1397	**Taxonomy** • Plumage • Museum Specimen • A. audax • Australia
1418	**Taxonomy** • Measure • Plumage • Photo Account • A. clanga • A. pomarina • Belgium
1451	**Taxonomy** • Distribution • Raptors • A. chrysaetos • North America
1505	**Taxonomy** • Dimorphism • Sexing • Museum Specimen • Measure • Raptors • A. chrysaetos • North America
1552	**Taxonomy** • Status • Raptors • A. chrysaetos • A. heliaca • A. nipalensis • A. clanga • United Soviet Socialists Republic
1553	**Taxonomy** • A. clanga • A. pomarina • A. nipalensis
1599	**Taxonomy** • Prey • A. nipalensis • A. clanga • Burma
1623	**Taxonomy** • A. clanga
1628	**Taxonomy** • Status • Raptors • A. rapax • A. nipalensis • A. verreauxi • Zambia
1639	**Taxonomy** • Dimorphism • Measure • Plumage • Distribution • Prey • Behavior • Management • Habitat • Territory • Mortality • Life History • Economics • Literature Review • A. chrysaetos • Oregon • Washington
1652	**Taxonomy** • Field Id • A. rapax • A. nipalensis
1665	**Taxonomy** • Status • Raptors • A. chrysaetos • A. heliaca • A. clanga • A. rapax • A. pomarina • India
1682	**Taxonomy** • A. gurneyi • A. audax • New Guinea • Australia
1683	**Taxonomy** • Anatomy • Osteology • Museum Specimen • Raptors • A. chrysaetos • A. rapax
1684	**Taxonomy** • Anatomy • Osteology • Museum Specimen • Raptors • A. chrysaetos • A. rapax
1685	**Taxonomy** • Anatomy • Osteology • Museum Specimen • Raptors • A. chrysaetos • A. rapax
1686	**Taxonomy** • Anatomy • Osteology • Museum Specimen • Raptors • A. chrysaetos • A. rapax
1697	**Taxonomy** • Plumage • Status • Distribution • Etymology • Field Id • Literature Review • Eagles • A. pomarina • A. clanga • France • Belgium • Switzerland • Italy
1760	**Taxonomy** • Measure • A. chrysaetos • Japan
1765	**Taxonomy** • Plumage • A. clanga • Europe
1784	**Taxonomy** • Collect • A. chrysaetos • Tunisia
1801	**Taxonomy** • Distribution • Raptors • A. nipalensis • A. rapax • United Soviet Socialists Republic
1808	**Taxonomy** • Etymology • Raptors • A. chrysaetos • A. heliaca • A. clanga • Yugoslavia
1874	**Taxonomy** • Plumage • Status • Photo Account • A. chrysaetos • A. clanga • A. nipalensis • France
1913	**Taxonomy** • A. clanga
1932	**Taxonomy** • Raptors • A. rapax • India
2055	**Taxonomy** • A. chrysaetos • A. clanga • Australia
2078	**Taxonomy** • Raptors • A. chrysaetos • North America
2118	**Taxonomy** • A. nipalensis • A. rapax • Africa • Asia
2120	**Taxonomy** • Descriptive Summary • Distribution • Plumage • Hunting Behavior • Raptors • A. chrysaetos • A. nipalensis • A. rapax • A. clanga • Saudi Arabia
2148	**Taxonomy** • Distribution • Field Id • Management • Habitat • Nest Survey • Prey • Breeding Biology • Photo Account • Eagles • A. pomarina • Europe
2149	**Taxonomy** • Breeding Biology • Cain/Abel • Nest • Prey • Photo Account • Popular Monograph • A. pomarina • West Germany
2354	**Taxonomy** • A. chrysaetos • North America
2355	**Taxonomy** • Collect • Plumage • Raptors • A. rapax • Kenya
2418	**Taxonomy** • Rarity • Raptors • A. nipalensis • A. rapax • A. clanga • A. pomarina • A. heliaca • Great Britain • Europe
2421	**Taxonomy** • Plumage • A. clanga
2469	**Taxonomy** • Status • Nest • Raptors • A. chrysaetos • A. pomarina • A. clanga • Yugoslavia
2516	**Taxonomy** • Collect • Raptors • A. pomarina • A. clanga • A. chrysaetos • A. nipalensis • A. heliaca • Poland
2552	**Taxonomy** • Descriptive Summary • Distribution • Plumage • Habitat • Measure • Raptors • A. gurneyi • New Guinea
2635	**Taxonomy** • Raptors • A. nipalensis • South Africa
2661	**Taxonomy** • A. clanga • A. pomarina
2666	**Taxonomy** • Plumage • A. clanga
2718	**Taxonomy** • Plumage • Raptors • A. chrysaetos • A. heliaca • A. clanga • Spain
2770	**Taxonomy** • A. heliaca • A. rapax • Spain
2783	**Taxonomy** • A. pomarina
2791	**Taxonomy** • A. chrysaetos • Europe
2822	**Taxonomy** • Paleontology • Eagles • A. chrysaetos
2823	**Taxonomy** • Paleontology • Eagles • A. chrysaetos • Oregon
2852	**Taxonomy** • Roadside Survey • Prey • Monograph • Density • Habitat • Interspecific Competition • Vocalization • Hunting Behavior • Nest Chronology • Photo Account • Raptors • A. wahlbergi • A. rapax • A. nipalensis • A. heliaca • Kenya
3028	**Taxonomy** • Plumage • Molt • A. chrysaetos • A. verreauxi • A. heliaca • A. rapax • A. clanga
3044	**Taxonomy** • Plumage • A. chrysaetos
3045	**Taxonomy** • Museum Specimen • Rarity • Measure • A. rapax • Rumania
3067	**Taxonomy** • Egg • Anatomy • Egg Measure • Raptors • A. chrysaetos • A. heliaca • A. pomarina • A. clanga • A. nipalensis
3068	**Taxonomy** • Egg • Anatomy • Egg Measure • Raptors • A. chrysaetos • A. heliaca • A. pomarina • A. clanga • A. nipalensis

Master No.	Keywords
3072	**Taxonomy** • Collect • Raptors • A. clanga • Siberia
3089	**Taxonomy** • Captivity • Behavior • Plumage • A. rapax
3148	**Taxonomy** • A. chrysaetos • India
3149	**Taxonomy** • A. chrysaetos • Nepal
3195	**Taxonomy** • Egg Measure • Photo Account • Raptors • A. chrysaetos • A. heliaca • A. rapax
3234	**Taxonomy** • Raptors • A. rapax • A. wahlbergi • A. verreauxi • Kenya • Uganda
3250	**Taxonomy** • Collect • Raptors • A. clanga • Hungary
3257	**Taxonomy** • Descriptive Summary • Distribution • Habitat • Raptors • A. rapax • A. nipalensis • A. heliaca • A. clanga • A. pomarina • A. verreauxi
3353	**Taxonomy** • Raptors • A. wahlbergi • Africa
3380	**Taxonomy** • A. chrysaetos • Europe
3405	**Taxonomy** • Distribution • Plumage • Subspecies • A. chrysaetos
3449	**Taxonomy** • Raptors • A. rapax • Somalia
3452	**Taxonomy** • (Title Only) • A. clanga • A. pomarina • Ukraine
764	**Telemetry** • Eagles • A. chrysaetos • North America
767	**Telemetry** • Protection • Nest • Falconry • Behavior • Pesticides • Photo Account • Popular Monograph • A. chrysaetos • United States
925	**Telemetry** • Raptors • A. chrysaetos • North America
959	**Telemetry** • Egg • Tech. • Equipment • A. chrysaetos • North America
1075	**Telemetry** • Rehabilitation • Raptors • A. chrysaetos • Washington
1387	**Telemetry** • Time Lapse • Prey • A. chrysaetos • Texas
1431	**Telemetry** • Poisoning • Toxicology • Habitat • Raptors • Cmpd. 1080 • A. chrysaetos • California
1944	**Telemetry** • Marking • Abnormal Behavior • Courtship • Nest Failure • Raptors • A. chrysaetos • Wyoming • Idaho
1945	**Telemetry** • Mining • Disturbance • Reproductive Success • Nest Survey • Raptors • A. chrysaetos • Montana • Wyoming
2085	**Telemetry** • Territory • Display • Agonistic Behavior • Behavior • Nest • A. chrysaetos • Wyoming • Montana
2684	**Telemetry** • Physiology • Temperature • A. chrysaetos • Idaho
2800	**Telemetry** • Molt • Captivity • Eagles • A. chrysaetos • Washington
3207	**Telemetry** • Refuge • Habitat • Nest Survey • Density • Prey • Predator/Prey • Disturbance • Territory • Reproductive Success • Raptors • SRBPNA • Photo Account • A. chrysaetos • Idaho
3239	**Telemetry** • Egg • Temperature • Measure • Incubation • Tech. • A. chrysaetos • United States
1155	**Temperature** • Behavior • Physiology • Nest • A. verreauxi • Rhodesia
2684	**Temperature** • Physiology • Telemetry • A. chrysaetos • Idaho
3021	**Temperature** • Captivity • Growth • Food Requirements • Measure • Molt • Pellet • Photo Account • Raptors • A. chrysaetos • California
3239	**Temperature** • Egg • Telemetry • Measure • Incubation • Tech. • A. chrysaetos • United States
544	**Terminology** • Taxonomy • Status • Field Id • Flight • Habitat • Distribution • Mortality • Behavior • Life History • Longevity • Territory • Photo Account • Raptors • A. chrysaetos • A. clanga • Great Britain • Ireland • Scotland
955	**Terminology** • Behavior • Pesticides • Parental Behavior • Vocalization • Disturbance • Measure • Breeding Biology • Parasite • Disease • Growth • A. chrysaetos • Montana
957	**Terminology** • Monograph • Behavior • Growth • Pesticides • Parental Behavior • Vocalization • Disturbance • Measure • Breeding Biology • Photography • Captivity • Parasite • Disease • Photo Account • A. chrysaetos • Montana
245	**Territory** • Raptors • A. audax • Australia
324	**Territory** • Reproductive Success • Food Requirements • Abandonment • Mortality • Cain/Abel • Density • Dispersal • Breeding Biology • Nest Survey • A. chrysaetos • Idaho
372	**Territory** • Habitat • Nest • Wintering Population • Hunting Behavior • A. chrysaetos • France
373	**Territory** • Reproductive Success • Sub-Adult • Shooting • Photo Account • A. chrysaetos • France
544	**Territory** • Taxonomy • Status • Field Id • Flight • Habitat • Terminology • Distribution • Mortality • Behavior • Life History • Longevity • Photo Account • Raptors • A. chrysaetos • A. clanga • Great Britain • Ireland • Scotland
551	**Territory** • Monograph • Conservation • Taxonomy • Courtship • A. chrysaetos • A. heliaca • A. clanga • A. rapax • A. verreauxi • A. wahlbergi • A. audax • A. gurneyi • A. nipalensis • A. pomarina
671	**Territory** • Interspecific Competition • Eagles • A. chrysaetos • France • Morocco
895	**Territory** • Nest • Egg • Measure • Intraspecific Behavior • Courtship • Interspecific Behavior • Hunting Behavior • A. chrysaetos • California
1007	**Territory** • Nest Survey • Reproductive Success • A. chrysaetos • Italy
1040	**Territory** • Monograph • Breeding Biology • Prey • Vocalization • Taxonomy • Migration • Mortality • Molt • Falconry • A. chrysaetos • A. verreauxi • A. audax
1118	**Territory** • Raptors • Aptors • A. clanga • United Soviet Socialists Republic
1150	**Territory** • Density • Dispersal • Nest • Survey • Intraspecific Behavior • Marking • Reproductive Success • Display • A. verreauxi • Rhodesia
1151	**Territory** • Reproductive Success • Density • Population Dynamics • Prey • Nest • Climatology • Interspecific Competition • Mortality • Clutch Size • Pesticides • Eagles • A. verreauxi • Rhodesia

Master No.	Keywords
1639	**Territory** • Taxonomy • Dimorphism • Measure • Plumage • Distribution • Prey • Behavior • Management • Habitat • Mortality • Life History • Economics • Literature Review • A. chrysaetos • Oregon • Washington
1778	**Territory** • Nest Survey • Nest Chronology • Prey • Reproductive Success • Abandonment • Marking • Roadside Survey • Population Dynamics • Raptors • SRBPNA • A. chrysaetos • Idaho
1780	**Territory** • Nest Survey • Nest Chronology • Prey • Reproductive Success • Abandonment • Marking • Roadside Survey • Population Dynamics • Raptors • SRBPNA • A. chrysaetos • Idaho
1781	**Territory** • Nest Survey • Reproductive Success • Marking • Habitat • Nest Chronology • Behavior • Prey • Banding • Population Dynamics • Raptors • SRBPNA • A. chrysaetos • Idaho
1934	**Territory** • Breeding Biology • Nest • Density • A. heliaca • Kazakhstan
1935	**Territory** • Distribution • A. heliaca • Kazakhstan
2085	**Territory** • Telemetry • Display • Agonistic Behavior • Behavior • Nest • A. chrysaetos • Wyoming • Montana
2239	**Territory** • Survey • Nest Failure • Nest • Migration • Habitat • Management • Reproductive Success • Prey • Roadside Survey • Status • Winter Survey • Uinta Nf • Wasatch Mt. • Raptors • A. chrysaetos • Utah
2854	**Territory** • Population Dynamics • Nest • Density • Prey • Breeding Biology • Distribution • Habitat • Directional Exposure • Reproductive Success • Mortality • Abandonment • Disturbance • Interspecific Behavior • A. chrysaetos • Utah
2856	**Territory** • Breeding Biology • Nest Survey • Habitat • Hunting Behavior • Interspecific Behavior • Prey • Raptors • A. chrysaetos • Utah
2871	**Territory** • Marking • Home Range • Dispersal • Raptors • Photo Account • A. rapax • A. wahlbergi • South Africa
3160	**Territory** • Wintering Population • Survey • Field Id • Age Structure • Supplemental Feeding • Illus. • A. chrysaetos • Sweden
3207	**Territory** • Refuge • Habitat • Nest Survey • Density • Prey • Predator/Prey • Disturbance • Telemetry • Reproductive Success • Raptors • SRBPNA • Photo Account • A. chrysaetos • Idaho
3259	**Territory** • Nest Survey • Habitat • Intraspecific Behavior • Density • Nest Chronology • Prey • Directional Exposure • Raptors • A. verreauxi • Rhodesia
3304	**Territory** • Nest Survey • Density • Directional Exposure • Reproductive Success • A. chrysaetos • Scotland
3323	**Territory** • Cain/Abel • Incubation • A. pomarina • West Germany
1387	**Time Lapse** • Telemetry • Prey • A. chrysaetos • Texas
1942	**Time Lapse** • Food Habits • Nest • Behavior • Economics • Animal Damage • Reproductive Success • Status • A. chrysaetos • Texas
1943	**Time Lapse** • Aerial Survey • Animal Damage • Nest Survey • Prey • Habitat • Winter Survey • Capture • Age Structure • Food Requirements • Shooting • A. chrysaetos • Texas
2465	**Time Lapse** • Habitat • Mining • Nest Survey • Disturbance • Marking • Prey • Raptors • Proj. Out. • A. chrysaetos • Wyoming • Montana
313	**Toxicology** • Poisoning • Autopsy • Captivity • Residues • Thal. Sulfate • A. chrysaetos
768	**Toxicology** • Abnormal Behavior • Raptors • A. chrysaetos • Great Britain
836	**Toxicology** • Pesticides • Mortality • Autopsy • Eagles • Popular Monograph • A. chrysaetos • North America
1431	**Toxicology** • Poisoning • Telemetry • Habitat • Raptors • Cmpd. 1080 • A. chrysaetos • California
3186	**Toxicology** • Pesticides • Comp. 1080 • Strychnine • Raptors • A. chrysaetos • United States
3299	**Toxicology** • Poisoning • Gophacide • Raptors • A. chrysaetos • United States
164	**Trade** • Economics • Protection • Raptors • A. rapax • Great Britain • India
403	**Trade** • Etymology • Raptors • A. chrysaetos • Japan
631	**Trade** • Conservation • Story • Trapping • Falconry • Rehabilitation • Wintering Population • Captivity • Raptors • Popular Monograph • A. chrysaetos • New York
973	**Trade** • Captivity • Raptors • A. rapax • A. clanga • A. pomarina • Great Britain
1609	**Trade** • Capture • A. chrysaetos • Tennessee
1845	**Trade** • Collect • Raptors • A. chrysaetos • China
2084	**Trade** • Collect • Prey • A. chrysaetos • California
2520	**Trade** • Raptors • A. chrysaetos
2573	**Trade** • Status • Symbolism • Raptors • A. chrysaetos • South Dakota
2645	**Trade** • Conservation • Shooting • Poisoning • Egg Collection • Nest Robbing • Protection • Trapping • Raptors • A. chrysaetos • Great Britain
2730	**Trade** • Symbolism • Eagles • A. chrysaetos • Pennsylvania
2731	**Trade** • A. heliaca • California
2755	**Trade** • Raptors • A. audax • Australia
1253	**Translocation** • Popular Monograph • (Title Only) • A. chrysaetos • Scotland
2156	**Translocation** • Cain/Abel • Management • Clutch Size • Eagles • Tech. • A. pomarina • A. verreauxi • A. rapax • A. chrysaetos • A. heliaca
2324	**Translocation** • Capture • Animal Damage • Banding • Management • Prey • Sexing • Marking • Measure • Captive Care • Tech. • Eagles • A. chrysaetos • Montana
2325	**Translocation** • Capture • Banding • Marking • Animal Damage • Sexing • Management • Prey • Survey • Measure • Mortality • Captive Care • Aging • A. chrysaetos • Montana
2326	**Translocation** • Capture • Banding • Marking • Animal Damage • Sexing • Management • Prey • Measure • Mortality • Captive Care • Aging • Food Requirements • A. chrysaetos • Montana

Master No.	Keywords
2759	**Translocation** • Animal Damage • Shooting • Conservation • Punishment • Attitude • Reward • Popular Monograph • A. chrysaetos • Texas
7	**Trapping** • Status • Shooting • Eagles • A. chrysaetos • Virginia
123	**Trapping** • Flight • Captivity • Photo Account • A. chrysaetos • North America
150	**Trapping** • Banding • Raptors • A. audax • Australia
159	**Trapping** • Mortality • Capture • Raptors • A. chrysaetos • Nevada
284	**Trapping** • A. chrysaetos • Ireland
286	**Trapping** • Survey • (Title Only) • A. chrysaetos • South Dakota
393	**Trapping** • Measure • A. chrysaetos • New York
496	**Trapping** • Rarity • A. nipalensis • South Africa
616	**Trapping** • Nest • Habitat • Egg • Prey • Vocalization • Growth • Poisoning • Behavior • Lift • Photo Account • A. chrysaetos • Montana
623	**Trapping** • A. chrysaetos • Louisiana
624	**Trapping** • A. chrysaetos • Louisiana
631	**Trapping** • Conservation • Story • Falconry • Rehabilitation • Trade • Wintering Population • Captivity • Raptors • Popular Monograph • A. chrysaetos • New York
703	**Trapping** • Eagles • A. chrysaetos • Mississippi
704	**Trapping** • Animal Damage • Measure • A. chrysaetos • North Carolina
902	**Trapping** • Raptors • A. chrysaetos • Colorado
935	**Trapping** • Status • A. chrysaetos • Prince Edward Island
951	**Trapping** • Measure • A. chrysaetos • Ontario
1067	**Trapping** • A. audax • Australia
1131	**Trapping** • Folklore • Hunting Behavior • Animal Damage • Protection • Eagles • A. chrysaetos • Tennessee
1246	**Trapping** • Behavior • Hunting Behavior • Lift • Attack • Photography • Behavior • Photo Account • Popular Monograph • A. chrysaetos • Scotland
1274	**Trapping** • Nesting Behavior • Photo Account • Popular Monograph • A. chrysaetos • Scotland
1296	**Trapping** • Prey • Status • Rarity • A. chrysaetos • Georgia
1299	**Trapping** • Measure • Raptors • A. chrysaetos • Alaska
1368	**Trapping** • A. chrysaetos • California
1626	**Trapping** • Migration • Etymology • Measure • A. chrysaetos • Alaska
1670	**Trapping** • Descriptive Summary • Etymology • Measure • Raptors • A. chrysaetos • Washington
1727	**Trapping** • A. chrysaetos • West Germany
1879	**Trapping** • Electrocution • Management • Eagles • A. chrysaetos • United States
1883	**Trapping** • Albinism • Measure • A. audax • Australia
2140	**Trapping** • Status • A. chrysaetos • Quebec
2200	**Trapping** • Prey • Animal Damage • Economics • Nest Survey • A. chrysaetos • New Mexico • Texas
2232	**Trapping** • Captivity • A. chrysaetos • Massachusetts
2280	**Trapping** • Protection • Economics • Raptors • A. chrysaetos • Nevada
2281	**Trapping** • Prey • Measure • A. chrysaetos • Maine
2304	**Trapping** • Prey • Attitude • Raptors • A. audax • Australia
2474	**Trapping** • Historical Account • Survey • A. chrysaetos • California
2554	**Trapping** • Conservation • Nest • Behavior • Interspecific Behavior • Prey • Animal Damage • Rehabilitation • Photo Account • Popular Monograph • A. chrysaetos • Wyoming
2576	**Trapping** • Rehabilitation • Shooting • Disease • Anaesthesia • Blood Sample • Raptors • A. chrysaetos • United States
2645	**Trapping** • Conservation • Shooting • Poisoning • Egg Collection • Nest Robbing • Protection • Trade • Raptors • A. chrysaetos • Great Britain
2711	**Trapping** • Shooting • Rehabilitation • Banding • A. chrysaetos • Tennessee
2756	**Trapping** • Status • Shooting • Collect • Photo Account • A. chrysaetos • Wisconsin
2907	**Trapping** • Wintering Population • Plumage • Measure • Molt • A. chrysaetos • Tennessee
2908	**Trapping** • Wintering Population • Measure • Plumage • A. chrysaetos • Tennessee
2909	**Trapping** • Nest • Field Id • Eagles • Popular Monograph • A. chrysaetos • Wyoming
2982	**Trapping** • A. chrysaetos • Virginia
3007	**Trapping** • Rehabilitation • A. chrysaetos • North Carolina
3034	**Trapping** • Shooting • Measure • A. chrysaetos • Pennsylvania
3097	**Trapping** • Senses • Raptors • A. chrysaetos • Oregon
3293	**Trapping** • Electrocution • Raptors • A. chrysaetos • Vermont
3312	**Trapping** • Conservation • Raptors • A. chrysaetos • United States
3315	**Trapping** • A. chrysaetos • Scotland
3370	**Trapping** • Hunting Behavior • Eagles • A. chrysaetos • Arizona
3377	**Trapping** • Rarity • Raptors • A. chrysaetos • Ireland
3415	**Trapping** • A. chrysaetos • Ireland
3426	**Trapping** • Measure • A. chrysaetos • Ireland
736	**Vaccination** • Captivity • Disease • Raptors • A. rapax • Africa
271	**Vocalization** • Behavior • Nest • Egg Collection • Behavior • Photo Account • Popular Monograph • A. chrysaetos • California
335	**Vocalization** • Distribution • Nest • Captivity • Prey • Raptors • A. chrysaetos • Oregon
351	**Vocalization** • Hunting Behavior • Measure • Life History • Courtship • Behavior • Distribution • Plumage • Raptors • Photo Account • A. chrysaetos • North America
464	**Vocalization** • Nest • Habitat • Egg Collection • Behavior • Animal Damage • A. audax • Australia
534	**Vocalization** • Habitat • Density • Nest Survey • Breeding Biology • Interspecific Competition • Prey • Reproductive Success • Etymology • Plumage • Nest Adornment • Raptors • A. verreauxi • A. wahlbergi • Kenya
587	**Vocalization** • Nest • Prey • Behavior • Growth • Nest Adornment • Photo Account • Illus. • A. chrysaetos • Switzerland

Master No.	Keywords
616	**Vocalization** • Nest • Habitat • Egg • Prey • Growth • Trapping • Poisoning • Behavior • Lift • Photo Account • A. chrysaetos • Montana
618	**Vocalization** • Prey • Nest • Attitude • Behavior • Breeding Biology • Flight • Poisoning • Molt • Hunting Behavior • Folklore • Lift • Photo Account • A. chrysaetos • Montana
698	**Vocalization** • Prey Behavior • Hunting Behavior • A. chrysaetos • California
770	**Vocalization** • Field Id • Descriptive Summary • Monograph • Distribution • Status • Migration • Prey • Plumage • Measure • Etymology • A. pomarina • A. clanga • A. nipalensis • A. heliaca • A. chrysaetos • A. verreauxi • Africa • Europe
900	**Vocalization** • Descriptive Summary • Prey • Hunting Behavior • Nest • Interspecific Behavior • Mortality • Raptors • Mt. Mckinley • A. chrysaetos • Alaska
955	**Vocalization** • Behavior • Pesticides • Parental Behavior • Disturbance • Measure • Breeding Biology • Parasite • Disease • Terminology • Growth • A. chrysaetos • Montana
957	**Vocalization** • Monograph • Behavior • Growth • Pesticides • Parental Behavior • Disturbance • Measure • Breeding Biology • Photography • Captivity • Parasite • Disease • Terminology • Photo Account • A. chrysaetos • Montana
963	**Vocalization** • Growth • Nest • Plumage • Prey • Behavior • Photo Account • A. audax • Australia
994	**Vocalization** • A. chrysaetos • Scotland
1040	**Vocalization** • Monograph • Breeding Biology • Prey • Territory • Taxonomy • Migration • Mortality • Molt • Falconry • A. chrysaetos • A. verreauxi • A. audax
1059	**Vocalization** • Field Id • Raptors • A. chrysaetos • A. heliaca • A. rapax • A. nipalensis • A. clanga • A. pomarina • Nepal
1174	**Vocalization** • Nest • Prey • Behavior • Photo Account • Popular Monograph • A. pomarina • West Germany
1186	**Vocalization** • Plumage • Illus. • A. pomarina • Yugoslavia
1187	**Vocalization** • Protection • Habitat • Nest • Raptors • A. chrysaetos • A. heliaca • Yugoslavia
1228	**Vocalization** • Monograph • Distribution • Prey • Plumage • Id Key • A. pomarina • A. chrysaetos • A. nipalensis • Europe
1344	**Vocalization** • Behavior • A. chrysaetos • West Germany
1564	**Vocalization** • Nest • Behavior • Hunting Behavior • Raptors • A. verreauxi • South Africa
1993	**Vocalization** • A. chrysaetos • Scotland
2048	**Vocalization** • Plumage • A. chrysaetos • Tunisia
2102	**Vocalization** • A. chrysaetos • Scotland
2144	**Vocalization** • Nest • Parental Behavior • Prey • Nest Adornment • Photo Account • A. chrysaetos • West Germany
2214	**Vocalization** • Nest • Habitat • Flight • A. chrysaetos • Idaho
2245	**Vocalization** • Nest • Prey • Etymology • Raptors • A. verreauxi • A. rapax • A. wahlbergi • Rhodesia
2463	**Vocalization** • Collect • Status • Behavior • Raptors • A. rapax • Somalia
2760	**Vocalization** • Nest • Prey • Behavior • A. chrysaetos • West Germany
2761	**Vocalization** • Breeding Biology • Prey • Directional Exposure • A. chrysaetos • West Germany
2830	**Vocalization** • Breeding Biology • Habitat • Cain/Abel • Nest Adornment • Growth • Plumage • Prey • Hunting Behavior • Illus. • A. pomarina • West Germany
2852	**Vocalization** • Roadside Survey • Prey • Monograph • Density • Habitat • Interspecific Competition • Taxonomy • Hunting Behavior • Nest Chronology • Photo Account • Raptors • A. wahlbergi • A. rapax • A. nipalensis • A. heliaca • Kenya
2949	**Vocalization** • Nest • Status • Prey • Behavior • Photo Account • A. chrysaetos • Switzerland
2969	**Vocalization** • Nest Survey • Habitat • Prey • Breeding Biology • Reproductive Success • Growth • Field Id • Migration • Behavior • Eagles • Photo Account • A. wahlbergi • A. rapax • Rhodesia
3022	**Vocalization** • Cain/Abel • Growth • Behavior • Pellet • Feeding Behavior • Raptors • A. chrysaetos • California
3061	**Vocalization** • Nest • Collect • Eagles • A. heliaca • Hungary
3065	**Vocalization** • Nest • Clutch Size • A. pomarina • Austria
3235	**Vocalization** • Growth • Nest • Photography • Prey • Behavior • Hunting Behavior • Nest Adornment • Egg • Raptors • Popular Monograph • A. rapax • Kenya
3368	**Vocalization** • Nest • Sub-Adult • A. chrysaetos • Norway
3369	**Vocalization** • Nest Survey • Egg • Habitat • Behavior • Hunting Behavior • Photo Account • A. chrysaetos • Arizona
2500	**Wing Loading** • Measure • Anatomy • Raptors • A. chrysaetos • North America
173	**Winter Survey** • Field Id • Measure • Protection • Reward • Eagles • A. chrysaetos • Arkansas
758	**Winter Survey** • Nest Survey • Density • Reproductive Success • Wintering Population • Status • Aerial Survey • A. chrysaetos • Colorado
760	**Winter Survey** • Eagles • A. chrysaetos • Colorado
762	**Winter Survey** • Roadside Survey • Raptors • A. chrysaetos • Idaho
968	**Winter Survey** • Roadside Survey • Raptors • A. chrysaetos • Colorado
1016	**Winter Survey** • Wintering Population • Eagles • A. chrysaetos • Illinois • Iowa • Kentucky • Missouri • Nebraska
1017	**Winter Survey** • Eagles • A. chrysaetos • Illinois • Kentucky • Tennessee • Indiana
1133	**Winter Survey** • Wintering Population • Eagles • A. chrysaetos • Tennessee
1501	**Winter Survey** • Aerial Survey • Age Structure • Eagles • A. chrysaetos • Wyoming
1675	**Winter Survey** • Roadside Survey • Raptors • A. chrysaetos • Colorado
1692	**Winter Survey** • Eagles • A. chrysaetos • South Dakota

Master No.	Keywords
1773	**Winter Survey** • Distribution • Population Dynamics • Prey • Age Structure • Wintering Population • Eagles • A. chrysaetos • Washington
1776	**Winter Survey** • Nest Survey • Refuge • Density • Electrocution • Shooting • Mortality • Prey • Predator/Prey • Raptors • SRBPNA • A. chrysaetos • Idaho
1796	**Winter Survey** • Raptors • A. chrysaetos • A. heliaca • A. rapax • Pakistan
1943	**Winter Survey** • Aerial Survey • Time Lapse • Animal Damage • Nest Survey • Prey • Habitat • Capture • Age Structure • Food Requirements • Shooting • A. chrysaetos • Texas
1973	**Winter Survey** • Poisoning • Roost • Refuge • Photo Account • Popular Monograph • A. chrysaetos • Wyoming
2239	**Winter Survey** • Survey • Territory • Nest Failure • Nest • Migration • Habitat • Management • Reproductive Success • Prey • Roadside Survey • Status • Uinta Nf • Wasatch Mt. • Raptors • A. chrysaetos • Utah
2621	**Winter Survey** • Raptors • A. chrysaetos • Kansas
2656	**Winter Survey** • Conservation • Shooting • Poisoning • Protection • Reward • Photo Account • Eagles • Popular Monograph • A. chrysaetos • Wyoming
2961	**Winter Survey** • Roadside Survey • Nest • Raptors • A. chrysaetos • Kansas
3418	**Winter Survey** • Eagles • A. chrysaetos • Wyoming
3419	**Winter Survey** • Aerial Survey • Population Estimate • Eagles • A. chrysaetos • Wyoming
65	**Wintering Population** • Museum Specimen • Migration • Descriptive Summary • Status • A. chrysaetos • Iowa
113	**Wintering Population** • Collect • A. pomarina • Southwest Africa
151	**Wintering Population** • Poisoning • Mortality • Conservation • Attitude • Punishment • A. chrysaetos • Wyoming
171	**Wintering Population** • Field Id • Raptors • A. clanga • A. pomarina • A. heliaca • Kenya
195	**Wintering Population** • Nest Survey • Habitat • Distribution • Attitude • Raptors • A. chrysaetos • Arizona
196	**Wintering Population** • Nest Survey • Habitat • Attitude • Falconry • Raptors • A. chrysaetos • Arizona
233	**Wintering Population** • Attitude • Reproductive Success • Density • Food Requirements • Habitat • Management • Aerial Survey • Nest Survey • Raptors • A. chrysaetos • Montana
268	**Wintering Population** • Historical Account • Status • A. chrysaetos • Kentucky
273	**Wintering Population** • Feeding Behavior • Raptors • A. chrysaetos • A. nipalensis • Aden
330	**Wintering Population** • Status • Raptors • A. chrysaetos • Saskatchewan
372	**Wintering Population** • Habitat • Nest • Hunting Behavior • Territory • A. chrysaetos • France
392	**Wintering Population** • Prey • Electrocution • A. chrysaetos • Denmark
423	**Wintering Population** • Aerial Survey • Nest Survey • Banding • Roadside Survey • Electrocution • A. chrysaetos • North America
425	**Wintering Population** • Aerial Survey • Management • Animal Damage • A. chrysaetos • Texas • New Mexico
427	**Wintering Population** • Aerial Survey • Age Structure • Nest Survey • Prey • Banding • A. chrysaetos • North America
428	**Wintering Population** • Reproductive Success • Status • Aerial Survey • Proj. Out. • A. chrysaetos • Colorado • Idaho • Montana • Utah • New Mexico • Wyoming
435	**Wintering Population** • Predator/Prey • Raptors • A. chrysaetos • Spain
490	**Wintering Population** • Flight Id • Aging • Eagles • A. nipalensis • A. pomarina • A. rapax • A. wahlbergi • A. verreauxi • Africa
495	**Wintering Population** • Prey • Distribution • Hunting Behavior • Age Structure • A. nipalensis • Africa
498	**Wintering Population** • Taxonomy • Field Id • Molt • Measure • Museum Specimen • Behavior • Interspecific Competition • Conservation • Eagles • Photo Account • A. nipalensis • A. pomarina • Africa
543	**Wintering Population** • Status • Habitat Loss • Raptors • AF • Africa
609	**Wintering Population** • Banding • Capture • Animal Damage • Eagles • A. chrysaetos • Tennessee
631	**Wintering Population** • Conservation • Story • Trapping • Falconry • Rehabilitation • Trade • Captivity • Raptors • Popular Monograph • A. chrysaetos • New York
654	**Wintering Population** • Nest • Plumage • Raptors • A. heliaca • A. chrysaetos • Spain
670	**Wintering Population** • A. chrysaetos • France
680	**Wintering Population** • Distribution • Field Id • Raptors • A. chrysaetos • A. heliaca • A. nipalensis • A. pomarina • Iraq
758	**Wintering Population** • Nest Survey • Density • Reproductive Success • Winter Survey • Status • Aerial Survey • A. chrysaetos • Colorado
763	**Wintering Population** • Roadside Survey • Nest • Survey • Raptors • A. chrysaetos • Idaho
823	**Wintering Population** • Distribution • Survey • A. chrysaetos • New Mexico
912	**Wintering Population** • Raptors • A. nipalensis • Zambia
969	**Wintering Population** • Aerial Survey • Eagles • A. chrysaetos • Colorado
999	**Wintering Population** • Conservation • Refuge • Management • Raptors • A. chrysaetos • Great Britain
1016	**Wintering Population** • Winter Survey • Eagles • A. chrysaetos • Illinois • Iowa • Kentucky • Missouri • Nebraska
1133	**Wintering Population** • Winter Survey • Eagles • A. chrysaetos • Tennessee
1134	**Wintering Population** • Raptors • A. chrysaetos • Tennessee
1336	**Wintering Population** • Habitat • Raptors • A. clanga • Thailand
1460	**Wintering Population** • Prey • Raptors • A. chrysaetos • North Dakota
1558	**Wintering Population** • Mortality • Prey • Raptors • A. chrysaetos • Alberta
1593	**Wintering Population** • Management • A. chrysaetos • North Carolina

Master No.	Keywords
1615	**Wintering Population** • Status • Roost • Raptors • A. chrysaetos • Kansas
1754	**Wintering Population** • Distribution • Nest • Habitat • Status • Raptors • A. chrysaetos • Colorado
1773	**Wintering Population** • Winter Survey • Distribution • Population Dynamics • Prey • Age Structure • Eagles • A. chrysaetos • Washington
1774	**Wintering Population** • Distribution • Prey • Raptors • Popular Monograph • A. chrysaetos • Washington
2057	**Wintering Population** • Educational Material • Raptors • A. chrysaetos • Minnesota
2058	**Wintering Population** • Roadside Survey • Raptors • A. chrysaetos • Nebraska
2126	**Wintering Population** • Status • Nest Survey • Mortality • Rehabilitation • Raptors • A. chrysaetos • A. heliaca • A. pomarina • A. nipalensis • A. clanga • Israel
2128	**Wintering Population** • Pesticides • Mortality • Prey • Rehabilitation • Azodrin • Raptors • A. heliaca • A. clanga • A. pomarina • Israel
2273	**Wintering Population** • Shooting • Animal Damage • Measure • Eagles • A. chrysaetos • Illinois
2394	**Wintering Population** • Raptors • A. chrysaetos • Siberia
2414	**Wintering Population** • Nest • Eagles • A. chrysaetos • Maine
2435	**Wintering Population** • Rarity • Raptors • A. clanga • A. pomarina • A. heliaca • A. nipalensis • Kenya
2436	**Wintering Population** • Raptors • A. heliaca • A. nipalensis • A. pomarina • Kenya
2437	**Wintering Population** • Field Id • A. pomarina • A. nipalensis • Kenya
2510	**Wintering Population** • Prey • Raptors • A. chrysaetos • A. clanga • A. heliaca • Albania
2512	**Wintering Population** • Interspecific Behavior • Eagles • A. chrysaetos • East Germany
2527	**Wintering Population** • Status • Raptors • A. chrysaetos • Russia
2591	**Wintering Population** • Roadside Survey • Animal Damage • Age Structure • Prey • Raptors • A. chrysaetos • New Mexico
2675	**Wintering Population** • Collect • Prey • Etymology • Status • Raptors • A. chrysaetos • A. clanga • Yugoslavia
2792	**Wintering Population** • Migration • Eagles • A. chrysaetos • Russia
2860	**Wintering Population** • Field Id • Migration • A. nipalensis • A. pomarina • A. clanga • A. heliaca • Africa
2907	**Wintering Population** • Trapping • Plumage • Measure • Molt • A. chrysaetos • Tennessee
2908	**Wintering Population** • Trapping • Measure • Plumage • A. chrysaetos • Tennessee
2929	**Wintering Population** • Raptors • A. chrysaetos • Colorado
2931	**Wintering Population** • Hunting Behavior • Raptors • A. chrysaetos • Colorado
2932	**Wintering Population** • Roadside Survey • Habitat • Raptors • A. chrysaetos • Colorado
2980	**Wintering Population** • Nest • Raptors • A. chrysaetos • Wisconsin
2997	**Wintering Population** • Distribution • A. pomarina • Europe
3080	**Wintering Population** • Cain/Abel • Distribution • Educational Material • Field Id • Habitat • Hunting Behavior • Prey • Artificial Nest • Nest • Incubation • Conservation • Photo Account • Eagles • A. chrysaetos • A. wahlbergi • A. verreauxi • A. nipalensis • A. pomarina • South Africa
3126	**Wintering Population** • Migration • Interspecific Competition • Raptors • A. nipalensis • A. pomarina • A. clanga • A. rapax • A. heliaca • Africa
3133	**Wintering Population** • Habitat • Prey • Density • Raptors • A. wahlbergi • A. rapax • A. nipalensis • A. pomarina • Cambodia
3160	**Wintering Population** • Survey • Territory • Field Id • Age Structure • Supplemental Feeding • Illus. • A. chrysaetos • Sweden
3161	**Wintering Population** • Aerial Survey • Eagles • A. chrysaetos • Arizona
3205	**Wintering Population** • Management • Animal Damage • Status • Distribution • A. chrysaetos • United States
3229	**Wintering Population** • Eagles • A. chrysaetos • Oklahoma
3231	**Wintering Population** • Prey • Raptors • A. chrysaetos • Washington
3236	**Wintering Population** • Rehabilitation • Falconry • Shooting • Photo Account • A. chrysaetos • Missouri
3298	**Wintering Population** • Raptors • A. chrysaetos • Nebraska

Chapter 5

Species Index of Citations

A. pomarina - Lesser spotted eagle - 9, 10, 11, 16, 21, 26, 27, 43, 45, 46, 52, 55, 56, 58, 68, 74, 79, 80, 88, 94, 112, 113, 128, 132, 139, 140, 142, 145, 171, 179, 188, 194, 203, 220, 221, 234, 242, 247, 248, 252, 263, 264, 265, 295, 298, 299, 300, 303, 304, 305, 311, 312, 318, 356, 386, 406, 408, 409, 410, 415, 429, 430, 442, 447, 455, 467, 490, 491, 492, 493, 498, 509, 510, 512, 513, 514, 515, 517, 531, 541, 545, 546, 547, 548, 551, 558, 564, 566, 567, 582, 583, 620, 626, 636, 665, 666, 680, 684, 685, 686, 689, 712, 717, 722, 724, 769, 770, 773, 774, 779, 791, 804, 816, 825, 843, 846, 861, 862, 864, 871, 879, 880, 881, 882, 883, 901, 906, 910, 911, 915, 916, 919, 923, 937, 938, 939, 953, 973, 981, 986, 987, 989, 1005, 1008, 1018, 1026, 1058, 1059, 1061, 1063, 1083, 1129, 1135, 1136, 1153, 1166, 1171, 1174, 1176, 1177, 1186, 1190, 1194, 1201, 1211, 1213, 1228, 1231, 1233, 1234, 1235, 1236, 1237, 1238, 1289, 1297, 1305, 1306, 1329, 1330, 1341, 1364, 1375, 1376, 1377, 1384, 1390, 1391, 1394, 1402, 1418, 1430, 1432, 1433, 1445, 1452, 1476, 1490, 1491, 1518, 1526, 1530, 1533, 1553, 1555, 1560, 1569, 1586, 1596, 1597, 1618, 1632, 1633, 1637, 1638, 1641, 1642, 1645, 1655, 1664, 1665, 1668, 1697, 1701, 1709, 1711, 1715, 1716, 1721, 1722, 1723, 1724, 1731, 1735, 1744, 1746, 1755, 1762, 1766, 1785, 1786, 1787, 1790, 1805, 1813, 1814, 1817, 1820, 1821, 1825, 1826, 1827, 1828, 1830, 1831, 1833, 1843, 1856, 1860, 1882, 1910, 1911, 1915, 1919, 1924, 1931, 1959, 1965, 1978, 2005, 2006, 2010, 2023, 2027, 2029, 2030, 2045, 2059, 2060, 2061, 2062, 2065, 2072, 2073, 2075, 2121, 2125, 2126, 2128, 2129, 2131, 2136, 2137, 2145, 2146, 2148, 2149, 2150, 2155, 2156, 2161, 2189, 2201, 2209, 2221, 2243, 2246, 2279, 2288, 2308, 2318, 2319, 2321, 2330, 2331, 2338, 2361, 2362, 2366, 2407, 2418, 2426, 2428, 2435, 2436, 2437, 2442, 2443, 2456, 2468, 2469, 2476, 2487, 2495, 2503, 2505, 2508, 2511, 2516, 2517, 2535, 2548, 2557, 2574, 2583, 2589, 2592, 2610, 2619, 2637, 2646, 2647, 2661, 2665, 2677, 2680, 2685, 2686, 2690, 2697, 2698, 2699, 2744, 2751, 2752, 2757, 2766, 2769, 2782, 2783, 2830, 2833, 2846, 2847, 2848, 2849, 2860, 2865, 2888, 2889, 2941, 2942, 2944, 2948, 2979, 2997, 3001, 3017, 3039, 3040, 3042, 3043, 3046, 3048, 3056, 3059, 3062, 3063, 3064, 3065, 3066, 3067, 3068, 3074, 3080, 3098, 3109, 3112, 3113, 3114, 3122, 3126, 3128, 3129, 3130, 3133, 3134, 3135, 3137, 3150, 3159, 3162, 3165, 3166, 3173, 3174, 3176, 3177, 3190, 3216, 3217, 3218, 3220, 3222, 3227, 3228, 3230, 3232, 3237, 3248, 3257, 3258, 3265, 3266, 3268, 3269, 3281, 3322, 3323, 3324, 3325, 3326, 3327, 3344, 3404, 3429, 3430, 3433, 3445, 3450, 3452, 3454

A. clanga - Spotted eagle - 2, 5, 21, 23, 26, 27, 43, 45, 46, 52, 54, 68, 70, 74, 80, 81, 88, 92, 95, 132, 133, 140, 145, 171, 179, 188, 205, 212, 216, 220, 227, 232, 242, 258, 262, 263, 266, 299, 304, 305, 307, 312, 318, 320, 334, 355, 358, 359, 379, 386, 406, 407, 408, 415, 419, 429, 442, 447, 448, 463, 477, 505, 506, 509, 510, 513, 514, 515, 516, 518, 529, 541, 544, 545, 546, 547, 548, 551, 558, 564, 566, 567, 582, 591, 595, 596, 620, 638, 643, 645, 660, 664, 665, 666, 678, 682, 685, 686, 688, 691, 693, 694, 709, 712, 715, 717, 770, 774, 776, 777, 780, 788, 797, 799, 801, 802, 804, 814, 815, 843, 845, 848, 849, 854, 861, 864, 871, 877, 881, 882, 882, 883, 891, 901, 903, 904, 907, 910, 911, 915, 916, 917, 918, 923, 924, 937, 939, 965, 967, 973, 980, 984, 985, 987, 988, 989, 1006, 1008, 1010, 1011, 1012, 1015, 1018, 1039, 1058, 1059, 1061, 1076, 1083, 1095, 1114, 1116, 1117, 1118, 1119, 1120, 1121, 1123, 1124, 1129, 1130, 1135, 1136, 1169, 1173, 1181, 1185, 1197, 1198, 1210, 1213, 1214, 1220, 1229, 1231, 1239, 1242, 1276, 1280, 1289, 1302, 1303, 1305, 1305, 1306, 1308, 1314, 1318, 1324, 1328, 1329, 1330, 1336, 1343, 1349, 1352, 1375, 1376, 1377, 1384, 1390, 1394, 1418, 1430, 1445, 1446, 1452, 1456, 1459, 1493, 1494, 1512, 1519, 1520, 1528, 1529, 1531, 1532, 1537, 1552, 1553, 1569, 1586, 1589, 1595, 1597, 1598, 1599, 1600, 1613, 1617, 1620, 1621, 1623, 1632, 1634, 1638, 1645, 1665, 1668, 1671, 1693, 1697, 1701, 1703, 1709, 1712, 1713, 1720, 1724, 1738, 1745, 1749, 1761, 1765, 1766, 1785, 1786, 1787, 1792, 1800, 1808, 1813, 1818, 1819, 1820, 1821, 1822, 1823, 1824, 1825, 1826, 1828, 1830, 1832, 1833, 1834, 1836, 1840, 1842, 1847, 1860, 1869, 1873, 1874,

A. heliaca - Imperial eagle - 2, 13, 21, 23, 24, 26, 27, 43, 45, 46, 54, 55, 57, 59, 78, 80, 145, 147, 171, 178, 179, 188, 190, 192, 194, 216, 218, 220, 242, 243, 248, 250, 259, 262, 263, 298, 299, 300, 303, 304, 312, 317, 318, 331, 350, 353, 359, 366, 386, 387, 399, 400, 406, 407, 408, 410, 413, 415, 429, 430, 432, 442, 447, 456, 505, 507, 508, 510, 516, 517, 541, 545, 546, 547, 548, 551, 564, 566, 567, 574, 575, 576, 582, 585, 596, 653, 654, 660, 664, 666, 678, 680, 682, 683, 685, 686, 697, 705, 712, 717, 769, 770, 773, 779, 780, 784, 795, 797, 798, 799, 815, 843, 849, 854, 864, 871, 879, 880, 881, 882, 883, 890, 904, 907, 908, 910, 911, 915, 917, 923, 924, 937, 965, 975, 987, 988, 998, 1006, 1008, 1013, 1021, 1022, 1023, 1024, 1025, 1046, 1053, 1059, 1063, 1109, 1115, 1126, 1128, 1129, 1135, 1136, 1137, 1159, 1160, 1162, 1163, 1164, 1169, 1173, 1180, 1187, 1199, 1213, 1215, 1242, 1276, 1289, 1305, 1310, 1317, 1322, 1323, 1325, 1327, 1328, 1330, 1332, 1333, 1349, 1376, 1384, 1394, 1395, 1428, 1433, 1445, 1449, 1461, 1475, 1537, 1552, 1586, 1589, 1595, 1598, 1600, 1613, 1632, 1634, 1665, 1668, 1671, 1693, 1695, 1701, 1703, 1707, 1719, 1720, 1731, 1747, 1748, 1749, 1761, 1766, 1767, 1785, 1790, 1796, 1808, 1825, 1826, 1827, 1828, 1830, 1833, 1841, 1842, 1843, 1847, 1857, 1860, 1906, 1916, 1917, 1918, 1924, 1925, 1931, 1933, 1934, 1935, 1936, 1952, 1953, 1961, 1969, 2009, 2020, 2029, 2030, 2034, 2038, 2042, 2045, 2050, 2059, 2060, 2063, 2064, 2076, 2094, 2095, 2111, 2112, 2116, 2119, 2121, 2123, 2125, 2126, 2128, 2129, 2131, 2132, 2150, 2151, 2152, 2153, 2154, 2155, 2156, 2157, 2158, 2159, 2160, 2161, 2183, 2189, 2191, 2194, 2197, 2198, 2205, 2207, 2209, 2221, 2240, 2241, 2242, 2243, 2279, 2300, 2308, 2312, 2313, 2318, 2320, 2321, 2333, 2338, 2352, 2361, 2362, 2364, 2378, 2388, 2393, 2418, 2424, 2426, 2429, 2435, 2436, 2441, 2442, 2447, 2448, 2449, 2456, 2476, 2480, 2481, 2482, 2483, 2495, 2502, 2503, 2504, 2505, 2510, 2516, 2577, 2583, 2589, 2619, 2647, 2665, 2676, 2678, 2679, 2699, 2703, 2704, 2716, 2717, 2718, 2719, 2731, 2754, 2770, 2773, 2782, 2784, 2815, 2833, 2834, 2836, 2838, 2850, 2852, 2853, 2857, 2858, 2860, 2861, 2864, 2882, 2883, 2888, 2889, 2923, 2944, 2948, 2979, 2995, 3001, 3003, 3004, 3008, 3009, 3011, 3013, 3028, 3029, 3030, 3039, 3042, 3046, 3048, 3052, 3061, 3067, 3068, 3069, 3070, 3071, 3074, 3075, 3076, 3078, 3086, 3087, 3108, 3111, 3126, 3135, 3138, 3142, 3145, 3147, 3151, 3152, 3162, 3166, 3181, 3195, 3209, 3212, 3218, 3220, 3225, 3226, 3232, 3240, 3241, 3242, 3243, 3245, 3246, 3249, 3251, 3257, 3265, 3270, 3277, 3294, 3297, 3341, 3342, 3343, 3344, 3357, 3366, 3395, 3404, 3421, 3430, 3436, 3437, 3443, 3450, 3454

A. wahlbergi - Wahlberg's eagle - 106, 108, 134, 180, 189, 203, 224, 261, 341, 345, 348, 383, 405, 406, 418, 446, 480, 481, 482, 484, 490, 494, 534, 535, 537, 540, 541, 545, 546, 547, 549, 550, 551, 553, 562, 592, 604, 640, 641, 692, 816, 882, 896, 952, 953, 998, 1032, 1033, 1088, 1093, 1182, 1305, 1330, 1334, 1337, 1340, 1354, 1393, 1402, 1422, 1429, 1526, 1547, 1640, 1641, 1690, 1733, 1734, 1735, 1736, 1737, 1806, 1849, 1857, 1891, 1892, 1893, 1905, 1911, 1931, 1958, 1982, 2005, 2006, 2039, 2099, 2142, 2170, 2181, 2186, 2219, 2245, 2287, 2306, 2308, 2341, 2403, 2404, 2406, 2471, 2485, 2535, 2618, 2634, 2636, 2637, 2707, 2772, 2773, 2774, 2796, 2802, 2807, 2816, 2842, 2852, 2870, 2871, 2894, 2969, 2970, 2972, 2976, 2983, 2989, 3001, 3046, 3054, 3055, 3079, 3080, 3084, 3108, 3119, 3120, 3121, 3122, 3123, 3124, 3125, 3127, 3128, 3130, 3131, 3133, 3134, 3136, 3172, 3174, 3177, 3187, 3211, 3234, 3237, 3238, 3262, 3268, 3278, 3285, 3352, 3353, 3384, 3385, 3386, 3388, 3404, 3411

A. chrysaetos - Golden eagle—Cont'd.

Chapter 6

Geographic Index of Citations

AFRICA

ASIA

AUSTRALIA

EUROPE

NORTH AMERICA

Chapter 7

List of Citations of Occurrences

CANADA

Irving, L. *1972.* Arctic Life of Birds and Mammals, Including Man. Springer-Verlag, Berlin, Heidelberg, New York. 192 pp.

Alberta

Brooks, W.S. and S. Cobb. *1911.* Notes from eastern Alberta. Auk 28(4):465-469.

Horsbraugh, C.B. *1915.* Ornithological notes from the Alix and Buffalo Lake Districts, Province of Alberta, Canada, 1914. Ibis (10th Ser.) 3(4):670-689.

Stansell, S.S.S. *1912.* Raptores of central Alberta. Oologist 29(2):231-232.

Taverner, P.A. *1919.* The birds of the Red Deer River, Alberta. Auk 36(1/2):1-21,248-265.

British Columbia

Alford, C.E. *1928.* Field-notes on the birds of Vancouver Island. Ibis 4(2):181-210.

British Columbia Provincial Museum. *1904.* Catalogue of British Columbia Birds. Victoria, B.C. 69 pp.

Brooks, A. *1917.* Birds of the Chilliwack District, B.C. Auk 34(1):28-50.

Brooks, A. *1925.* A distributional list of the birds of Britsh Columbia. Pac. Coast Avifauna 17:1-158.

Kelso, J.E.H. *1926.* Birds of the Arrow Lakes, west Kootenay District, British Columbia. Ibis 2(4):689-723.

Laing, H.M. *1942.* Birds of the coast of central British Columbia. Condor 44(4):175-181.

Laing, H.M. *1956.* Nesting of golden eagle on Vancouver Island. Can. Field-Nat. 70(2):95.

Munro, J.A. *1929.* August bird life on Liumchin Mountain, British Columbia. Murrelet 10(1):7-9.

Munro, J.A. *1935.* Bird life at Horse Lake, British Columbia. Condor 37(4):185-193.

Munro, J.A. *1945.* The birds of the Cariboo Parklands, British Columbia. Can. J. Res. 23(3):17-103.

Pearse, T. *1943.* Golden eagle on Vancouver Island, B.C. Can. Field-Nat. 57(4/5):97.

Racey, K. *1926.* Notes on the birds observed in the Alta Lake region, B.C. Auk 43(3):319-325.

Racey, K. *1948.* Birds of the Alta Lake region, British Columbia. Auk 65(3):383-401.

Rhoads, S.N. *1893.* The birds observed in British Columbia and Washington during spring and summer, 1892. Proc. Acad. Nat. Sci. Philadelphia 45:21-67.

Swarth, H.S. *1922.* Birds and mammals of the Stikine River region of northern British Columbia and southeastern Alaska. Univ. Calif. Publ. Zool. 24(2):125-314.

Swarth, H.S. *1924.* Birds and mammals of the Skeena River region of northern British Columbia. Univ. Calif. Publ. Zool. 24(3):315-394.

Swarth, H.S. *1926.* Report on a collection of birds and mammals from the Atlin region, northern British Columbia. Univ. Calif. Publ. Zool. 30(4):51-162.

Weeden, R.B. *1960.* The birds of Chilkat Pass, British Columbia. Can. Field-Nat. 74(2):119-129.

Labrador

Weiz, S. *1866.* List of vertebrates observed at Ada, Labrador. Proc. Boston Soc. Nat. Hist. 10:264-269.

Manitoba

Cooke, F., R.K. Ross, R.K. Schmidt and A.J. Pakulak. *1975.* Birds of the tundra biome at Cape Churchill and La Perouse Bay. Can. Field-Nat. 89(4):413-422.

Jehl, J.R., Jr. and B.A. Smith. *1970.* Birds of the Churchill Region, Manitoba. Manitoba Mus. Man and Nature, Spec. Publ. No. 1.

Seton, E.T. *1886.* The birds of western Manitoba. Auk 3(2):145-156.

Skinner, K.L. *1935.* The birds of Churchill, Manitoba. Ool. Rec. 15(1):18-24.

Taverner, P.A. and G.M. Sutton. *1934.* The birds of Churchill, Manitoba. Ann. Carnegie Mus. 23:1-83.

Thompson, E.E. *1890.* The birds of Manitoba. Proc. U.S. Natl. Mus. 13:457-643.

New Brunswick

Chamberlain, M. *1882.* A catalogue of the birds of New Brunswick. Bull. Nat. Hist. Soc. New Brunswick 1:23-68.

Pettingill, O.S., Jr. *1937-1938.* The bird life of the Grand Manan Archipelago. Proc. Nova Scotia Inst. Sci. 19(4):293-372.

New Brunswick —Cont'd.

Townsend, C.W. *1923.* Notes on the birds of Grand Manan, New Brunswick. Can. Field-Nat. 37(8):141-144.

Newfoundland

Tuck, L.M. *1968.* Recent Newfoundland bird records. Auk 85(2):304-311.

Northwest Territory

Blakiston, C. *1863.* On the birds of the interior of British North America. Ibis 5(17):39-87.

Clarke, C.H.D. *1940.* A biological investigation of the Thelon Game Sanctuary. Natl. Mus. Canada, Biology Ser. 25, Bull. No. 96. p. 1-135.

Hohn, O.E. *1959.* Birds of the mouth of the Anderson River and Liverpool Bay, Northwest Territories. Can. Field-Nat. 73(2):93-114.

L.G.L., Ltd. *1972.* Surveys of terrestrial bird populations in Alaska, Yukon Territories, Northwest Territories, and northern Alberta. Part One. Report Prepared for Northern Engineering Services, Ltd. 171 pp.

McEwen, E.H. *1957.* Birds observed at Bathurst Inlet, Northwest Territories. Can. Field-Nat. 71(3):109-115.

Preble, E.A. *1902.* A biological investigation of the Hudson Bay region. U.S.D.A., Div. Biology Survey, N. Am. Fàuna 22. p. 1-40.

Seton, E.T. *1908.* Bird records from Great Slave Lake region. Auk 25(1):68-74.

Sutton, G.M. *1931.* Notes on birds observed along the west coast of Hudson bay. Condor 33(4):154-159.

Sutton, G.M. *1932.* The birds of Southampton Island. Mem. Carnegie Mus. 12(2):1-275.

Nova Scotia

Gilpin, J.B. *1880-1881.* On the birds of prey of Nova Scotia. Proc. and Trans. Nova Scotia Inst. Nat. Sci. 5(3):255-268.

Tufts, R.W. *1956.* Annotated list of Nova Scotia birds. Nova Scotia Mus. Sci., Newsletter 1(4):1-22.

Ontario

Fleming, J.H. *1901.* A list of the birds of the districts of Parry Sound and Muskoka, Ontario. Auk 18:33-45.

Fleming, J.H. and H. Lloyd. *1920.* Ontario bird notes. Auk 37(3):429-439.

Hitchcock, H.B. *1943.* Yellow-headed blackbird and golden eagle in Middlesex County, Ontario. Can. Field-Nat. 57(4/5):88.

James, R.D., P.L. McLaren and J.C. Barlow. *1976.* Annotated checklist of the birds of Ontario. Royal Ontario Mus., Life Sci. Misc. Publ. 75 pp.

Jones, L. *1909.* The birds of Cedar Point and vicinity. Wilson Bull. 21(4):187-204.

Lloyd, H. *1923.* The birds of Ottawa, 1923- continued. Can. Field-Nat. 37(8):151-156.

Lloyd, H. *1949.* Bird records for the Ottawa District. Can. Field-Nat 63(1):31-34.

Manning, T.H. *1952.* Birds of the west James Bay and southern Hudson Bay coast. Natl. Mus. of Canada, Biology Ser. No. 43, Bull. No. 125. p. 1-114.

Ricker, W.E. *1939.* The birds of the vicinity of Lake Nipissing, Ontario. Royal Ontario Mus. Zool., Contrib. No. 16.

Saunders, W.E. *1937.* Golden eagle and Richardson's owl in Brant County, Ontario. Can. Field-Nat. 51(3):44.

Snyder, L.L. *1940.* Golden eagle reported nesting in Ontario. Auk 57(4):565-566.

Snyder, L.L., E.B.S. Logier, T.B. Kurata, F.A. Urquart and J.F. Brimley. *1941.* A faunal investigation of Prince Edward Island County, Ontario. Royal Ontario Mus. Zool., Contrib. No. 19. (Reprint from Univ. of Toronto Studies, Biology Ser. No. 48.)

Wood, A.A., G.M. Stirrett and D.A. Arnott. *1941.* Observations on some interesting birds in Kent County, Ontario. Can. Field-Nat. 55(2):15.

Quebec

Braund, F.W. and E.P. McCullagh. *1940.* The birds of Anticosti Island, Quebec. Wilson Bull. 52(2):96-123.

Eifrig, C.W.G. *1906.* Notes on some northern birds. Auk 23:313-318.

Lloyd, H. *1953.* Dead golden eagle at Perkins Mills, Quebec. Can. Field-Nat. 67(1):44.

Saskatchewan

Bent, A.C. *1907.* Summer birds of southwestern Saskatchewan. Auk 24(4):407-430.

Blakiston, C. *1861.* On birds collected and observed in the British North America. Ibis 3(12):314-320.

Mitchell, H.H. *1924.* Birds of Saskatchewan. Can. Field-Nat. 38(6):101-115.

Yukon Territory

Bishop, L.B *1900.* Birds of the Yukon region, with notes on other species. Pages 1-100. *in:* W.H. Osgood, Ed. Results of A Biological Reconnaisance of the Yukon River Region. N. American Fauna No. 19.

L.G.L., Ltd. *1972.* Surveys of terrestrial bird populations in Alaska, Yukon Territories, Northwest Territories, and northern Alberta. Part One. Report Prepared for Northern Engineering Services, Ltd. 171 pp.

Rand, A.L. *1946.* List of Yukon birds and those of the Canol Road. Natl. Mus. of Canada, Biology Ser. No. 33, Bull. No. 105. p. 1-76.

White, C.M. and J.R. Haugh. *1969.* Recent data on summer birds of the Upper Yukon River, Alaska, and adjacent part of the Yukon Territory, Canada. Can. Field-Nat. 83(3):257-271.

MEXICO

Banks, R.C. *1963.* Birds of Cerralvo Island, Baja California. Condor 65(4):300-312.

Davis, J. *1953.* Birds of the Tzitzio region Michoacan, Mexico. Condor 55(2):90-98.

Ely, C.A. *1962.* The birds of southeastern Coahuila, Mexico. Condor 64(1):34-39.

Huey, L.M. *1935.* February bird life of Punta Penascosa, Sonora, Mexico. Aux 52(3):249-256.

Mengel, R.M. and D.W. Warner. *1948.* Golden eagles in Hidalgo, Mexico. Wilson Bull. 60(2):122.

Price, W.W. *1899.* Some winter birds of the lower Colorado Valley. Condor 1(5):89-93.

Short, L.L., Jr. *1967.* Notes on the avifauna of northwestern Baja California. Trans. San Diego Soc. Nat. Hist. 14(20):281-300.

Vuilleumier, F. and J.E. Williams. *1964.* Notes on some birds from the Rio, Chihuahua, Mexico. Condor 66(6):515-516.

Webster, J.D. *1958.* Further ornithological notes from Zacatecas, Mexico. Wilson Bull. 70(3):243-256.

UNITED STATES

Bailey, F.M. *1927.* Handbook of birds of the western United States. Houghton Mifflin Comp., Boston and New York. 590 pp.

Alabama

Holt, E.G. *1924.* Golden eagle *(Aquila chrysaetos)* in Alabama. Auk 41(4):601-602.

Howell, A.H. *1928.* Birds of Alabama. Dept. Game and Fisheries of Alabama. 384 pp.

Alaska

Anthony, A.W. *1906.* Stray notes from Alaska. Auk 23:179-184.

Bailey, A.M. *1926.* A report on the birds of northwestern Alaska and regions adjacent to Bering Strait. Part IX. Condor 28(3):121-126.

Bailey, A.M. *1927.* Notes on the birds of southeastern Alaska. Auk 44(1/2):1-23,184-205.

Bailey, A.M. *1948.* Birds of arctic Alaska. Colo. Mus. Nat. Hist. Pop. Series No. 8. 317 pp.

Bailey, A.M., C.D. Brower and L.B. Bishop. *1933.* Birds of the region of Point Barrow, Alaska. Prog. Activities of Chicago Acad. Sci. 4(2):15-39.

Brokaw, H.P., T.C.T. Brokaw, N.V.L. Brokaw and J.J. Brokaw. *1975.* Birds seen in the Arctic National Wildlife Range, July 1970. Delmarva Ornithol. 10(2):42-45.

Cahn, A.R. *1947.* Notes on the birds of the Dutch Harbor area of the Aleutian Islands. Condor 49(2):78-82.

Clark, A.H. *1911.* The birds collected and observed during the cruise of the U.S. fisheries steamer 'Albatross" in the North Pacific Ocean, and in the Bering, Okhotsk, Japan and Eastern seas. Proc. of the U.S. Natl. Mus. 38(1727):25-74.

Dall, W.H. *1873.* Notes on the avifauna of the Aleutian Islands, from Unalaska eastward. Proc. Calif. Acad. Sci. 5(1):25-35.

Dice, L.R. *1920.* Notes on some birds of interior Alaska. Condor 22(5):176-185.

Friedmann, H. *1937.* Bird bones from archeological sites in Alaska. J. Wash. Acad. Sci. 27(10):431-438.

Grinnell, J. *1910.* Miscellaneous records from Alaska. Condor 12(1):41-43.

Hasselborg, A.E. *1918.* Bird notes from Admiralty Island, southeastern Alaska. Condor 20(5):191.

Hines, J.Q. *1963.* Birds of the Noatak River, Alaska. Condor 65(5):410-425.

Irving, L. *1960.* Birds of Anaktuvuk Pass, Kobuk, and Old Crow. U.S. Natl. Mus. Bull. 217. 409 pp.

Irving, L. *1972.* Arctic Life of Birds and Mammals, Including Man. Springer-Verlag, Berlin, Heidelberg, New York. 192 pp.

Kessel, B. *1953.* A study of the birds of the Colville River. Off. Naval Res., Rep. 8 pp.

Kessel, B. and T.J. Cade. *1958.* Birds of the Colville River, northern Alaska. Univ. of Alaska, Biology Papers, No. 2. 83 pp.

Kessel, B. and G.B. Schaller. *1960.* Birds of the Upper Sheenjek Valley, northeastern Alaska. Univ. of Alaska, Biology Papers No. 4. 59 pp.

Krog, J. *1953.* Notes on the birds of Amchitka Island, Alaska. Condor 55(6):299-304.

L.G.L., Ltd. *1972.* Surveys of terrestrial bird populations in Alaska, Yukon Territories, Northwest Territories, and northern Alberta. Part One. Report Prepared for Northern Engineering Services, Ltd. 171 pp.

Maher, W.J. *1959.* Habitat distribution of birds breeding along the upper Kaolak River, northern Alaska. Condor 61(5):351-368.

Manuwal, D.A. and N.J. Manuwal. *1979.* Habitat utilization and migration of land birds on the Barren Islands, Alaska. West. Birds 10(4):201-213.

Alaska —Cont'd.
Murie, A. *1942.* The wilds where the caribou roam. Audubon Mag. 44(1):2-11.
Osgood, W.H. *1904.* A biological reconnaissance of the base of the Alaska Peninsula. N. American Fauna, No. 24. p. 1-86.
Sage, B.L. *1975.* Recent observations in the Wrangell Mountains, Alaska. Condor 77(2):206-207.
Shortt, T.M. *1939.* The summer birds of Yakutat Bay, Alaska. Royal Ontario Mus. Zool., Contrib. No. 17.
Turner, L.M. *1885.* Notes on the birds of the Nearer Islands, Alaska. Auk 2(2):154-159.
Willett, G. *1927.* Notes on the occurrence and distribution of some southeastern Alaska birds. Condor 29(1):58-60.

Arizona

Arizona Game and Fish Department. *1977.* Wildlife surveys and investigations. Nongame investigations. Special Performance Report, Proj. No. W-53-R-27. 72 pp.
Arizona Game and Fish Department. *1978.* Wildlife surveys and investigations. Nongame investigations. Special Performance Report, Proj. No. W-53-R-28. 12 pp.
Bailey, F.M. *1923.* Birds recorded from the Santa Rita Mountains in southern Arizona. Pac. Coast Avifauna 15:1-60.
Bruner, S.C. *1926.* Notes on the birds of the Baboquivari Mountains, Arizona. Condor 28(5):231-238.
Bryant, H.C. *1942.* Golden eagles visit northern Arizona desert. Condor 44(1):41.
Coues, E. *1866.* List of the birds of Fort Whipple, Arizona. Proc. Acad. Nat. Sci. Philadelphia 1866(1):39-100.
Coues, E. *1868.* List of birds collected in southern Arizona by Dr. E. Palmer; with remarks. Proc. Acad. Nat. Sci. Philadelphia 1868(1):81-85.
Fisher, A.K. *1903.* A partial list of the birds of Keam Canyon, Arizona. Condor 5(2):33-36.
Huey, L.M. *1936.* Notes on summer and fall birds of the White Mountains, Arizona. Wilson Bull. 48(2):119-130.
Huey, L.M. *1939.* Birds of the Mount Trumbull region, Arizona. Auk 56(3):320-325.
Huey, L.M. *1942.* A vertebrate faunal survey of the Organ Pipe Cactus National Monument, Arizona. Trans. of the San Diego Soc. Nat. Hist. 9(32):353-376.
Ligon, J.D. and R.P. Balda. *1968.* Recent data on summer birds of the Chiricahua Mountains area, southeastern Arizona. Trans. San Diego Soc. Nat. Hist. 15(2):41-50.
Lincoln, F.C. *1917.* Some notes of the birds of Rock Canyon, Arizona. Wilson Bull. 29(2):65-73.
Mearns, E.A. *1890.* Observation on the avifauna of portions of Arizona. Auk 7(1):45-55.
Monson, G. *1944.* Notes on birds of the Yuma region. Condor 46(1):19-22.
Osgood, W.H. *1903.* A list of birds observed in Cochise County, Arizona. Condor 5(5):128-131.
Rhoads, S.N. *1892.* The birds of southeastern Texas and southern Arizona observed during May, June and July, 1891. Proc. Acad. Sci. Philadelphia 44:98-126.
Scott, W.E.D. *1886.* On the avi-fauna of Pinal County, with remarks on some birds of Prima and Gila Counties, Arizona. Auk 3(2/3/4):249-258,383-389,421-432.
Stevenson, J.O. *1936.* Bird notes from the Haulapi Mountains, Arizona. Condor 38(6):244-245.
Sutton, G.M. and A.R. Phillips. *1942.* June bird life of the Papago Indian Reservation, Arizona. Condor 44(2):57-65.
Swarth, H.S. *1904.* Birds of the Huachuca Mountains, Arizona. Pac. Coast Avifauna 4:1-70.
Swarth, H.S. *1908.* Some fall migration notes from Arizona. Condor 10(3):107-116.
Swarth, H.S. *1914.* A distributional list of the birds of Arizona. Pac. Coast Avifauna 10:1-133.
Swarth, H.S. *1924.* Fall migration notes from the San Francisco Mountain region, Arizona. Condor 26(5):183-190.
Swarth, H.S. *1929.* The faunal areas of southern Arizona: a study in animal distribution. Proc. Calif. Acad. Sci. 18(12):267-383.
Van Rossem, A.J. *1946.* Two new races of birds from the Harquahala Mountains, Arizona. Auk 63(4):560-563.

Arkansas

Howell, A.H. *1911.* Birds of Arkansas. U.S.D.A. Biol Survey Bull. No. 38. 100 pp.
Smith, A.P. *1915.* Birds of the Boston Mountains, Arkansas. Condor 17(1):41-57.

California

Anonymous. *1936.* Minutes of the Cooper Club meetings. Condor 38(2):92-95.
Anthony, A.W. *1893.* Birds of San Pedro Martir, Lower California. Zoe 4(3):228-247.
Anthony, A.W. *1895.* Birds of San Fernando, lower California. Auk 12(2):134-143.
Barlow, C. *1901.* A list of the land birds of the Placerville - Lake Tahoe stage road. Central Sierra Nevada Mountains, Cal. Condor 3(6):151-184.
Belding, L. *1879.* A partial list of the birds of central California. Proc. U.S. Natl. Mus. 2:388-449.
Bloeker, J.C. von, Jr. *1937.* Some winter field notes from Monterey County, California. Condor 39(3):128-129.
Bradley, R.A. *1980.* Avifauna of the Palos Verdes Peninsula, California. West. Birds 11(1):1-24.
Butterfield, A.D. *1883.* Golden eagle captured. Ornithol. and Ool. 8(5):35.
Cooper, J.G. *1874.* Animal life of the Cuyamaca Mountains. Am. Nat. 8(1):14-18.
Cooper, J.G. *1887.* Additions to the birds of Ventura County, California. Auk 4(2):85-94.
Dawson, W.L. *1916.* A personal supplement to the distributional list of the birds of California. Condor 18(1):22-30.
De Groot, D.S. *1934.* Field observations from Echo Lake, California. Condor 36(1):6-9.
Dixon, J.S. *1943.* Birds of the Kings Canyon National Park area of California. Condor 45(6):205-219.

California —Cont'd.

Edwards, H.A. *1914.* Bird notes from the Sierra Madre Mountains, southern California. Condor 16(5):207-210.
Ferry, J.J. *1908.* Notes from the diary of a naturalist in northern California. Condor 10(1):30-44.
Fisher, W.K. *1900.* A list of birds observed on Mt. St. Helena, California. Condor 2(6):135-138.
Grinnell, J. *1902.* Check-list of California birds. Pac. Coast Avifauna 3:1-98.
Grinnell, J. *1902.* Birds of the Little Sur River, Monterey County. Condor 4(6):125-128.
Grinnell, J. *1905.* Summer birds of Mt. Pinos, California. Auk 22:378-391.
Grinnell, J. *1908.* The biota of the San Bernadino Mountains. Univ. Calif. Publ. Zool. 5(1):1-170.
Grinnell, J. *1915.* A distributional list of the birds of California. Pac. Coast Avifauna 11:1-217.
Grinnell, J. *1928.* A distributional summation of the ornithology of lower California. Univ. Calif. Publ. Zool. 32(1):1-300.
Grinnell, J. *1934.* Further observations upon the bird life of Death Valley. Condor 36(2):67-72.
Grinnell, J. and H.S. Swarth. *1913.* An account of the birds and mammals of the San Jacinto area of southern California. Univ. Calif. Publ. Zool. 10(10):197-406.
Grinnell, J. and M.W. Wythe. *1927.* Directory to the bird-life of the San Francisco Bay region. Pac. Coast Avifauna 18:1-160.
Grinnell, J., J. Dixon and J.M. Linsdale. *1930.* Vertebrate natural history of a section of northern California through the Lassen Peak region. Univ. Calif. Publ. Zool. 35:1-594.
Harwell, C.A. *1935.* Golden eagle flight. Yosemite Nat. Notes 14(3):24.
Hill, H.M. and I.L. Wiggins. *1948.* Ornithological notes from lower California. Condor 50(4):155-161.
Huey, L.M. *1926.* Notes from northwestern lower California, with the description of an apparently new race of the screech owl. Auk 43(3):347-362.
Lamb, C. and A.B. Howell. *1913.* Notes from Buena Vista Lake and Fort Tejon. Condor 15(3):115-120.
Linsdale, J.M. *1938.* Environmental responses of vertebrates in the Great Basin. Am. Midland Nat. 19(1):1-206.
Linsdale, J.M. *1947.* A ten-year record of bird occurrence on the Hastings Reservation. Condor 49(6):236-241.
Mailliard, J. *1900.* Land birds of Marin County, Cal. Condor 2(3):62-68.
Mailliard, J. *1927.* The birds and mammals of Modoc County, California. Proc. Calif. Acad. Sci. 16(10):1-359.
McGregor, R.C. *1901.* A list of the land birds of Santa Cruz County, California. Pac. Coast Avifauna 2:1-22.
Michael, C.W. *1925.* Additional information concerning the birds of Yosemite Valley. Condor 27(3):109-113.
Monson, G. *1944.* Notes on birds of the Yuma region. Condor 46(1):19-22.
Pemberton, J.R. and H.W. Carriger. *1915.* A partial list of the summer resident land birds of Monterey County, California. Condor 17(5):189-201.
Prill, A.G. *1925.* A trip into the realms of nature. Oologist 42(1):2-8.
Ray, M.S. *1903.* A list of the land birds of Lake Valley, central Sierra Nevada Mountains, California. Auk 20:180-193.
Ray, M.S. *1913.* Some further notes on Sierran field-work. Condor 15(6):198-203.
Ray, M.S. *1918.* Six weeks in the high Sierras in nesting time. Condor 20(2):70-78.
Richards, E.B. *1924.* A list of the land birds of the Grass Valley District, California. Condor 26(3):98-104.
Ridgway, R. *1874.* Lists of birds observed at various localities contiguous to the Central Pacific Railroad, from Sacramento City, California, to Salt Lake City, Utah. Bull. Essex Inst. 6:169-174.
Ridgway, R. *1875.* List of the birds observed at various localities contiguous to the Central Pacific Railroad, from Sacramento City, California, to Salt Lake City, Utah. Bull. Essex Inst. 7:10-35.
Sharp, C.S. *1907.* The breeding birds of Escondido. Condor 9(3):84-91.
Sherwood, J. *1936.* A California bird list. Oologist 53(8):101-104.
Silliman, O.P. and J.C. Von Bloeker. *1937.* Some winter field notes from Monterey County, California. Condor 39(3):128.
Sumner, E.L., Jr. *1929.* Golden eagle in Death Valley. Condor 31(3):127.
Swarth, H.S. *1900.* Avifauna of a 100-acre ranch. Condor 2(1):14-16.
Thompson, C.S. *1900.* Notes from San Luis Obispo County, Cal. Condor 2(4):89-90.
Townsend, C.H. *1887.* Field-notes on the mammals, birds and reptiles of northern California. Proc. U.S. Natl. Mus. 10:159-241.
Tyler, J.G. *1913.* Some birds of the Fresno District, California. Pac. Coast Avifauna 9:1-114.
Van Rossem, A. *1911.* Winter birds of the Salton Sea region. Condor 13(4):129-137.
Wauer, R.H. *1962.* A survey of the birds of Death Valley. Condor 64(3):220-233.
Wilder, H.E. *1916.* Some distributional notes on California birds. Condor 18(3):127-128.
Wilkinson, G.S. and K.R. Debban. *1980.* Habitat preferences of wintering diurnal raptors in the Sacramento Valley. West. Birds 11(1):25-34.
Willett, G. *1908.* Summer birds of the upper Salinas Valley and adjacent foothills. Condor 10(4):137-139.
Willett, G. *1909.* Bird notes from the coast of San Luis Obispo County. Condor 11(6):185-187.
Willett, G. *1919.* Bird notes from southeastern Oregon and northeastern California. Condor 21(5):194-207.
Willett, G. *1933.* A revised list of the birds of southwestern California. Pac. Coast Avifauna 21:1-204.
Williams, L. *1938.* Birds observed in the vicinity of Twenty-nine Palms, California. Condor 40(6):258.
Wilson, I.D. *1945.* Birds of the Kettleman Hills area, California. Condor 47(4):149-153.

Colorado

Aiken, C.E.H. *1937.* Birds of the Southwest. Colo. Coll. Publ., Gen. Ser. No. 212. 73 pp.
Allen, J.A. *1871.* Notes of an ornithological reconnaissance of portions of Kansas, Colorado, Wyoming and Utah. Bull. Mus. Comp. Zool. 3(6):113-183.
Bergtold, W.H. *1927.* Denver birds. Auk 44(3):432-433.
Betts, N.D. *1935.* Birds of the Boulder County, Colorado. Univ. Colo. Studies 10(4):177-232.

Colorado —Cont'd.

Cooke, W.W. *1897.* The birds of Colorado. Colorado State Agric. College, Agric. Expt. Stat. Bull. No. 37, Tech. Ser. 2. 143 pp.
Cory, C.B. *1881.* Description of a new species of the family Procellaridae. Bull. Nuttall Ornithol. Club 6(2/3):85-91,138-143.
Dille, F.M. *1903.* Nesting dates for birds in the Denver District, Colorado. Condor 5(3):73-74.
Gilman, M.F. *1907.* Some birds of southwest Colorado. Condor 9(5):152-158.
Hendee, R.W. *1929.* Notes on birds observed in Moffat County, Colorado. Condor 31(1):24-32.
Hersey, L.J. and R.B. Rockwell. *1909.* An annotated list of the birds of the Barr Lake District, Adams County, Colorado. Condor 11(4):109-122.
Hubbard, R. *1920.* Colorado notes. Condor 22(1):37-38.
Knox, D.A. *1944.* Summer birds of the Gothic area, Gunnison County, Colorado. Auk 61(1):19-30.
Lincoln, F.C. *1920.* Birds of the Clear Creek District, Colorado. Auk 37(1):60-77.
Lowe, W.P. *1894.* A list of the birds of the Wet Mountains, Huerfano County, Colorado. Auk 11(4):266-270.
Morrison, C.F. *1888.* A list of some birds of La Plata County, Col., with annotations. Ornithol. and Ool. 13(9):139-140.
Packard, F.M. *1945.* The birds of Rocky Mountain National Park, Colorado. Auk 62(3):371-394.
Packard, F.M. *1950.* The birds of Rocky Mountain National Park. Rocky Mountain Nature Assoc., Co. 81 pp.
Rockwell, R.B. *1908.* An annotated list of the birds of Mesa County, Colorado. Condor 10(4):152-180.
Sclater, W.L. *1908.* The winter birds of Colorado. Ibis (9th Ser.) 2(7):443-450.
Warren, E.R. *1908.* Northwestern Colorado bird notes. Condor 10(1):18-26.
Warren, E.R. *1909.* Notes on the birds of southwestern Montrose County, Colorado. Condor 11(1):11-17.
Warren, E.R. *1910.* Bird notes from Salida, Chaffee County, Colorado. Auk 27(2):142-151.
Warren, E.R. *1915.* Some Park County, Colorado, bird notes. Condor 17(2):90-95.
Warren, E.R. *1916.* Notes on the birds of the Elk Mountain region, Gunnison County, Colorado. Auk 33(3):292-317.
White, C.M. and W.H. Behle. *1961.* Birds of Navajo Reservoir Basin in Colorado and New Mexico, 1960. Univ. of Utah Anthropol. Papers 55:130-154.

Connecticut

Anonymous. *1886.* Golden eagle in Connecticut. Forest and Stream 27(19):362.
Anonymous. *1909.* The last of C.L. Rawson's quail trap eagles. Oologist 26(4):58.
Bishop, L.B. *1910.* Notes from Connecticut. Auk 27(4):462-464.
Bishop, L.B. *1921.* Notes from Connecticut. Auk 38(4):582-589.
Brewer, T.M. *1875.* Catalogue of the birds of New England, with brief notes indicating the manner and character of their presence. Proc. Boston Soc. Nat. Hist. 17:436-454.
Hopkins, D.A. and G.S. Mesereau. *1977.* Hawk migration, 1976 report. New England Hawk Watch. 14 pp.
Howes, P.G. *1928.* Notes of the birds of Stamford, Connecticut and vicinity. Oologist 45(6):70-96.
Moffatt, E.E. *1932.* Duck hawk nest. Oologist 44(8):93-94.
Sage, J.H. *1893.* Notes on some Connecticut birds. Auk 10(2):206-207.
W., J.M. *1887.* The golden eagle in Connecticut. Ornithol. and Ool. 12(12):206.

District of Columbia

Cooke, M.T. *1929.* Birds of the Washington, D.C. region. Proc. Biology Soc. Wash. 42:1-80.
Coues, E. and D.W. Prentiss. *1861.* List of birds ascertained to inhabit the District of Columbia, with the times of arrival and departure of such as are non-residents, and brief notices of habits, etc. Smithsonian Inst. Ann. Report 1861. p. 399-421.
Swales, B.H. *1920.* Records of several rare birds from near Washington, D.C. Proc. Biology Soc. Wash. 33:181-182.

Florida

Fisher, G.C. *1910.* Another golden eagle taken in west Florida. Auk 27(2):206.
Fisher, G.C. *1910.* Golden eagle taken in west Florida. Auk 27(1):80.
Sprunt, A., Jr. *1954.* Florida bird life. Coward-Mc Cann, Inc., NY. 527 pp.
Weston, F.M. *1965.* A survey of the birdlife of northwestern Florida. Bull. of Tall Timbers Res. Station 5:1-147.

Georgia

Bell, G.W. *1940.* Random notes from about state. Oriole 5(3):34.
Bell, G.W. *1944.* The golden eagle in northeast Georgia. Oriole 9(3/4):37.
Greene, E.R. *1931.* Golden eagle captured in Georgia. Auk 48(1):118.
Hebard, F.V. *1945.* Recent new or interesting winter records from interior southeastern Georgia. Oriole 10(1):4-6.
Hopkins, M., Jr. *1968.* Golden eagle in Irwin County, Georgia. Oriole 33(4):50.
Sciple, G.W., Jr. and B. Griffith. *1939.* Golden eagle in Georgia in summer. Oriole 4(3):31.
Stevenson, H.M. *1950.* Distribution of certain birds in the southeastern United States. Am. Midland Nat. 43(3):605-626.
Stoddard, H.L. *1940.* The golden eagle in southwest Georgia. Oriole 5(1):8.

Idaho

Arvey, M.D. *1947.* A check-list of the birds of Idaho. Univ. Kansas Publ. Mus. Nat. Hist. 1(10):193-216.
Hand, R.L. *1941.* Birds of the St. Joe National Forest, Idaho. Condor 43(5):220-232.
Jewett, S.G. *1912.* Some birds of the Saw-tooth Mountains, Idaho. Condor 14(5):191-194.
Johnston, D.W. *1949.* Populations and distribution of summer birds of Latah County, Idaho. Condor 51(3):140-149.
Levy, S.H. *1950.* Summer birds in southern Idaho. Murrelet 31(1):2-8.
Marshall, W.H. *1945.* Winter bird observations in the Boise National Forest, Idaho. Condor 47(4):170-172.
Merrill, J.C. *1897.* Notes on the birds of Fort Sherman, Idaho. Auk 14(4):347-357.
Oring, L.W. *1962.* Observations on the birds of southeastern Idaho. Murrelet 43(3):40-50.
Rust, H.J. *1916.* Additional notes on the birds of Kootenai County, Idaho. Condor 18(2):81-82.
Rust, H.J. *1917.* An annotated list of the birds of Fremont County, Idaho, as observed during the summer of 1916. Condor 19(2):29-43.
Snyder, J.O. *1900.* Notes on a few species of Idaho and Washington birds. Auk 17(3):242-245.

Illinois

Bellrose, F. *1936.* Notes from north central Illinois. Auk 53(3):348.
Black, C.T. *1935.* Notes form northeastern and east central Illinois. Auk 52:464.
Brodkorb, P. *1934.* Notes from central Illinois. Auk 51(3):403.
Graber, R.R. and J.S. Golden. *1960.* Hawks and owls: population trends from Illinois Christmas counts. Illinois Nat. Hist. Survey, Biology Notes No. 41. 24 pp.
Holder, R.H. *1861.* Birds of Illinois. Trans. Illinois Nat. Hist. Soc. 1(1):1-194.
Ridgway, R. *1874.* The lower Wabash Valley, considered in its relation to the faunal districts of the eastern region of North America: with a synopsis of its avian fauna. Proc. Boston Soc. Nat. Hist. 16:304-332.
Ridgway, R. *1881.* A revised catalogue of the birds ascertained to occur in Illnois. Illinois State Lab. Nat. Hist. Bull. No. 4. p. 163-208.
Southern, W.E. *1963.* Winter populations, behavior, and seasonal dispersal of bald eagles in northwestern Illinois. Wilson Bull. 75(1):42-55.

Indiana

Butler, A.W. *1897.* The birds of Indiana. 22nd Report Dept. Geol. and Nat. Resources of Indiana.
Carter, L. *1933.* Golden eagles in central Indiana. Indiana Audubon Year Book 1933:67.
Evermann, B.W. *1888.* Birds of Carroll County, Indiana. Auk 5(4):344-351.
Fite, C.F. *1894.* The golden eagle in Indiana. Oologist 11(5):184.
Perkins, S.E., III. *1928.* The golden eagle in Indiana. Wilson Bull. 40(3):196.
Phillips, R.E. *1953.* Indiana hawks and owls. Indiana Dept. of Conserv., Div. of Fish and Game. 47 pp.
Ridgway, R. *1874.* The lower Wabash Valley, considered in its relation to the faunal districts of the eastern region of North America: with a synopsis of its avian fauna. Proc. Boston Soc. Nat. Hist. 16:304-332.

Iowa

Allert, O.P. *1939.* Notes on certain raptores in Allamakee, Clayton and Dubuque Counties, Iowa. Iowa Bird Life 9(3):34-36
Brown, W.H. *1962.* Golden eagle in Polk County. Iowa Bird Life 32(4):85.
Bryant, E. *1967.* Golden eagle in Plymouth Co. Iowa Bird Life 37(1):23.
Jaques, H.E. *1931.* The golden eagle in southeastern Iowa. Wilson Bull. 43(2):58.
Lee, Z.C. *1943.* Golden eagle in northwest Iowa. Iowa Bird Life 13(4):64-65.
Morrissey, T.J. *1968.* Notes of birds in the Davenport area. Iowa Bird Life 38:67-80.
Parker, H.W. *1871.* Iowa birds. Am. Nat. 5(3):168-170.
Pierce, F.J. *1930.* Birds of Buchanan County, Iowa. Wilson Bull. 42(4):253-285.
Spurrell, J.A. *1917.* Annotated list of the water birds, game birds and birds of prey of Sac County, Iowa. Wilson Bull. 29(3):141-160.
Wolden, B.O. *1947.* Eagles in Emmet County in 1944. Iowa Bird Life 17(1):16.
Youngworth, W. *1961.* Golden eagle in Monona County. Iowa Bird Life 31(1):19.

Kansas

Allen, J.A. *1871.* Notes of an ornithological reconnaissance of portions of Kansas, Colorado, Wyoming and Utah. Bull. Mus. Comp. Zool. 3(6):113-183.
Douthitt, B.P. *1919.* Migration records for Kansas birds. Wilson Bull. 31(1):6-20.
Goodrich, A.L. *1946.* Birds in Kansas. Kansas Board Agriculture, Topeka. 340 pp.
Harrish, H. *1919.* Birds of the Kansas City region. Trans. Acad. of Sci. St. Louis 23(8):213-271.
Isely, D. *1912.* A list of the birds of Sedgwick County, Kansas. Auk 29(1):25-44.
Lantz, D.E. *1899-1900.* A list of birds seen in Dickinson County, Kansas from August 1898, to August, 1900. Trans. Kansas Acad. Sci. 17:116-121.
Linsdale, J. *1927.* Notes on summer birds of southwestern Kansas. Auk 44(1):47-58.

Kansas —Cont'd.

Long, W.S. *1940.* Check-list of Kansas birds. Trans. Kansas Acad. Sci. 43:432-456.
Love, G. *1916.* A golden eagle's nest in Decator County, Kansas. Oologist 33:12.
Wells, W. *1940.* Some changes in forty years. Oologist 57(1):9-12.

Kentucky

Alden, L. and E. Risser. *1963.* Golden eagle at Nerstrand Woods. Flicker 35(4):139.
Anonymous. *1967.* The 1967 winter eagle count in Kentucky. Ky. Warbler 43(2):51-52.
Feck, W. *1965.* Sight record of golden eagle in Jessamine County. Ky. Warbler 41(1):16.
Peterson, C.T. *1962.* Eagles in Kentucky Woodlands National Wildlife Refuge. Ky. Warbler 38(3):43-44.
Peterson, C.T. *1963.* The eagles of Hematite Lake. Ky. Warb. 39(3):35-45.
Wilson, G. *1922.* Birds of Bowling Green, Kentucky. Auk 39(2):233-243.
Wilson, G. *1969.* Additions to "Birds of south-central Kentucky." Ky. Warb. 45(2):32-40.

Louisiana

Beyer, G.E., A. Allison and H.H. Kopman. *1908.* List of the birds of Louisiana. Part V. Auk 25(4):439-448.
Holt, E.G. *1933.* Golden eagle in Louisiana. Auk 50(3):355.
Oberholser, H.C. *1938.* Bird life of Louisiana. Louisiana Dept. of Conserv., Bull. No. 289. 834 pp.
Tanner, J.T. *1939.* Observations in Madison Parish, Louisiana. Auk 56(1):90.

Maine

Boardman, G.A. *1862-1863.* Catalogue of the birds found in the vicinity of Calais, Maine, and about the islands at the mouth of the Bay of Fundy. Proc. Boston Soc. Nat. Hist. 9:122-132.
Brewer, T.M. *1875.* Catalogue of the birds of New England, with brief notes indicating the manner and character of their presence. Proc. Boston Soc. Nat. Hist. 17:436-454.
Brock, H.H. *1896.* Recent capture of the golden eagle near Portland, Maine. Auk 13(3):256.
Knight, O.W. *1896.* The golden eagle in Maine. Auk 13(1):82.
Palmer, R.S. *1946.* Birds of the Mt. Katahdin region of Maine. Auk 63(3):299-314.
Sweet, D.W. *1906.* Migration reports, 1905. J. Maine Ornithol. Soc. 8(2):35-39.
Verrill, A.E. *1864.* Catalogue of the birds found at Norway, Oxford Co., Maine. Proc. Essex Inst. 3:136-153.

Maryland

Buckalew, J.A. *1950.* Records from the Del-Mar-Va Peninsula. Auk 67(2):250-252.
Fisher, W.H. *1896.* Raptores of Maryland. Nidologist 3(10/11):119-123.
Kirkwood, F.C. *1895.* A list of the birds of Maryland. Trans. Maryland Acad. Sci. 1:241-382.
Stewart, R.E. and C.S. Robbins. *1947.* Recent observations on Maryland birds. Auk 64(2):266-274.

Massachusetts

Bagg, A.C. and S.A. Eliot, Jr. *1973.* Birds of the Connecticut Valley in Massachusetts. Hampshire Bookshop, MA. 813 pp.
Brewer, T.M. *1875.* Catalogue of the birds of New England, with brief notes indicating the manner and character of their presence. Proc. Boston Soc. Nat. Hist. 17:436-454.
Brewster, W. *1884.* Notes on the summer birds of Berkshire County, Massachusetts. Auk 1(1):5-16.
Brewster, W. *1887.* The golden eagle in eastern Massachusetts. Auk 4(1):75.
Eliot, S.A., Jr. *1939.* Golden eagle at Chilmark, Massachusetts. Auk 56:75.
Eliot, S.A., Jr. *1942.* Golden eagle visits Vineyard. Bull. Mass. Audubon Soc. 26(6):142.
Falk, L.L. *1973.* Winter birding in northeastern Massachusetts. Delmarva Ornithol. 8(2):68-73.
Griscom, L. and D.E. Snyder. *1955.* The birds of Massachusetts. Peabody Mus., Salem. 295 pp.
Hopkins, D.A. and G.S. Mesereau. *1977.* Hawk migration, 1976 report. New England Hawk Watch. 14 pp.
Howe, R.H. and G.M. Allen. *1901.* Birds of Massachusetts. Publ. by Subscription, Cambridge, Mass. 154 pp.
Howe, R.H., Jr. *1908.* Massachusetts records. Auk 25(3):323.
Kneeland. S., Jr. *1855.* Young golden eagle. Proc. Boston Soc. Nat. Hist. 5:272.
Kneeland, S., Jr. *1855.* Young male golden eagle. Proc. Boston Soc. Nat. Hist. 5:272-273.
Morris, R.O. *1917.* Golden eagle *(Aquila chrysaetos)* at Springfield, Mass. Auk 34(4):477.
Stubbs, A.P. and S.G. Emilio. *1931.* A list of the birds of Essex County, Massachusetts. Essex County Ornithol. Club Bull. 13:13-52.

Michigan

Blain, A.W., Jr. *1902.* Michigan bird notes, 1901. Auk 19(2):209-210.

Michigan —Cont'd.

Judson, R.F. *1892.* A golden eagle taken at Kalamazoo, Mich. Ornithol. and Ool. 17(1):9.
Lorge, L.L. *1966.* Summer sighting of the golden eagle in Michigan. Jack-Pine Warb. 44(1):22.
Purdy, J.B. *1902.* Snowy owl and golden eagle at Plymouth, Michigan. Auk 19(1):80.
Scolopax. *1893.* Raptores of Michigan. Oologist 10(11):294-297.
Swales, B.H. *1903.* Notes on the winter birds of Wayne County, Michigan. Wilson Bull. 15(1):20-24.

Minnesota

Barrett, L.L. *1947.* Eagles in Minnesota. Flicker 19(2):53-54.
Brown, A.D. *1890.* Specimen of golden eagle. Ornithol. and Ool. 15(5):79.
Cantwell, G.G. *1890.* A list of the birds of Minnesota. Ornithol. and Ool. 15(9):129-137.
Farnes, C.A. *1981.* Birds of the St. Croix River Valley: Minnesota and Wisconsin. U.S.D.I. Fish and Wildlife Service, N. American Fauna, No. 73.
Green, J.C. *1965.* 1965 fall hawk migration, northwestern Minnesota. Loon 37(4):126-130.
Green, J.C. *1965.* 1964 fall hawk migration, Duluth. Loon 37(2):88-89.
Greenwood, O.M. *1915.* Golden eagle. Oologist 32(6):98.
Hofslund, P.B. *1954.* The hawk pass at Duluth. Flicker 26(3):96-99.
Hofslund, P.B. *1966.* Hawk migration over the western tip of Lake Superior. Wilson Bull. 78(1):79-87.
Rustad, O.A. *1976.* Birds of Rice County, Minnesota (Part 1). Loon 48(4):136-149.

Mississippi

Burleigh, T.D *1944.* The bird life of the Gulf Coast region of Missippi. Louisiana St. Univ., Mus. Zool. Occas. Pap. No. 20. p. 331-490.

Missouri

Bennitt, F. *1932.* Check-list of the birds of Missouri. Univ. of Missouri Studies, Quarterly of Res. 7(3):1-81.
Bolt, B.J. *1912.* Capture of a golden eagle at Kansas City, Mo. Auk 29(1):102.
Widmann, O. *1907.* A preliminary catalog of the birds of Missouri. Trans. Acad. Sci. of St. Louis. 296 pp.

Montana

Allen, J.A. *1874.* Notes on the natural history of portions of Dakota and Montana Territories. Proc. Boston Soc. Nat. Hist. 17:33-86.
Bosson, C. *1936.* Ornithological sidelights of a motor trip through the west. Essex County Ornithol. Club Bull. 17:29-31.
Grinnell, G.B. *1876.* List of mammals and birds. Pages 59-92. *in:* W. Ludlow. Report of A Reconnaissance from Carroll, Montana Territory, on the Upper Missouri, to the Yellowstone National Park, and Return, Made in the Summer of 1875. War Dept., Wash., D.C.
Johnson, R.E. *1966.* Alpine birds of the Little Belt Mountains, Montana. Wilson Bull. 78(2):225-227.
Pattie, D.L. and N.A.M. Verbeek. *1966.* Alpine birds of the Beartooth Mountains. Condor 68(2):167-176.
Richmond, C.W. and F.H. Knowlton. *1894.* Birds of south-central Montana. Auk 11(4):298-308.
Saunders, A.A. *1911.* A preliminary list of the birds of Gallatin County, Montana. Auk 28(1):26-49.
Saunders, A.A. *1912.* Some birds of southwestern Montana. Condor 14(1):22-32.
Saunders, A.A. *1914.* The birds of Teton and northern Lewis and Clark Counties, Montana. Condor 16(3):124-144.
Saunders, A.A. *1921.* A distributional list of the birds of Montana. Pac. Coast Avifauna 14:1-194.
Silloway, P.M. *1901.* Summer birds of Flathead Lake. Univ. of Montana, Bull. No. 3. Biology Ser. No. 1. 83 pp.
Thompson, L.S. *1978.* Species abundance and habitat relations of an insular montane avifauna. Condor 80(1):1-14.
Thorne, P.M. *1895.* List of birds observed in the vicinity of Fort Keogh, Montana, from July, 1888, to September, 1892. Auk 12(3):211-219.
Weydemeyer, W. *1975.* Half-century record of the breeding birds of Fortine area, Montana: nesting data and population status. Condor 77(3):281-287.

Nebraska

Fiala, K.L. *1970.* The birds of Gage County, Nebraska. Nebr. Bird Rev. 38(3):43-70.
Haecker, F.W., R.A. Moser and J.B. Swenk. *1945.* Check-list of the birds of Nebraska. Nebr. Bird Rev. 13(1):1-40.
Patton, F. *1951.* Golden eagles seen in Gage County. Nebr. Bird Rev. 19(2):31.
Rapp, W.F., Jr. *1951.* Twenty-five year summary of bird migration in Nebraska. Nebr. Bird Rev. 19(1):3-8.
Taylor, W.E. and A.H. Van Vleet. *1889.* Notes on Nebraska birds. Ornithol. and Ool. 14(11):163-165.
White, C.A. *1893.* The raptores of Omaha and vicinity. Oologist 10(5):138-140.
Zimmer, J.T. *1911.* Some notes on the winter birds of Dawes County. Proc. Nebr. Ornithol. Union 5(2):19-30.

Nevada

Alcorn, J.R. *1946.* The birds of Lahontan Valley, Nevada. Condor 48(3):129-138.
Baldwin, G.C. *1944.* Uncommon birds of the Boulder Dam area, Nevada. Condor 46(1):35.
Bond, R.M. *1940.* New bird records for Lincoln County, Nevada. Condor 42(4):220-222.
Cottam, C. *1942.* Records from extreme northeastern Nevada. Condor 44(3):127-128.
Gabrielson, I.N. *1949.* Bird notes from Nevada. Condor 51(4):179-187.
Gullion, G.W., W.M. Pulich and F.G. Evenden. *1959.* Notes on the occurence of birds in southern Nevada. Condor 61(4):278-297.
Linsdale, J.M. *1936.* The birds of Nevada. Pac. Coast Avifauna 23:1-145.
Linsdale, J.M. *1951.* A list of the birds of Nevada. Condor 53(5):228-249.
Taylor, W.P. *1912.* Field notes on amphibians, reptiles and birds of northern Humboldt County, Nevada, with a discussion of some of the faunal features of the region. Univ. Calif. Publ. Zool. 7(10):319-436.
Van Rossem, A.J. *1936.* Birds of the Charleston Mountains, Nevada. Pac. Coast Avifauna 24:1-65.

New Hampshire

Brewer, T.M. *1875.* Catalogue of the birds of New England, with brief notes indicating the manner and character of their presence. Proc. Boston Soc. Nat. Hist. 17:436-454.
Richards, T. *1958.* The distribution of hawks and owls in New Hampshire. New Hampshire Bird News 11(4):18-26.
Shelley, L.O. *1931.* The golden eagle *(Aquila chrysaetos)* in Cheshire County, New Hampshire. Auk 48(1):118.

New Jersey

Braislin, W.C. *1896.* The golden eagle in New Jersey. Auk 13(1):81-82.
Chapman, F.M. *1898.* Golden eagle in New Jersey. Auk 15(1):54.
Clark, W.S. *1971.* Migration trapping of hawks (and owls) at Cape May Point, N.J. E.B.B.A. 34(4):160-169.
Clark, W.S. *1973.* Cape May Point raptor banding station - 1972 results. E.B.B.A. News 36(3):150-164.
Dunne, P. *1977-1978.* Summary of the 1977 hawk flight at Cape May Point, N.J. Cassinia 57:21.
Miller, W.D. *1919.* Four rare birds in Sussex County, New Jersey. Auk 36(2):293.
Sage, J.H. *1895.* The golden eagle in New Jersey. Auk 12(2):179.

New Mexico

Bailey, F.M. *1904.* Additional notes on the birds of the upper Pecos. Auk 21:349-363.
Bergtold, W.H. *1912.* October birds of the headwaters of the Gila River, New Mexico. Auk 29(3):327-336.
Gilman, M.F. *1908.* Birds on the Navajo Reservation in New Mexico. Condor 10(4):146-152.
Henry, T.C. *1859.* Catalogue of the birds of New Mexico as compiled from notes and observations made while in that Territory, during a residence of six years. Proc. of the Acad. Nat. Sci. Philadelphia. 11:104-109.
Henshaw, H.W. *1886.* List of birds observed in summer and fall on the upper Pecos River, New Mexico. Auk 3(1):73-80.
Huey, W.S. and J.R. Travis. *1961.* Burford Lake, New Mexico, revisted. Auk 78(4):607-626.
Hunn, J.T.S. *1906.* Notes on birds of Silver City, New Mexico. Auk 23:418-425.
Tatschl, J.L *1967.* Breeding birds of the Sandia Mountains and their ecological distributions. Condor 69(5):479-490.
Wetmore, A. *1920.* Observations on the habits of birds at Lake Burford, New Mexico. Auk 37(2/3):221-247,393-412.
White, C.M. and W.H. Behle. *1961.* Birds of Navajo Reservoir Basin in Colorado and New Mexico, 1960. Univ. of Utah Anthropol. Papers 55:130-154.

New York

Bagg, E. *1897.* Some new records from central New York. Auk 14(2):226-227.
Beardslee, C.S. *1938.* The golden eagle in western New York. Prothonotary 4(10):2.
Carleton, G. *1951.* Bird notes from Essex County. Kingbird 1(4):86-88.
Chahoon, G. *1905.* Notes on some Adirondack birds. Warbler 1(1):4-14.
Crumb, D.W. and B. Peebles. *1977.* To find a fall hawk lookout. Kingbird 27(2):82-85.
Gardner, G.B. *1918.* Notes from Canandaigua, N.Y. Bird Lore 20(5):351-352.
Haugh, J.R. and T.J. Cade. *1966.* The spring hawk migration around the southeastern shore of Lake Ontario. Wilson Bull. 78(1):88-110.
Hopkins, D.A. and G.S. Mesereau. *1977.* Hawk migration, 1976 report. New England Hawk Watch. 14 pp.
Kibbe, A.E. *1911.* A golden eagle record. Oologist 28(3):50.
Single, E.W. *1975.* Hook Mountain: highway for hawks. Kingbird 25(3):124-131.
Spofford, W.R. *1959.* Some golden eagles in New York. Kingbird 9(1):23-24.
Thomas, S. *1975.* Hook Mountain, New York. Pages 19-21. *in:* Proc. of the N. American Hawk Migration Conf., 1974. Syracuse, N.Y. Hawk Migration Assoc. of North America.

North Carolina

Brewster, W. *1886.* An ornithological reconnaissance in western North Carolina. Auk 3(1):94-112.
Brown, W., Jr. and J.A. Lee. *1974.* Golden eagle in eastern North Carolina. Chat 38(3):71.
Cairns, J.S. *1887.* A list of birds of Buncombe Co., North Carolina. Ornithol. and Ool. 12(1):3-6.
Cairns, J.S. *1889.* The summer birds of Buncombe County, North Carolina. Ornithol. and Ool. 14(2):17-23.
Clark, A.H. and L.F. Clark. *1934.* The golden eagle in North Carolina. Auk 51(1):80.
De Vore, J.E. *1973.* Golden eagle in the Unicoi Mountains. Chat 37(2):50.
Herndon, L.R. *1949.* Golden eagle at Hump Mountain, Tenn.-N.C. Migrant 20(1):17.
Holmes, R., III. *1954.* Notes on North Carolina colonies. Chat 18(1):23-24.
Mattocks, J. *1959.* Golden eagles. Chat 23(2):35.
McCulloch, N.B., Jr. *1952.* Golden eagle at Pea Island Refuge. Chat 16(1):24.
Moore, C.F. *1917.* Golden eagle in Rutherford, County, N.C. Oologist 34(11):193.
Quay, T.L. *1956.* A record of the golden eagle in North Carolina. Chat 20(3):59.
Rudolph, R. *1959.* Untitled. Chat 23(2):36.
Simpson, M.B., Jr. *1976.* Birds of the Plott Balsam Mountains of North Carolina. Chat 40(3):53-62.
Stupka, A. *1963.* Notes on the birds of the Great Smoky Mountains National Park. Univ. of Tennessee Press, Knoxville, TN. 242 pp.
Wayne, A.T. *1910.* The cerulean warbler *Dendroica cerulea* breeding in Burke County, North Carolina. Auk 27(1):84-85.

North Dakota

Larson, A. *1928.* Birds of eastern McKenzie County, North Dakota. Wilson Bull. 40(1):37-48.
Monson, G.W. *1934.* The birds of Berlin and Hardwood Townships, Cass County, North Dakota. Wilson Bull. 46(1):37-58.
Williams, H.V. *1926.* Birds of the Red River Valley of northeastern North Dakota. Wilson Bull. 38(1):17-33.
Wood, N.A. *1923.* A preliminary survey of the bird life of North Dakota. Univ. Mich. Mus. of Zool., Misc. Publ. No. 10. p. 1-94.

Ohio

Baird, R.L. *1905.* Bird migration at Oberlin, Ohio. Wilson Bull. 17(3):75-83.
Davie, O. *1882.* Eagles in Ohio. Ornithol. and Ool. 7(18):142.
Fisher, G.C. *1907.* A golden eagle in Miami County, Ohio. Wilson Bull. 19(1):30-31.
Goodpaster, W. *1941.* A list of the birds and mammals of southwestern Ohio. J. Cinn. Soc. Nat. Hist. 22(3):1-47.
Henninger, W.F. *1902.* A preliminary list of the birds of middle southern Ohio. Wilson Bull. 9(3):77-93.
Henninger, W.F. *1906.* A preliminary list of the birds of Seneca County, Ohio. Wilson Bull. 18(2):47-60.
Jones, L. *1902.* Winter birds. Wilson Bull. 14(4):129-132.
Langdon, F.W. *1879.* A revised list of Cininnati birds. J. Cinn. Soc. Nat. Hist. 1(4):167-193.
McConnell, H.B. *1908.* Golden eagle *(Aquila chrysaetos)* at Cadiz, Ohio. Wilson Bull. 20(4):214.
Oberholser, H.C. *1892.* The golden eagle in Ohio. Auk 9(2):200.
Roads, K.M. *1938.* Four Ohio records of golden eagle. Wilson Bull. 50(1):64.
Swales, B.H. *1915.* Golden eagle *(Aquila chrysaetos)* in Oho *(sic).* Wilson Bull. 27(1):288.

Oklahoma

Sutton, G.M. *1934.* Notes on the birds of the western panhandle of Oklahoma. Ann. Carnegie Mus. 24:1-50.
Tyler, J.D. *1979.* Birds of southwestern Oklahoma. Univ. of Oklahoma, Stovall Mus., Contrib. No 2.

Oregon

Browning, M.R. *1975.* The distribution and occurrence of the birds of Jackson County, Oregon, and surrounding areas. U.S.D.I. Fish and Wildlife Service, N. American Fauna, No. 70. 69 pp.
Gabrielson, I.N. *1924.* Notes on the birds of Wallowa County, Oregon. Auk 41(4):552-565.
Gabrielson, I.N. *1931.* The birds of the Rogue River Valley, Oregon. Condor 33(3):110-121.
Green, C.A. *1978.* Summer birds of the Alvord Basin, Oregon. Murrelet 59(2):59-69.
Gullion, G.W. *1951.* Birds of the southern Willamette Valley, Oregon. Condor 53(3):129-149.
Jewett, S.G. *1909.* Some birds of Baker County, Oregon. Auk 26(1):5-9.
Laval, J. *1947.* White-tailed kites in Oregon. Murrelet 28(3):41.
McAllister, T.H. and D.B. Marshall. *1945.* Summer birds of the Fremont National Forest, Oregon. Auk 62(2):177-189.
Merrill, J.C. *1888.* Notes on the birds of Fort Klamath, Oregon. Auk 5(2):139-140,145-146.
Pedersen, R.J., A.W. Adams and L.D. Bryant. *1975.* Observations on birds of the Blue Mountains. Murrelet 56(1):7-10.
Prill, A.G. *1922.* Nesting birds of Lake County, Oregon (with special reference to Warner Valley). Wilson Bull. 34(3):131-140.
Walker, A. *1924.* Notes on some birds from Tillamook County, Oregon. Condor 26(5):180-182.
Willett, G. *1919.* Bird notes from southeastern Oregon and northeastern California. Condor 21(5):194-207.

Pennsylvania

Anonymous. *1973.* Three records set at Hawk Mountain. Penn. Game News 44(7):11.

Barnard, V. *1860.* A catalogue of the birds of Chester County, Pennsylvania, with their times of arrival in spring, from observations annually for ten successive years. Smithsonian Inst. Ann. Rep. 1860:434-438.

Barnes, R.M. *1931.* Golden eagle. Oologist 48(5):70.

Bergey, A. *1974-1975.* Hawk migration behavior at the Susquehanna River. Cassinia 55:37-38.

Burns, F.L. *1909.* Golden eagle *(Aquila chrysaetos).* Wilson Bull. 21(1):48.

Frock, R.F., Jr. *1970.* Golden eagle in Bucks County, Pennsylvania. Cassinia 52:40.

Heintzelman, D.S. *1973.* The 1972 autumn hawk count at Tott's Gap, Pennsylvania. N.J. State Mus., Sci. Notes No. 129. 3 pp.

Hofman, M.J. *1917.* Eagles in Pennsylvania. Oologist 34(1):11.

Leberman, R.C. *1976.* The birds of the Ligonier Valley. Carnegie Mus. Nat. Hist., Spec. Publ. No. 3. 67 pp.

Ressel, C.R. *1889.* Birds of Chester County, Penn. Ornithol. and Ool. 14(7):97-101.

Street, P.B. *1954.* Birds of the Pocono Mountains, Pennsylvania. Cassinia 41:3-76.

Street, P.B. *1974-1975.* Birds of the Pocono Mountains, 1955-1975. Cassinia 55:3-16.

Rhode Island

Brewer, T.M. *1875.* Catalogue of the birds of New England, with brief notes indicating the manner and character of their presence. Proc. Boston Soc. Nat. Hist. 17:436-454.

Dexter, N. *1887.* Golden eagle in Rhode Island. Forest and Stream 28(6):106.

Howe, R.H. and E. Sturtevant. *1903.* A supplement to the birds of Rhode Island. Privately Published. Middletown, RI. 111 pp.

South Carolina

Edwards, R.D. *1962.* Golden eagle in Sumter County, South Carolina. Chat 26(1):19.

Hudson, G.E. and F. Sherman. *1936.* Some miscellaneous notes on South Carolina birds. Auk 53(3):311-315.

Pickens, A.L. *1927.* Supplementary records for upper South Carolina. Auk 44(3):428-429.

Pickens, A.L. *1928.* Birds of upper South Carolina: a study in geographical distribution. Wilson Bull. 40(3):182-191.

Pickens, A.L. *1928.* Additions to the list of birds of the Piedmont region of South Carolina. Auk 45(1):66-69.

Wayne, A.T. *1906.* A contribution to the ornithology of South Carolina, chiefly the coast region. Auk 23(1):56-68.

Wayne, A.T. *1910.* Birds of South Carolina. Contrib. from Charleston Museum, Charleston, SC. 254 pp.

South Dakota

Agersborg, G.S. *1885.* The birds of southeastern Dakota. Auk 2(3):276-289.

Allen, J.A. *1874.* Notes on the natural history of portions of Dakota and Montana Territories. Proc. Boston Soc. Nat. Hist. 17:33-86.

Blankenship, L.H., I.G. Bue and W.H. Marshall. *1953.* Spring migration and nesting data in Stanley County, South Dakota. S. Dakota Bird Notes 5(1):10-13.

Cary, M. *1901.* Birds of the Black Hills. Auk 18:231-233.

Hoffman, W.J. *1875.* List of birds observed at Grand River Agency, Dakota Terr., from October 7th, 1872, to June 7th, 1873. Proc. Boston Soc. Nat. Hist. 18:169-175.

Krumm, K. *1951.* Winter notes from southwestern S.D. S. Dakota Bird Notes 3(1):11-12.

Lamster, E.R. *1951.* Eagles near Pierre. S. Dakota Bird Notes 3(1):10.

Mallory, W.B. *1950.* Golden eagle near Canton, S.D. S. Dakota Bird Notes 2(1):14.

Morse, G.W. *1925.* An oologist's outing. Oologist 42:108-116.

Patton, F.A. *1926.* Our trip to the eagle's nest. Oologist 43(3):30-31.

Visher, S.S. *1909.* A list of the birds of western South Dakota. Auk 26(2):144-153.

Visher, S.S. *1911.* Annotated list of the birds of Harding County, northwestern South Dakota. Auk 25(1):5-16.

Visher, S.S. *1915.* A list of the birds of Clay County, southeastern South Dakota. Wilson Bull. 37(2):321-335.

Youngworth, W. *1935.* The birds of Fort Sisseton, South Dakota, a sixty year comparison. Wilson Bull. 47(3):209-235.

Tennessee

De Vore, J.E. *1973.* Golden eagle in the Unicoi Mountains. Chat 37(2):50.

Ganier, A.F. *1932.* Facts about eagles in Tennessee. J. Tenn. Acad. Sci. 6(2):49-57.

Ganier, A.F. *1951.* Mississippi kite and golden eagle in Overton Co., Tenn. Migrant 22:29.

Herndon, L.R. *1949.* Golden eagle at Hump Mountain, Tenn.-N.C. Migrant 20(1):17.

Ogden, J.C. *1962.* Golden eagle in Knox County. Migrant 33(1):20.

Rhoads, S.N. *1895.* Contributions to the zoology of Tennessee. No. 2. Birds. Proc. Acad. Sci. Philadelphia. 1895 (Part 3):463-501.

Stupka, A. *1963.* Notes on the birds of the Great Smoky Mountains National Park. Univ. of Tennessee Press, Knoxville, TN. 242 pp.

Wetmore, A. *1939.* Notes on the birds of Tennessee. Proc. U.S. Natl. Mus. 86:175-243.

Texas

Benners, G.B. *1887.* A collecting trip in Texas. Ornithol. and Ool. 12(5):65-69.
Carroll, J.J. *1900.* Notes on the birds of Refugio County, Texas. Auk 17(4):337-348.
Griscom, L. and M.S. Crosby. *1925.* Birds of the Brownsville region, southern Texas. Auk 42(3/4):432-440,519-537.
Kumlien, L. *1877.* Ornithological notes from Texas. Field and Forest 2(8):127-132.
Lacey, H. *1912.* Additions to birds of Kerrville, Texas. Auk 29(2):254.
Lawrence, R.B. *1926.* Golden eagle at Houston, Texas. Auk 43(3):369.
Montgomery, T.H., Jr. *1905.* Summer resident birds of Brewster County, Texas. Auk 22:12-15.
Rhoads, S.N. *1892.* The birds of southeastern Texas and southern Arizona observed during May, June and July, 1891. Proc. Acad. Sci. Philadelphia 44:98-126.
Smith, A.P. *1912.* Recent records from the Valley of the Lower Rio Grande. Auk 29(2):254-255.
Smith, A.P. *1917.* Some birds of the Davis Mountains, Texas. Condor 19(5):161-165.
Stevenson, J.O. *1942.* Birds of the central panhandle of Texas. Condor 44(3):108-115.
Van Tyne, J. and G.M. Sutton. *1937.* The birds of Brewster County, Texas. Univ. of Mich. Mus. of Zool., Misc. Publ. No. 37. p. 1-119.

Utah

Allen, J.A. *1871.* Notes of an ornithological reconnaissance of portions of Kansas, Colorado, Wyoming and Utah. Bull. Mus. Comp. Zool. 3(6):113-183.
Behle, W.H. *1944.* Check-list of the birds of Utah. Condor 46(2):67-87.
Behle, W.H. *1955.* The birds of the Deep Creek Mountains of central western Utah. Univ. of Utah, Biology Ser. 11(4):1-34.
Behle, W.H. *1960.* The birds of southeastern Utah. Univ. of Utah, Biology Ser. 12(1):1-56.
Behle, W.H., J.B. Bushman and C.M. Greenhalgh. *1958.* Birds of the Kanab area and adjacent high plateaus of southern Utah. Univ. of Utah, Biology Ser. 11(7):1-92.
Hardy, R. *1940.* An annotated check-list of the birds of Washington County, Utah. Proc. Utah Acad. Sci., Arts and Letters 17:95-111.
Knopf, F.L. and V.C. Bachman. *1977.* Birds of Gunnison and Cub Islands, Great Salt Lake, Utah. Great Basin Nat. 37(2):247-251.
Ridgway, R. *1875.* List of the birds observed at various localities contiguous to the Central Pacific Railroad, from Sacramento City, California, to Salt Lake City, Utah. Bull. Essex Inst. 7:10-35.
Stanford, J.S. *1931.* Notes on hawks and owls in Sevier Co., Utah. Auk 48(4):618-620.
Tanner, V.M. *1927.* Notes on birds collected in the Virgin River Valley of Utah. Condor 29(4):196-200.
Twomey, A.C. *1942.* The birds of the Uinta Basin, Utah. Ann. Carnegie Mus. 28:341-490.
Wauer, R.H. *1965.* Birds of Zion National Park and vicinity. Zion Nat. Hist. Assoc., in Coop. With the National Park Service. 92 pp.
White, C.M. *1969.* Population trends in Utah raptors. Pages 359-362. *in:* J.J. Hickey, Ed. Peregrine Falcon Populations; Their Biology and Decline. Univ. of Wisconsin Press, Madison, WI.
Woodbury, A.M., C. Cottam and J.W. Sugden. *1949.* Annotated checklist of the birds of Utah. Bull. of the Univ. of Utah, Biology Ser. 11(2):1-40.

Vermont

Brewer, T.M. *1875.* Catalogue of the birds of New England, with brief notes indicating the manner and character of their presence. Proc. Boston Soc. Nat. Hist. 17:436-454.
Kirk, G.L. *1908.* Golden eagle in Rutland County. Vermont Bird Club Bull. 3:20.
Ross, L.H. *1913.* Birds observed at Bennington, Vermont. Auk 30(3):436-437.

Virginia

Burford, F.C. *1966.* Golden eagle at Peaks of Otter. Raven 37(4):89.
Murray, J.J. *1933.* Additions to the Virginia avifauna since 1890. Auk 50(2):190-200.
Murray, J.J. *1936.* The land birds of Rockbridge County, Vorginia *(sic).* Oologist 53(3):26-35.
Murray, J.J. *1953.* First revision of the Virginia 1952 "check-list." Raven 24(5/6):34-45.
Shaver, B.F. *1916.* A golden eagle in Virginia. Bird-Lore 18(5):319-320.
Smyth, E.A., Jr. *1912.* Birds observed in Montgomery County, Virginia. Auk 29(4):508-530.
Surber, T. *1894.* Capture of the golden eagle at Covington, Virginia. Auk 11(1):77.
Toms, H. and S. Cromer. *1973.* Golden eagles in southwest Virginia. Raven 44(4):98.

Washington

Bowles, J.H. *1906.* A list of the birds of Tacoma, Washington, and vicinity. Auk 23(2):138-148.
Brown, D.E. *1934.* Washington bird notes- fall of 1934. Murrelet 15(3):79-80.
Burke, T. and J. Nisbet. *1979.* Birds of the Colville National Forest. U.S.D.A. Forest Service, Colville National Forest. 8 pp.
Dice, L.R. *1918.* The birds of Walla Walla and Columbia Counties, southeastern Washington. Auk 35(1):40-51.
Edson, J.M. *1908.* Birds of the Bellingham Bay region. Auk 25(4):425-439.

Washington —Cont'd.

Edson, J.M. *1926.* Ten days among east-side birds and mammals. Murrelet 7(1):3-7.

Fielder, P.C. *1976.* The impact of the Grand Coulee third powerhouse extension on fish and wildlife populations. Columbia River, Washington. Wash. Dept. of Game, Appl. Res. Info. Rep. 63 pp.

Hall, F.S. *1934.* Studies in the history of ornithology in the State of Washington (1792-1932) with special reference to the discovery of new species. Murrelet 15(1):3-19.

Harris, S.W. and C.F. Yocom. *1952.* Birds of the Lower Grand Coulee and potholes area, Grant County, Washington. Murrelet 33(2):18-28.

Hudson, G.E. *1947.* Birds observed in the Blue Mountains area of Washington during 1946. Murrelet 28(1):2-5.

Hurley, J.B. *1921.* Birds of Yakima County, Wash. Oologist 38(2):14-17.

Johnsgard, P.A. *1954.* Birds observed in the potholes region during 1953-1954. Murrelet 35(2):25-31.

Kitchin, E.A. *1939.* A distributional check-list of the birds of Mount Ranier National Park. Murrelet 20(2):27-37.

Lawrence, R.H. *1892.* A preliminary list of the birds of the Gray's Harbor region, Washington. Auk 9(1):39-47.

Mattocks, P.W., E.S. Hunn and T.R. Wahl. *1976.* A checklist of the birds of Washington State, with recent changes annotated. West. Birds 7(1):1-24.

Palmer, R.H. *1927.* Birds of the Olympic Peninsula of Washington. Part II. Murrelet 8(3):59-70.

Rathburn, S.F. *1902.* A list of the land birds of Seattle, Washington, and vicinity. Auk 19(2):131-141.

Rhoads, S.N. *1893.* The birds observed in British Columbia and Washington during spring and summer, 1892. Proc. Acad. Nat. Sci. Philadelphia 45:21-67.

Snyder, J.O. *1900.* Notes on a few species of Idaho and Washington birds. Auk 17(3):242-245.

Taylor, W.P. and W.T. Shaw. *1927.* Mammals and birds of Mount Rainier National Park. U.S.D.I. National Park Service. 249 pp.

U.S. Department of the Interior. *1977.* Bird list, Lewis and Clark and Columbian White-tailed Deer National Wildlife Refuges. U.S.D.I. Fish and Wildlife Service. 6 pp.

U.S. Department of the Interior. *1978.* Birds of the... Toppenish National Wildlife Refuge, Washington. U.S.D.I. Fish and Wildlife Service. 8 pp.

U.S. Department of the Interior. *1979.* Birds: Ridgefield National Wildlife Refuge, Washington. U.S.D.I. Fish and Wildlife Service. 8 pp.

U.S. Department of the Interior. *1980.* Birds: Turnbull National Wildlife Refuge, Washington. U.S.D.I. Fish and Wildlife Service. 6 pp.

U.S. Department of the Interior. *1980.* Birds of the... Willapa National Wildlife Refuge. U.S.D.I. Fish and Wildlife Service. 15 pp.

Wick, W.A. *1958.* A nine year bird list from Eliza and Protection Islands, Washington. Murrelet 39(1):1-9.

Wing, L. *1944.* Summer bird studies in the Okanogan Cascades. Murrelet 25(1):3-8.

Yocom, C.F. *1958.* Records of birds observed in the Okanogan Highlands and the Selkirk Mountains of northern Washington. Murrelet 39(2):15-18.

West Virginia

Anonymous. *1936.* Golden eagle in West Virginia. Cardinal 4(4):96.

Brooks, M. *1936.* Notes on some West Virginia birds. Cardinal 4(5):118-120.

Glover, G.A. *1947.* Golden eagle observed in Cranberry Glades. Redstart 14(7):41.

Tinsley, J.C., Jr. *1971.* Hawks over Peters Mountain. Raven 42(2):23-25.

Wetmore, A. *1937.* Observations on the birds of West Virginia. Proc. U.S. Natl. Mus. 84:401-441.

Young, C.M. *1956.* Golden eagle at Bluestone Reservoir. Redstart 23(4):73.

Wisconsin

Barry, C. *1854.* Ornithological fauna of Wisconsin. Proc. Boston Soc. Nat. Hist. 5:1-13.

Farnes, C.A. *1981.* Birds of the St. Croix River Valley: Minnesota and Wisconsin. U.S.D.I. Fish and Wildlife Service, N. American Fauna, No. 73.

Kumliem, L. and N. Hollister. *1903.* Birds of Wisconsin. Wisc. Nat. Hist. Soc. Bull. 3:1-143.

Kumlien, L. and N. Hollister. *1949.* The birds of Wisconsin. Pass. Pigeon 11(3):114-124.

Lesher, F. *1966.* Golden eagles in Houston County. Loon 38(8):144-145.

Robbins, S., C. Kemper and F.G. Daley. *1963.* Golden eagle observations. Pass. Pigeon 25(1):5.

Vos Burg, G.W. *1939.* Some rare birds I have seen in Wisconsin. Oologist 56(7):77.

Vos Burgh, G.W.H. *1912.* Some rareties. Oologist 29(6):295.

Wyoming

Allen, J.A. *1871.* Notes of an ornithological reconnaissance of portions of Kansas, Colorado, Wyoming and Utah. Bull. Mus. Comp. Zool. 3(6):113-183.

Finzel, J.E. *1964.* Avian populations of four herbaceous communities in southeastern Wyoming. Condor 66(6):496-510.

Kemsies, E. *1930.* Birds of the Yellowstone National Park, with some recent additions. Wilson Bull. 42(3):198-210.

Marler, G.D. *1951.* Birds of the geyser basins of the Firehole. Yellowstone Nat. Notes 25(1):1-7.

McCreary, O.C. and A.B. Mickey. *1935.* Bird migration records from southeastern Wyoming. Wilson Bull. 47(2):129-157.

Neilson, J.A. *1926.* Bird notes from Wheatland, Wyoming. Condor 28(2):99-102.

Pattie, D.L. and N.A.M. Verbeek. *1966.* Alpine birds of the Beartooth Mountains. Condor 68(2):167-176.

APPENDIX A

DICTIONARY OF PRIMARY KEYWORDS

ABANDONMENT	Interruption in the normal nesting sequence, usually caused by a disturbance.
ABNORMAL BEHAVIOR	Behavior contrary to that known to be normal for a species in courtship, parenting, hunting, brooding, incubation, etc.
AERIAL SURVEY	Any survey technique utilizing aircraft. (See "Nest Survey", "Survey", "Winter Survey")
AGE STRUCTURE	Ratio of various age classes in a population.
AGING	Standards of morphology or anatomy which can be used to determine the age of a raptor.
AGONISTIC BEHAVIOR	Hostile activities or displays.
AIRCRAFT	Fixed-wing aircraft and helicopters and/or their effects on observed birds.
AIRCRAFT COLLISION	Collision of a raptor and an aircraft.
ALBINISM	Albinism in a raptor species.
ALTITUDE	Observations on the altitude at which a raptor is observed.
ANAESTHESIA	Anaesthesia used for raptors.
ANATOMY	Morphology and/or structure of a raptor.
ANIMAL DAMAGE	Depredation on species valuable from an economic or recreational standpoint. (See "Economics")
ANTHROPOLOGY	The role of birds of prey in human culture.

ART	Art work relating to a raptor.
ARTIFICIAL INCUBATION	Incubation of eggs in an incubator.
ARTIFICIAL INSEMINATION	The act of artificially implanting semen into a raptor.
ARTIFICIAL NEST	Nest site provided by man, intentionally or not.
ATTACK	An account of a raptor attack directed toward a human, provoked or unprovoked. (See "Folklore")
ATTITUDE	Human attitudes toward birds of prey.
AUTOPSY	Examination of a dead raptor to determine causes of mortality.
BANDING	Banding activities and techniques. Includes band-return information. (See "Marking")
BEHAVIOR	What a bird does and how it responds to its environment. (See "Abnormal Behavior", "Agonistic Behavior", "Feeding Behavior", "Hunting Behavior", "Interspecific Behavior", "Intraspecific Behavior", "Nest Behavior" "Parental Behavior")
BENEFICIAL EFFECT	Beneficial effects of raptors on man's activities, such as agriculture.
BIOLOGY	The physical characteristics, life processes, habits, etc. of a raptor.
BLOOD SAMPLE	Blood sample analysis.
BOUNTY	Money paid in return for proof of raptor being killed.
BREEDING BIOLOGY	Breeding biology of a raptor, includes physical and behavioral.
CACHE	The act of storing food for subsequent use.

CAIN/ABEL	The act of an older chick killing its younger sibling.
CAPTIVE BREEDING	Practice of breeding raptors in captivity.
CAPTIVE CARE	Care of raptors kept in captivity. (See "Rehabilitation")
CAPTIVITY	Raptors kept in captivity for purposes other than captive breeding and falconry.
CAPTURE	Any technique or device to live capture raptors. (See "Trapping")
CHILDREN'S	Information directed toward children.
CLIMATOLOGY	Science dealing with climate and meteorological changes as they influence birds of prey.
CLUTCH SIZE	Number of eggs per nesting attempt.
COLLECT	Act of collecting raptors or raptor parts for scientific purposes.
CONSERVATION	Articles dealing with raptor conservation.
COURTSHIP	Reproductive behavior related to pair formation, stimulation of ovulation and timing of sexual readiness.
DENSITY	Number of raptors per unit area at a specific time.
DESCRIPTIVE SUMMARY	Brief or regional account covering a variety of factors dealing with a raptor, such as range, biology, habits, size, etc.
DIGESTION	Act or process of digesting food.
DIMORPHISM	Differentiation of sexes by plumage and/or size.
DIRECTIONAL EXPOSURE	The relationship of the exposure of a nest and the compass direction that a raptor nest faces.
DISEASE	Deviation from the normal or healthy state of a raptor.

DISPERSAL	Long-distance movement of individuals or groups of birds which is not necessarily characterized as migration.
DISPLAY	A territorial display of a raptor, usually aerial. (See "Courtship")
DISTRIBUTION	The range of a species. Could be specific to particular political boundaries.
DISTURBANCE	Activity which adversely affects a raptor.
ECONOMICS	Refers to studies which attempt to quantify the economic value of birds of prey and/or economic losses attributable to them. (See "Animal Damage", "Beneficial Effect")
EDUCATIONAL MATERIAL	Reference is of a general, usually non-scientific nature. Also used when reference discusses educational materials which are available.
EGG	A description or discussion of a raptor's egg. (See "Egg Measure")
EGG COLLECTION	Reference to egg collecting or egg collections.
EGG MEASURE	Thickness or size measurements for detection of pollutant/pesticide effects.
ELECTROCUTION	Electrocution of a raptor by powerlines, etc. (See "Mortality")
ENERGY DEVELOPMENT	The effects of energy development on birds of prey.
ENVIRONMENTAL IMPACT	References dealing with the effect of environmental modifications on raptors.
ETYMOLOGY	The origin and development of a raptor name.
FALCONRY	Birds of prey used for falconry or maintained by using falconry techniques.
FAT	Presence of body fat, and measurements of the amounts of fat.

FEEDING BEHAVIOR	Characteristic behavioral patterns associated with food consumption.
FIELD ID	Description of field marks to be used in the identification of a raptor species.
FLEDGING	Used to denote period after leaving the nest when young are still dependent on adults.
FLEDGLING	A young raptor which has left the nest.
FLIGHT	Characteristics of flight, including wing-loading, and other physical characteristics.
FLIGHT SPEED	Recorded or estimated speed of a raptor in flight.
FOLKLORE	Stories and often unfounded tales about raptors, including stories of eagles carrying off babies. Also includes debunking articles.
FOOD HABITS	Summary of feeding habits and food types.
FOOD REQUIREMENTS	Nutritional requirements of a raptor.
FORESTRY	Management of forest lands; includes logging and silvicultural practices and their influence on birds of prey.
FOSTERING	Transfer of a nestling raptor into the nest of another raptor. Includes cross fostering.
GROWTH	Development and physical growth of a raptor.
HABITAT	Description of habitat type, species of vegetation, etc. in relation to use by raptors (wintering and nesting primarily).
HABITAT LOSS	Loss of habitat area for birds of prey, generally due to man's influences.
HABITAT PROTECTION	Incidences of protecting habitat through land purchase, mitigation and/or other protective methods.

HACKING	Falconry technique used to release captive-raised, rehabilitated, or young wild birds under controlled conditions.
HEAVY METALS	Includes copper, lead, mercury, cadmium and other heavy metals and their effects on raptors.
HISTORICAL ACCOUNT	Historical record of raptors.
HISTORY	History as it involves raptor species.
HOME RANGE	Area in which a raptor spends most or all of its time.
HUNTING BEHAVIOR	Techniques and behavioral patterns associated with hunting.
HYPERTHERMIA	Heat stress.
ID KEY	A key to the identification of raptors in the hand, based on plumage, size and other physical characteristics.
INCUBATION	Discussion and description of incubation behavior, time period, etc.
INFERTILITY	Infertility of a raptor.
INTERSPECIFIC BEHAVIOR	Behavioral interaction between species.
INTERSPECIFIC COMPETITION	Articles dealing with competition for food, nest sites or other needed resources, between two or more species.
INTRASPECIFIC BEHAVIOR	Interactions between individuals of the same species.
INVERTEBRATES	Invertebrates associated with raptors.
KARYOTYPE	General appearance, including size, number and shape of the set of diploid chromosomes.
KNOWLEDGE STATUS	Articles summarizing the extent of our knowledge of raptor species.

LIFE HISTORY	Information detailing the sequence of events in the life of a bird of prey.
LIFT	Accounts of weight lifting capabilities of a raptor.
LITERATURE REVIEW	A review of the literature on a raptor species.
LONGEVITY	Life spans of birds of prey.
MANAGEMENT	Management techniques which can be applied to raptor species.
MARKING	Methodologies for the identification of individual raptors by marking with patagial wing markers, tarsal flags, feather coloring, etc.
MEASURE	Measurement data on the size and weight of raptors.
MELANISM	Melanism in a raptor.
MIGRATION	Movement of a raptor from one locality or climate to another for breeding, wintering and/or feeding purposes.
MINING	Effects of mining activities on birds of prey.
MITIGATION	Process of minimizing or alleviating potential impacts of an activity on raptors.
MOBBING	The reaction of some bird species to the presence of a raptor by attacking as a group.
MOLT	Timing, or sequence of, feather loss and replacement.
MONOGRAPH	A lengthy paper or book dealing with most or all aspects of a raptor species or one or more aspects in great detail. (Includes keywords of many other categories.)
MORTALITY	Characterization of one or more cases of death with causative factors usually identified. Also includes mortality rates. (See "Shooting,", "Electrocution", "Poisoning")
MUSEUM DISPLAY	A display or diorama of raptors in a museum.

MUSEUM SPECIMEN	Reference to locations where specimens are stored, or details of specimen characteristics.
NEST	Articles indicating casual accounts, description of nest parameters or discussion of factors dealing with nesting.
NEST ADORNMENT	Accounts of greenery (and other items) brought to the nest by raptors.
NEST BEHAVIOR	Behavior associated with nesting activities.
NEST CHRONOLOGY	The nesting sequence of a raptor.
NEST CONSTRUCTION	The physical process of building a nest.
NEST FAILURE	Documentation of unsuccessful nest(s) and/or discussion of reason(s) for failure. (See "Abandonment")
NEST ROBBING	The illegal act of removing eggs or young from the nest of a raptor.
NEST SURVEY	A survey of nesting raptors to determine reproductive success, etc. (See "Aerial Survey")
NESTLING	Raptor still in the nest and under parental care.
NICHE	The functional role of a raptor in a community.
OSTEOLOGY	The study of the skeletal structure of raptors.
PAIRED OVARIES	Occurrence of paired ovaries in a raptor.
PALEONTOLOGY	Science dealing with raptors of past geological periods as known from fossil remains.

PARASITE	A plant or animal that lives on or in another organism and derives sustenance or protection without benefit to the host and usually doing harm.
PARENTAL BEHAVIOR	Behavior related to the raising of young.
PELLET	Undigestible prey material which is regurgitated.
PESTICIDES	Primarily organochlorine pesticides, but could include organophosphates and carbamates. Includes residues remaining in tissues of a raptor resulting from pesticide ingestion.
PHOTOGRAPHY	Refers to photographic techniques including blinds, etc. (See "Time Lapse")
PHYSIOLOGY	The functions and vital processes of a raptor.
PIRACY	Account of a raptor pirating prey.
PLUMAGE	Description of juvenile, immature, or adult feathering of a raptor.
POETRY	Poetry which involves a raptor.
POISONING	Poisoning which results from poisoned bait or secondary poisoning by consuming animals that have become contaminated with lead, strychnine, thallium sulfate, etc.
POPULAR MONOGRAPH	A generalized account relating to a raptor, written in popular style.
POPULATION DECLINE	Decline of a raptor population over a period of time.
POPULATION DYNAMICS	Analysis of mechanisms of population regulation.
POPULATION ESTIMATE	An estimate of the number of raptors in a geographic area, at a specific time.
PREDATOR/ PREY	Relationship between predators and prey, numbers, etc.

PREY BEHAVIOR	Behavior of prey species in response to the presence or behavioral actions of a predator.
PROTECTION	Laws or acts designed to protect birds of prey.
PUNISHMENT	Fines and/or imprisonment for crimes against raptors.
QUESTIONNAIRE	Questionnaire used to gather data or information about raptors.
RARITY	The rare occurrence of a raptor in an area or at a time when it is not normally encountered.
REFUGE	Used to denote any sanctuary where the major emphasis is on the protection of wildlife, (specifically birds of prey).
REHABILITATION	Organized efforts to systematically rehabilitate birds of prey. Also used with references that deal with specific raptor rehabilitation techniques.
REPRODUCTIVE SUCCESS	Data provided on the number of successful nesting attempts relative to total attempts; also used when relative data are provided on number of young successfully produced.
RESIDUES	Residual levels of toxics, pesticides, and/or heavy metals in a raptor.
REWARD	A reward provided for information leading to conviction of person(s) responsible for illegally killing a raptor.
ROADSIDE SURVEY	Survey which involves counting raptors from a vehicle.
ROOST	Use of a location by a raptor or concentration of raptors during a non-active period.
SENSES	Any of the five faculties receiving sensations and the organs and nerves associated with them. (See "Anatomy" and "Physiology")
SEXING	Standards of morphology and anatomy which can be used to determine the sex of a raptor.

SEX RATIO	The ratio of males to females in a population.
SHOOTING	Shooting of raptors.
STATISTICS	Statistical techniques for the analysis of data.
STATUS	Used in conjunction with references that emphasize a discussion on the status of birds of prey; also used when several years of survey data are presented.
STORY	Popular account of human interactions with raptors.
SUB-ADULT	Account of a raptor not yet in adult plumage.
SUPPLEMENTAL FEEDING	Artificially providing a food source, usually carrion, for raptors.
SURVEY	Methods of quantifying numbers and distribution of raptors. (see "Roadside Survey", "Aerial Survey", "Nest Survey", "Winter Survey")
SYMBOLISM	Using a raptor to express or represent the intangible.
SYNONOMY	Scientific names used in different systems of nomenclature to designate the same species, genus.
TALON GRAPPLING	Account of raptors joining feet during flight.
TAXIDERMY	Art of preparing and mounting the skins of animals.
TAXONOMY	Classification of birds of prey into naturally related groups based on some factor(s) common to each, such as structure, embryology, biochemistry, etc.
TELEMETRY	Use of radio transmitting and tracking equipment. Includes radar tracking.
TEMPERATURE	Temperature of a raptor or raptor egg.
TERMINOLOGY	Discussion of the use or misuse of terms and efforts to try to establish standard terminology.

TERRITORY	An area defended by a bird against individuals of its own species.
TIME LAPSE	Use of time lapse photography in a raptor study.
TOXICOLOGY	Science dealing with poisons, their effects and antidotes.
TRADE	Buying or selling live or dead birds of prey, or parts of birds of prey (e.g. feathers).
TRANSLOCATION	Intentional moving of a raptor by man to a different area, nest, etc. as in a reintroduction project.
TRAPPING	Incidental and/or accidental trapping of raptors when traps were set for other species such as furbearers.
VACCINATION	Protection of a raptor against disease, usually in captivity.
VOCALIZATION	Instances of vocal sound production.
WING LOADING	Ratio of weight to wing area of a raptor.
WINTER SURVEY	Used to describe organized efforts to document numbers of raptors on specific wintering grounds, often repeated each year in a systematic fashion.
WINTERING POPULATION	Used with casual observations and/or accounts of raptors on their wintering grounds; numbers or other documentation often not provided.